# MATERIALS THERMODYNAMICS

With Emphasis on Chemical Approach

# MATERIALS THERMODYNAMICS

## With Emphasis on Chemical Approach

HAE-GEON LEE

POSTECH, Korea

World Scientific

NEW JERSEY • LONDON • SINGAPORE • BEIJING • SHANGHAI • HONG KONG • TAIPEI • CHENNAI

*Published by*

World Scientific Publishing Co. Pte. Ltd.
5 Toh Tuck Link, Singapore 596224
*USA office:* 27 Warren Street, Suite 401-402, Hackensack, NJ 07601
*UK office:* 57 Shelton Street, Covent Garden, London WC2H 9HE

**British Library Cataloguing-in-Publication Data**
A catalogue record for this book is available from the British Library.

**MATERIALS THERMODYNAMICS**
**With Emphasis on Chemical Approach**
**(With CD-ROM)**



ISBN-13 978-981-4368-05-6
ISBN-10 981-4368-05-9

To my Lord

To my parents

To the late chairman
Chang, Sang-Tae

To my wife and children
Myoung-Hi, Hanna, Rebecca, Chris

# Preface

This book is the expanded edition of the first book entitled "Chemical Thermodynamics for Metals and Materials."

This book presents thermodynamics of materials with emphasis on the chemical approach. This expanded edition keeps most of the contents of the first edition, but with substantial revisions. It also includes a number of additional topics of thermodynamics which are relevant to the wide spectrum of materials. This book is thus suitable to students in materials science and metallurgical engineering, and also to students in related fields such as physical chemistry and chemical engineering.

As thermodynamics is a key discipline in most science and engineering fields, a great number of books, each of which claims originality and uniqueness in presentation and approach, have been published on the subject. However, thermodynamics remains a subject that is difficult to grasp for many uninitiated students and an "easy to forget" one for graduate engineers.

After many years of teaching thermodynamics at different universities and conducting actual application of it at the industry level, I have come to the following conclusions:

- each of the thermodynamic topics must be explained so as to make a clear connection to its physical significances, and
- mathematical derivation of thermodynamic relationships must be easy to follow in its sequential flow.

This book has been written with such insights as guiding principles. As such, it has unique features as described below:

- Classical thermodynamics is amalgamated with statistical approach, wherever it provides a better understanding. While this book is essentially based on classical thermodynamics, it is supported by statistical treatments with molecular-level models, wherever it helps to make understanding of a particular concept easier. This combination should help readers obtain a solid grasp of physical significance, which may otherwise remain as concepts that are difficult to comprehend.

- In contrast with most books on this subjects in which mathematical derivation of thermodynamic relationships is explained line-by-line in a largely one-dimensional way, this book adopts a two-dimensional approach in which the derivation flows in two ways: vertically and horizontally. This flowchart-like format should provide readers with clear reasoning of the derivation flow.

A number of worked examples are included in each chapter, which are designed to emphasize extended principles and to present the strategies of applications. Many exercise questions which helps develop skills necessary to handle numerical problems are also included in each chapter.

A computer-aided learning package is included with this book in an accompanying CD-ROM, which simulates the way a lecturer would teach this subject in a classroom setting. Unlike the typical non-interactive presentation found in most textbooks, this multimedia package provides the users with helpful interactive learning tools. The package includes much of the contents contained in the book as well as the solutions to the exercise questions.

As this book has developed from my teaching materials, it unavoidably includes contributions of many other authors. I wholeheartedly acknowledge their contributions.

I am indebted to my teacher, Professor Y.K. Rao for introducing me to the world of thermodynamics. I am especially grateful to my former colleague, Professor Peter Hayes, at The University of Queensland, Australia, for making many useful comments and giving me constant encouragement. I am also thankful for the assistance provided by my colleague, Professor Youn-Bae Kang in proof-reading parts of the book, and providing constructive comments.

I dedicate this small achievement to the late chairman Chang Sang-Tae of Dongkuk Steel Co., who offered me unconditional support and encouragement for my academic career. He was indeed my mentor for my whole life, having taught me the ways to harmonize knowledge with wisdom.

I wish to acknowledge generous support of POSCO, which has enabled me to apply thermodynamics to the industry level.

Finally, I am deeply thankful to my wife and children for the love, support and encouragement they have given to me.

Hae-Geon Lee
POSTECH
Pohang

# Contents

Preface vii

**1. Introduction** 1

1.1 Energy 2

1.2 Heat and Work 4

**2. The First Law of Thermodynamics** 7

2.1 The First Law of Thermodynamics 7

2.2 Enthalpy and heat Capacity 15

2.3 Enthalpy Change 23

**3. The Second Law of Thermodynamics** 27

3.1 Grades of Energy 27

3.2 Heat Engines and Entropy 32

3.3 Energy Dispersion and Entropy 53

3.4 Equilibrium Criterion and Entropy 62

3.5 Entropy Changes and Entropy Productions 73

3.6 Statements of the Second Law of Thermodynamics 94

**4. Free Energy Functions** 97

4.1 Energy Functions 97

4.2 Helmholtz Energy 99

4.3 Gibbs Energy 101

4.4 Effect of Pressure on Gibbs Energy Change 104

4.5 Effect of Temperature on Gibbs Energy Change 106

4.6 Some Useful Equations 107

**5. The Third Law of Thermodynamics** 111

5.1 Statements of the Third Law 111

5.2 Corollaries to the Third Law 112

5.3 Absolute Entropies 115

5.4 Molecular Interpretation 116

**6. Enthalpy and Gibbs Energy Changes** 119

6.1 Standard States 119
6.2 Heat of Formation 120
6.3 Heat of Reaction 122
6.4 Adiabatic Flame Temperature 127
6.5 Gibbs Energy Changes 129

**7. Behavior of Gases** 133

7.1 Ideal Gases 133
7.2 Real Gases and Fugacity 136

**8. Thermodynamic Functions of Mixing** 141

8.1 Activity 141
8.2 Partial Properties 148

**9. Behavior of Solutions** 157

9.1 Ideal Solutions 157
9.2 Non-ideal Solutions and Excess Properties 161
9.3 Dilute Solutions 167
9.4 Gibbs-Duhem Equation 171
9.5 Solution Models 175

**10. Reaction Equilibria** 185

10.1 Equilibrium Constants 185
10.2 Criteria of Reaction Equilibrium 190
10.3 Effect of Temperature on Equilibrium Constant 201
10.4 Effect of Pressure on Equilibrium Constant 203
10.5 Le Chatelier's Principle 205
10.6 Alternative Standard States 207
10.7 Interaction Coefficients 215
10.8 Ellingham Diagram 218
10.9 Adsorption Equilibria 227

**11. Phase Equilibria** 239

11.1 Phase rule 239
11.2 Phase Transformations 248
11.3 Phase Equilibria and Gibbs Energies 255
11.4 Influence of Interfaces on Equilibrium 271

**12. Phase Diagrams** 287

12.1 One Component (Unary) Systems 287
12.2 Binary Systems 297
12.3 Thermodynamic Models 326

12.4 Ternary Systems 335
12.5 Predominance Diagrams 363

**13. Electrochemistry** 367
13.1 Basic Electrochemical Concepts 367
13.2 Electrochemical Cell Thermodynamics 370
13.3 Electrochemical Cells and Electrodes 378
13.4 Concentration Cells 390
13.5 Activities in Aqueous Solutions 403
13.6 Solubility Products 411
13.7 Pourbaix Diagrams 416

**Appendices** 427
Appendix I Heats of formation, standard entropies and heat capacities 429
Appendix II Standard Gibbs energies of formation 436
Appendix III Properties of selected elements 449
Appendix IV Standard half-cell potentials in aqueous solutions 452

**Index** 455

# Chapter 1

# Introduction

Thermodynamics is a science concerned with the transfer and conversion of energy associated with various chemical and physical processes. It is temperature that plays the central role in thermodynamics. Temperature is a fundamental notion since it is not expressible in terms of basic quantities like mass, length and time. This makes thermodynamics unique and different from other branches of science. The processes which are subject to thermodynamic considerations include all of the phenomena that occur in nature, and also include the chemical and physical processes that are manipulated intentionally.

Thermodynamics is limited only by completely general restrictions on macroscopic energy transfer in nature and hence exceedingly general in its applicability. It deals with the bulk (macroscopic) properties of matter and does not concern itself with whether or not there are atoms or molecules. In fact, thermodynamics does not care whether or not there are atoms and molecules. In this regard, it is sometimes called "*classical thermodynamics*." On the other hand, *statistical the rmodynamics* which is footed on quantum mechanics deals with the individual particles of matter, and provides means for calculating the properties of bulk material (macroscopic samples) from the properties of the atoms and molecules which comprise the material.

| ***Classical Thermodynamics*** | ***Statistical Thermodynamics*** |
|---|---|
| The macroscopic approach to the study of thermodynamics which does not require knowledge of the behavior of individual particles is called classical thermodynamics.<br>Example: The pressure of a gas in a container is the result of momentum transfer of the particles (atoms or molecules) to the wall of the container, but the behavior of individual particles does not need to be known in classical thermodynamics. A pressure gage is sufficient to measure the pressure. | The microscopic approach based on the average behavior of large groups of individual particles is called statistical thermodynamics. It provides a framework for relating the microscopic properties of individual atoms and molecules to the macroscopic or bulk properties of materials that can be observed. This ability to make macroscopic predictions based on microscopic properties is the main advantage of statistical thermodynamics. |

***Chemical thermodynamics*** in particular is the study of the interrelation of heat and work with chemical reactions or with physical changes of state. For example, it deals with determination of feasibility of transformation from one phase to another. It also establishes a criterion as to whether or not a particular chemical process can take place under any given conditions. How much of energy is required for the process and how far the process will go (the maximum yield) can also be determined.

The laws of thermodynamics are the results of concise summaries of experimental observations. Prerequisite to the understanding of the laws of thermodynamics is a clear understanding of some basic terms like heat, work and energy.

## 1.1 Energy

Energy can be viewed as the ability to cause changes. Energy is an attribute of a substance as a consequence of its structure (atomic, molecular or aggregate). Matter is made up of atoms and molecules (groupings of atoms) and the properties of the matter depend on the behavior of these particles. Energy causes the atoms and molecules to always be in motion.

Since a chemical transformation involves a change in the structure, it invariably causes the increase or decrease of energy of the substances involved. Energy can also be defined as the ability to do work. While there are many forms of energy, they can be grouped into two categories: ***potential energy***, or stored energy; and ***kinetic energy***, or energy of motion. All of these energies are ***quantized***: *i.e.*, it is not allowed to take any arbitrary value of energy, but only certain discrete numerical values of energy can be taken.

### Potential Energy

The potential energy is the energy that atoms and molecules possess in the form of bonds associated with the intermolecular attractive forces. A form of potential energy which is of prime importance in chemical thermodynamics is ***chemical energy*** which is related to chemical bonds for the structural arrangement of atoms or molecules. Chemical reactions occur by breaking the bonds and the corresponding chemical energy is released or absorbed. When released (exothermic), it is either reused in forming new bonds with other atoms or energizes atoms or molecules, or enters the surroundings as heat.

### Kinetic Energy

All atoms or molecules at temperatures above absolute zero Kevin are in a state of motion and hence possess the kinetic energy associated with their various motions. What kinds of motions are possible? The motions of individual atoms or molecules are random and chaotic. The motions are classified into three different modes; namely, ***translation***, ***rotation*** and ***vibration***. The kinetic energy is the sum of the energy due to these three motions. The sum total of all of this microscopic-scale randomized kinetic energy within a body is sometimes given a special name, ***thermal energy***. Translation is the movement from one location to another. Monatomic molecules store the kinetic energy through the translational motion, as it is the virtually only possible motion for them. For solids and liquids, translational motions of molecules are restricted to very short distances. But solids and liquids, and gas molecules consisting of two or more atoms can undergo internal vibration and rotation.

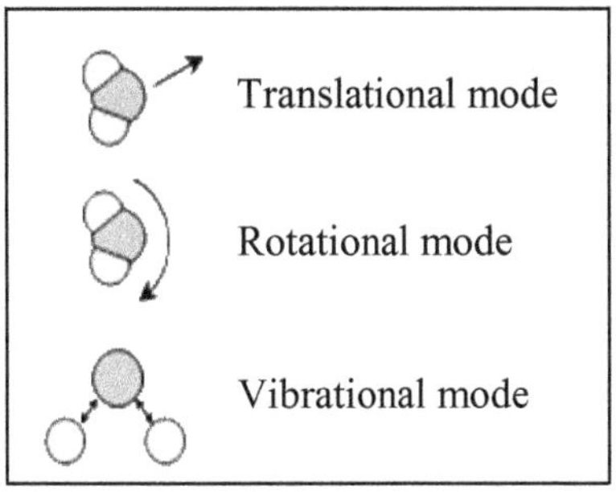


**Internal energy**

Internal energy, $U$, is the sum of the kinetic and potential energies of all the particles (atoms or molecules) making up a body, and hence it involves energy on the microscopic scale. Atoms and molecules play the role of storage and transport of energy, and conversion of energy from one form to another by uptake or release of heat. The internal energy does not include the kinetic and potential energies of the system as a whole. For instance, the kinetic energy related to the motion of a massive body such as a baseball or a car that is moving along a uniform trajectory is not included in the internal energy of the body. It excludes any potential energy that a body may have because of its location in external gravitational or electrostatic field.

> ***Temperature***
>
> Temperature is a measure of the average kinetic energy of particles in a substance, which is related to how hot or cold that substance is. Temperature is not directly proportional to the internal energy since temperature measures only the kinetic energy, but do not tell anything directly about the potential energy of the substance. Two objects with the same temperature do not in general have the same internal energy.

The following diagram shows a schematic expression of the internal energy of two different materials at the same temperature. It is noted that at the same temperature, the average kinetic energy of the molecules is the same, but not the potential energy.

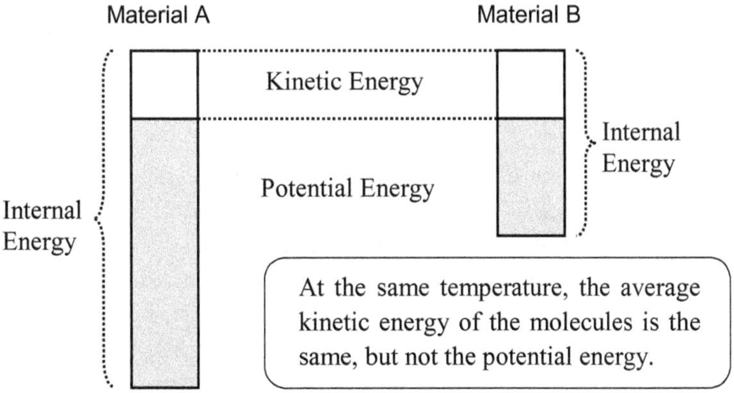

When energy is transferred into a substance, it can be used to increase the kinetic energy of particles (atoms or molecules), which causes increase in temperature. The energy can also be used to increase the potential energy of the particles, which does not cause increase in temperature.

Internal energy is a property of a system that depends only on the current state of the system, not on how the state is reached. In thermodynamics this kind of property is referred to as a ***state function***.

*Example 1.1*

The internal energy ($U$) of a monatomic ideal gas is given as

$$U = \tfrac{1}{2} Nm < v^2 >$$

where $N$ is the number of atoms and $m$ is the atomic mass. According to the kinetic theory of gases, the mean-square speed of the atoms, $<v^2>$, is given as,

$$< v^2 > = 3PV / Nm$$

1) Find out how $U$ depends on temperature.
2) Determine how $U$ depends on pressure or volume at a constant temperature.

1) By removing $<v^2>$ from the two equations above, and knowing $PV = nRT$,

$$U = \frac{3}{2} PV = \frac{3}{2} nRT$$

   Thus, $U$ is directly proportional to $T$.

2) From the above equation, $U$ is independent of both $P$ and $V$ at constant $T$.

## 1.2 Heat and Work

Heat and work are the only two mechanisms by which energy can be transferred to or from a body. It is important to understand clearly the difference between work and heat.

### Work

Work is a form of energy transfer to or from the body that is due to a change of the external macroscopic parameters of the body (for example, expansion of the volume of a system against an external pressure, and driving of a piston-head out of a cylinder against an external force.) By work energy is transferred due to organized motion of particles (*i.e.*, atoms, molecules, electrons, etc.). For instance, when a piston-head is driven to do work, all the atoms in the piston-head moved together in the same direction.

## Heat

Heat is the transfer of thermal energy. The energy transfer by heat occurs from a body at a higher temperature to one at a lower temperature. In molecular terms heat is the transfer of energy via microscopic thermal (chaotic, random, disorganized) motion of atoms and molecules. When energy is added to a system in the form of heat, it is stored as kinetic and potential energy of the atoms and molecules making up the system.

Heat transfer into the system and work done on the system can increase thermal energy of the system. There is no way to detect any qualitative difference in the thermal energy resulting from any of these sources. For example, work in the form of violent agitation can increase thermal energy of a system, but it is indistinguishable from that produced by direct heat transfer into it from some source at a higher temperature.

| Units of energy |
|---|
| 1 joule (J) = $1\ kg\ m^2s^{-2}$ |
| 1 erg = $1\ g\ cm^2s^{-2} = 10^{-7}$ J |
| 1 cal = 4.184 J |
| 1 liter atm = 101.325 J |
| 1 kWH = $3.6 \times 10^6$ J |
| 1 eV = $1.6022 \times 10^{-19}$ J |

Neither heat nor work is thermodynamic property of a system. A system cannot contain or store either heat or work. They are two different forms of energy transfer.

The ***sign convention*** is that,

- Heat ($q$) is *positive* when it flows to the system from the surroundings and *negative* when it flows from the system to the surroundings.
- Work ($w$) is *positive* when the surroundings does work on the system, and *negative* when the system does work on the surroundings.

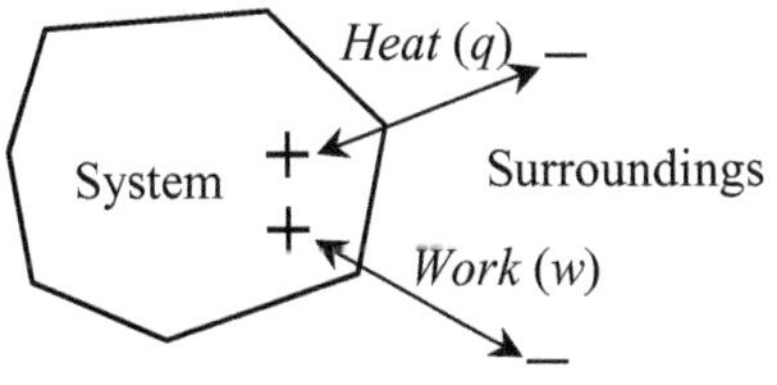


***System***: a portion of the universe that is chosen for thermodynamic discussion
***Surroundings***: the remainder of the universe.

*Caution*: Some texts adopt different sign convention: *i.e.*, $w$ is positive when work is done *by* the system

# Chapter 2

# The First Law of Thermodynamics

## 2.1 The First Law of Thermodynamics

The First Law of Thermodynamics is really a statement of *the principle of conservation of energy*:

- Energy can neither be created, nor destroyed.
- Energy can be transported or converted from one form to another, but cannot be either created or destroyed.
- Chemical and/or physical changes are accompanied by changes in energy.

Energy may be converted from one form to another, but it cannot be created or destroyed. This principle is referred to as the *First Law of Thermodynamics*. The internal energy of the system ($U$) increases, if heat ($q$) is supplied to the system, or work ($w$) is done on the system. The internal energy of the system ($U$) decreases if heat is extracted from the system, or the system does work ($w$) to the surroundings.

Net change in the internal energy ($\Delta U$) is then

$$\boxed{\Delta U = q + w}$$

*Example 2.1*

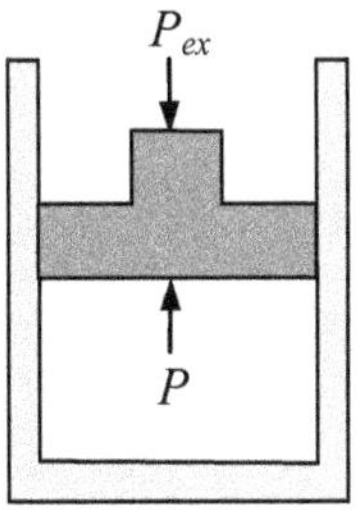


Work can be expressed in terms of a force and the displacement of its point of action. If the gas inside the cylinder shown expands and pushes the piston against the external pressure $P_{ex}$, can the force ($F$) exerted by the gas on the piston be represented by the following equation?

$F = AP_{ex}$ where $A$ is the cross sectional area of the piston.

As *force* = *area* × *pressure*, the expression is correct. If the gas expands against the external pressure $P_{ex}$ from $L_1$ to $L_2$ in the figure, the work done by the system (*i.e.*, the gas) is

$$w = -\int_{L_1}^{L_2} AP_{ex}dL = -\int_{V_1}^{V_2} P_{ex}dV$$

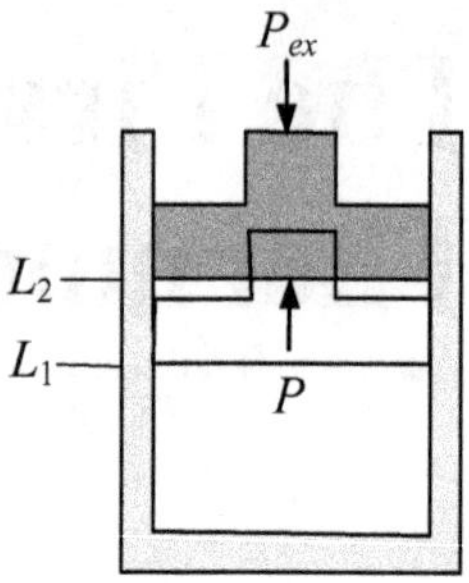


If the external pressure is continuously adjusted so that it is kept the same as the internal pressure P,

$$w = -\int_{V_1}^{V_2} PdV$$

Work under these conditions is called ***reversible work.***

*Example 2.2*

Fine steel fiber is contained in a cylinder filled with pure oxygen. The cylinder is fitted with a frictionless piston which maintains the oxygen pressure at 1 atm. The iron in the steel fiber reacts with oxygen very slowly to form $Fe_2O_3$. Heat generated by the reaction is removed during the process so as to keep the temperature constant at 25°C. For the reaction of 2 moles of iron, 831.08 kJ of heat is removed:

$$2\text{Fe} + \frac{3}{2}\text{O}_2 = \text{Fe}_2\text{O}_3$$

Calculate heat ($q$), work ($w$) and internal energy change ($\Delta U$).

Assumptions:

(1) The volume of oxygen = the total volume of the cylinder (Fe and $Fe_2O_3$ are both solid and occupy a negligible volume compared with the gaseous $O_2$.)
(2) Oxygen is an ideal gas.

$$PV = nRT$$

↓ $\Delta V$ : volume change due to reaction; $\Delta n$ : number of moles of $O_2$ reacted

$$P\Delta V = (\Delta n)RT$$

↓ $w = - P\Delta V$

$$w = -(\Delta n)RT$$

↓ Given: $\Delta n$ = -1.5 moles and $T$ = 298 K, Knowing R = 8.314 J $mol^{-1}K^{-1}$

$w = 3.72$ kJ

↓ $w$>0 →work is done on the cylinder (system)
$\Delta U = q + w$ : First law
$q$ = -831.08 kJ (given)

$\Delta U = -827.36$ kJ

$\Delta U < 0 \rightarrow$ Decrease in bond energy (chemical energy) caused by chemical reaction. Bonds of reactants (Fe and $O_2$) break and new bonds of the products ($Fe_2O_3$) form. As the molecular bond energy of the products is smaller than that of the reactants, the balance of the bond energy is released as heat, and the internal energy decreases.

*Example 2.3*

A system can change from one state to another in many different ways. Suppose a system changes from the initial state (A) in the figure to the final state (B). Determine the work done by the system for each of the following paths:

Path 1 : A→C→B
Path 2 : A→E→B
Path 3 : A→D→B

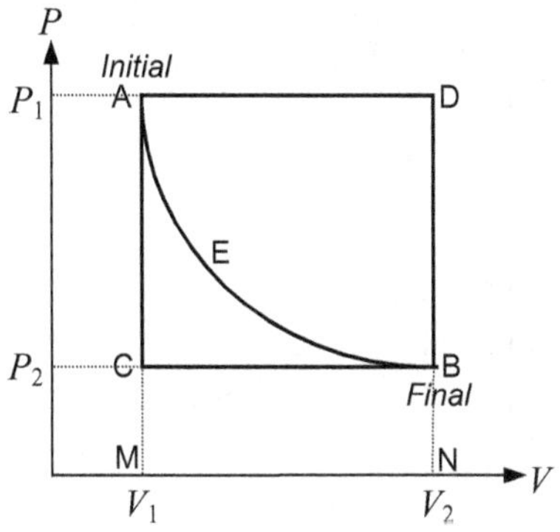


Recall that

$$w = -\int_{V_1}^{V_2} P\,dV$$

<u>*Path 1*</u>: Initially the pressure decreased from $P_1$ to $P_2$ (*i.e.*, A→C) at the constant volume $V_1$ by decreasing the temperature. In this process no work has been done as there was no volume change. Next, the volume of the system expands from $V_1$ to $V_2$ (*i.e.*, C→B) at the constant pressure $P_2$. The amount of work done in this process is represented by the area CBNM. This is the total work done if the system follows the path 1.

<u>*Path 2*</u>: If the system follows the path A→E→B, work done by the system is represented by the area AEBNM.

<u>*Path 3*</u>: Similarly, the amount of work done by the system is given by the area ADNM. The amount of work done by the system depends on the path taken, and hence cannot be evaluated without a knowledge of the path.

*Example 2.4*

The gas inside the cylinder shown expands and pushes the piston against the external pressure $P_{ex}$. Determine the work done by the cylinder for each of the following paths:

(1) $P_{ex}$ suddenly drops to $P_f$
(2) $P_{ex}$ drops to $P_2$ and then to $P_f$
(3) $P_{ex}$ drops in sequence to $P_1$, $P_2$, $P_3$ and $P_f$.
(4) $P_{ex}$ decreases in such a manner that it always infinitesimally smaller than the pressure inside the cylinder.

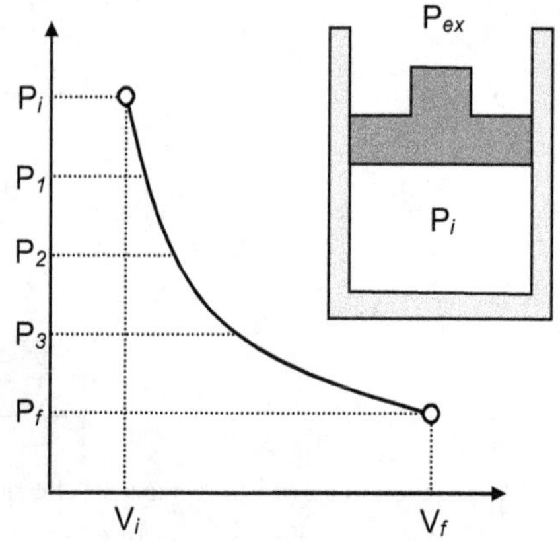


The shaded area in each case is the amount of work done by the cylinder (system) to the surroundings by volume expansion.

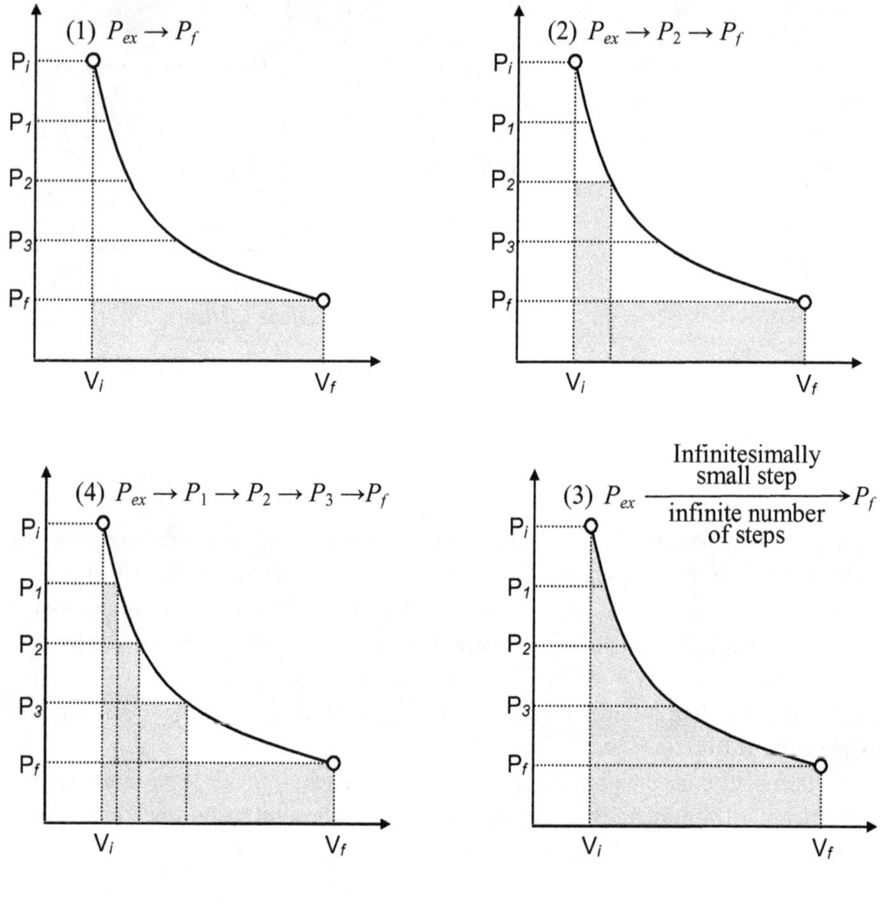

It is noted that, as the number of steps increases, the amount of work done *by the system* also increases for the *same* initial and final states. When the external pressure decreases by infinitesimally small steps, the work done by the system becomes maximum (the case 4). This represents the ***reversible process***. All others are ***irreversible processes***.

*Example 2.5*

The gas inside the cylinder shown below is compressed by pushing the piston against the internal pressure $P_i$. Determine the work done on the cylinder for each of the following paths:

(1) $P_{ex}$ is suddenly increased to $P_f$.
(2) $P_{ex}$ is increased to $P_2$ and then to $P_f$.
(3) $P_{ex}$ is increased in sequence to $P_1$, $P_2$, $P_3$ and $P_f$.
(4) $P_{ex}$ is increased in such a manner that it always infinitesimally larger than the pressure inside the cylinder.

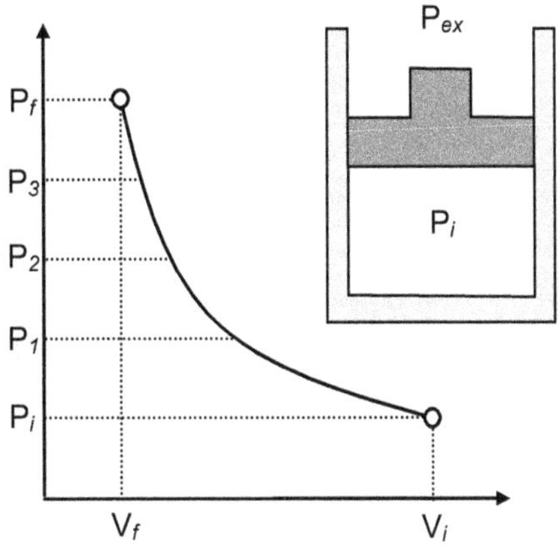


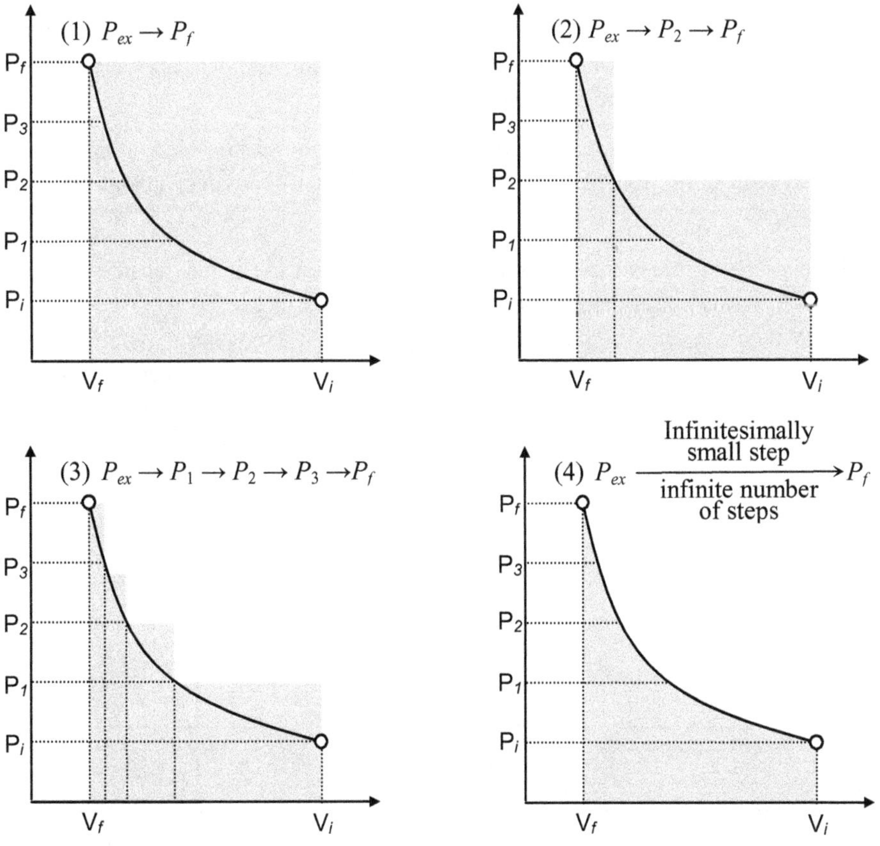

It is noted in this compression process that, as the number of steps increases, the amount of *work done on the system* decreases for the same initial and final states. When the external pressure increases by infinitesimally small steps, the work done on the system becomes minimum (the case 4). This represents the *reversible process*. All others are *irreversible processes*.

It is thus seen from the above two examples that ***reversibility*** is approached when the expansion and compression processes are carried out in such a manner that the external pressure is never more than infinitesimally different from the internal pressure.

*More about reversible work*

Work is a mode of energy transfer which occurs due to the existence of imbalance of forces between the system and the surroundings. When the forces are *infinitesimally* unbalanced throughout the process in which energy is transferred as work, then the process is said to be reversible.

A reversible change may also be defined as one carried out in such a way that, when undone, both the system and surroundings remain unchanged. In other words, a process is called reversible if, after the process has occurred, both the system and its surroundings can be wholly restored by any means to their respective initial states.

Heat can be transferred reversibly between two bodies by changing the temperature difference between them in infinitesimal steps each of which can be undone by reversing the temperature difference.

Any process that proceeds in infinitesimal steps would take infinitely long to occur, so thermodynamic reversibility is an idealization that is never achieved in real processes.

But if no real process can take place reversibly, what use is the consideration of a reversible process? Although a real process is impossible to be achieved in a truly reversible manner, this idealized pathway provides a crucially useful means to determining the change of a state function like internal energy and other thermodynamic properties, as the change of the state function is independent of the pathway from the initial to final states: the same net change via any pathway.

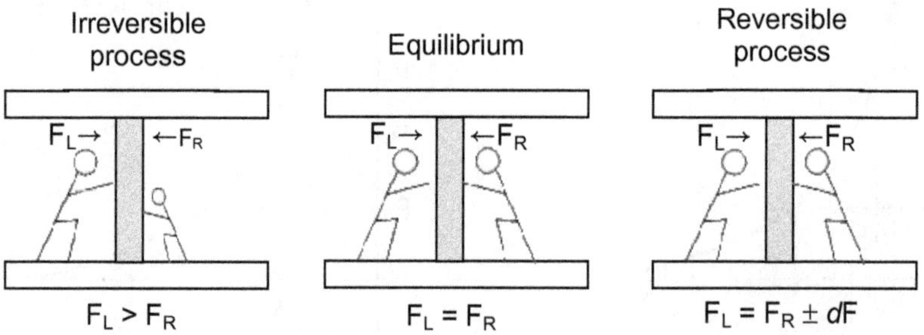

*Example 2.6*

A substance in state A undergoes a change to state B via state 1, and then comes back to state A via states 2 and 3. Is it possible that the gain of internal energy in the forward process (A→1→B) can be different from the loss in the backward process (B→2→3→A)?

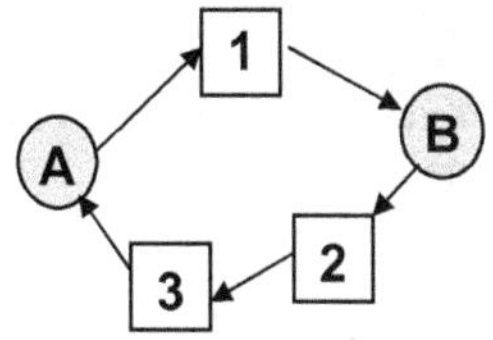


No. It should be the same. If different, the system will return to the initial state A with a net gain of internal energy. In other words, if different, the system will gain even more energy from nowhere just by repeating the process. This is contrary to the First Law of Thermodynamics. Therefore the internal energy gained in the forward process must be equal to the energy lost in the return process.

We have seen here that internal energy ($U$) differs from heat ($q$) and work ($w$) in that it depends only on the state of the system, not on the path it takes. A function which depends only on the initial and final states and not on path is called ***state function***.

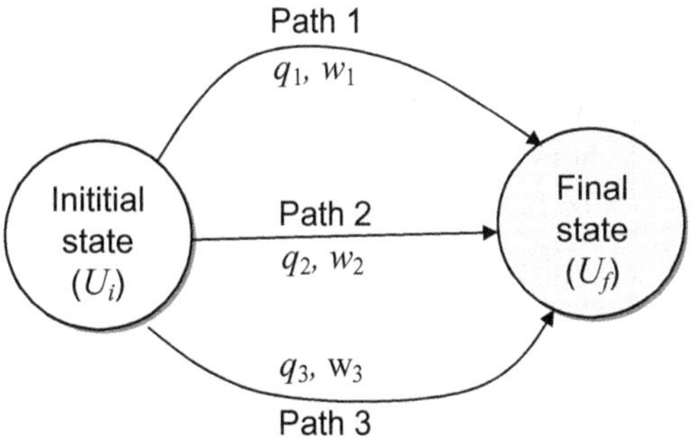


$U$: State Function

$\Delta U = U_f - U_i$

$q_1 \neq q_2 \neq q_3$,

$w_1 \neq w_2 \neq w_3$,

but, $q_1 + w_1 = q_2 + w_2 = q_3 + w_3 = \Delta U$

Therefore, $\Delta U = q + w$

*Example 2.7*

Which of the following thermodynamic terms are state functions?

Temperature ($T$), Pressure ($P$), Heat ($q$), Work ($w$), Volume ($V$)

State functions: $T, P, V$ Non-state functions: $q, w$

State functions depend on the mass of material which are called ***extensive properties*** (*e.g.*, $U$, $V$). On the other hand some state functions are independent of the amount of materials. These are called ***intensive properties*** (e.g., $P$, $T$).

- Thermodynamics is largely concerned with the relations between state functions which characterize systems.
- A state function can be integrated between the initial (A) and final (B) states, being independent of integration path.

$$\text{(Example)} \quad \Delta U = \int_{A}^{B} dU$$

- An exact differential can be written in terms of partial derivatives. For instance, as $U = f(T,V)$,

$$dU = \left(\frac{\partial U}{\partial T}\right)_V dT + \left(\frac{\partial U}{\partial V}\right)_T dV$$

- The order of differentiation of a state function is immaterial.

$$\left[\frac{\partial}{\partial V}\left(\frac{\partial U}{\partial T}\right)_V\right]_T = \left[\frac{\partial}{\partial T}\left(\frac{\partial U}{\partial V}\right)_T\right]_V$$

- The First Law of Thermodynamics may be summarized by the following equation:

$$\boxed{\Delta U = q + w}$$

↓ For infinitesimal change of state

$$\boxed{dU = dq + dw}$$

↓ But $dq$ and $dw$ are not exact differential because they depend on the path. To remind us of this, they are customarily written as $\delta q$ and $\delta w$.

$$\boxed{dU = \delta q + \delta w}$$

*Exercises*

2.1 Calculate the work done by one mole of an ideal gas when it isothermally expands from 1 $m^3$ to 10 $m^3$ at 300 K.

2.2 A system moves from state A to state B as shown in the figure. When the system takes path 1, 500J of heat flow into the system and 200J of work done by the system.

(1) Calculate the change of the internal energy.

(2) If the system takes path 2,100 J of work is done by the system. How much heat flows into the system?

(3) Now the system returns from state B to state A via path 3,100 J of work is done on the system. Calculate the heat flow.

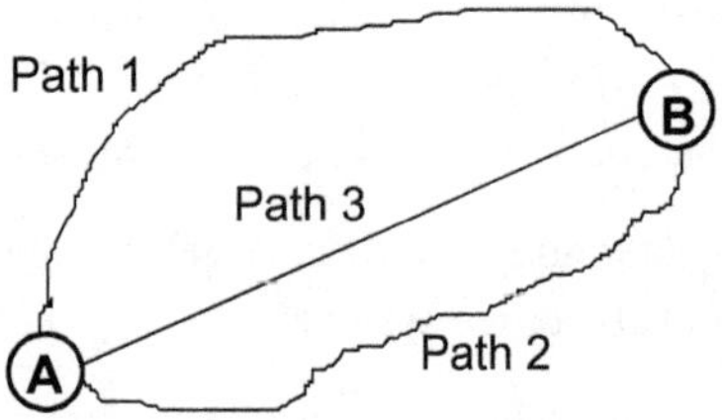

## 2.2 Enthalpy and Heat Capacity

### Enthalpy

If a process takes place at *constant volume*,

$$\boxed{U = q + w}$$

At constant volume,
$w = P\Delta V = 0$

$$\boxed{\Delta U = q}$$

Therefore, the increase or decrease in internal energy of the system is equal to the heat absorbed or released, respectively, at constant volume.

If a process is carried out at a *constant pressure* rather than at a constant volume, then the work done by the system as a result of the volume change is

$$w = -\int_1^2 PdV = -P\int_1^2 dV = -P(V_2 - V_1)$$

From the first law of thermodynamics,

$$\boxed{\Delta U = q + w}$$

$\Delta U = U_2 - U_1$ $\quad$ $w = -P(V_2 - V_1)$

$$\boxed{(U_2 + PV_2) - (U_1 + PV_1) = q}$$

The function $U + PV$ occurs frequently in chemical thermodynamics and hence it is given a special name, ***enthalpy*** with the symbol $H$.

$$\boxed{H = U + PV}$$

$\Delta H = H_2 - H_1$

$$\boxed{\Delta H = q}$$

Therefore, for a system at constant pressure,

- Increase in enthalpy ( $\Delta H > 0$) : Heat is absorbed by the system (***endothermic***)
- Decrease in enthalpy ( $\Delta H < 0$) : Heat is released by the system (***exothermic***)

For changes at other than constant pressure, $\Delta H$ still has a definite value, but $\Delta H \neq q$.

*More about enthalpy*

Although enthalpy is defined as a matter of convenience because it often occurs in thermodynamic discussions, it also carries its own physical significance. Suppose that a system is newly created. The energy required for the creation of the system is the internal energy ($U$) of the system and an additional energy to provide room to place the system by pushing the surroundings ($PV$). Enthalpy is the sum of these two energies.

## Heat Capacity

Heat capacity of a system is defined as the amount of heat required to raise one unit of temperature of a unit mass of homogeneous material, provided that no phase or chemical changes occur during the transfer of heat.

Thus, when the temperature of a unit mass of a material is raised by $\Delta T$ by absorbing heat of $q$,

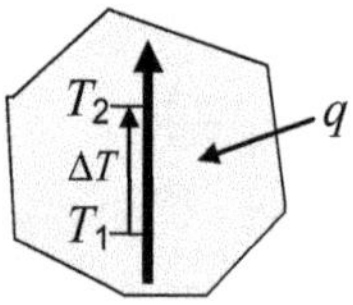


$$C = \frac{q}{\Delta T}$$ where $C$ is the *heat capacity*.

For an infinitesimal change in $T$,

$$C = \frac{\delta q}{dT}$$

Materials with large heat capacities, like water, hold thermal energy well - their temperature do not rise much for a given amount of heat transferred, whereas materials with small heat capacities, like copper, do not hold thermal energy well, and hence temperature rises quickly.

Recall that heat $q$ is not a state function, so that the change in $q$ depends on the other variable, for instance, $V$ or $P$, in addition to $T$. Therefore,

At constant volume

$$\delta q = dU \longrightarrow C_V = \frac{dU}{dT}$$

and,

At constant pressure

$$\delta q = dH \longrightarrow C_P = \frac{dH}{dT}$$

where $C_V$ is the heat capacity at constant volume, and $C_P$ is the heat capacity at constant pressure.

*More about heat capacity*

Heat capacity is directly related to temperature, and in turn the temperature is the result of the average total kinetic energy of particles in material. Hence heat capacity must be closely related to the kinetic energy portion of the internal energy of the material (recall that the internal energy is the sum of the kinetic and potential energies). Heat is transfer of thermal energy. Thermal energy is stored as kinetic energy and also as potential energy. This distribution of thermal energy contributes to the heat capacity. The more thermal energy goes to the potential energy, the higher the heat capacity is.

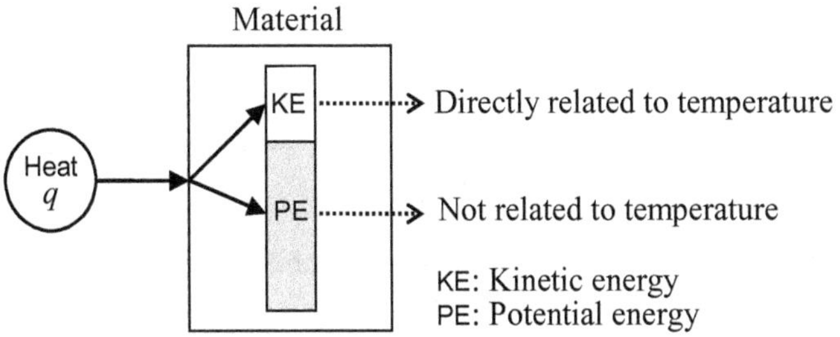


The variation with temperature of the heat capacity, $C_P$, for a substance is often given by an expression of the form:

$$C_P = a + bT + cT^{-2}$$

where $a$, $b$ and $c$ are constants to be determined empirically.

*Some examples,*

$$C_{P,\mathrm{Al_2O_3}} = 106.6 + 17.8\times10^{-3}T - 28.5\times10^{5}T^{-2}, \mathrm{J\ mol^{-1}K^{-1}}$$

$$C_{P,\mathrm{CO_2(g)}} = 44.1 + 9.04\times10^{-3}T - 8.54\times10^{5}T^{-2}, \mathrm{J\ mol^{-1}K^{-1}}$$

*Example 2.8*

(1) Prove the following statements:
   (a) $\Delta U$ and $\Delta H$ are usually very similar to each other for processes involving solids or liquids.
   (b) If gases are involved in a process, these may be significantly different.
   (c) Derive the relationship of $C_V$ and $C_P$ for an ideal gas.

(2) If a reaction involves an increase of 1 mole of gases in the system, calculate the difference $\Delta H$ - $\Delta U$ at 298 K.

(1) (a) $PV$ work of condensed phases is normally negligibly small:

$$\Delta H = \Delta U + \Delta(PV) \cong \Delta U$$

$$C_V = \left(\frac{\partial U}{\partial T}\right)_V \qquad C_P = \left(\frac{\partial H}{\partial T}\right)_P$$

↓

$$C_P = C_V$$

(b) If gases are involved in a process,

$$\Delta H = \Delta U + \Delta(PV) \xrightarrow[\text{for a perfect gas}]{PV = nRT} \Delta H = \Delta U + \Delta(n)RT$$

Therefore, if there is a change in the total number of moles of the gas phase, $\Delta H$ may be significantly different from $\Delta U$.

(c) For ideal gases,

$$H = U + PV$$

↓ $PV = RT$

$$H = U + RT$$

↓ Differentiating

$$dH = dU + RdT$$

↓ $U = f(T)$ only for ideal gases.

$$\frac{dH}{dT} = \frac{dU}{dT} + R$$

↓

$$C_P - C_V = R$$

(2) $\Delta H - \Delta U = \Delta(n)RT = (1)(8.314 \text{ J mol}^{-1} \text{ K}^{-1})(298K) = 2.48 \text{ kJ mol}^{-1}$

*Example 2.9*

$C_P$ and $C_V$ for argon gas are $C_P = 20.8 \text{ J mol}^{-1}\text{K}^{-1}$ and $C_V = 12.5 \text{ J mol}^{-1}\text{K}^{-1}$. Calculate $C_P - C_V$, and discuss significance of the value calculated.

$C_P - C_V = 20.8 - 12.5 = 8.3$ J mol$^{-1}$K$^{-1}$.
This value is very close to the gas constant $R$, which verifies the relationship $C_P - C_V = R$.

*Example 2.10*

Substances usually expand with increase in temperature at constant pressure. Is $C_P$ usually larger than $C_V$?

When thermal energy (heat) is supplied to a substance,

At constant $P$,

Heat $q$

- Part of the heat may be used for work of volume expansion,
- The rest will produce a rise in temperature.

At constant $V$,

Heat $q$

- All the heat supplied produces a rise in temperature.

Therefore, $C_p$ is larger than $C_v$.

More rigorous analysis:

$$C_P - C_V = \left(\frac{\partial H}{\partial T}\right)_P - \left(\frac{\partial U}{\partial T}\right)_V$$

from the definition of $C_P$ and $C_V$.

$H = U + PV$

$$C_P - C_V = \left(\frac{\partial U}{\partial T}\right)_P + P\left(\frac{\partial V}{\partial T}\right)_P - \left(\frac{\partial U}{\partial T}\right)_V$$

$U = f(T, V)$

$$dU = \left(\frac{\partial U}{\partial V}\right)_T dV + \left(\frac{\partial U}{\partial T}\right)_V dT$$

$$\left(\frac{\partial U}{\partial T}\right)_P = \left(\frac{\partial U}{\partial V}\right)_T \left(\frac{\partial V}{\partial T}\right)_P + \left(\frac{\partial U}{\partial T}\right)_V$$

$$C_P - C_V = P\left(\frac{\partial V}{\partial T}\right)_P + \left(\frac{\partial U}{\partial V}\right)_T \left(\frac{\partial V}{\partial T}\right)_P$$

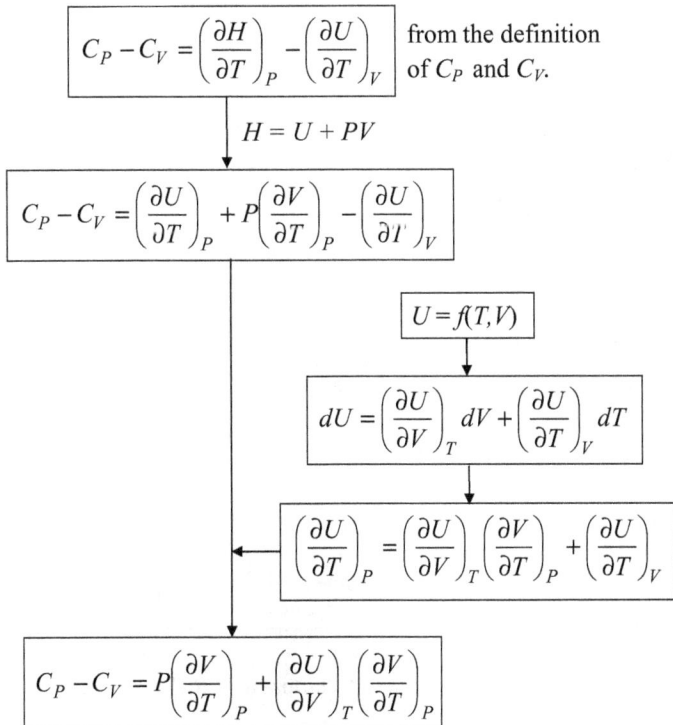

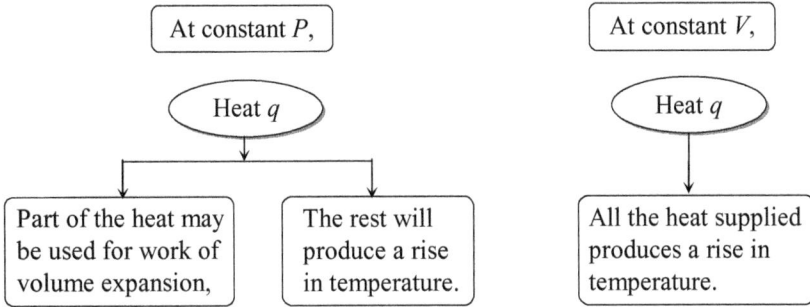

$$C_P - C_V = P\left(\frac{\partial V}{\partial T}\right)_P + \left(\frac{\partial U}{\partial V}\right)_T\left(\frac{\partial V}{\partial T}\right)_P$$

The contribution to $C_P$ through the change in the volume of the system due to the increase in temperature against the constant external pressure.

The contribution from the energy required for the change in volume against the internal cohesive forces acting between the constituent particles of a substance. For liquids and solids, which have strong internal cohesive forces, the term $(\partial U/\partial V)_T$ is large. For gases this term is usually small compared with $P$. An ideal gas is a gas consisting of non-interacting particles, and hence this term is zero.
$(\partial U/\partial V)_T = 0$ for ideal gases

$$C_P - C_V = P\left(\frac{\partial V}{\partial T}\right)_P$$

$PV = RT$

$$C_P - C_V = R$$

*Example 2.11*

In a reversible, adiabatic process of a system comprising of one mole of an ideal gas, prove the following relationships:

$$dU = \delta w$$

$$C_V dT = -PdV$$

(1) $dU = \delta q + \delta w$, but $\delta q = 0$ in an adiabatic process.

(2)

$$dU = \delta w$$

$dU = C_V dT$
$\delta w = -PdV$

$$C_V dT = -PdV$$

From this equation, a useful relationship can be derived, as given below:

$$C_V dT = -PdV$$

↓ $PV = RT$ or $P = RT/V$

$$C_V \frac{dT}{T} = -R\frac{dV}{V}$$

↓ On integrating between states 1 and 2

$$C_V ln\left(\frac{T_2}{T_1}\right) = Rln\left(\frac{V_1}{V_2}\right) \rightarrow \left(\frac{T_1}{T_2}\right)^{C_V} = \left(\frac{V_1}{V_2}\right)^{R}$$

↓

$$\frac{T_2}{T_1} = \left(\frac{V_1}{V_2}\right)^{\frac{R}{C_V}}$$

$$C_P - C_V = R$$

↓

$$\frac{C_P}{C_V} - 1 = \frac{R}{C_V}$$

↓ Put $C_P/C_V = \gamma$

$$\gamma - 1 = R/C_V$$

↓

$$\frac{T_2}{T_1} = \left(\frac{V_1}{V_2}\right)^{\gamma-1}$$

↓ $PV = RT$ or $T = PV/R$

$$P_1V_1^{\gamma} = P_2V_2^{\gamma} = \text{constant}$$

*Example 2.12*

For reversible adiabatic expansion of an ideal gas, we have seen

$$PV^{\gamma} = c \text{ (constant)}$$

When a system comprising of one mole of an ideal gas changes its state from $(P_1, V_1, T_1)$ to $(P_2, V_2, T_2)$, prove that work done by the system is

$$w = \frac{P_2V_2 - P_1V_1}{\gamma - 1} = C_V(T_2 - T_1)$$

$$\delta w = -PdV$$

↓ $PV^{\gamma} = c$

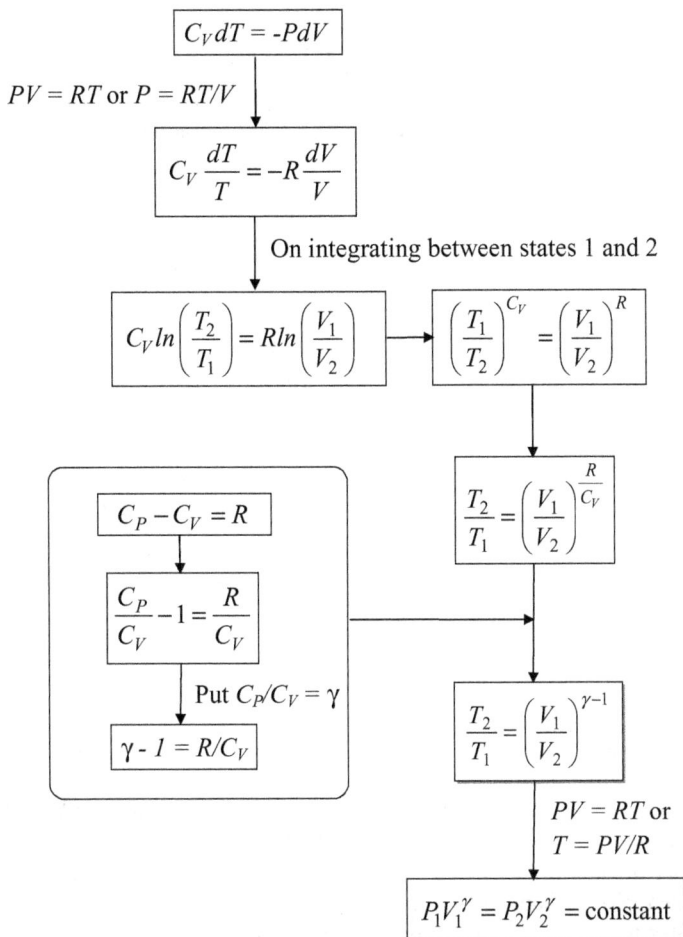

$$\delta w = -c\frac{dV}{V^{\gamma}}$$

↓ Integrating

$$w = \frac{-c}{1-\gamma}\left(V_2^{(1-\gamma)} - V_1^{(1-\gamma)}\right)$$

↓ $c = P_1V_1^{\gamma} = P_2V_2^{\gamma}$

$$w = \frac{P_2V_2 - P_1V_1}{\gamma - 1}$$

↓ For an ideal gas, $PV = RT$ and knowing $C_P$ - $C_V = R$, and $C_P/C_V = \gamma$

$$w = C_V\left(T_2 - T_1\right)$$

*Example 2.13*

Phase transitions between solid and liquid, and liquid and vapor involve large amounts of energy compared to the heat capacity. The diagram below shows schematically the change of temperature of $H_2O$ as heat is added at a given pressure. When phase changes (ice-to-water and water-to-vapor) occur, the diagram shows plateaus, *i.e.*, the temperature stays unchanged until the phase changes are complete even though heat continues to be added. What is the role of the thermal energy added during the phase transitions?

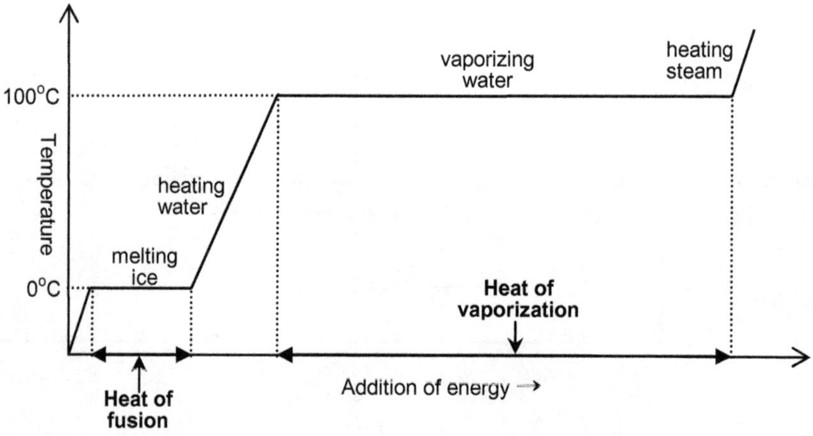

When thermal energy (heat) goes into material, one of two things can happen: The temperature of the material rises, or the material changes the state. When ice is heated, its temperature rises until the temperature reaches the melting point (the transition temperature from ice to water).

At the melting point (0°C), ice melts into water, but the temperature stays at 0°C until ice has melted completely. During the transition, the average kinetic energy of ice is the same as the average kinetic energy of water. Thermal energy absorbed does not go to the kinetic energy of the molecules (not speed up the molecules), but instead it is used to change (loosen) intermolecular bonds and hence stored at potential energy which does not affect the temperature.

The energy required to change a liquid into the gaseous state at the boiling point is called ***heat of vaporization***. This extra energy needed to turn water into vapor is used to break down the intermolecular attractive bonds, and also to provide the energy necessary to expand the gas (the $PV$ work).

*Exercises*

2.3 Metallic vapors generally have a monatomic constitution. According to the kinetic theory of gases, only three translational degrees of freedom need to be considered for a monatomic gas and hence the translational kinetic energy is given by

$$U = \tfrac{3}{2} nRT$$

(1) Calculate $\Delta H$ when the temperature of 3 moles of the gas is raised from 700 to 1000 K.
(2) Calculate $C_V$ for the gas.
(3) Calculate $C_P$ for the gas.

## 2.3 Enthalpy Change

For a substance of fixed composition, the enthalpy change with change in temperature at constant pressure $P$ can be calculated as follows:

From the definition,

$$C_P = \frac{dH}{dT}$$

↓

$$dH = C_P dT$$

↓ On integration from $T_1$ to $T_2$,

$$\Delta H = \int_{T_1}^{T_2} C_P \mathrm{d}T$$

$$C_P = a + bT + cT^{-2}$$

$$\Delta H = \int_{T_1}^{T_2} (a + bT + cT^{-2})\mathrm{d}T$$

The enthalpy change associated with a chemical reaction or phase change at constant pressure and temperature can be calculated from the enthalpy of each species involved in the process. When species A undergoes phase transformation from α to β,

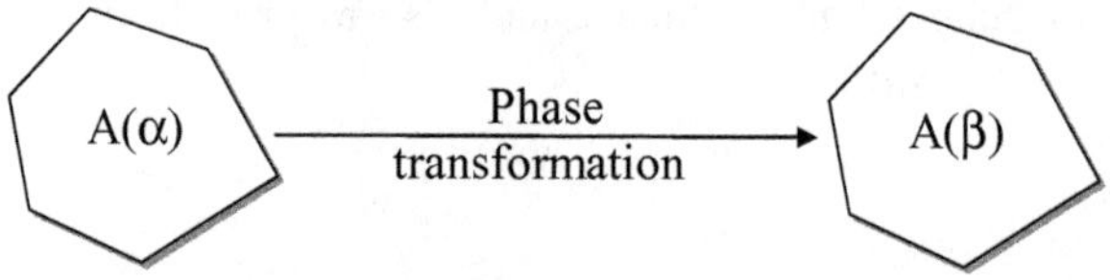


$$\Delta H_t = H_{A(\beta)} - H_{A(\alpha)}$$

The enthalpy change due to chemical reaction ($\Delta H$) is the difference between the sum of enthalpies of the products and the sum of enthalpies of the reactants:

$$\Delta H = \sum H_{products} - \sum H_{reactants}$$

(Example) $Fe_2O_3 + 2Al = Al_2O_3 + 2Fe$: $\Delta H = (H_{Al_2O_3} + 2H_{Fe}) - (H_{Fe_2O_3} + 2H_{Al})$

*Example 2.14*

Pure copper melts at 1,084°C. Calculate the enthalpy change when 1 mole of copper is heated from 1,000°C to 1,100°C. ($C_{P,Cu(l)} = 31.4$ J mol$^{-1}$K$^{-1}$, $C_{P,Cu(s)} = 22.6 + 6.28 \times 10^{-3}\ T$, J mol$^{-1}$K$^{-1}$, Heat of fusion ($\Delta H_t$) : 13,000 J mol$^{-1}$)

Total enthalpy change = Enthalpy change associated with heating of solid copper to the melting temperature

+ Heat of fusion at the melting temperature

+ Enthalpy change associated with heating of liquid copper to the temperature of 1,100°C.

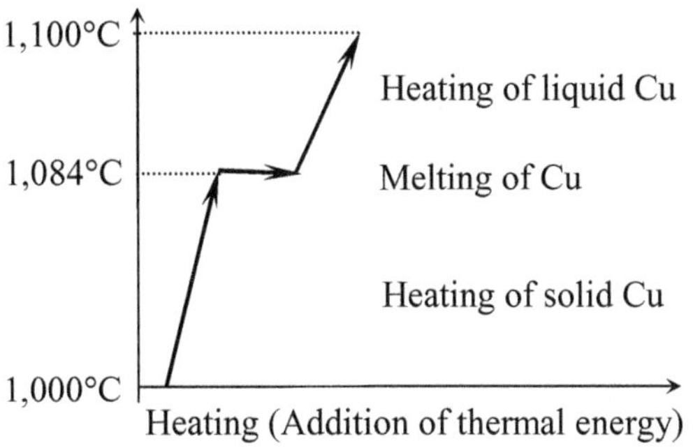


Heating of liquid Cu $\Delta H_L = \int_{1357}^{1373} C_{P,\mathrm{Cu}(l)} \mathrm{d}T = 502 \text{ J mol}^{-1}$

Melting of Cu $\Delta H_t = 13{,}000 \text{ J mol}^{-1}$

Heating of solid Cu $\Delta H_S = \int_{1273}^{1357} C_{P,\mathrm{Cu}(s)} \mathrm{d}T = 2{,}592 \text{ J mol}^{-1}$

Thus, the total enthalpy change: $\Delta H = \Delta H_S + \Delta H_t + \Delta H_L = 16{,}094 \text{ J mol}^{-1}$

> *Example 2.15*
>
> Molten copper is supercooled to 5°C below its true melting point (1,084°C). Nucleation of solid copper then takes place and solidification proceeds under adiabatic conditions. Calculate the percentage of the solid copper.

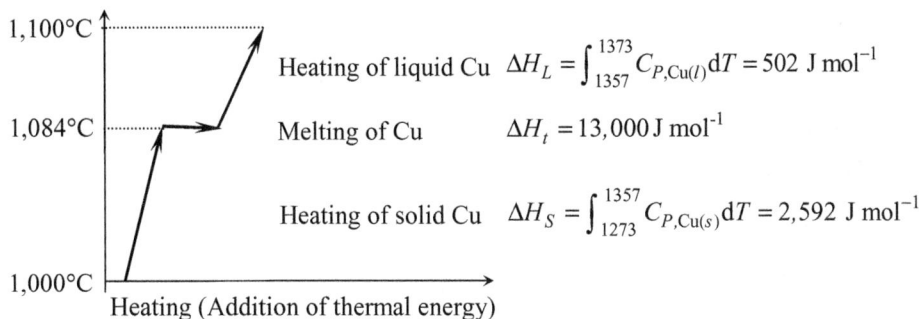

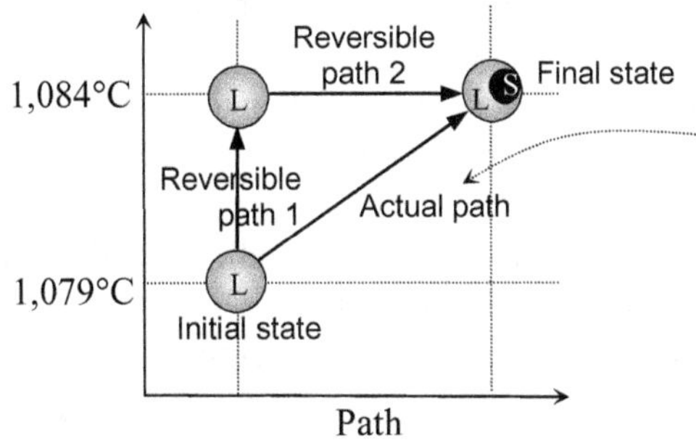


The actual path is not convenient for thermodynamic calculations in this case, because copper solidifies during the process and the heat of fusion at temperatures other than the true melting point is not readily known.

Instead, it is more convenient to take imaginary paths:

(1) Liquid copper is heated from 1,079°C to 1,084°C (Path 1): $\Delta H_{path1} > 0$.

(2) And then portion of the liquid copper solidifies at the true melting point (Path 2): $\Delta H_{path2} < 0$.

(3) As the process is adiabatic, $\Delta H_{path1} + \Delta H_{path2} = 0$.

Since enthalpy is a state function, the enthalpy change along this imaginary paths should be the same as that along the true path.

$$\Delta H_{path1} = \int_{1352}^{1357} C_{P,\mathrm{Cu}(l)} \mathrm{d}T = 31.4(1357 - 1352) = 157 \text{ J mol}^{-1}$$

↖ $31.4 \text{ J mol}^{-1}\text{K}^{-1}$

$$\Delta H_{path2} = -x\Delta H_t = -13{,}000x$$

where $x$ = fraction of copper solidified, and

$\Delta H_t$ = heat of fusion of copper (13,000 J mol$^{-1}$)

As solidification proceeds under adiabatic conditions,

$$\Delta H_{whole\ process} = 0 = \Delta H_{path1} + \Delta H_{path2}$$

Therefore, $x = 0.012$: fraction of solid copper

*Example 2.16*

The reaction between methane and oxygen produces carbon dioxide and water vapor.

Reaction path 1 : $CH_4 + 2O_2 \rightarrow CO_2 + 2H_2O$ $\quad \Delta H_1$

Reaction path 2 : $CH_4 \rightarrow C + 2H_2$ $\quad \Delta H_{2a}$

$\quad\quad \rightarrow 2H_2 + O_2 \rightarrow 2H_2O$ $\quad \Delta H_{2b}$

$\quad\quad \rightarrow C + O_2 \rightarrow CO_2$ $\quad \Delta H_{2c}$

Prove that $\Delta H_1 = \Delta H_{2a} + \Delta H_{2b} + \Delta H_{2c}$.

Because enthalpy is a state property, the enthalpy change depends on the initial and final states only, not on the path the process follows. As the sum of all the reactions in path 2 results in the same reaction as the one in path 1, the enthalpy change should be the same for both paths.

The additive properties of enthalpy is known as ***Hess's Law***. According to this law,

- The enthalpy change associated with a given chemical reaction is the same whether it takes place in one or several steps.
- Enthalpies or enthalpy changes may be added or subtracted in parallel with the same manipulations performed on their respective components or reactions.

The above is in fact a different expression of the state property of enthalpy.

*Exercises*

2.4 The melting point of $CaTiSiO_5$ is 1,400°C and the heat of fusion at the normal melting point is 123,700 $J\ mol^{-1}$. Calculate the heat of fusion at 1,300°C.

$C_{P,solid} = 177.4 + 23.2\times10^{-3}T - 40.3\times10^{5}T^{-2}$, J mol$^{-1}$K$^{-1}$

$C_{P,liquid} = 279.6$ J mol$^{-1}$K$^{-1}$

2.5 Enthalpy changes resulting from temperature change can be represented on an enthalpy-temperature diagram as shown in the figure. Express on the diagram the answers to the following questions:

(1) Enthalpy change when solid A melts at $T_m$

(2) Enthalpy change when liquid A is supercooled from $T_m$ to $T_1$, and then solidifies.

(3) Enthalpy change when solid A is superheated from $T_m$ to $T_2$ and then melts

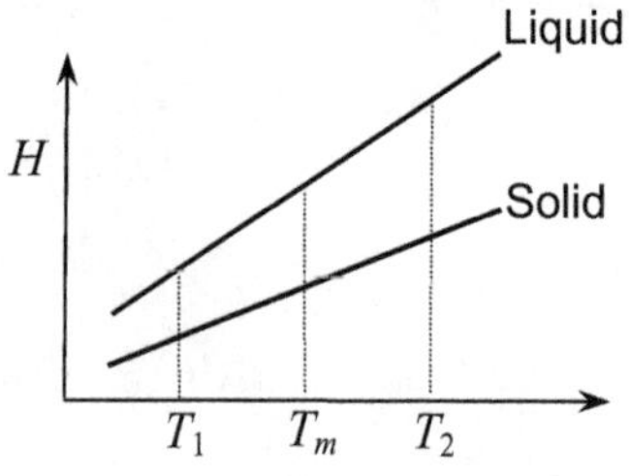

# Chapter 3

# The Second Law of Thermodynamics

## 3.1 Grades of Energy

Thermodynamics deals with energy. The first law is concerned with the conservation of energy in energy transfer. Energy may be transferred from one place to the other or transformed in one form to another, but the total energy of the universe (the system + the surroundings) is always conserved.

Now a question arises as to in which direction energy flows in a natural process, and if there is any law which governs the direction of the energy flow. Thus, the primary interest in thermodynamics is to predict the direction of chemical or physical processes, in particular the spontaneous, natural direction. Some examples we observe in our everyday life are,

- Energy always flows spontaneously from a higher temperature to a lower temperature, but not the reverse (thermal energy).
- Energy always flows naturally from a higher pressure to a lower pressure (mechanical energy).
- Energy always flows from a higher voltage potential to a lower voltage potential (electrical energy).
- Wood burns spontaneously in air if ignited, but the reverse process, *i.e.*, the spontaneous recombination of the combustion products to wood and oxygen in air, has never been observed in nature (chemical energy).
- Ice at 1 *atm.* pressure and a temperature above 0°C always melts spontaneously, but water at 1 *atm.* pressure and a temperature above 0°C never freezes spontaneously in nature (phase transition).

There will be numerous examples of processes which proceed in one direction, but not in the other direction spontaneously. The above examples strongly support the view that energy contained in the source from which the energy is drawn must be different in ***quality*** or ***grade*** from the energy contained in the sink into which the energy flows.

Thermal energy in water

Suppose we have two buckets of water of the same amounts, but one at temperature of 90°C, and the other at 40°C (Refer to the figure below.) Now thermal energy (heat) (Q) is taken out of the hot water by cooling from 90°C to 80°C ($\Delta T$ = -10°C). The energy so extracted is transferred to the cold water, and heats the cold water from 40°C to 50°C ($\Delta T$ = +10°C), assuming that there is no heat loss and the heat capacity of water is constant. The energy Q which was originally in the hot water is now in the cold water. Here a question arises as to whether it would be possible to let the same amount of energy Q return to the hot water *spontaneously* so that the temperature of the hot water recovers its original temperature, 90°C ($\Delta T$ = +10°C) – see the following figure. Our everyday experience disproves it. If it were the case, the hot water could get hotter while the cold water becomes colder. Even though we talk about the same amount of energy Q, say, 1,000 joules, the energy in the hot water and that in the cold water are different in terms of *ability* or *usefulness*. This example clearly reveals that energy has quantity, and also *quality* as well.

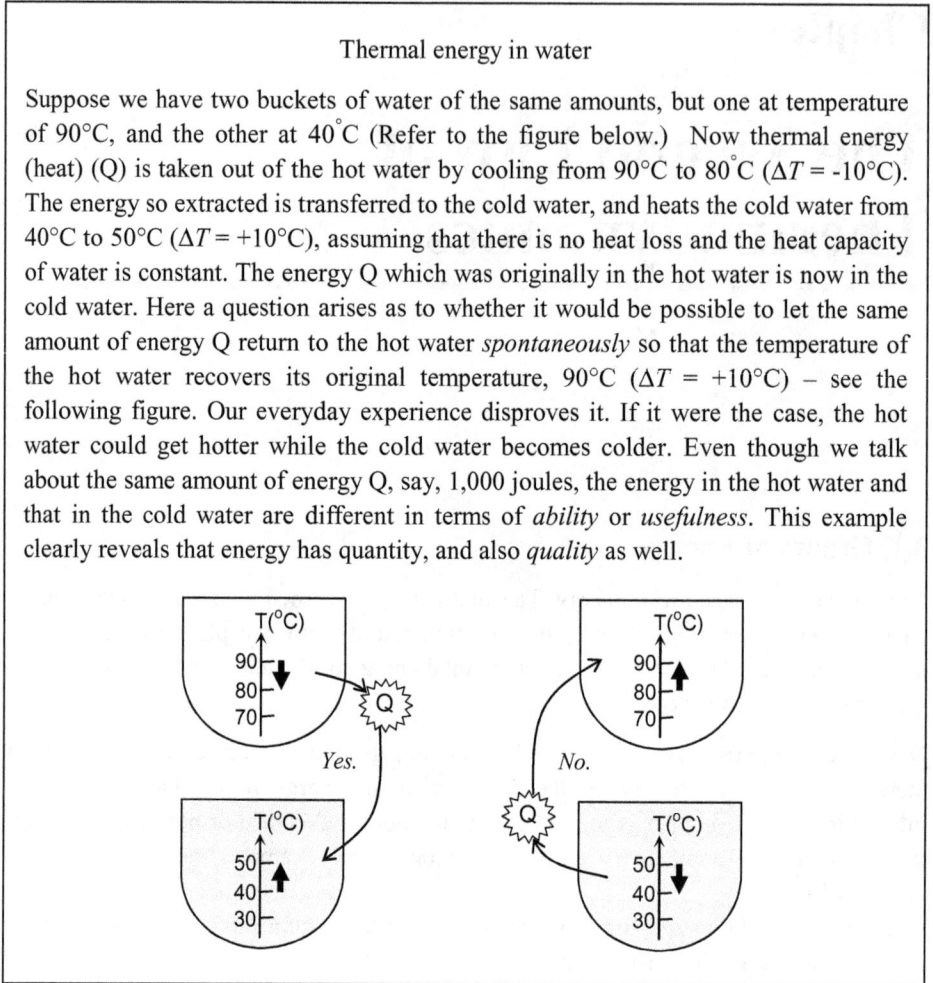

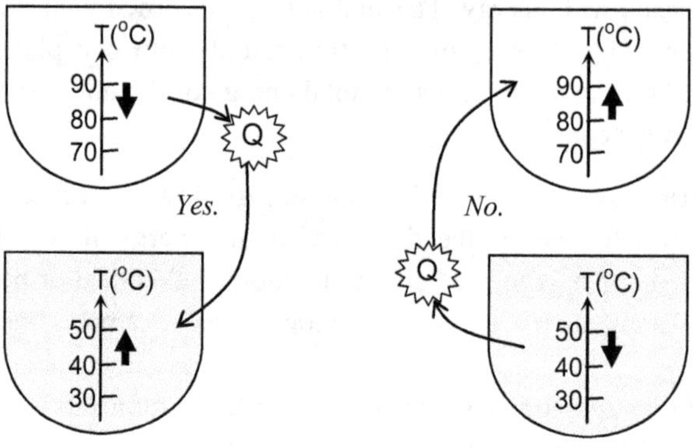


It would be very useful if we could find *a function or property of a system by which the quality* (*or grade*) of *energy is quantitatively determined*. As it determines the direction of its flow, the energy grade of a system must be a function of the thermodynamic driving forces of the system which provides the impetus for the transfer of the energy from the system. When the thermodynamic driving forces in the system decrease, the energy in the system is degraded. In this regard, the energies contained in the sources in the above examples are at the higher energy grades than those in the sinks. For example, the energy contained in a higher temperature system is at the higher energy grade than that in a lower temperature system.

When two systems exchange thermal energy in the mode of heat, the energy grades of both systems will be changed. Suppose two systems at different temperatures are placed in contact with each other. Then thermal energy will be transferred in the mode of heat from the high temperature system to the low temperature one. The following diagram schematically represents the changes of the energy and temperature of the systems:

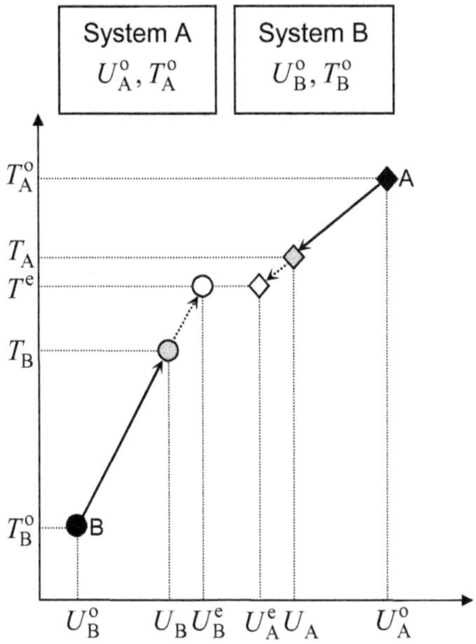


- System A (high temperature) and System B (low temperature) are brought into contact with each other.
- As the thermal energy is transferred, System A experiences decrease in the amount of energy and also its temperature, whereas in System B both the amount of energy and its temperature increase.
- The transfer of thermal energy continues until the temperatures of both systems arrive at the same temperature of $T^e$.
- During the thermal process, the energies of both systems change:

System A: $U_A^o \rightarrow U_A \rightarrow U_A^e$ (energy loss)

System B: $U_B^o \rightarrow U_B \rightarrow U_B^e$ (energy gain)

- According to the first law, the energy loss of System A is equal to the energy gain of System B, and thus the total energy is conserved:

$$U_A^o + U_B^o = U_A + U_B = U_A^e + U_B^e$$

- The temperatures of both systems change:

$$\underline{\text{System A}} \qquad\qquad \underline{\text{System B}}$$
$$T_A^o \rightarrow T_A \rightarrow T^e \leftarrow T_B \leftarrow T_B^o$$
$$\text{(decrease)} \qquad\qquad \text{(increase)}$$

- Note that the energy in System A has experienced decrease in quantity and also in quality (grade) as the temperature of A has decreased, whereas the energy in System B is increased in quantity and also quality.
- The net effect of the above heat transfer process is that the total energy is conserved, but System A has left with a permanent ***degradation*** of its initial energy. The energy in System B has been upgraded, but the so-upgraded energy is not useful (or available) for the purpose of returning the so-degraded energy of System A back to the level of its initial quality. The only way to bring the energy grade of System A back to the initial value by means of thermal energy is to transfer energy from a system which is at the temperature higher than $T_A^o$.

From the above example, it can be concluded that,

- the increase of the energy grade or energy quality of a system by heat transfer is possible, but only at the expense of lowering the energy grade of another system whose energy grade is higher than that of the system, and

- the so-upgraded energy of the system is of no use for recovering the energy grade of the other so-degraded system, and
- the universe (the sum of both systems in the present case) has suffered from a permanent degradation of the energy.

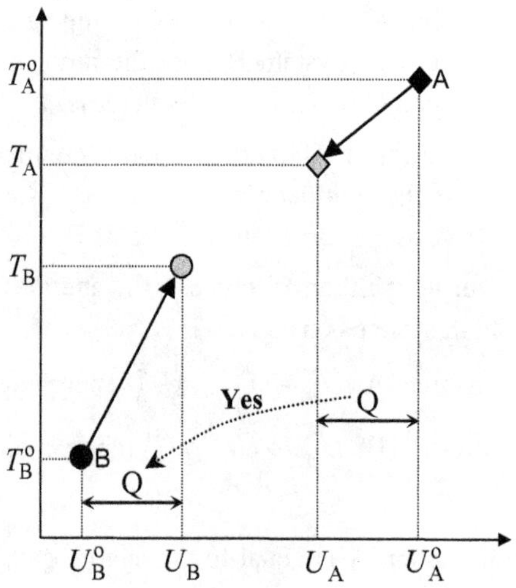


Thermal energy (Q) released from System A by cooling can be transferred to System B directly and thus raise the temperature of System B.

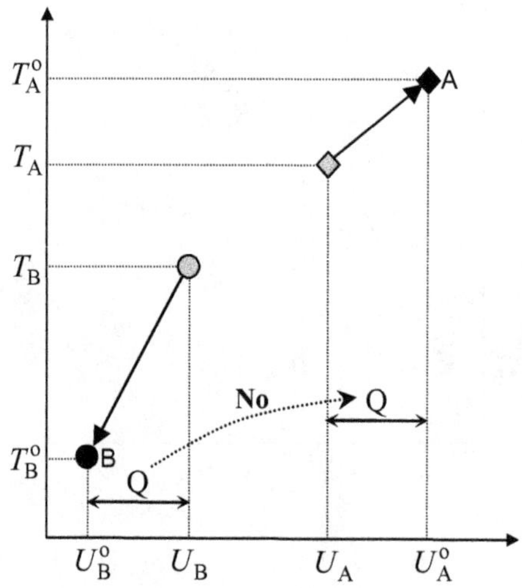


Thermal energy (Q) released from System B by cooling cannot be transferred to System A directly and is unavailable for raising the temperature of System B.

Through the above discussions, it is now clear that factors which determine the extent of change in the grade of energy of a system is not just the *quantity of energy* transferred, but also the *temperature* of the system at which the energy transfer occurs.

Consider the following example in which thermal energy is transferred by heat flow:

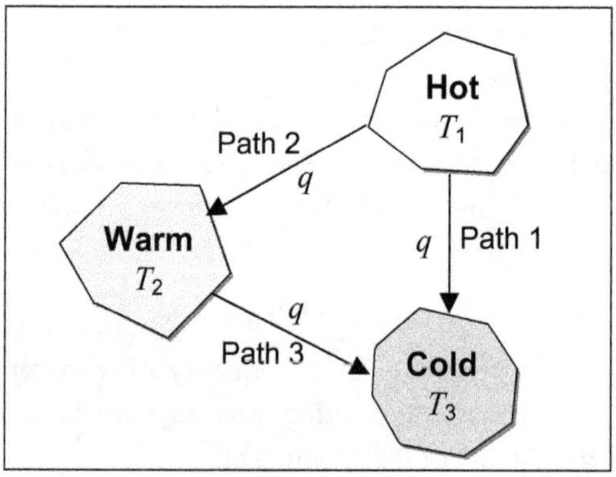

Thermal energy $q$ is to flow in the mode of heat from the hot body to the cold body. The heat flow may occur either

(1) directly from the hot to cold bodies, *i.e.*, Path 1, or
(2) from the hot to warm, and then to cold bodies, *i.e.*, Path 2 + Path 3.

The degradation or decrease in quality of the thermal energy $q$ is greater in Path 1 than in Path 2, as in the latter path further degradation should occur through Path 3 to reach the same final state. Thus,

- Degradation in Path 1 > Degradation in Path 2
- Degradation in Path 1 > Degradation in Path 3

In other words,

*Path 1 is more irreversible than either Path 2 or Path 3.*

From the view point of the conversion of thermal energy into work ($w$), the above analysis can be interpreted as follows:

- Once the thermal energy $q$ is transferred to a lower temperature, say, from $T_1$ to $T_3$, it loses the capacity to do work which might have been done if a chance were given during the course of the transfer.
- The loss of capacity to do work is greater in Path 1 than in Path 2, as, after Path 2, there is a chance to do work during further cooling from $T_2$ to $T_3$ (Path 3).
- The loss of capacity to do work and the degradation (or decrease in quality) of thermal energy shares the same origin; that is, the transfer of thermal energy.

Therefore, the quantification of degradation of thermal energy or the loss of capacity to do work must include both the amount of thermal energy transferred in the mode of heat, $q$, and the temperature $T$ at which the transfer occurs.

- The larger the amount of heat flow, the greater the extent of degradation.
- The lower the temperature of the body to which heat flows, the higher the degree of degradation.

Therefore,

$$\text{Extent of degradation} \propto \frac{q}{T}$$

*Example 3.1*

Thermodynamics may be defined as a physical science concerned with the transfer of heat and the performance of work accompanying various physical and chemical processes. The first law of thermodynamics is the law of energy conservation. Explain, using an example, why the first law is not enough to answer the following comments:

"Some things happen spontaneously; some other things don't."

Refer to the following two figures:

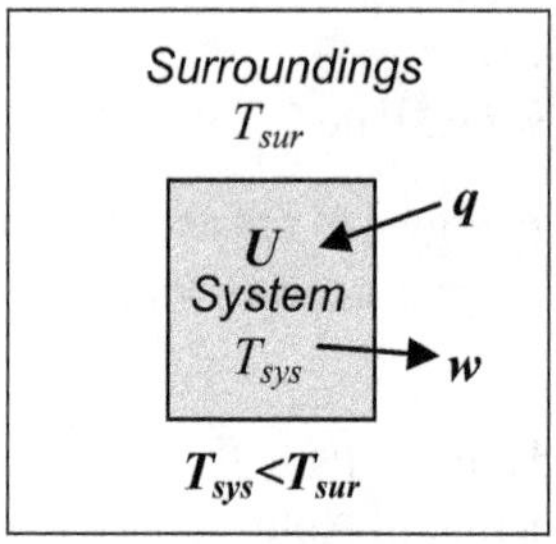


The process satisfies $\Delta U = q + w$, and occurs spontaneously as indicated by the arrows.

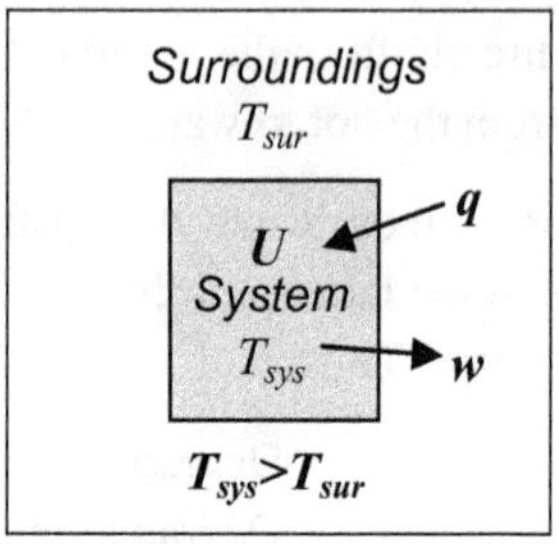


The process satisfies $\Delta U = q + w$, but does not occur spontaneously as indicated by the arrows.

The above example demonstrates that the first law of thermodynamics alone is not enough to completely deal with processes which occur naturally. We now search in several different ways a function or property which can be a barometer of the direction of a natural or spontaneous process.

## 3.2 Heat Engines and Entropy

A heat engine is a device for converting heat (thermal energy) into work (*e.g.*, steam engine, internal combustion engine). The figure on the right is the schematic representation of a heat engine.

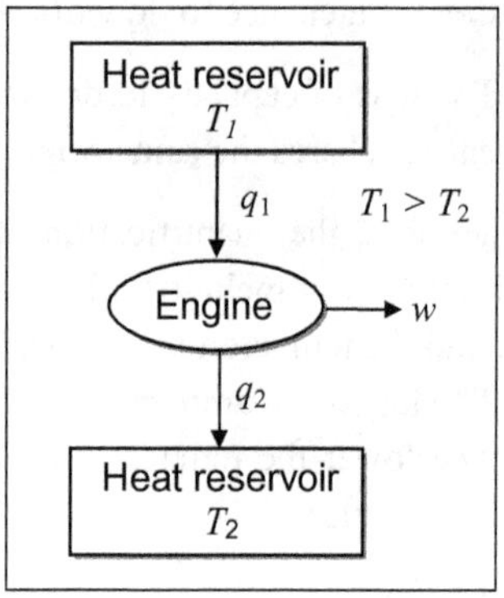


- Each cycle takes thermal energy of heat ($q_1$) from the high temperature heat reservoir, and
- uses some of it to generate work ($w$), and
- rejects the unused portion ($q_2$, thermal energy) to the low temperature heat reservoir

The ***efficiency*** of the engine is defined as

$$\varepsilon = \frac{\text{The amount of thermal energy expended to conduct work}}{\text{The amount of thermal energy supplied to the engine}} = \left|\frac{w}{q_1}\right|$$

Where "| |" is to ensure $\varepsilon$ to be positive.

The prime interest we may have in dealing with a heat engine is to know the maximum amount of work that can be obtained from each cycle of the heat engine. We already know that the maximum work can be obtained when the process is conducted in a reversible manner (refer to Chapter 2.1). It is possible to devise a cyclic process which consists of processes which are all reversible.

## Carnot Cycle

The ***Carnot cycle*** is the cyclic operation of an idealized (ideal gas) engine which runs in four steps: two *reversible* isothermal steps and two *reversible* adiabatic steps. After completing a cycle, the system returns to its initial state so that all of the state functions restore their initial values.

The figure below shows the Carnot engine, in which heat $q_1$ is taken in from the high temperature heat reservoir, work $w$ is done through running a complete cycle of the Carnot engine, and heat $q_2$ is rejected to the low temperature heat reservoir.

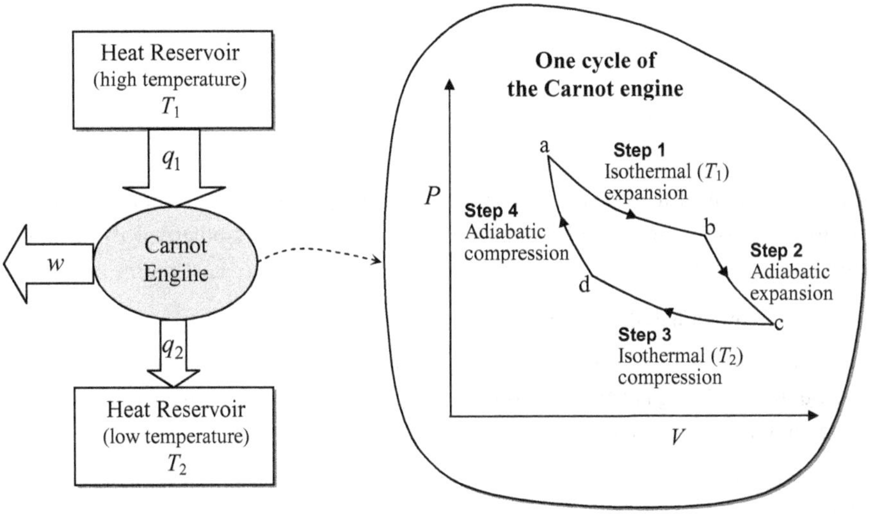


It is clear from the figure that work is done *by* the Carnot engine through Step 1 and Step 2 as the gas expands, and work is done *on* the engine through Step 3 and Step 4 as the gas is compressed. The net amount of work ($w$) done by the engine is the balance of the work done by the engine and the work done on the engine.

Now we discuss each step of the Carnot cycle one by one.

**Step 1**: Reversible isothermal ($T_1$) expansion:

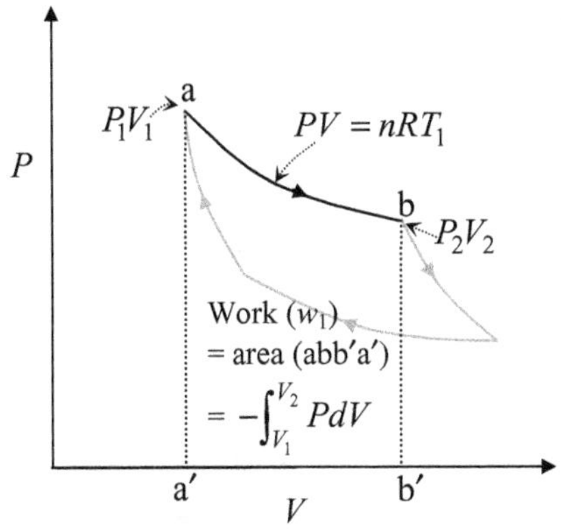


$$w = -\int_{V_1}^{V_2} PdV$$

$$\downarrow \; PV = nRT_1$$

$$w_1 = -nRT_1 \int_{V_1}^{V_2} \frac{1}{V} dV$$

$$\downarrow$$

$$w_1 = nRT_1 \ln \frac{V_1}{V_2}$$

The engine takes heat $q_1$ from the high temperature ($T_1$) heat reservoir, and the gas in the engine expands isothermally and reversibly. The figure shown above explains how to obtain the amount of work done by the engine in Step 1.
Since the gas is at the constant temperature $T_1$, the internal energy ($U$) stays unchanged:

$$\Delta U_1 = 0$$

$$\Delta U_1 = q_1 + w_1$$

$$w_1 = nRT_1 \, ln \frac{V_1}{V_2}$$

$$q_1 = nRT_1 \, ln \frac{V_2}{V_1}$$

This is the amount of heat taken in from the high temperature heat reservoir to do work $w_1$ by the reversible isothermal expansion. Note that in this isothermal expansion the temperature of the gas is the same as that of the high temperature heat reservoir ($T_1$).

**Step 2**: Reversible adiabatic expansion:

The engine at this step is now *insulated* so that no heat is allowed to be exchanged with the surroundings. The gas expands from $V_2$ to $V_3$ and the pressure changes from $P_2$ to $P_3$, as depicted in the figure below.

Since the expansion is conducted under the adiabatic condition, the temperature of the gas decreases. The expansion continues until the temperature has become the same as the temperature of the low temperature heat reservoir ($T_2$).

For the adiabatic change of the gas, the relationship of $PV = nRT$ does not hold, but instead the relationship of $PV^\gamma = c$ (constant) applies (see Example 2.11).

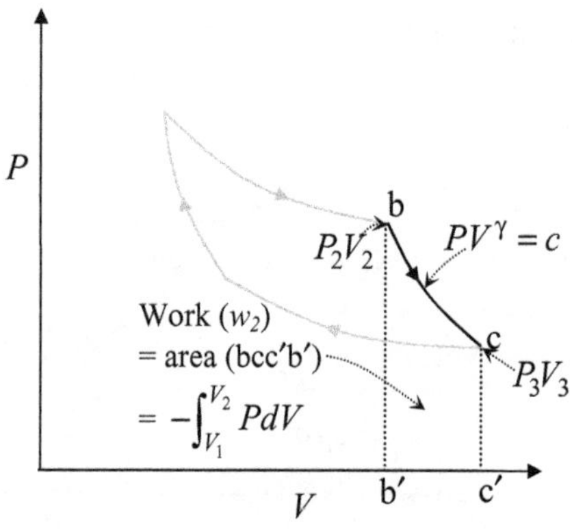


$$w_2 = -\int_{V_2}^{V_3} PdV$$

$$PV^\gamma = c$$

$$w_2 = -c\int_{V_2}^{V_3} \frac{1}{V^\gamma} dV$$

$$w_2 = -c \frac{V_3^{1-\gamma} - V_2^{1-\gamma}}{1-\gamma}$$

$$c = P_2V_2^\gamma = P_3V_3^\gamma$$

$$w_2 = -\frac{P_3V_3 - P_2V_2}{1-\gamma}$$

This is the work done by the engine through the reversible adiabatic expansion.
From the ideal gas law, we know the relationships of $P_2V_2 = nRT_1$ and $P_3V_3 = nRT_2$, and thus the last equation becomes

$$w_2 = -nR\frac{(T_2 - T_1)}{1-\gamma}$$

$$\gamma = \frac{C_P}{C_V}$$

$$C_P - C_V = R$$

$$w_2 = nC_V(T_2 - T_1)$$

$$q = 0 \text{ (adiabatic)}$$

$$\Delta U = q + w$$

$$\Delta U_2 = nC_V(T_2 - T_1)$$

As can be seen above, by the reversible adiabatic expansion, the engine does work $w_2$ against the external pressure, and accordingly the internal energy decreases by $\Delta U_2$ and the temperature of the gas decreases from $T_1$ to $T_2$.

In order to bring the engine back to the initial starting point, the gas which has expanded through the above two steps must now be compressed to the initial volume.

**Step 3**: Reversible isothermal compression:

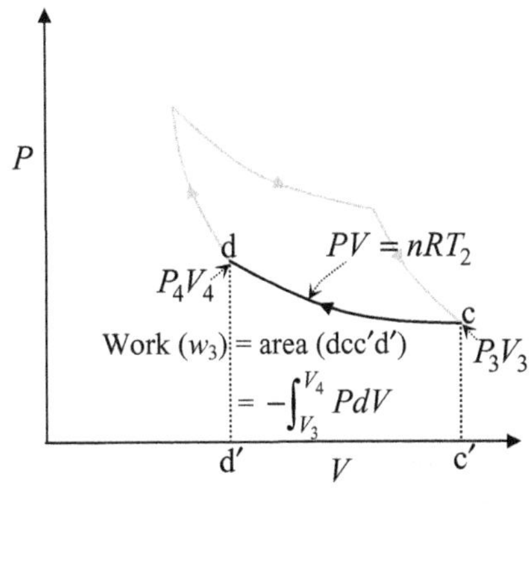


$$w_3 = -\int_{V_3}^{V_4} PdV$$

$$PV = nRT_2$$

$$w_3 = -nRT_2\int_{V_3}^{V_4} \frac{1}{V}dV$$

$$w_3 = nRT_2 \ln\frac{V_3}{V_4}$$

This is the amount of work done *on* the engine through the compression of the gas. Since the gas is at the constant temperature $T_2$, the internal energy ($U$) stays unchanged:

$$\Delta U_3 = 0$$

$$\Delta U_3 = q_2 + w_3$$

$$w_3 = nRT_2 \, ln\frac{V_3}{V_4}$$

$$q_2 = nRT_2 \, ln\frac{V_4}{V_3}$$

This is the amount of heat ($q_2 < 0$) rejected by the gas to the low temperature heat reservoir. Note that in this isothermal compression the temperature of the gas is the same as that of the low temperature heat reservoir ($T_2$).

**Step 4**: Reversible adiabatic compression:

Now the engine is insulated again so that no heat exchange with the surroundings is allowed. The gas is compressed from $V_4$ to $V_1$ and the pressure changes from $P_4$ to $P_1$. The process is graphically represented in the figure below.

Since the compression is conducted under the reversible adiabatic condition, the temperature of the gas rises. The compression continues until the temperature of the gas reaches that of the high temperature heat reservoir ($T_2$) which is also the temperature of the gas at the beginning of the cycle. As seen in Step 2, for the reversible adiabatic change of the gas, the relationship of $PV = nRT$ does not hold, but instead the relationship of $PV^\gamma = c$ (constant) applies.

In a similar way to Step 2,

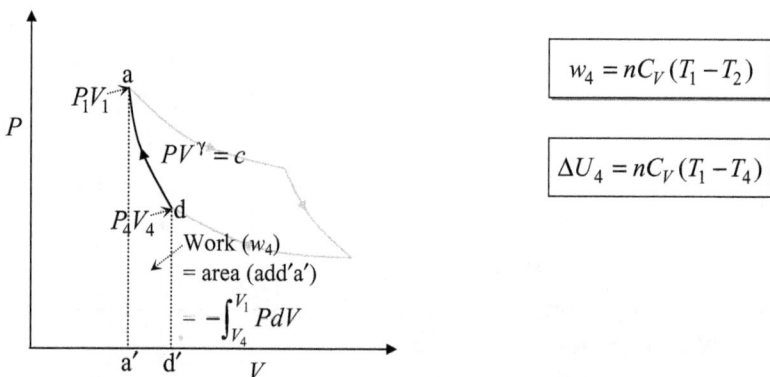

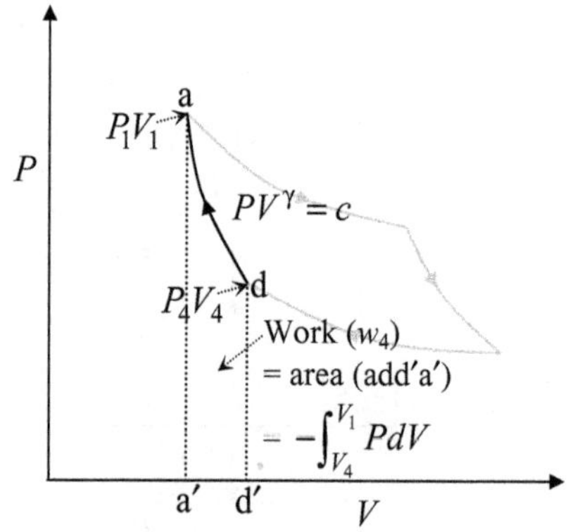


$$w_4 = nC_V(T_1 - T_2)$$

$$\Delta U_4 = nC_V(T_1 - T_4)$$

As seen above, by the reversible adiabatic compression, the work $w_4$ is done *on* the engine against the internal pressure, and accordingly the internal energy increases by $\Delta U_4$ and the temperature of the gas rises from $T_2$ to $T_1$.

The complete Carnot cycle is summarized in the following figure:

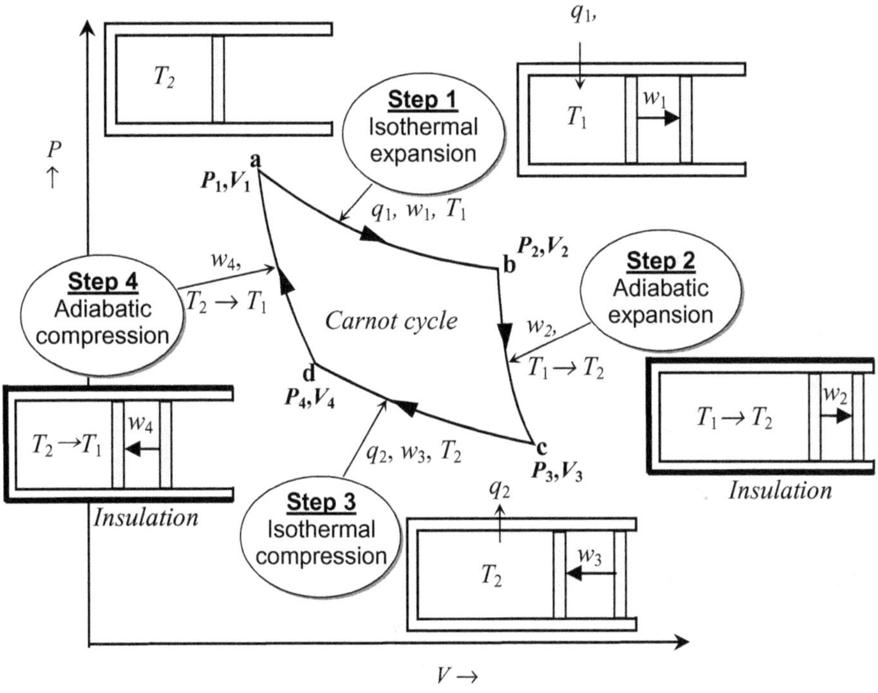


In a complete cycle, the gas follows the path "abcd". During the cycle,

- the amount of heat taken in from the high temperature heat reservoir: $q_1 = nRT_1 \, ln\dfrac{V_2}{V_1}$
- the amount of heat rejected to the low temperature heat reservoir: $q_2 = nRT_2 \, ln\dfrac{V_4}{V_3}$
- the amount of work done by the engine:

$$w = w_1 + w_2 + w_3 + w_4$$

$$w_1 = nRT_1 \, ln\frac{V_1}{V_2} \qquad w_2 = nC_V(T_2 - T_1)$$

$$w_3 = nRT_2 \, ln\frac{V_3}{V_4} \qquad w_4 = nC_V(T_1 - T_2)$$

$$w = nRT_1 \ln\frac{V_1}{V_2} + nRT_2 \ln\frac{V_3}{V_4}$$

$$q_1 = nRT_1 \ln\frac{V_2}{V_1}$$

$$q_2 = nRT_2 \ln\frac{V_4}{V_3}$$

$$w = -(q_1 + q_2)$$

For further consideration on the last equation, let us employ the equations for adiabatic conditions given in Step 2 and Step 4.

For Step 2,

$$P_2V_2^{\gamma} = P_3V_3^{\gamma}$$

$$P_2V_2 = nRT_1,\ P_3V_3 = nRT_2$$

$$T_1V_2^{\gamma-1} = T_2V_3^{\gamma-1}$$

For Step 4, in a similar way,

$$T_2V_4^{\gamma-1} = T_1V_1^{\gamma-1}$$

Combination of the last two equations yields,

$$\frac{V_2}{V_1} = \frac{V_3}{V_4}$$

This last result is useful to bring a very important relationship when it is combined with $q_1$ and $q_2$ in the Carnot cycle.

$$q_1 = nRT_1 \ln\frac{V_2}{V_1} \qquad q_2 = nRT_2 \ln\frac{V_4}{V_3}$$

$$\frac{V_2}{V_1} = \frac{V_3}{V_4}$$

$$\frac{q_1}{T_1} + \frac{q_2}{T_2} = 0$$

This equation is resulted from the reversible cyclic heat engine which is a closed system, in which heat $q_1$ is taken in at $T_1$ and heat $q_2$ is removed at $T_2$, both of which are under the isothermal conditions. Since the sum of changes of *any* state property must be zero in a cyclic process, this last equation suggests that there exists a state property which is related to $q/T$. The change of this property during the expansion is $q_1/T_1$ and its change during the compression is $q_2/T_2$, and the sum of these two changes for a complete cycle is zero, which is the necessary and sufficient condition for a state property. This property is termed ***entropy***. The concept of entropy will be further refined after discussing the efficiency of the Carnot engine.

$$\varepsilon = \frac{\text{The amount of thermal energy expended to conduct work}}{\text{The amount of thermal energy supplied to the engine}} = \left|\frac{w}{q_1}\right|$$

$$w = -(q_1 + q_2)$$

$$\varepsilon = 1 + \frac{q_2}{q_1} \quad \xrightarrow{\frac{q_1}{T_1} + \frac{q_2}{T_2} = 0} \quad \varepsilon = 1 - \frac{T_2}{T_1}$$

This is a remarkable result:

*The efficiency depends only on the temperatures of the reservoirs, and is independent of the nature of the engine, working substance, or the type of work performed.*

The above two equations indicate that only a fraction of the heat taken in from the high temperature reservoir is converted to work. This fraction is

$$\frac{q_1 + q_2}{q_1} \quad \text{or} \quad \frac{T_1 - T_2}{T_1}$$

It is obvious from the above relationship that complete conversion of heat into work is possible only if $q_2$ is equal to zero which means that $T_2$ is equal to $0K$, and thus it indicates that the complete conversion of heat into work is impossible in practice.

*Example 3.2*

Refer to the diagram of the Carnot cycle in the text. An engine operates between 1,200°C ($T_1$) and 300°C ($T_2$), and Step 1 (isothermal expansion) involves an expansion where the pressure of the gas drops from $6\times10^5$ N m$^{-2}$ to $4\times10^4$ N m$^{-2}$. The working substance is one mole of an ideal gas.

(1) Calculate the efficiency of the heat engine.
(2) Calculate heat absorbed in Step 1.
(3) Calculate the amount of heat rejected in Step 3 (isothermal compression).

(1) $\varepsilon = 1 - \frac{T_2}{T_1}$ $\xrightarrow[T_2 = 573K]{T_1 = 1473K}$ $\varepsilon = 0.611$ or 61.1%

(2)

$$q_1 = nRT_1 \, ln\frac{V_2}{V_1}$$

↓ $PV = nRT$

$$q_1 = nRT_1 \, ln\frac{P_1}{P_2}$$

↓
$P_1 = 6\times10^5 \, N\,m^{-2}$
$P_2 = 4\times10^4 \, N\,m^{-2}$
$n = 1mole$
$R = 8.314J\,mol^{-1}K^{-1}$
$T_1 = 1473K$

$$q_1 = 33{,}164 \, J$$

(3)

$$\frac{q_1}{T_1} + \frac{q_2}{T_2} = 0$$

↓
$q_1 = 33{,}164 \, J$
$T_1 = 1473K$
$T_2 = 573K$

$$q_1 = -12{,}900J$$

**Entropy**

Let us return to the important relationship obtained for a reversible cyclic process:

$$\frac{q_1}{T_1} + \frac{q_2}{T_2} = 0$$

↓ Since the heat exchanges are conducted reversibly,

$$\frac{q_{1,rev}}{T_1} + \frac{q_{2,rev}}{T_2} = 0$$

↓ If heat exchanges are conducted in an infinitesimal amount,

$$\frac{\delta q_{1,rev}}{T_1} + \frac{\delta q_{2,rev}}{T_2} = 0$$

↓ Since the operation is a cyclic process,

$$\oint \frac{\delta q_{i,rev}}{T_i} = 0$$

This last equation applies not only to the Carnot cycle, but also to any cycle which is reversible (later in Example 3.5 we will see that this equation is equally valid for any cyclic process which includes irreversible steps).

A reversible cycle can be approximated by a series of Carnot cycles as seen on the right in the following figure. In the figure, a number of curved lines are drawn for the isothermal and adiabatic processes. The entropy change of each step of individual Carnot cycles is cancelled by the neighboring Carnot cycle, except all outmost steps indicated by thick lines in the figure. The sum of the entropy changes of these surviving steps must also be zero. The approximation can be close enough to the real arbitrary cycle by making the individual Carnot cycles infinitesimally small.

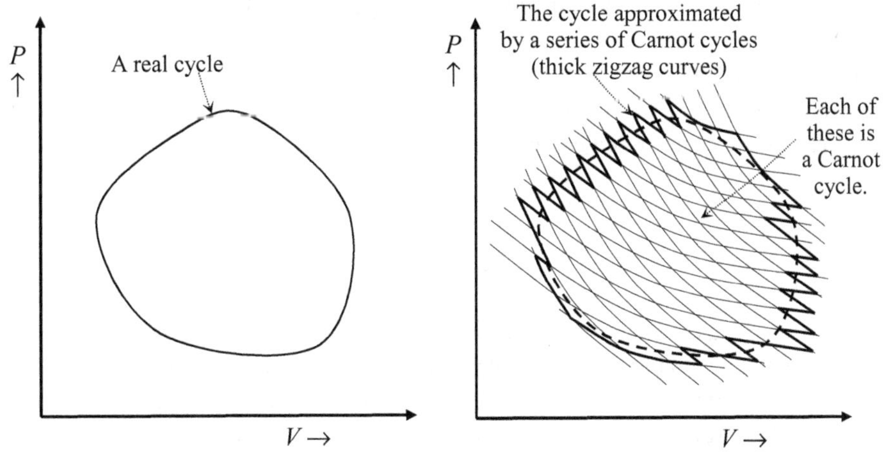


So far we have dealt with a cycle which is combination of reversible steps only. However, it is reminded that the equation in the above is equally valid for a cycle which involves irreversible steps. This is discussed in *Example 3.5*.

Now it is in order to examine whether or not each term in the above equation, *i.e.*, $q_{i,rev}/T_i$ represents the change of a state property of the system. If it is really the case, its value for all processes which share the same initial and final states must be the same, irrespective of whether the process is reversible or irreversible.
The following figure shows two reversible cycles, *i.e.*, cycles "1a2r1" and "1b2r1".

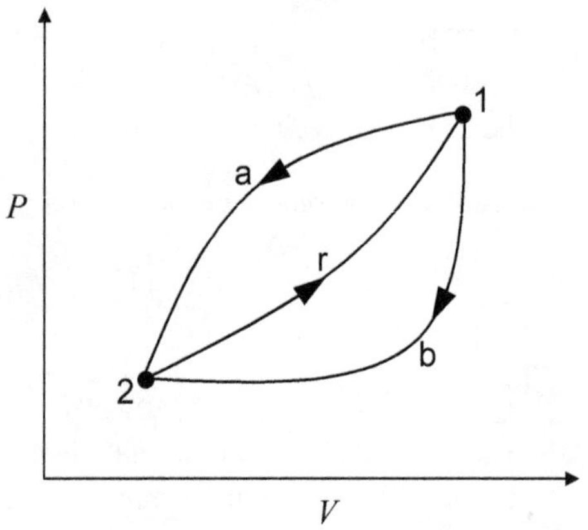


For the cycle "1a2r1",

$$\oint \frac{\delta q_{i,rev}}{T_i} = 0 = \int_1^2 \left(\frac{\delta q}{T}\right)_a + \int_2^1 \left(\frac{\delta q}{T}\right)_r$$

For the cycle "1b2r1",

$$\oint \frac{\delta q_{i,rev}}{T_i} = 0 = \int_1^2 \left(\frac{\delta q}{T}\right)_b + \int_2^1 \left(\frac{\delta q}{T}\right)_r$$

Combination of the above two equations yields,

$$\int_1^2 \left(\frac{\delta q}{T}\right)_a = \int_1^2 \left(\frac{\delta q}{T}\right)_b$$

In the above equation, both integral terms indicate the change of the property, *q/T*, when the state is changed from State 1 to State 2 in the figure. The term on the left side is the change along the path "1a2" and the term on the right side is along the path "1b2", and both are the same. This result ensures that the quantity of *q/T* for a reversible path is independent of the choice of the path as long as the initial and final states are not altered. Thus it is a function of the state. In other words it is the change of a property of the system. This property is termed ***entropy*** with the symbol of S.

$$dS \equiv \frac{\delta q_{rev}}{T}$$

Since the entropy is a state function, its change for a path must be the same irrespective of the path being reversible or irreversible. That is, the above relationship holds for both reversible and any irreversible processes, even though the equation includes $q_{rev}$ which is the heat flow for a reversible process (refer to Example 3.6 and page 77).

If a system undergoes a change in state, say, from State 1 to State 2, the entropy change by the change of the state is given by integrating the above equation:

$$\Delta S = \int_1^2 \frac{\delta q_{rev}}{T}$$

This equation can be integrated by expressing $q$ as a function of $T$.

*Example 3.3*

For a system which undergoes the process of the Carnot cycle, calculate the entropy change of the system, the surroundings and the universe (the system + the surroundings).

1. Entropy change of the system ( $\Delta S_{sys}$ ):

   (1) Step 1: Isothermal expansion : $\Delta S_{sys,1} = \frac{q_1}{T_1}$

   (2) Step 2: Adiabatic expansion: $\Delta S_{sys,2} = 0$, as $q = 0$.

   (3) Step 3: Isothermal compression: $\Delta S_{sys,3} = \frac{q_2}{T_2}$

   (4) Step 4: Adiabatic compression: $\Delta S_{sys,4} = 0$, as $q = 0$.

   Thus,

   $\Delta S_{sys} = \Delta S_{sys,1} + \Delta S_{sys,2} + \Delta S_{sys,3} + \Delta S_{sys,4} = \frac{q_1}{T_1} + \frac{q_2}{T_2} = 0$ (see page 38)

2. Entropy change of the surroundings ( $\Delta S_{sur}$ )

   (1) Step 1: Isothermal expansion of the system: $\Delta S_{sur,1} = \frac{-q_1}{T_1}$ ( heat loss)

   (2) Step 2: Adiabatic expansion of the system: $\Delta S_{sur,2} = 0$, as $q = 0$.

   (3) Step 3: Isothermal compression of the system: $\Delta S_{sur,3} = \frac{-q_2}{T_2}$ (heat gain)

   (4) Step 4: Adiabatic compression of the system ( $\Delta S_{sur,4}$ ): $\Delta S_{sur,4} = 0$, as $q = 0$.

   Thus,

   $$\Delta S_{sur} = \Delta S_{sur,1} + \Delta S_{sur,2} + \Delta S_{sur,3} + \Delta S_{sur,4} = -\left(\frac{q_1}{T_1} + \frac{q_2}{T_2}\right) = 0$$

3. Entropy change of the universe ( $\Delta S_{uni}$ )

   $$\Delta S_{uni} = \Delta S_{sys} + \Delta S_{sur} = 0$$

In summary, for a reversible cyclic process the entropies of the system, the surroundings and the universe all remain unaltered.

> *Example 3.4*
>
> In the previous example (Example 3.3), when we consider the step of isothermal expansion only, what are the entropy changes of the system, surroundings and the universe?

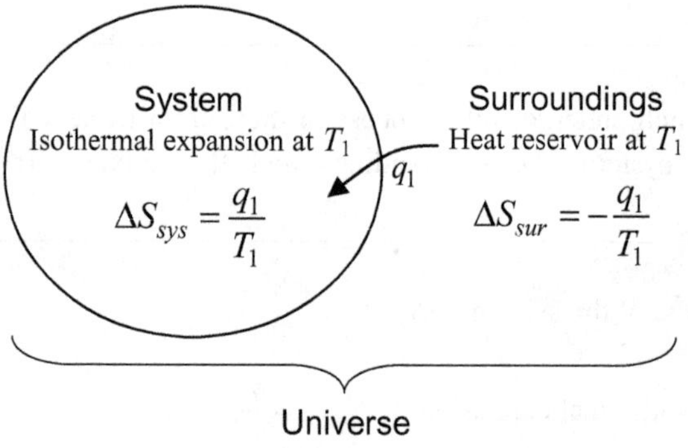


$$\Delta S_{uni} = \Delta S_{sys} + \Delta S_{sur} = 0$$

It can be seen that the increase in the entropy of the system is exactly the same in numerical value as the decrease in the entropy of the surroundings, and thus, the total entropy (the entropy of the universe) is unaltered.
In the reversible process, thus, we may say that:

*Entropy of the amount of* $\frac{q_1}{T_1}$ *has been transferred from the surroundings to the system.*

> *Example 3.5*
>
> An irreversible step is inserted into the Carnot cycle as shown in the figure below. The irreversible step added is an adiabatic free expansion. Determine the entropy changes of the system (the revised cycle), the surroundings and the universe. Assume that the working substance is $n$ moles of ideal gas. Determine also the efficiency ε.

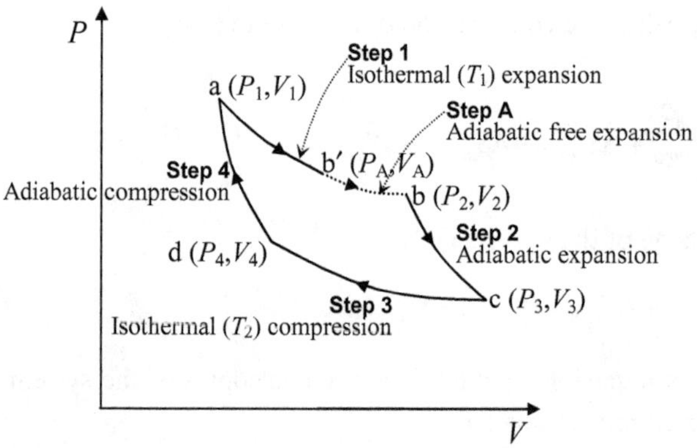

Step 1: Reversible isothermal expansion (a→b′)

(1) Heat : $q_1 = nRT_1 \, ln \frac{V_A}{V_1} (> 0)$

(2) Work : $w_1 = nRT_1 \, ln \frac{V_1}{V_A} (< 0)$

(3) Entropy change : $\Delta S_1 = \frac{q_1}{T_1} = nR \, ln \frac{V_A}{V_1}$

Step A: Adiabatic free expansion (b′→b)

(1) Heat : $q_A = 0$ (adiabatic)

(2) Work : $w_A = 0$ (free expansion)

(3) Entropy change

- The internal energy does not change for an adiabatic free expansion ($\Delta U = q + w = 0 + 0 = 0$).
- Thus there will be no change in temperature ($T_1$)
- However, the entropy change is not equal to zero. We may find the entropy change by reversing the process since the entropy change for Step b′→b must be the same in numerical value as the entropy change for the Step b→b′, but in opposite sign.
- The entropy is a state function, and thus we may choose the isothermal compression for Step b→b′ (Temperature: constant at $T_1$).
- $\Delta S_{b \to b'} = nR \, ln \frac{V_A}{V_2}$, and thus $\Delta S_A = \Delta S_{b \to b'} = nR \, ln \frac{V_2}{V_A}$
- Refer to the further discussion given in Example 3.13.

Step 2: Reversible adiabatic expansion (b→c)

(1) Heat : $q = 0$

(2) Work : $w_2 = nC_V (T_2 - T_1) (< 0)$

(3) Entropy change : $\Delta S_2 = 0$ since $q = 0$. (Note that T = constant at $T_1$ for adiabatic free expansion, but $T_1 \to T_2$ for adiabatic reversible expansion.)

Step 3: Reversible isothermal compression (c→d)

(1) Heat : $q_2 = nRT_2 \, ln \frac{V_4}{V_3} (< 0)$

(2) Work : $w_3 = nRT_2 \, ln \frac{V_3}{V_4} (> 0)$

(3) Entropy change : $\Delta S_3 = \frac{q_2}{T_2} = nR \, ln \frac{V_4}{V_3} (< 0)$

Step 4: Reversible adiabatic compression (d→a)

(1) Heat : $q = 0$

(2) Work : $w_4 = nC_V(T_1 - T_2)\,(>0)$

(3) Entropy change : $\Delta S_4 = 0$

<u>Entropy change of the system</u> ($\Delta S_{sys}$ : Total entropy change for the modified cycle)

$$\Delta S_{sys} = \Delta S_1 + \Delta S_A + \Delta S_2 + \Delta S_3 + \Delta S_4$$

$$\Delta S_{sys} = nR\,ln\frac{V_A}{V_1} + nR\,ln\frac{V_2}{V_A} + 0 + nR\,ln\frac{V_4}{V_3} + 0$$

$$\Delta S_{sys} = nR\,ln\left[\left(\frac{V_2}{V_1}\right)\left(\frac{V_4}{V_3}\right)\right]$$

From the two adiabatic steps,

$\frac{V_2}{V_1} = \frac{V_3}{V_4}$ (Refer to earlier discussion)

$$\Delta S_{sys} = nR\,ln 1 = 0$$

Recall that the modified cycle includes an irreversible step, but the entropy change of the system for the cycle is zero. This is another evidence that entropy is a state function.

Next, let us examine the entropy change of the surroundings.

Step 1: Reversible isothermal expansion

(1) Heat : $q_{1,sur} = -q_1 = -nRT_1\,ln\frac{V_A}{V_1}$

(2) Entropy change : $\Delta S_{1,sur} = \frac{q_{1,sur}}{T_1} = -nR\,ln\frac{V_A}{V_1}$

Step A: Adiabatic free expansion

(1) Heat : $q_{A,sur} = 0$

(2) Entropy change : $\Delta S_{A,sur} = 0$

Step 2: Reversible adiabatic expansion

(1) Heat : $q_{2,sur} = 0$

(2) Entropy change : $\Delta S_{2,sur} = 0$

Step 3: Reversible isothermal compression

(1) Heat : $q_{3,sur} = -q_3 = -nRT_2\,ln\frac{V_4}{V_3}$

(2) Entropy change : $\Delta S_{3,sur} = \dfrac{q_{3,sur}}{T_2} = -nR\,ln\dfrac{V_4}{V_3}$

Step 4: Reversible adiabatic compression

(1) Heat : $q_{4,sur} = 0$

(2) Entropy change : $\Delta S_{4,sur} = 0$

Entropy change of the surroundings ( $\Delta S_{sur}$ )

$$\Delta S_{sur} = \Delta S_{1,sur} + \Delta S_{A,sur} + \Delta S_{2,sur} + \Delta S_{3,sur} + \Delta S_{4,sur}$$

$$\downarrow$$

$$\Delta S_{sur} = -nR\,ln\frac{V_A}{V_1} - nR\,ln\frac{V_4}{V_3}$$

From the two adiabatic steps,

$$\frac{V_2}{V_1} = \frac{V_3}{V_4}$$

$$\downarrow$$

$$\Delta S_{sur} = nR\,ln\frac{V_2}{V_A} \; (> 0)$$

Entropy change of the universe ( $\Delta S_{uni}$ )

$$\Delta S_{uni} = \Delta S_{sys} + \Delta S_{sur}$$

$$\downarrow$$

$$\Delta S_{uni} = nR\,ln\frac{V_2}{V_A} \; (> 0)$$

The above analysis of a cyclic process may be summarized as follows:

- The entropy change of the system (the engine) in a cyclic process is zero (no entropy change) irrespective of the cycle being reversible or irreversible.
- The entropy change of the surroundings in a cyclic process is zero if the entire process is reversible, but greater than zero if the process includes any irreversible step.
- The entropy change of the universe in a cyclic process is zero if the process is reversible, but greater than zero if the process includes any irreversible step. Thus the entropy of the universe increases each time the cycle is completed.

The efficiency of the revised cycle:

(1) Work done by the modified cycle ($w_{sys}$)

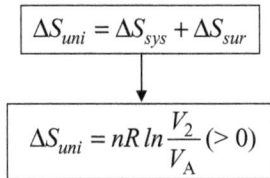

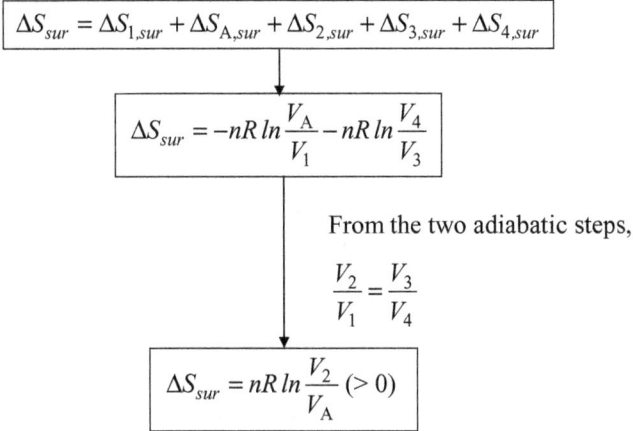

$$w_{sys} = nRT_1\, ln\frac{V_1}{V_A} + nC_V(T_2 - T_1) + nRT_2\, ln\frac{V_3}{V_4} + nC_V(T_1 - T_2)$$
$$= nRT_1\, ln\frac{V_1}{V_A} + nRT_2\, ln\frac{V_3}{V_4} \quad (< 0)$$

(2) Heat taken in from the surroundings: ($q_{sys}$)

$$q_{sys} = q_1 = nRT_1\, ln\frac{V_A}{V_1} \quad (> 0)$$

(3) Efficiency of the modified cycle: (ε)

$$\varepsilon = \frac{-w_{sys}}{q_{sys}}$$

$$w_{sys} = nRT_1\, ln\frac{V_1}{V_A} + \left(nRT_2\, ln\frac{V_1}{V_A} - nRT_2\, ln\frac{V_1}{V_A}\right) + nRT_2\, ln\frac{V_3}{V_4}$$

and we know $\frac{V_2}{V_1} = \frac{V_3}{V_4}$

Then $w_{sys} = nR\, ln\frac{V_1}{V_A}(T_1 - T_2) + nRT_2\, ln\frac{V_2}{V_A}$

$$\varepsilon = 1 - \frac{T_2}{T_1} - \left(\frac{T_2\, ln\frac{V_2}{V_A}}{T_1\, ln\frac{V_A}{V_1}}\right)$$

When all steps are reversible

$$\varepsilon_{rev} = 1 - \frac{T_2}{T_1}$$

$$\varepsilon = \varepsilon_{rev} - \left(\frac{T_2\, ln\frac{V_2}{V_A}}{T_1\, ln\frac{V_A}{V_1}}\right)$$

$V_2 > V_A$ and $V_A > V_1$.
Thus the second term on the right side is positive.

$$\varepsilon < \varepsilon_{rev}$$

The above result demonstrates that the efficiency of a cyclic engine which involves an irreversible step is lower than that of the Carnot cycle in which all steps are reversible.

That is, the efficiency of the Carnot engine is the maximum that a cyclic engine can obtain, and all practical engines which involve irreversible step(s) have efficiencies lower than that of the Carnot engine.

*Example 3.6*

Let us consider a partial process of the Carnot cycle as shown in the figure below. The process starts at State "a" and ends at State "c" via States "b′" and "b". Determine the entropy changes of the system, the surroundings, and the universe.

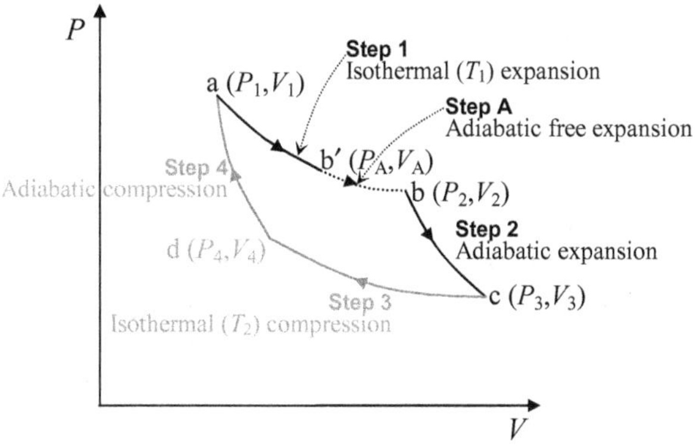


From the results in Example 3.5,

$$\Delta S_{sys,irr} = nR\,ln\frac{V_A}{V_1} + nR\,ln\frac{V_2}{V_A} = nR\,ln\frac{V_2}{V_1}\ (= \Delta S_{sys,rev})$$

$$\Delta S_{sur,irr} = -nR\,ln\frac{V_A}{V_1}\ (cf,\ \Delta S_{sur,rev} = -nR\,ln\frac{V_2}{V_1})$$

$$\Delta S_{uni} = nR\,ln\frac{V_2}{V_A}\ (>0)$$

Note from the above result that the entropy change (increase) of the system is the same for both the reversible and irreversible processes, whereas the entropy change (decrease) of the surroundings for the irreversible process is less than that for the reversible process. The net result is that the total entropy, *i.e.*, the entropy of the universe, increases due to the irreversible step involved. The entropy change of the system can be viewed as follows:

| Entropy change of the system | = | Entropy transferred from the surroundings | + | Entropy produced (generated) inside the system |
|---|---|---|---|---|
| $(\Delta S_{sys})$ | | $(nR\,ln\frac{V_A}{V_1})$ | | $(nR\,ln\frac{V_2}{V_A})$ |

The sum is always the same as the entropy change for the reversible process.

**Example 3.7**

When the Carnot engine completes its cycle, the amount of thermal energy entered into the system during the expansion is not the same as that left during the compression. The former is larger than the latter. We know that the difference has been converted into work. That is, the amount of thermal energy of the system ($q$) alone is not conserved. However, the quantity of $q/T$ is conserved. As a proof of this, calculate the change of entropy of the system for following two reversible paths:

1) Path A: An isothermal expansion followed by an adiabatic expansion
2) Path B: An adiabatic expansion followed by an isothermal expansion

Suppose that the following two figures represent Path A and Path B, respectively"

Path A

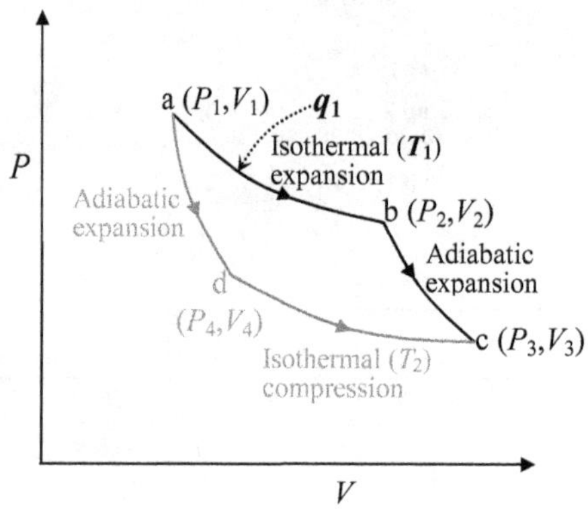


Path B

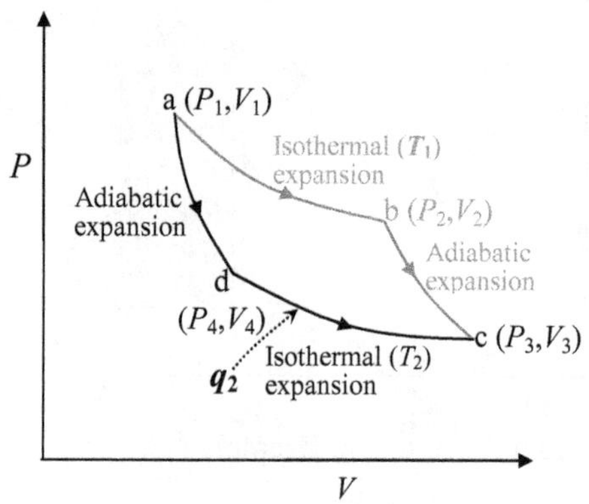


Note that Path B is the reverse of the compression paths in the Carnot cycle. Note also that both paths share the same starting point "a" and the finishing point "c", and the processes are fully reversible in both paths. However, the amount of thermal energy taken in by the reversible Path A ($q_{rev} = q_1$) is not the same as that taken in by the reversible Path B ($q_{rev} = q_2$). When the thermal energy supplied is divided by the temperature at which it is delivered, it becomes the entropy change of the process:

$$\Delta S_A = \frac{q_1}{T_1} \text{ and } \Delta S_B = \frac{q_2}{T_2}$$

But we know that $\Delta S_A = \Delta S_B$ as the entropy is a state function. Therefore,

$$\frac{q_1}{T_1} = \frac{q_2}{T_2}$$

This result assures that even though the thermal energy (heat, $q$) is not a state property (different for different paths), the entropy ($q/T$) is indeed a state property (independent of the path taken).

*Example 3.8*

A Carnot engine can be run in reverse and used to transfer energy in the mode of heat from a low temperature reservoir to a high temperature reservoir.

This type of a device is called ***a heat pump***, if it is used as a heat source.

It is called ***a refrigerator***, if it is used to remove heat.

Prove that, either for heat pump or for refrigerator, work must be done *on* the engine.

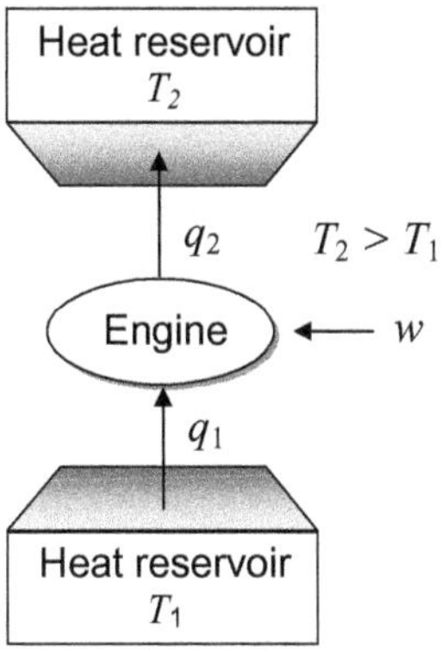


Just as for a Carnot engine running in the forward direction we have an engine running in the reverse direction.

For a complete cycle

$$\Delta S_{cycle} = \Delta S_1 + \Delta S_2 = 0$$

$$\Delta S_1 = \frac{q_1}{T_1} \qquad \Delta S_2 = \frac{q_2}{T_2}$$

$$\frac{q_1}{T_1} + \frac{q_2}{T_2} = 0$$

From the first law

$\Delta U = (q_1 + q_2) + w = 0$

$$w = -q_2\left(1 - \frac{T_1}{T_2}\right)$$

$$1 - \frac{T_1}{T_2} > 0 \qquad q_2 < 0$$

$$w > 0$$ Work is done on the engine.

The *coefficient of performance of a heat pump* (η) is defined as

$$\eta = \left|\frac{q_1}{w}\right|$$

$$\frac{q_1}{T_1}+\frac{q_2}{T_2}=0 \quad \text{from the earlier discussion.}$$

$$\downarrow$$

$$\boxed{\eta=\frac{T_1}{T_2-T_1}}$$

Note that the coefficient of performance of a heat pump of a refrigerator, unlike the efficiency of a heat engine, can be greater than unity.

*Exercises*

3.1 The following diagram shows the operation cycle of a Carnot refrigerator. The refrigerator operates between 25°C ($T_2$) and -10°C ($T_1$) and step 2 involves heat absorption of 500 J.

(1) Calculate the coefficient of the refrigerator.
(2) Calculate the total work done per cycle.

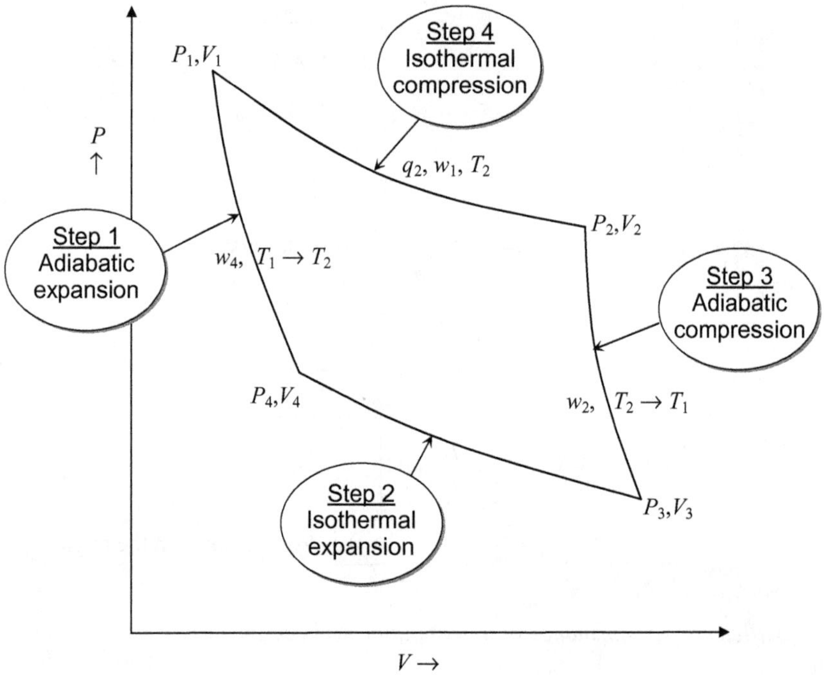

## 3.3 Energy Dispersion and Entropy

Transfer of thermal energy induces degradation or decrease in quality of energy, which in turn results in the loss of capacity to do work. Natural or spontaneous processes always occur in the direction by which thermal energy is degraded. Entropy is not energy, but a kind of the measure of the amount of energy in a physical system not available to do work. However, the definition of entropy, $\Delta S = q_{rev}/T$, which has been introduced earlier is in fact the definition of the *change* of entropy, not entropy itself. In order to define entropy itself explicitly, we need to consider the transfer of thermal energy in a microscopic point of view.

### Macrostates and Microstates

Discussions so far have been based on macroscopic matter, and made no reference to the microscopic nature of matter. As we know that atoms or molecules of matter provide storage to the energy transferred to the matter, a question arises as to what happens in the matter in the molecular point of view when thermal energy has been transferred to or from the matter. To answer this question, we first define two important terms: ***macrostates*** and ***microstates***.

Dictionaries define,

"*macro-*" as the prefix of the word "macroscopic" which designates a size scale very much larger than that of atoms and molecules, and

"*micro-*" as the prefix of the word "microscopic" which designates a size scale comparable to the subatomic particles, atoms, and molecules.

But a macrostate and a microstate in thermodynamics do not just designate something big and small in size. In thermodynamics, a microstate isn't just about a smaller amount of matter or a smaller size of matter, but it implies *a detailed look at the energy* that molecules or other particles in a system have.

Let us start our discussion with an example of a simple system.
Suppose that there is a system consisting of three molecules and three units of energy to share among them. The *macrostate* of the system is the one having the physical size of three molecules with the energy of three units. In classical thermodynamics, the macrostate is simply called the *state*. The macrostate or state does not care as to how the energy is distributed among the molecules. As long as the total number of molecules and the total units of energy are kept unaltered, the system is at the same macrostate.
As for the *microstate*, however, we need to look at the energy distribution in detail. What we need to consider is the number of ways the total energy can be distributed among the molecules (or atoms). For the present example, there are several different ways of distributing these three units of energy to the three molecules. One way is that the energy is shared equally and each molecule has one unit of energy. The other way is that one molecule has all three units of energy and the other two have none. There is still another way: two units to one molecule, one unit to the other one molecule, and none to the last molecule. The following table shows all possible combinations.

The system of (n = 3, ε = 3)

No. of macrostates= 1

No. of microstates = 10

| | | | | | | | | | | | |
|---|---|---|---|---|---|---|---|---|---|---|---|
| Energy level | 4 | | | | | | | | | | |
| | 3 | | Ⓐ | Ⓑ | Ⓒ | | | | | | |
| | 2 | | | | | Ⓐ | Ⓐ | Ⓑ | Ⓑ | Ⓒ | Ⓒ |
| | 1 | ⒶⒷⒸ | | | | Ⓑ | Ⓒ | Ⓐ | Ⓒ | Ⓑ | Ⓐ |
| | 0 | | ⒷⒸ | ⒶⒸ | ⒶⒷ | Ⓒ | Ⓑ | Ⓒ | Ⓐ | Ⓐ | Ⓑ |
| Microstates | | 1 | 2 | 3 | 4 | 5 | 6 | 7 | 8 | 9 | 10 |
| Configurations | | I | II | | | III | | | | | |

These ten possible combinations are referred to as ***microstates*** of the system. In other words, the system has ten different microstates and thus it takes *one of these ten microstates at any one instant in time*. The specific phrase, "one of these …at any instant in time" is important. The system cannot take two or more microstates at the same time, but only one microstate at a time. The system will be in one of the microstates at one instant, and in the next instant the system can immediately move to another microstate. The system has 10 different choices for its microstate, but no options for its macrostate; *i.e.*, it is under only one macrostate. The more number of microstates a system has, the more number of choices the system can take for the next instant.

Let us consider more about the above system.

First we define a "***configuration***" as the collection of those microstates that possess identical distributions of energy among the accessible energy levels without distinguishing individual molecules. According to this definition of configuration, the system above has three different configurations. If all microstates are equally probable, the probability of any one configuration is proportional to the number of microstates: that is, the system will be at Configuration I for 10% of the time, at Configuration II for 30% of the time, at Configuration III for 60% of the time.

As the number of molecules and the number of energy units increases, the number of accessible microstates will grow explosively; if a system has the size of 1000 molecules with the total energy of 1000 units, the number of available microstates will be around $10^{600}$- a number that greatly exceeds the number of atoms in the observable universe.

A more rigorous definition of the microstate is;

> "A microstate is one of the large numbers of different accessible arrangements of the molecules' motional energy (the translational, rotational, and vibrational modes of molecular motion) for a particular macrostate."

One microstate is something like an instantaneous snapshot of the energy of all the individual molecules in the macrostate. In the next instant the system immediately changes to another microstate. A single microstate of a system has all the energies of all the molecules on specific energy levels at one instant. All of the energy of a system can only be in one microstate at any time.

**Energy Dispersion and Entropy**

A macrostate is the thermodynamic state of a system that is exactly characterized by the system's properties such as $P$, $V$, $T$, and number of moles of each constituent. Thus, if its observable properties ($P$, $V$, $T$, ...) do not change, a macrostate does not change over time, but the microstate does.

The energy of a system becomes *more dispersed* when the number of microstates available to occupy, *i.e.*, the number of accessible microstates, becomes larger - there are more choices the system can take for distributing its energy at one instant.

*The bottom line is that the change in the number of accessible microstates is the determiner of the spontaneous direction of a natural process.*

Entropy is now defined to be the measure of the *spontaneous dispersal of energy* (in the previous sections it was called degradation of energy) for a system. As the number of microstates that are accessible for a system indicates all the different ways that the energy can be arranged in that system, the larger the number of accessible microstates, the greater the entropy of the system at a given temperature.

The correlation of the entropy with the number of microstates is shown by the ***Boltzmann equation*** which is one of the most celebrated equations in statistical thermodynamics:

$$\boxed{S = k \ln W}$$

where $W$ is the number of microstates, and $k$ is Boltzmann's constant ($1.38\text{x}10^{-23}$ J $K^{-1}$).

Then the change of entropy between two different states (the initial and final states) will be,

$$\boxed{\Delta S = k \ln\left(\frac{W_{final}}{W_{initial}}\right)}$$

How is this equation related to the following expression of the entropy change developed from the macroscopic approach of classical thermodynamics?

$$\Delta S = \frac{q_{rev}}{T}$$

These two expressions, the latter from macroscopic approach and the former from microscopic considerations, are indeed related to each other, and the relationship is clarified in Examples 3.14 and 3.17.

Why does $S$ depend not on $W$, but on the *logarithm* of $W$? Suppose we have two systems: System 1 (entropy of $S_1$ and the number of microstates of $W_1$) and System 2 (entropy of $S_2$ and the number of microstates of $W_2$). If we now redefine these two systems as a single system, then the entropy of the new system will be the sum of the two: $S = S_1 + S_2$. But the number of microstates will be the product $W_1 \times W_2$ because for each microstate of System 1, System 2 can be in any of $W_2$ numbers of microstates.

System 1

$\boxed{S_1, W_1}$

$S_1 = k \ln W_1$

System 2

$\boxed{S_2, W_2}$

$S_2 = k \ln W_2$

New System

$\boxed{S_1, W_1} \quad \boxed{S_2, W_2}$

$S, \quad W = W_1 W_2$

$$\boxed{S = S_1 + S_2}$$

$S_1 = k \ln W_1 \quad | \quad S_2 = k \ln W_2$

$$\boxed{S = k \ln W_1 + k \ln W_2}$$

$$\boxed{S = k \ln (W_1 W_2)}$$

$$\boxed{S = k \ln W}$$

$$W = W_1 W_2$$

$$\boxed{S = k \ln (W_1 W_2)}$$

$$\boxed{S = k \ln W_1 + k \ln W_2}$$

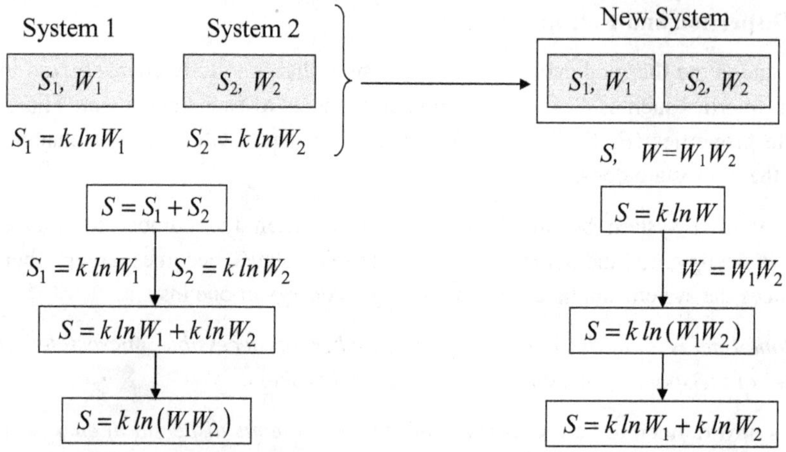

Now the entropy expressed by the logarithm of $W$ warrants the additive nature of entropy. Provided that we are able to calculate the number of microstates for a given macrostate, the Boltzmann equation gives us answers to the relation between molecular motion and entropy of a system, *i.e.*, the relationship between

| Molecules (or atoms or ions) constantly energetically speeding, colliding with each other, moving distances in space (or, just vibrating rapidly in solids) | ⟷ | What we define as its entropy of the system |

Suppose that we have two hypothetical metal blocks: Block 1 having mass of three atoms and energy of 6 units, and Block 2 having mass of three atoms and energy of 2 units. There will be a number of ways of distributing the energy to the atoms in each block. The following diagram shows the number of ways of distributing the energy between the atoms, *i.e.*, the number of microstates each block has:

*No energy transfer*

| **Block 1 ($n = 3$, $\varepsilon = 6$)** | ✳⟷ | **Block 2 ($n = 3$, $\varepsilon = 2$)** |

$W = 28$

$W = 6$

| Conf. | I | II | III | IV | V | VI | VII |
|-------|---|----|-----|----|---|----|-----|
| Micro. | 3 | 6 | 6 | 3 | 3 | 6 | 1 |

| Conf. | I | II |
|-------|---|----|
| Micro. | 3 | 3 |

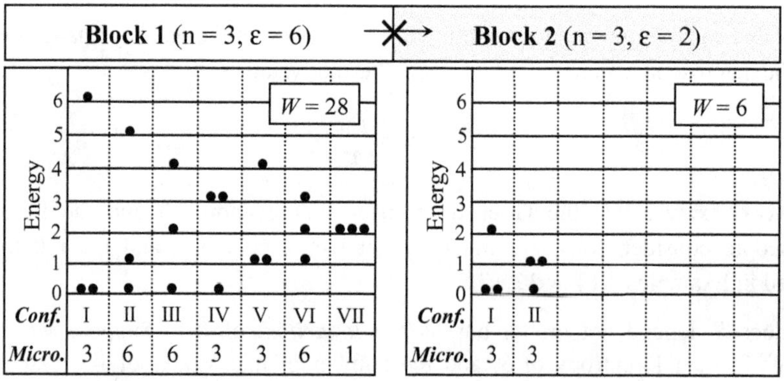

\* *Config.*: configuration number, *Micro.*: Number of microstates

Block 1 has 7 different arrangements (configurations) of atoms to the accessible energy levels, while Block 2 has 2 different configurations. Each of the configurations has its own number of microstates. For example, Configuration I of Block 1 has three different ways of arranging three atoms in those two energy levels. Let us name the atoms A, B and C, then the diagram on the right shows three different arrangements. In this way we can find that Block 1 has 28 microstates ($W = 28$) in total, and Block 2 has 6 ($W = 6$).

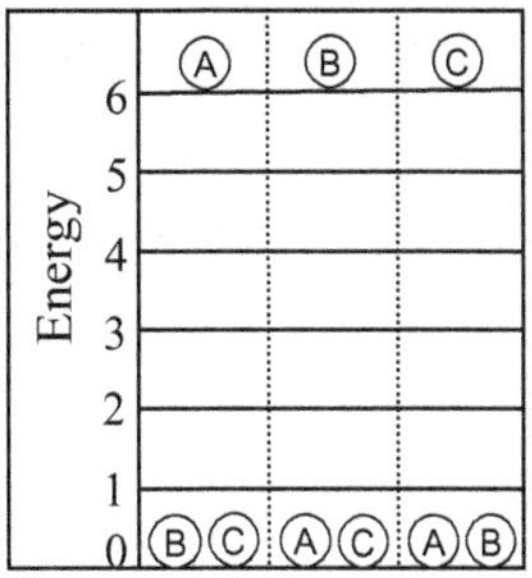


If the combination of Block 1 and Block 2 is considered **to** form one system, but without allowing energy transfer between them, then the number of microstates of the system will be $W = 168$ (= 28×6).

Now suppose that one unit of energy is transferred from Block 1 to Block 2 so that the energy of Block 1 is reduced from 6 to 5 units, while the energy of Block 2 is increased from 2 to 3 units. Microstates of the two blocks in the new system are shown in the diagram below. The number of microstates of the combined system (Block 1 + Block 2) is now 210 (= 21×10), which is larger than that of the previous case where it was 168. Therefore, transfer of one unit of energy from Block 1 to Block 2 results in increase of the number of microstates, and thus the transfer must be spontaneous or natural.

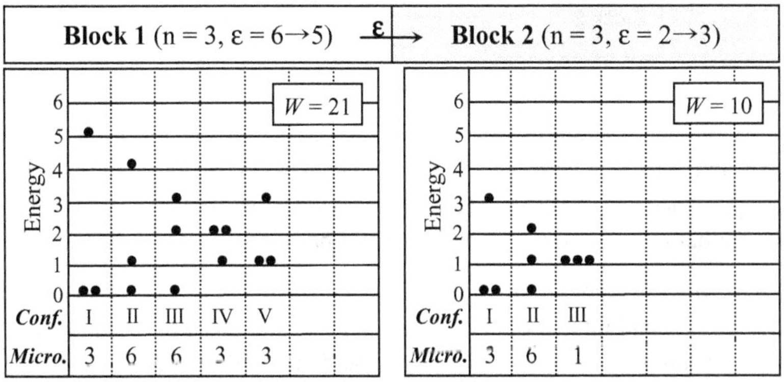


Further transfer of energy by one unit equalizes the energy of both blocks. The result is shown in the following figure:

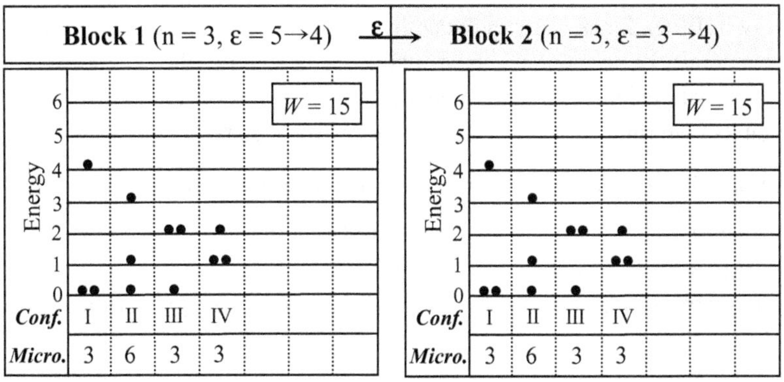

The number of microstates of the combined system is 225 (= 15×15), and thus this transfer of energy is also spontaneous.

Further transfer of energy will not occur spontaneously as it will decrease the number of microstates because it is the reverse of the above transfer sequence.

The above results are graphically shown in the following figure:

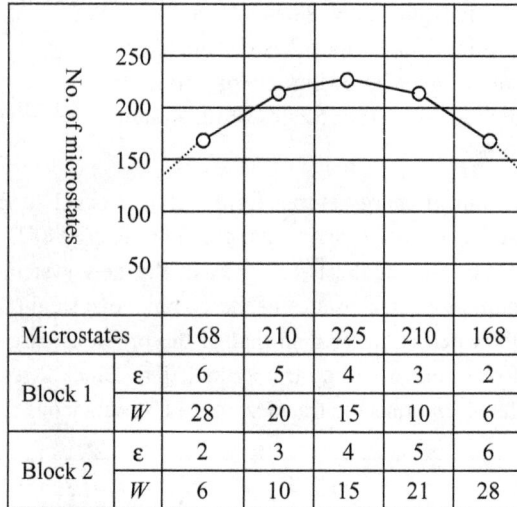

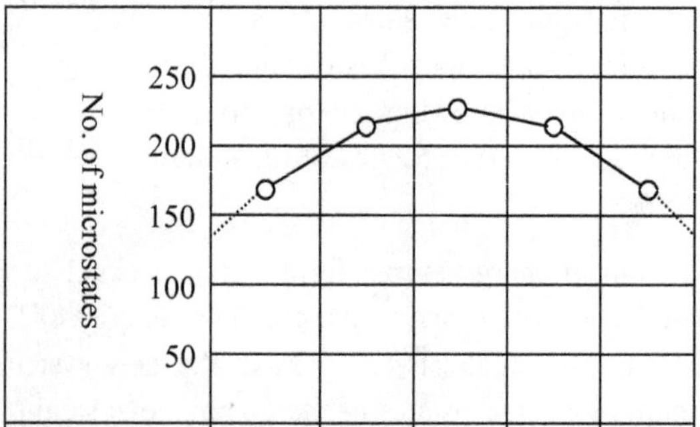


| | | | | | | |
|---|---|---|---|---|---|---|
| Microstates | | 168 | 210 | 225 | 210 | 168 |
| Block 1 | ε | 6 | 5 | 4 | 3 | 2 |
| | $W$ | 28 | 20 | 15 | 10 | 6 |
| Block 2 | ε | 2 | 3 | 4 | 5 | 6 |
| | $W$ | 6 | 10 | 15 | 21 | 28 |

Note that the number of microstates reaches the maximum when both blocks share the same energy in the present case (*i.e.*, the same temperature). No further transfer of energy will occur spontaneously. Note also that the energy disperses to become uniform, through which the number of microstates reaches the maximum.

As for the view of degradation of energy, let us look at the energy transfer again. For Block 1, it initially had 6 units of energy, but, after losing one unit of energy to Block 2, it is left with 5 units of energy. However, these 5 units of energy are not the same as before not only in the quantity of energy, but also in its quality. These remaining 5 units of the energy have been degraded, since these energies are not available for conducting all the work the initial 6 units of energy could do.

The energy in Block 2 has been upgraded by receiving one unit of energy from Block 1, but these upgraded energies are not available or useful for the energy of Block 1 to recover its initial grade. It is not possible for Block 2 to return one unit of energy back to Block 1 on its own right. If it could, the number of microstates would have decreased.

When we calculate the entropy of these three hypothetical systems,

- Before energy transfer : $S = k\,ln168$
- After energy transfer by one unit : $S = k\,ln210$
- After energy transfer by two units : $S = k\,ln225$ (uniform energy distribution)

It is clear from the above that the total energy disperses toward uniform distribution, *i.e.*, toward the maximum number of microstates, or toward the maximum entropy.

In the previous discussion, we have seen that a particular configuration can have several different arrangements of molecules (or atoms) to the accessible energy levels (microstates). Here we discuss in a systematic way as to how to determine the number of microstates for a given configuration.

Consider the number of ways of distributing total $n$ particles in a number of different energy levels in such a way that

$n_0$ particles in the $\varepsilon_0$ energy level,
$n_1$ particles in the $\varepsilon_1$ energy level,
$n_2$ particles in the $\varepsilon_2$ energy level,
......................................
$n_i$ particles in the $\varepsilon_i$ energy level.

Then the total number of ways to arrange $n$ particles is $n!$, but there are,

$n_0!$ ways of arranging $n_0$ particles in the $\varepsilon_0$ energy level: *not distinguishable*,
$n_1!$ ways of arranging $n_1$ particles in the $\varepsilon_1$ energy level: *not distinguishable*,
$n_2!$ ways of arranging $n_2$ particles in the $\varepsilon_2$ energy level: *not distinguishable*,
...........................................................................................
$n_i!$ ways of arranging $n_i$ particles in the $\varepsilon_i$ energy level: *not distinguishable*.

Therefore, not all n! ways are *distinguishable*, since the particles at a same energy level are merely changing their sequential positions at the same level. Thus the number of distinguishable ways of arrangements (the number of microstates), $W$, is given by

$$W = \frac{n!}{n_0!\,n_1!\,n_2!.....n_i!}$$

*Example 3.9*

Five units of energy are distributed among three distinguishable particles. Calculate the total number of accessible microstates in the system.

If $E$ units of energy are distributed among $n$ distinguishable particles, the total number of accessible microstates $W$ is given by

$$W = \frac{(n+E-1)!}{(n-1)!\,E!}$$

$n = 3$
$E = 5$

↓

$$W = \frac{(3+5-1)!}{(3-1)!5!}$$

$$W = 21$$

*Example 3.10*

Let us define the notation {a, b, c,..} in which the numbers from the left, a, b, c, …, represent the occupancy of the energy levels from the lowest one upwards. Calculate the number of microstates for a system consisting of five energy levels in a state with occupation of the energy levels {2,3,4,2,1}.

$$W = \frac{n!}{n_0!\,n_1!\,n_2!.....n_i!}$$

$$n = 2+3+4+2+1 = 12$$

$$W = \frac{12!}{2!\,3!\,4!2!1!} = 831{,}600$$

*Example 3.11*

Calculate the entropy of the system which consists of 12 particles in a state with occupation of the energy levels {2,3,4,2,1}.

$$S = k\,ln\,W$$

$W = 831{,}600$
$k = 1.38066 \times 10^{-23}$ J K$^{-1}$

$$S = 1.38066\times10^{-23}\,ln(831{,}600)\,\text{J K}^{-1}$$
$$= 1.88\times10^{-22}\,\text{J K}^{-1}$$

*Example 3.12*

A system consists of 10 particles distributed over four energy levels {5,3,2,0}. If a single particle is excited by one energy level, what would be the new distribution of particles in the energy levels which maximizes the entropy of the system?

Exciting a particle from the configuration {5,3,2,0} leads to the following three possibilities: {4,4,2,0}, {5,2,3,0} and {5,3,1,1}. Then,

$$\{4,4,2,0\} \rightarrow W = \frac{10!}{4!\,4!\,2!0!} = 3{,}150$$

$$\{5,2,3,0\} \rightarrow W = \frac{10!}{5!\,2!\,3!0!} = 2{,}520$$

$$\{5,3,1,1\} \rightarrow W = \frac{10!}{5!\,3!\,1!1!} = 5{,}040$$

The configuration which maximizes the entropy is thus {5,3,1,1}: *spreading out to the higher energy level.*

*Example 3.13*

In Example 3.5, an adiabatic *free* expansion step was inserted to know the effect of an irreversible step on the entropy change of the system. Although there was no heat transfer and no work done during the expansion, there was a change of entropy of the system. The entropy change was found in the Example by reversing the process; *i.e.*, by compressing the expanded gas back to its initial volume.
Explain the entropy change along with an adiabatic free expansion in view of the change in the number of microstates.

When the volume of a system is increased without change in energy, its energy levels become closer together, *i.e.*, more energy levels become accessible to molecules which are within the original energy range. Thus, while the original molecular motional energy is still the same in the larger volume (isothermal), the system has more energy levels, meaning that the number of accessible microstates increases.

This can be conceptually proved by employing quantum mechanics. The energy (translational energy) for a particle in a cubic box is given by

$$\varepsilon = \frac{h^2(n_x^2 + n_y^2 + n_z^2)}{2mV^{2/3}}$$

where $m$ is the mass of the particle, $h$ is the Plank constant, $n_x$, $n_y$ and $n_z$ are the quantum numbers which can have positive integer values, and $V$ is the volume of the box.

Upon expansion of the box (*i.e.*, increase in $V$), the energy (ε) of each quantum level, *i.e.*, a given ($n_x$, $n_y$, $n_z$), decreases as can be seen in the above equation. Since the total energy of the gas does not change upon expansion into vacuum, the gap between the energy levels narrows as the volume expands, and thus more accessible energy levels are provided, leading to increase in the number of accessible microstates.

The following figure schematically represents the above discussion:

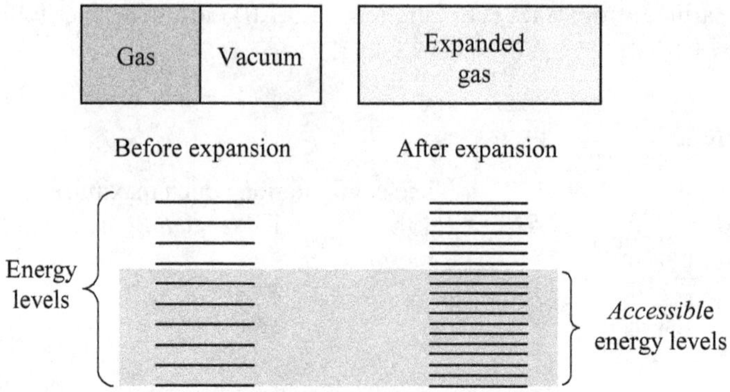


## 3.4 Equilibrium Criterion and Entropy

### Equilbirum Criterion

When a system is left to itself, it would either remain unchanged in its initial state, or move spontaneously to some other state. If the former is the case, the initial state is indeed the ***equilibrium*** state. If the latter is the case, however, the system is initially in a ***non-equilibrium*** state, and the system will spontaneously move to the equilibrium state.

From the discussions done so far, we may be able to draw the following summary:

- All real processes involve some degree of irreversibility and thus all real processes lead to an increase in the total entropy: $S_{tot} = S_{sys} + S_{sur}$, and $\Delta S_{tot} > 0$.
- The total entropy does not change in the reversible process ($\Delta S_{tot} = 0$).
- The reversible process is the succession of equilibrium states.
- At equilibrium, the total entropy or the number of accessible microstates is maximum.

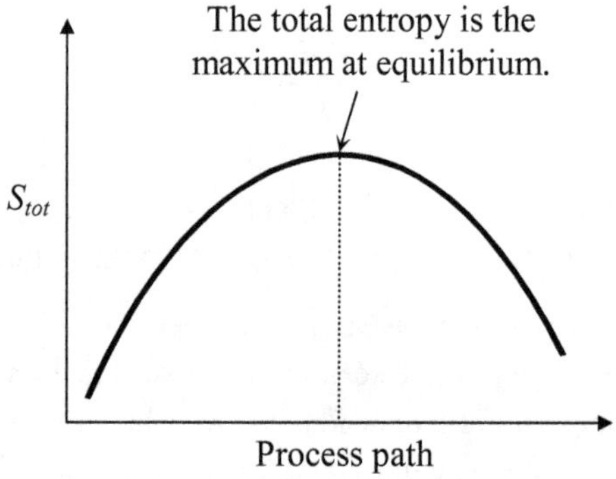


From the molecular point of view, the equilibrium state is the state at which thermal energy is dispersed to the maximum: *i.e.*, a state at which the number of accessible microstates is the maximum. Therefore, the following is the condition to be satisfied for a state to be at equilibrium:

$$\boxed{dW = 0}$$

where $W$ is the total number of microstates (accessible) which was given earlier as,

$$W = \frac{n!}{n_0!\,n_1!\,n_2!\ldots.n_i!}$$

*Example 3.14*

Recall that the number of microstates defined by above equation is in fact the total number of microstates *for a given configuration*, not for all configurations of the a system. Under what conditions can we assume the above equation to represent approximately the total number of microstates of the whole system?

Suppose that we have $n$ particles which are distinguishable from each other, but do not interact with each other. All $n$ particles are contained in a box. We are interested in the total number of possible distributions of the particles in the box whether each of them is on the left or right half. The number of distributions may be called the number of microstates. If the total number of particles is four, then the statistics will be like what we can see in the following:

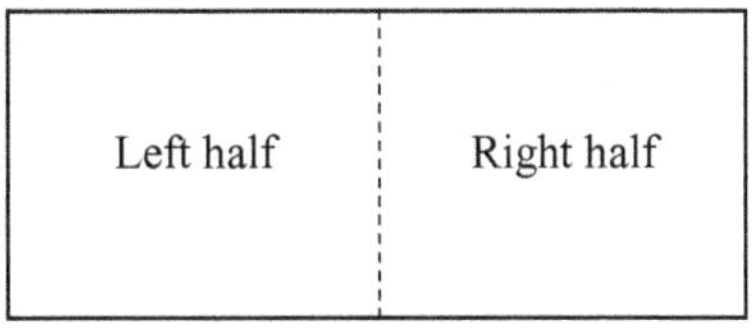


| Particles | | No. of microstates | Probability |
|---|---|---|---|
| Left | Right | | |
| 0 | 4 | 1 | 1/16 |
| 1 | 3 | 4 | 4/16 |
| 2 | 2 | 6 | 6/16 |
| 3 | 1 | 4 | 4/16 |
| 4 | 0 | 1 | 1/16 |
| Total | | 16 | 16/16(=1) |

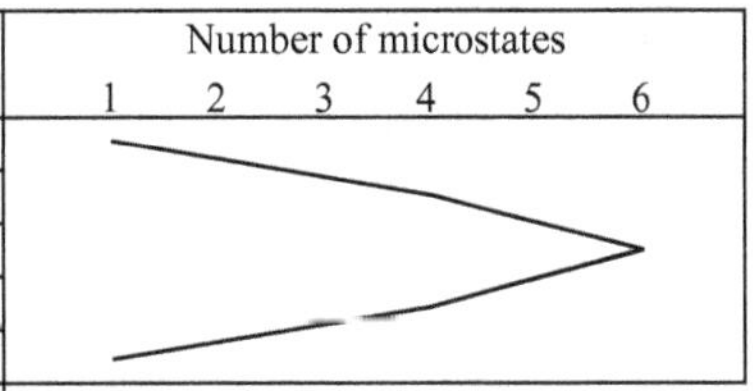


Note that the curve is symmetric about the configuration of equal distribution (2 and 2).

Suppose that we now increase the number of particles to distribute, and the following is the selected statistics with 10 particles. Note that the smallest number of microstates is still unity as with less number of particles, but the largest number has grown to 252, and the plot shows much sharper peak at the configuration having equal number of particles in both halves (*i.e.*, uniform distribution).

When we have 100 particles, the number of microstates of the most probable configuration (equal distribution of 50:50) will explosively grow as given below:

$$W = \frac{100!}{50!50!} = 10^{30}$$

| Particles | | No. of microstates | Probability |
|---|---|---|---|
| Left | Right | | |
| 0 | 10 | 1 | 1/814 |
| 1 | 9 | 10 | 10/814 |
| 2 | 8 | 45 | 45/814 |
| 3 | 7 | 120 | 120/814 |
| 4 | 6 | 210 | 210/814 |
| 5 | 5 | 252 | 252/814 |
| 6 | 4 | 210 | 210/814 |
| 7 | 3 | 120 | 120/814 |
| 8 | 2 | 45 | 45/814 |
| 9 | 1 | 10 | 10/814 |
| 10 | 0 | 1 | 1/814 |
| Total | | 814 | 1 |

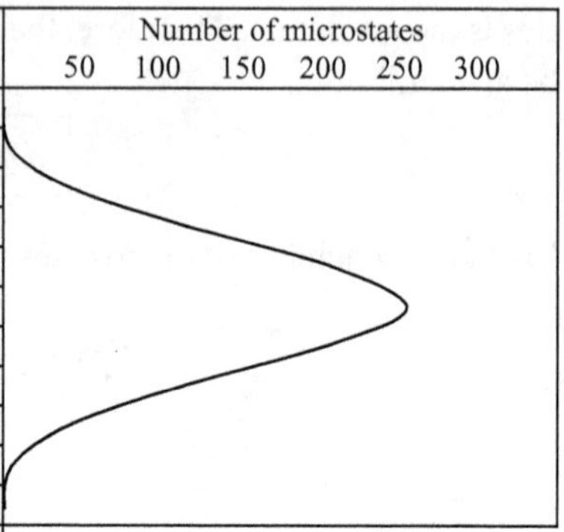


For 1,000 particles, the number of microstates of the most probable configuration becomes to be

$$W = \frac{1000!}{500!500!} = 10^{300}$$

As the number of particles is increased, the number of microstates of the most probable configuration becomes dominating and virtually represents the total number of microstates. That is,

$$W_{\text{(most probable configuration)}} \approx W_{\text{(total)}}$$

The systematic way of finding the maximum number of microstates for a configuration, which must be the most probable configuration, is to take derivative of the equation for the number of microstates and put it equal to zero:

$$W = \frac{n!}{n_0!\,n_1!\,n_2!.....n_i!}$$

$n_L$ = Number of particles at the left half

$n_R$ = Number of particles at the right half

$n_R = n - n_L$

$$W = \frac{n!}{n_L!(n-n_L)!}$$

Taking logarithm for convenience,

$$lnW = ln\,n! - ln\,n_L! - ln(n-n_L)!$$

Applying the Stirling's approximation, $ln\,n! = n\,ln\,n - n$ for large $n$

$$\ln W = n \ln n - n_L \ln n_L - (n - n_L) \ln (n - n_L)$$

↓ Taking derivative with respect to $n_L$,

$$\frac{d \ln W}{dn_L} = \ln \left( \frac{n - n_L}{n} \right) = 0$$

↓

$$n_L = \tfrac{1}{2} n \quad \rightarrow \quad n_R = \tfrac{1}{2} n$$

Now the procedure seen in the above example is extended to be applicable to more general cases:

$$W = \frac{n!}{n_0!\, n_1!\, n_2! \ldots. n_i!}$$

↓ Taking the natural logarithm,

$$\ln W = \ln n! - (\ln n_0! + \ln n_1! + \ln n_2! + \cdots)$$

↓ or

$$\ln W = \ln n! - \sum_i \ln n_i!$$

↓ Applying the Stirling's approximation $\ln x! = x \ln x - x$

$$\ln W = (n \ln n - n) - \sum_i (n_i \ln n_i - n_i)$$

↓ Taking derivative,

$$d \ln W = -\sum_i (n_i d \ln n_i + \ln n_i dn_i - dn_i)$$

↓

$$d \ln W = -\sum_i \left( n_i \frac{dn_i}{n_i} + \ln n_i dn_i - dn_i \right)$$

↓

$$d \ln W = -\sum_i (\ln n_i dn_i)$$

↓ At the maximum of *W*, $d \ln W = 0$

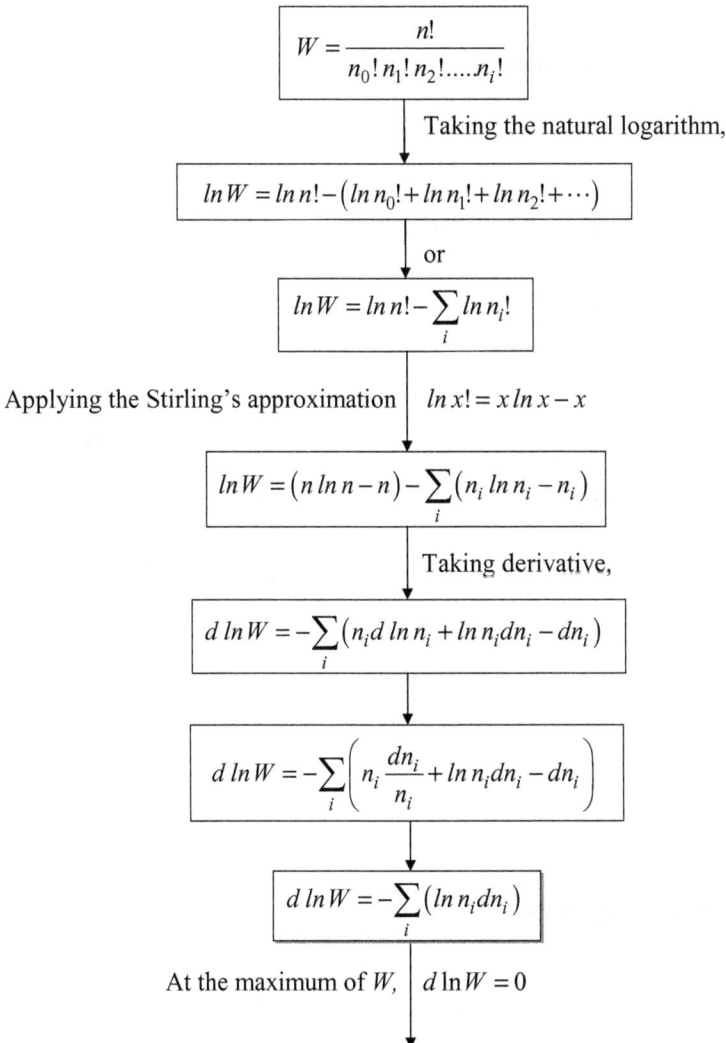

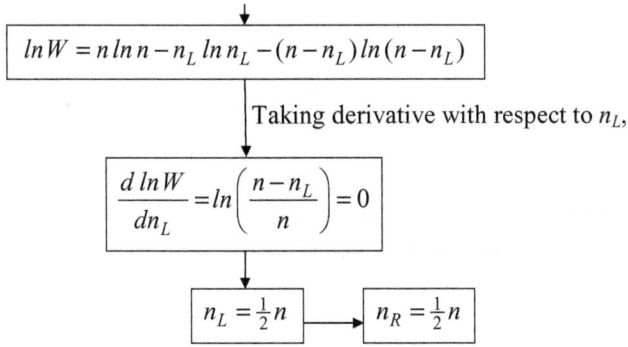

$$\sum_i (\ln n_i dn_i) = 0$$

This equation represents the condition for the maximum number of microstates of the most probable configuration, and in fact the total number of microstates of the system, provided that $n$ is sufficiently large. However, there are two constraints which must be met;

(1) The total energy ($U$) must be constant:

$$\sum_i n_i \varepsilon_i = U \quad \text{or in the differential form} \quad \sum_i \varepsilon_i dn_i = 0$$

(2) The total number of atoms (or molecules) must be constant:

$$\sum_i n_i = n \quad \text{or in the differential form} \quad \sum_i dn_i = 0$$

Applying so-called the ***undetermined multipliers***,

$$\sum_i (\ln n_i dn_i) = 0$$

$$\beta \sum_i \varepsilon_i dn_i = 0$$

$$\alpha \sum_i dn_i = 0$$

where $\alpha$ and $\beta$ are constant.

Summation of these three equations yields,

$$\sum_i (\ln n_i + \alpha + \beta\varepsilon_i) dn_i = 0$$

$dn_i \neq 0$
Therefore,

$$\ln n_i + \alpha + \beta\varepsilon_i = 0$$

or

$$n_i = e^{-\alpha} e^{-\beta\varepsilon_i}$$

Summation yields

$$\sum_i n_i = n = e^{-\alpha} \sum_i e^{-\beta\varepsilon_i}$$

Combining the last two equations,

$$\frac{n_i}{n} = \frac{e^{-\beta\varepsilon_i}}{\sum_i e^{-\beta\varepsilon_i}}$$

This equation gives the number of particles ($n_i$) which will be found at the energy level of $\varepsilon_i$, or the probability ($n_i / n$) that a particular particle will be found at the energy level of $\varepsilon_i$ for the maximum number of microstates ($W$), or for the maximum entropy ($S = k \ln W$), or at equilibrium ($d \ln W = 0$).

**Partition Function**

The constant β introduced in the above equation is related to temperature and given as*,

$$\beta = \frac{1}{kT}$$

* Readers who are interested in more discussion on this are suggested to refer to books on statistical thermodynamics.

where $k$ is Boltzmann's constant: $k = 1.38 \times 10^{-23}$ J K$^{-1}$

Substitution yields,

$$\frac{n_i}{n} = \frac{e^{-\frac{\varepsilon_i}{kT}}}{\sum_i e^{-\frac{\varepsilon_i}{kT}}}$$

***Boltzmann distribution***

$$Z = \sum_i e^{-\frac{\varepsilon_i}{kT}}$$

Definition of a function termed
***Partition function***

$$\frac{n_i}{n} = \frac{e^{-\frac{\varepsilon_i}{kT}}}{Z}$$

The ***partition function*** defined above is the sum of all the energy states allowed. The partition function is a function of temperature $T$ and microstate energies $\varepsilon_i$'s, and the microstate energies are determined by other thermodynamic variables such as the number of particles and the volume, as well as microscopic quantities like the mass of the constituent particles. The partition function is a particularly useful function which allows us to calculate all the thermodynamic properties of the system, once it is known. Partition function encodes how the probabilities are partitioned among the different microstates, based on their individual energies. If all states are equally probable (equal energies) the partition function is representative of the total number of possible states.

*Example 3.15*

Suppose that we have a hypothetical system in which $n$ atoms are distributed in two energy levels. Using the Boltzmann distribution, discuss the dependence of the entropy change on 1/T, *i.e.*,

$$\Delta S = \frac{q_{rev}}{T}$$

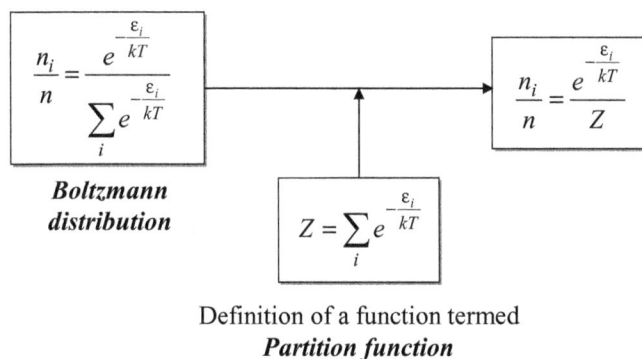

According to the Boltzmann distribution,

$$\frac{n_i}{n} = \frac{e^{-\frac{\varepsilon_i}{kT}}}{\sum_i e^{-\frac{\varepsilon_i}{kT}}}$$

For two energy levels 1 and 2

$$\frac{n_i}{n} = \frac{e^{-\frac{\varepsilon_i}{kT}}}{e^{-\frac{\varepsilon_1}{kT}} + e^{-\frac{\varepsilon_2}{kT}}}$$

$\varepsilon_1 = 0$ and $n_1$ for level 1

$\varepsilon_2 = \varepsilon$ and $n_2$ for level 2

$$\frac{n_1}{n} = \frac{1}{1+e^{-\frac{\varepsilon}{kT}}} \qquad \frac{n_2}{n} = \frac{e^{-\frac{\varepsilon}{kT}}}{1+e^{-\frac{\varepsilon}{kT}}}$$

Then the distribution of atoms in level 1 and level 2 can be represented by the ratio of the number of atoms in level 2 to that in level 1, which is given by

$$\frac{n_2}{n_1} = e^{-\frac{\varepsilon}{kT}}$$

It is seen that the ratio is dependent on the temperature. If the temperature approaches $0K$, the ratio becomes zero, meaning that $n_2 = 0$ and thus all atoms are in the ground state (level 1). On the other hand, if the temperature is increased very high, the ratio approaches unity, which means the atoms are evenly distributed between the two levels. As the temperature is increased, therefore, more particles are accessed to the higher energy level, and thus the number of accessible microstates increases, and then the entropy of the system increases. It can be said that the entropy is a kind of function of the above ratio:

$$S = f\left(\frac{n_2}{n_1}\right)$$

Suppose we have two systems; one at a cold temperature $T_c$, and the other at a higher temperature $T_h$, both of which have two energy levels and $n$ atoms. Then, the ratios ($\lambda$'s) are given by

$$\lambda_{T_c} = \left(\frac{n_2}{n_1}\right)_{T_c} = e^{-\frac{\varepsilon}{kT_c}} \qquad \lambda_{T_h} = \left(\frac{n_2}{n_1}\right)_{T_h} = e^{-\frac{\varepsilon}{kT_h}}$$

When the two systems are placed in contact with each other, the temperature will be equalized to the equilibrium temperature ($T_e$).

$$T_e = \frac{T_h + T_c}{2}$$

Then,

$$\lambda_{T_e} = \left(\frac{n_2}{n_1}\right)_{T_e} = e^{-\frac{\varepsilon}{kT_e}}$$

Let us assume that $T_h$ is higher than $T_c$ by $\alpha$ times, that is, $T_h = \alpha T_c$.

$$\lambda_{T_c} = e^{-\frac{\varepsilon}{kT_c}} \qquad \lambda_{T_h} = e^{-\left(\frac{1}{\alpha}\right)\frac{\varepsilon}{kT_c}} \qquad \lambda_{T_e} = e^{-\left(\frac{2}{1+\alpha}\right)\frac{\varepsilon}{kT_c}}$$

The number of microstates of the high temperature system will decrease by cooling from $T_h$ to $T_e$, and on the other hand the number of microstates of the cold temperature system will increase by heating from $T_c$ to $T_e$. The ratio of the microstates at $T_e$ to that at $T_h$ of the high temperature system will be given by,

$$\frac{\lambda_{T_e}}{\lambda_{T_h}} = e^{\left(\frac{1-\alpha}{\alpha(1+\alpha)}\right)\frac{\varepsilon}{kT_c}}$$

According to the Boltzmann equation for entropy, the entropy change due to the change in state is proportional to the logarithm of the ratio of the number of microstates of the initial and final states. Thus the entropy change by changing the temperature from $T_h$ to $T_e$ will be proportional to the logarithm of the above equation:

$$ln\frac{\lambda_{T_e}}{\lambda_{T_h}} = \left(\frac{1-\alpha}{\alpha(1+\alpha)}\right)\frac{\varepsilon}{kT_c}$$

The above equation clearly predicts that the entropy change is inversely proportional to the temperature for a given temperature difference (*i.e.*, for a given $\alpha$). The above equation also predicts decrease of the entropy upon cooling, since the value of $\alpha$ is greater than unity. In a similar way, for heating of the low temperature system to the equilibrium temperature,

$$ln\frac{\lambda_{T_e}}{\lambda_{T_c}} = \left(\frac{\alpha-1}{1+\alpha}\right)\frac{\varepsilon}{kT_c}$$

Note in the above equation that it predicts that the entropy is also inversely proportional to the temperature for a given temperature difference and it increases upon heating. The change of the total entropy may also be predicted by adding the above two equations:

$$ln\frac{\lambda_{T_e}}{\lambda_{T_h}} + ln\frac{\lambda_{T_e}}{\lambda_{T_c}} = \left(\frac{1-\alpha}{\alpha(1+\alpha)}\right)\frac{\varepsilon}{kT_c} + \left(\frac{\alpha-1}{1+\alpha}\right)\frac{\varepsilon}{kT_c} = \left(\frac{(\alpha-1)^2}{\alpha(1+\alpha)}\right)\frac{\varepsilon}{kT_c} > 0$$

The above proves that heat transfer always results in increase of the total entropy of the universe (for the present case, the sum of the high temperature and low temperature systems).

*Example 3.16*

The macrostate (or simply state) of a system can be fixed by fixing some of the variables like *U, V, n* and *T*. Not all of them are independent. When we fix *V* and *n*, for instance, *U* is dependent on *T*. Express internal energy in terms of partition function.

$$U = \sum n_i\varepsilon_i$$ Internal energy

$$\frac{n_i}{n} = \frac{e^{-\frac{\varepsilon_i}{kT}}}{Z}$$

$$U = \frac{n}{Z}\sum_i \varepsilon_i e^{-\frac{\varepsilon_i}{kT}}$$

For internal energy per mole ($u$) of particles

$$u = \frac{N_0}{n}U = \frac{N_0}{Z}\sum_i \varepsilon_i e^{-\frac{\varepsilon_i}{kT}}$$

$$Z = \sum_i e^{-\frac{\varepsilon_i}{kT}}$$ Partition function

Taking derivative

$$\frac{dZ}{dT} = \sum_i \frac{\varepsilon_i}{kT^2} e^{-\frac{\varepsilon_i}{kT}}$$

Multiplying $\frac{N_o kT^2}{Z}$ to both sides

$$\frac{N_o kT^2}{Z}\frac{dZ}{dT} = \frac{N_0}{Z}\sum_i \varepsilon_i e^{-\frac{\varepsilon_i}{kT}}$$

By comparison of the two equations

$$u = \frac{N_o kT^2}{Z}\frac{dZ}{dT}$$

$$\frac{1}{x}dx = d\,ln\,x$$

$$u = N_o kT^2 \frac{d\,ln\,Z}{dT}$$

As the volume is constant,

$$u = N_o kT^2 \left(\frac{d\,ln\,Z}{dT}\right)_V$$

*Example 3.17*

Express entropy in terms of partition function. Make use of the results obtained in the previous example.

Recall

$$ln\,W = n\,ln\,n - \sum n_i\,ln\,n_i$$

and

$$n_i = n\left(\frac{e^{-\frac{\varepsilon_i}{kT}}}{Z}\right)$$

Combining the two equations,

$$ln\,W = n\,ln\,n - \sum_i \frac{n}{Z} e^{-\frac{\varepsilon_i}{kT}}\,ln\left(\frac{n}{Z}e^{-\frac{\varepsilon_i}{kT}}\right)$$

$$ln\,W = n\,ln\,n - \frac{n}{Z}\sum_i e^{-\frac{\varepsilon_i}{kT}}\,ln\,n + \frac{n}{Z}ln\,Z\sum_i e^{-\frac{\varepsilon_i}{kT}} + \frac{n}{ZkT}\sum \varepsilon_i e^{-\frac{\varepsilon_i}{kT}}$$

$$Z = \sum_i e^{-\frac{\varepsilon_i}{kT}} \qquad u = \frac{N_o}{n}U = \frac{N_o}{Z}\sum_i \varepsilon_i e^{-\frac{\varepsilon_i}{kT}}$$

$$ln\,W = n\,ln\,n - n\,ln\,n + n\,ln\,Z + \frac{nu}{N_o kT}$$

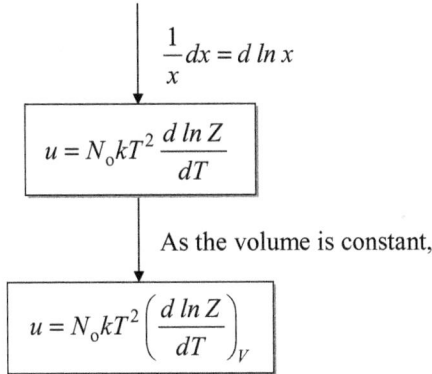

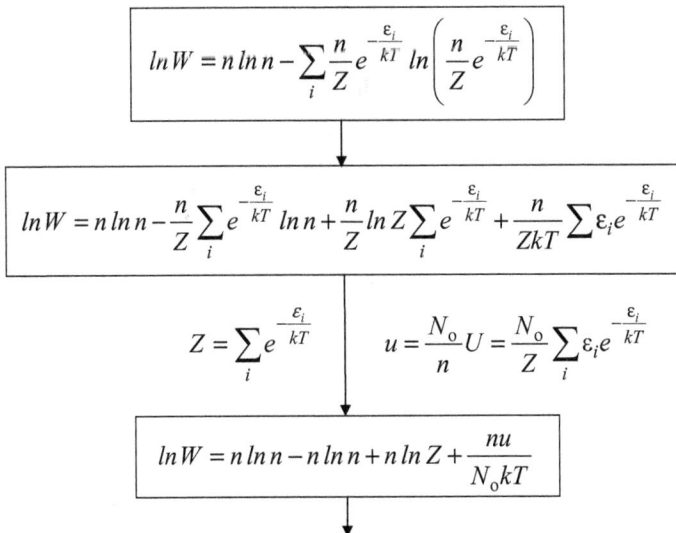

↓

$$\ln W = n \ln Z + \frac{nu}{N_o kT}$$

↓ $u = N_o kT^2 \left( \frac{d \ln Z}{dT} \right)_V$

$$\ln W = n \ln Z + nT \left( \frac{d \ln Z}{dT} \right)_V$$

↓ $S = k \ln W$

$$S = nk \ln Z + nkT \left( \frac{d \ln Z}{dT} \right)_V$$

↓ Per mole of particles,

$$s = N_o k \ln Z + N_o kT \left( \frac{d \ln Z}{dT} \right)_V$$

*Example 3.18*

Classical thermodynamics defines the change of entropy a system by

$$dS = \frac{dq_{rev}}{T}$$

Statistical thermodynamics, on the other hand, defines entropy by

$$S = k \ln W$$

or the change of entropy by

$$dS = k\, d \ln W\ .$$

Discuss how they are related to each other.

From the previous example,

$$\ln W = n \ln Z + \frac{nu}{N_o kT}$$

↓

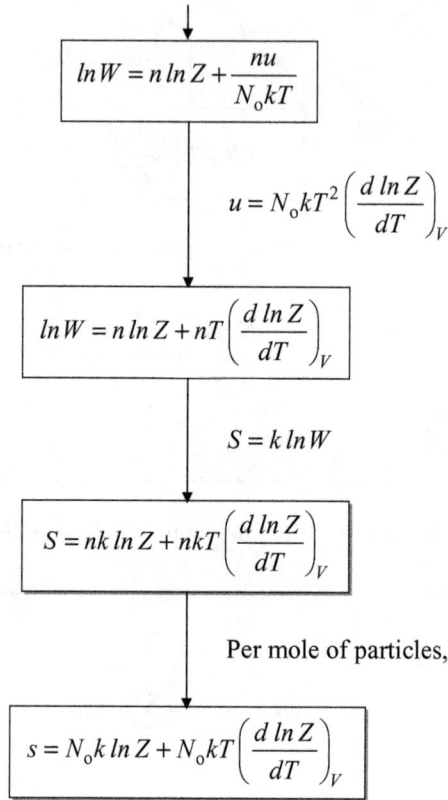

For constant temperature,

$$d\ln W = \frac{n}{N_o kT}\,du$$

$$u = \frac{N_o}{n}U$$

$$d\ln W = \frac{dU}{kT}$$

$dU = dq_{rev}$ at constant volume

$$d\ln W = \frac{dq_{rev}}{kT}$$

$$kd\ln W = \frac{dq_{rev}}{T}$$

$dS = kd\ln W$
From statistical
thermodynamics

$dS = \dfrac{dq_{rev}}{T}$
From classical
thermodynamics

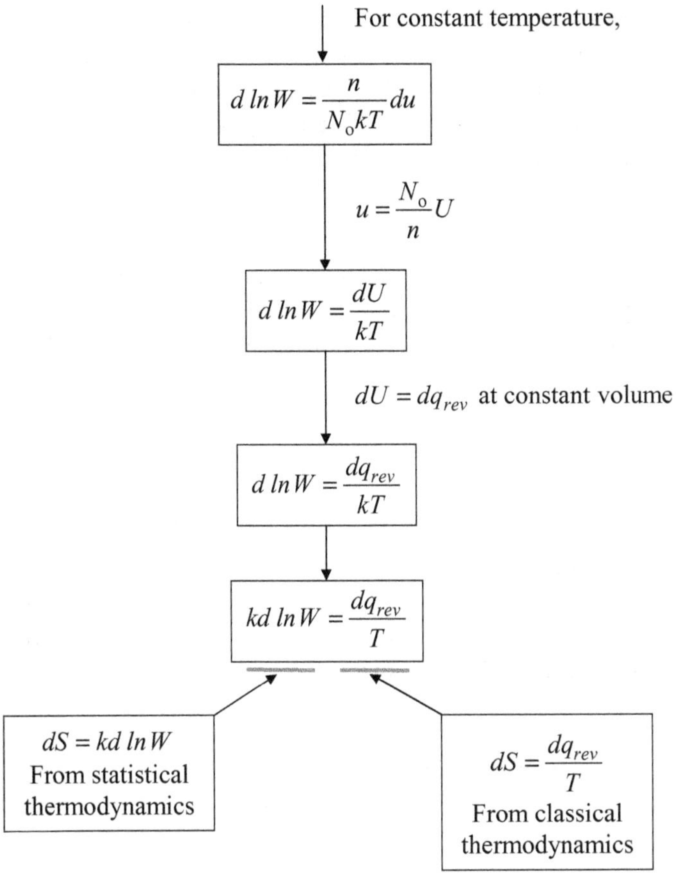

## 3.5  Entropy Changes and Entropy Productions

If we are faced with the problem of deciding whether a given process will proceed *spontaneously*, we might intuitively think whether there is enough energy available. We might suppose that it is perhaps the energy of the system that tends to become minimum and then would be tempted to suggest that,

- if the reaction is exothermic ($\Delta H < 0$), it takes place spontaneously, and
- if the reaction is endothermic ($\Delta H > 0$), it does not take place spontaneously.

However, there are numerous reactions that are endothermic (heat absorbing), and occur spontaneously. For example, the phase transformation of white tin (Sn) to gray tin is exothermic:

$$Sn(white, 298K) = Sn(gray, 298K) \qquad \Delta H = -2,100\ J\ mol^{-1}$$

but white tin is more stable at 298K. Thus energy alone then is not sufficient.
Another example is dissolution of solid $Na_2NO_3$ in water. It dissolves, even though the process is endothermic ($\Delta H > 0$).

Moreover, if the energy of the system decreases during a spontaneous change, its surroundings must experience increase in energy by the same amount in order to satisfy the first law of thermodynamics. This implies that the increase in energy of the surroundings is also spontaneous. Therefore energy alone does not provide a decisive clue as to whether or not a process of interest is spontaneous.

**Entropy Changes**

In order to clarify how a system changes of its state, we need to know not only the amount of energy in the system, but also the direction of distribution of the energy. We already discussed that it is entropy that satisfies the requirements, and the entropy change was defined as,

$$\Delta S = \frac{q_{rev}}{T}$$

Application of this equation is now discussed using a simplifed process.

Let us consider a cylinder which contains water and water vapor at temperature $T$. The cylinder is in thermal contact with a heat reservoir, and they are in thermal equilibrium with each other at temperature $T$. The system (cylinder) and the surroundings (heat reservoir) are schematically shown below:

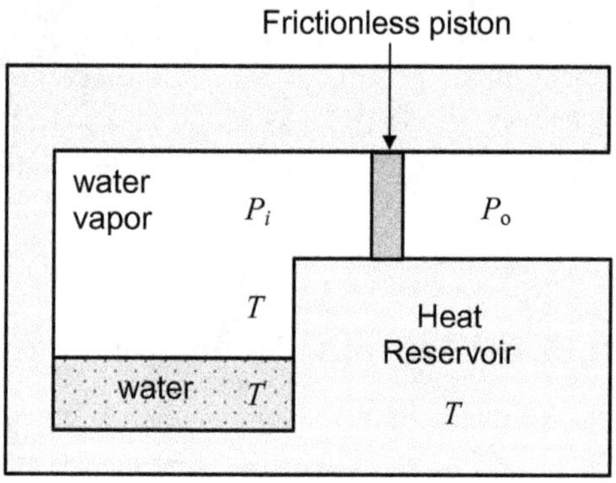


Now thermodynamic states of the above setting can be summarized as follows:

- The temperature inside the cylinder is the same as that of the heat reservoir at $T$.
- The pressure inside the cylinder is the saturation vapor pressure of water at $T$ ($P_i$).
- The external pressure ($P_o$) is kept the same as the internal pressure $P_i$, *i.e.*, $P_o = P_i$.
- The whole setting is in equilibrium and the piston does not move in either direction.

Next, the external pressure is suddenly decreased by $\Delta P$, and thus the piston moves out due to the pressure imbalance (Refer to the following figure). After one mole of water has vaporized, the external pressure is restored to the saturation vapor pressure of water ($P_i$) at $T$.

During the above process, the following events will occur:

- The volume inside the cylinder expands and the internal pressure decreases.
- Water vaporizes and heat flows from the reservoir as the vaporization is endothermic.
- The piston (frictionless) moves outwards and conducts work against the new external pressure ($P_{ext} = P_o - \Delta P$).
- After one mole of water has vaporized, the system (cylinder: water + water vapor) and the surroundings restores the new equilibrium at temperature $T$ and pressure $P_i$.
- But the amount of water vapor in the system at the new equilibrium state has been increased by one mole. Otherwise the system at the new equilibrium is the same as before (the same $T$ and $P$).

In the process described above, the system (cylinder) conducted work against the external pressure of $P_{ext} = P_o - \Delta P$, and hence the amount of work done by the system is

$$\boxed{w = -(P_o - \Delta P)V}$$

where $V$ = molar volume of water vapor. The negative sign in the equation (*i.e.*, the negative amount of work) warrants that the work was done by the system and thus the system has lost energy.

The amount of heat (thermal energy) transferred from the heat reservoir to the cylinder can be found from the first law:

$$\Delta U = q + w$$

Thus,

$$\boxed{q = \Delta U + (P_o - \Delta P)V}$$

As the reservoir (surroundings) has lost this amount of thermal energy as heat to the system, the entropy change of the surroundings (which is so large that its temperature virtually does not change by losing the heat of $q$) is given by,

$$\boxed{\Delta S_{sur} = -\frac{q}{T}} \quad \text{or} \quad \boxed{dS_{sur} = -\frac{dq}{T}} \quad \text{for an infinitesimal change,}$$

Note that the entropy change of the reservoir ($\Delta S_{sur}$) is not constant, but varies with the degree of irreversibility; that is, it changes with change in $\Delta P$:

$$\Delta S_{sur} = -\frac{q}{T} = -\frac{\Delta U + (P_o - \Delta P)V}{T}$$

The larger $\Delta P$ is, the less amount of work the cylinder (the system) performs, and the smaller the amount of thermal energy the surroundings loses to the system, and thus the smaller the entropy change (decrease as $q < 0$) of the surroundings becomes.

When the above process is conducted in a reversible manner by infinitesimally small change in $P$, *i.e.*, $\Delta P \rightarrow 0$, then the above equation will become,

$$\Delta S_{sur} = -\frac{q_{rev}}{T} = -\frac{\Delta U + P_o V}{T}$$

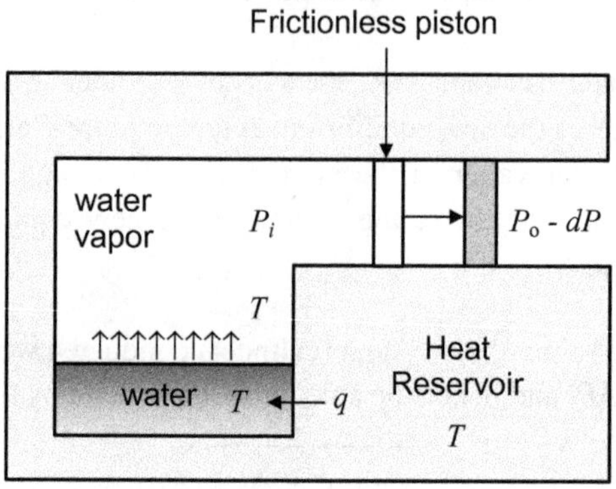


The entropy change of the surroundings as a function of the degree of irreversibility can be graphically represented as follows:

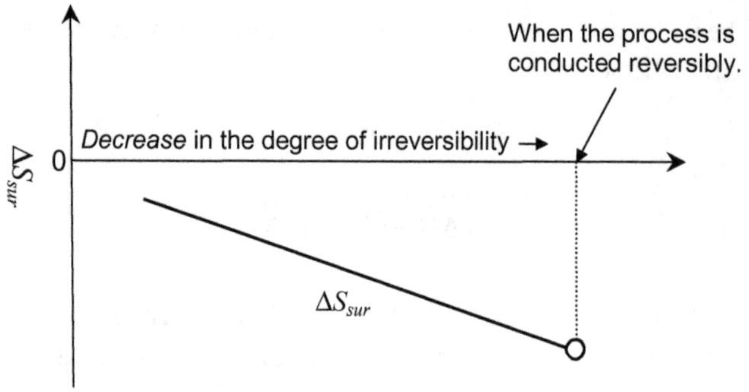


We are now ready to determine the entropy change of the system by applying our knowledge in two basic points:

(1) Entropy is a state function.

(2) When a process undergoes in a reversible manner, the sum of the entropies of the system and of the surroundings is equal to zero:

$$\Delta S_{sys} + \Delta S_{sur} = 0$$

*Reminder*: A reversible process is the process in which the system and surroundings can be restored to the initial state from the final state without producing any changes in the thermodynamics properties of the universe (the system + the surroundings).

We already know the change of the entropy of the surroundings when the process is reversible, and thus the entropy change of the system must be,

$$\Delta S_{sys} = \frac{q_{rev}}{T} \quad \text{or} \quad dS_{sys} = \frac{dq_{rev}}{T}$$

These are valid for any process, irrespective of the process being reversible or irreversible, since the entropy is a state function.

To help better understand the above equations as to why $q_{rev}$ is used even for an irreversible process, we now analyze in depth the transfer of thermal energy in the mode of heat and the conversion of thermal energy into the mode of work. The figure given below explains the transfer of thermal energy in an arbitrary irreversible process:

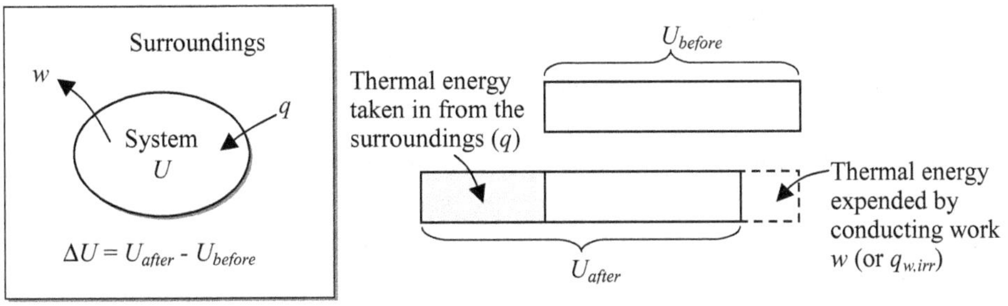


If the process is conducted in a reversible manner, the transfer and conversion of thermal energy can be represented by the following figure:

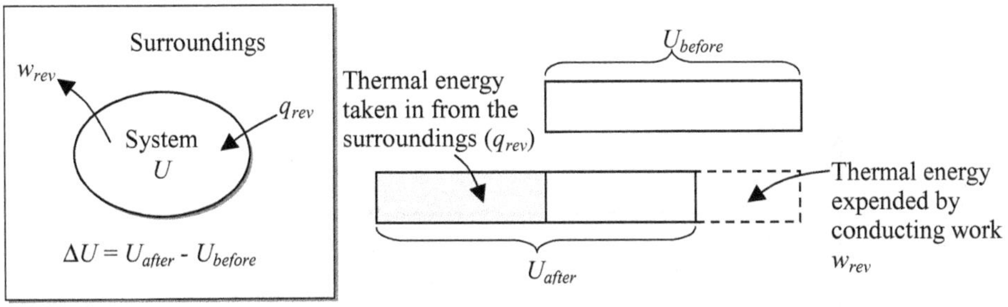


Note that, in the case of the reversible process both thermal energy taken in and that expended to conduct work are larger than those in the case of an irreversible process, but the change in the internal energy, $\Delta U$, is the same as it is a state property.

When the above two processes are compared,

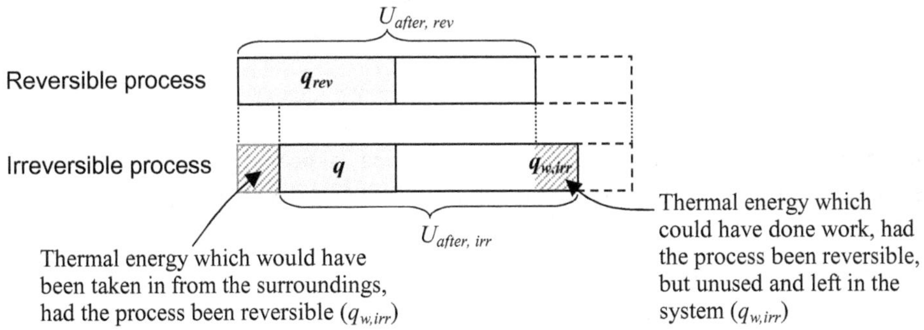

Two important points can be noted from the above:

(1) $U_{after,\,irr} = U_{after,rev}$ since the internal energy is a state property.

(2)

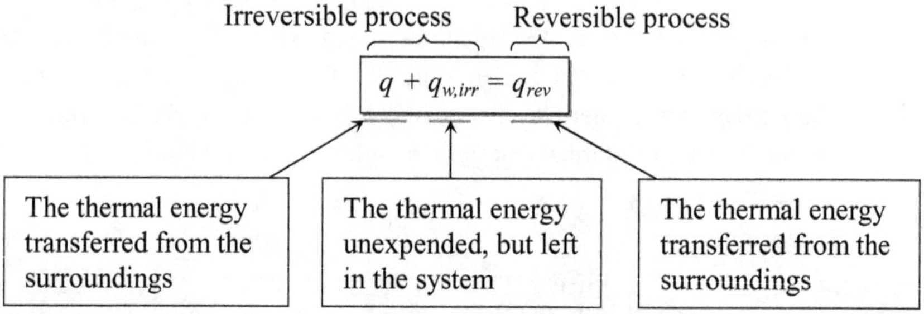


The change of thermal energy which the system experiences in an irreversible process is the sum of the energy transferred from the surroundings and the thermal energy unexpended because of less amount of work done due to the process being irreversible, and the sum is always the same as the thermal energy transferred in the reversible process.

The following figure shows schematically the entropy changes of the system, the surroundings and the universe as a function of the degree of irreversibility:

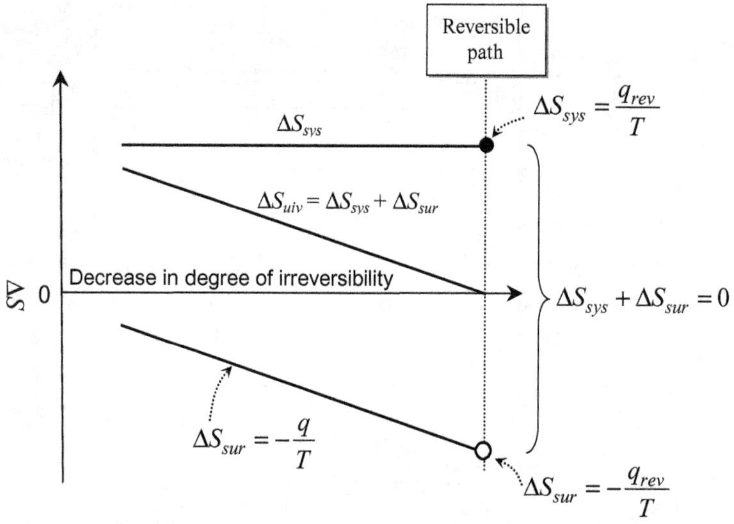


It is emphasized again that although the thermal energy appeared in the expression for the entropy change of the system is that for a reversible process, the application of the equation

$$\Delta S_{sys} = \frac{q_{rev}}{T}$$

is not limited to the reversible process only, but equally valid for any irreversible processes. The total entropy change is then,

$$\Delta S_{tot} = \Delta S_{sys} + \Delta S_{sur}$$

$$\Delta S_{sys} = \frac{q_{rev}}{T} \qquad \Delta S_{sur} = \frac{-q}{T}$$

$$\Delta S_{tot} = \frac{q_{rev}}{T} + \left(\frac{-q}{T}\right)$$

$$\Delta S_{tot} = \frac{q_{rev} - q}{T}$$

As $q_{rev} = q_{max}$, and thus $q_{rev} \geq q$

$$\Delta S_{tot} \geq 0$$

Thus, the entropy of the universe (the system + the surroundings: $\Delta S_{tot}$) increases in irreversible (or natural, or spontaneous) processes. Only if the process is reversible, in other words, if the process is a succession of equilibrium states, the entropy of the universe remains unchanged. In other words the entropy is merely transferred from the system to the surroundings, or vice versa in the reversible process.

The entropy change accompanied by the change in state can be found by integrating the differential expression of the entropy change between the two limiting states:

$$dS = \frac{dq_{rev}}{T}$$

Integration

$$\Delta S = \int_{State\,1}^{State\,2} \frac{dq_{rev}}{T}$$

Suppose we are interested in finding the entropy change due to the change in temperature from $T_1$ to $T_2$ at a constant pressure. Then,

$$\Delta S = \int_{State\,1}^{State\,2} \frac{dq_{rev}}{T}$$

State 1 = $T_1$, State 2 = $T_2$

At constant $P$,

$$dq_{rev} = dH$$

$$dH = C_P dT$$

$$dS = \frac{dq_{rev}}{T} = \frac{C_P}{T} dT$$

$$\Delta S = \int_{T_1}^{T_2} \frac{C_P}{T} dT$$

The above equation allows us to find the entropy change of a material due to the change of temperature.

> *Example 3.19*
>
> Entropy is defined as
>
> $$dS \geq \frac{\delta q}{T}$$
>
> for a system, where the equality sign holds for a reversible process and the inequality sign for an irreversible process. We know that $\delta q$ is not a state function, *i.e.*, it depends on the path the process takes. Using the case of an ideal gas, prove that entropy is a state function.

Consider one mole of a perfect gas.

$$dU = \delta q + \delta w$$ First law of thermodynamics

$dU = C_V dT$

$\delta q = dq_{rev}$ for a reversible process

$\delta w = -PdV$ for $PV$ work

$$C_V dT = dq_{rev} - PdV$$

$PV = RT$ for one mole of a perfect gas

$$\frac{dq_{rev}}{T} = C_V \frac{dT}{T} + R\frac{dV}{V}$$

$dS = dq_{rev}/T$ by definition

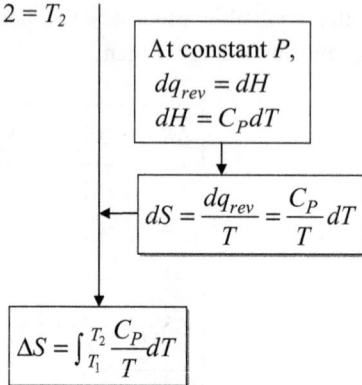

$$S_2 - S_1 = \int_{state\,1}^{state\,2} \frac{\delta q_{rev}}{T} = C_V \ln\left(\frac{T_2}{T_1}\right) + R \ln\left(\frac{V_2}{V_1}\right)$$

① This side depends only on the initial and final states. ($T_1, V_1$ and $T_2, V_2$)

② Therefore *S is a state function* for a perfect gas. Refer to the explanation given in page 42 for more general discussions.

*Example 3.20*

If one mole of a perfect gas undergoes an isothermal process, is the change of entropy independent of pressure?

From Example 3.19,

$$\Delta S = C_V \ln\left(\frac{T_2}{T_1}\right) + R \ln\left(\frac{V_2}{V_1}\right)$$

$T_1 = T_2 =$ constant

$$\Delta S = R \ln\left(\frac{V_2}{V_1}\right)$$

$P_1 V_1 = P_2 V_2$

$$\Delta S = R \ln\left(\frac{P_1}{P_2}\right)$$

The entropy change depends on the volume or pressure change. Refer to Example 3.13.

*Example 3.21*

When a system undergoes a process at a constant pressure, does the entropy change depend on the temperature?

As entropy is a state function, we are free to choose a path from the initial to final states. The path along which a process takes place *reversibly* would be most convenient for thermodynamic calculations, because the heat absorbed or released can be directly related to the entropy change:

$$dS = \frac{dq_{rev}}{T} \quad \text{for a reversible process}$$

$$C_P = \left(\frac{\delta q}{\mathrm{d}T}\right)_P$$

$$\mathrm{d}S = \frac{C_P}{T}\mathrm{d}T$$

$$\frac{\mathrm{d}S}{\mathrm{d}T} = \frac{C_P}{T} > 0 \quad \text{at constant pressure.}$$

This equation tells us that the entropy of a substance held at constant pressure increases when the temperature increases.

*Example 3.22*

Suppose that thermal energy $q$ is transferred as heat spontaneously from a system at a fixed temperature $T_1$ to a system at a fixed temperature $T_2$ without performing any work. Prove that the total entropy change of the process, $\Delta S_{tot}$, is positive.

System 1, $T_1$ → ($q$, $w = 0$) → System 2, $T_2$

$$\Delta S_1 = \frac{-q}{T_1}$$

$$\Delta S_2 = \frac{q}{T_2}$$

$$\Delta S_{tot} = \Delta S_1 + \Delta S_2 = q\left(\frac{1}{T_2} - \frac{1}{T_1}\right)$$

$$\Delta S_{tot} = q\left(\frac{T_1 - T_2}{T_1 T_2}\right) > 0$$

Whenever a system undergoes a change in state and the amount of work done by the system is less than the maximum possible amount of work, then there is a net increase in entropy.

*Example 3.23*

Prove the following statement:

"An isothermal change in phase (phase transformation) of a substance produced by input of energy as heat always leads to an increase in the entropy of the substance."

$$\Delta S = \frac{q_{rev}}{T}$$

$q_{rev} = \Delta H_t$ (heat of transformation) $> 0$
$T = T_t$ (transformation temperature)

$$\Delta S = \frac{\Delta H_t}{T_t} > 0$$

*Example 3.24*

Develop equations for entropy changes for the following three cases:

(1) Thermal energy $q$ is transferred from the heat reservoir I at temperature $T_{\mathrm{I}}$ to the heat reservoir II at temperature $T_{\mathrm{II}}$. Assume the heat transfer causes no temperature change in either reservoir.

(2) Thermal energy $q$ is transferred from the heat reservoir I to the finite mass II ($n_{\mathrm{II}}$ moles). Assume there is no temperature change in the heat reservoir I, but the temperature of the finite mass II rises from $T_{\mathrm{II}}$ to $T_{\mathrm{II}}'$.

(3) Thermal energy $q$ is transferred from the finite mass I ($n_{\mathrm{I}}$ moles) to another finite mass II ($n_{\mathrm{II}}$ moles). Assume the temperature of both masses changes: Mass I from $T_{\mathrm{I}}$ to $T_{\mathrm{I}}'$, and Mass II from $T_{\mathrm{II}}$ to $T_{\mathrm{II}}'$.

We may apply a number of equations developed so far to answer the above questions:

- When heat is transferred without changing temperature,

$$\Delta S = -\frac{q_{rev}}{T}$$

- When heat is transferred with accompanying the change in temperature,

$$\Delta S = \int_{T_1}^{T_2} \frac{dq_{rev}}{T}$$

- When the pressure is constant,

$$dq_{rev} = dH = C_P dT\,,$$

and thus

$$\Delta S = \int_{T_1}^{T_2} \frac{C_P}{T} dT$$

| | Reservoir I → Reservoir II | Reservoir I → Finite mass II | Finite mass I → Finite mass II |
|---|---|---|---|
| Processes | 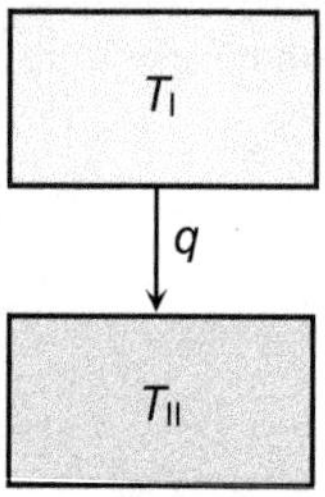 | 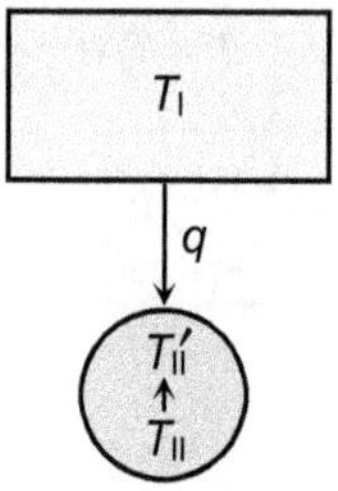 | 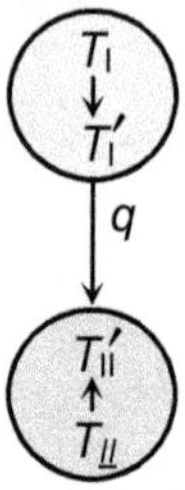 |
| Entropy change in I ($J\,K^{-1}$) | $\Delta S_{\text{I}} = -\frac{q}{T_{\text{I}}}$ | $\Delta S_{\text{I}} = -\frac{q}{T_{\text{I}}}$ | $\Delta S_{\text{I}} = n_{\text{I}} \int_{T_{\text{I}}}^{T'_{\text{I}}} \frac{C_{P(\text{I})}}{T} dT$ |
| Entropy change in II ($J\,K^{-1}$) | $\Delta S_{\text{II}} = \frac{q}{T_{\text{II}}}$ | $\Delta S_{\text{II}} = n_{\text{II}} \int_{T_{\text{II}}}^{T'_{\text{II}}} \frac{C_{P(\text{II})}}{T} dT$ | $\Delta S_{\text{II}} = n_{\text{II}} \int_{T_{\text{II}}}^{T'_{\text{II}}} \frac{C_{P(\text{II})}}{T} dT$ |
| Remarks | $dH = C_p dT \quad \rightarrow \quad \Delta H = \int_T^{T'} C_P dT = \frac{q}{n}$ : Find $T'$.<br>$dS = \frac{C_P}{T} dT \quad \rightarrow \quad \Delta S(J\,mol^{-1}K^{-1}) = \int_T^{T'} \frac{C_P}{T} dT$<br>$\Delta S(J\,K^{-1}) = n \int_T^{T'} \frac{C_P}{T} dT,$ | | |

*Example 3.25*

Develop a general equation for the entropy change of an ideal gas when a system undergoes the change in state from State 1 ($P_1$, $T_1$) to State 2 ($P_2$, $T_2$).

Since entropy is a state function, the change in entropy is independent of the path the process takes. In other words, we can take any path that is convenient, as long as the path connects the two states. Two convenient paths will be,

(1) Reversible, isothermal process ($T_1$, $P_1 \rightarrow P_2$) followed by reversible, isobaric process ($P_2$, $T_1 \rightarrow T_2$)
(2) Reversible, isobaric process ($P_1$, $T_1 \rightarrow T_2$) followed by reversible, isothermal process ($T_2$, $P_1 \rightarrow P_2$).

Let us take the first path. The following is the graphical expression of the first path:

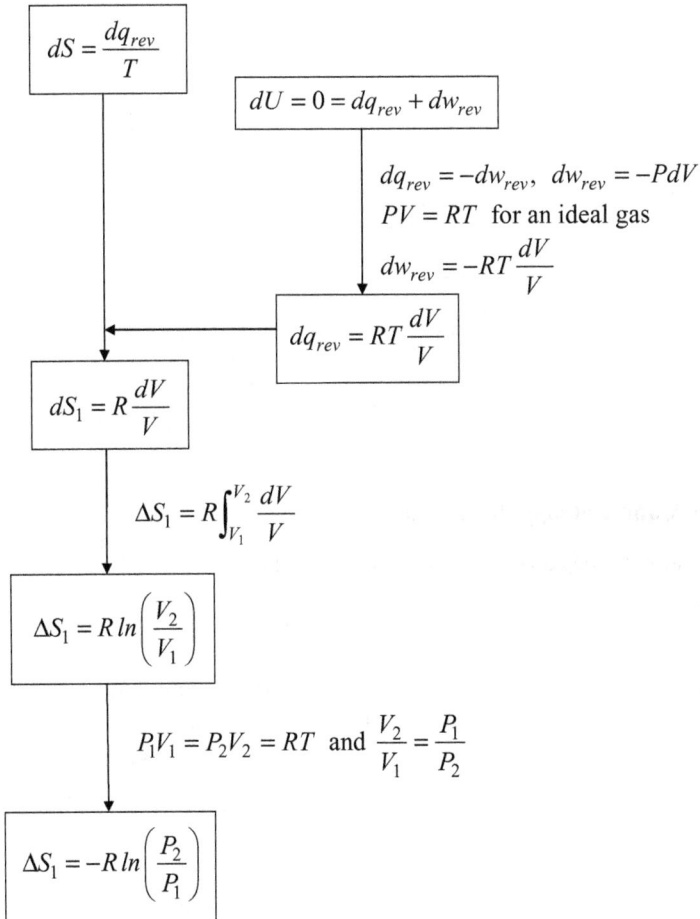

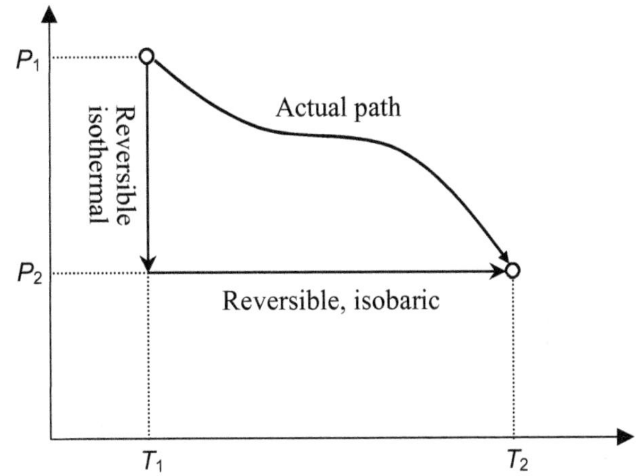


Step 1: Reversible, isothermal process

$$dS = \frac{dq_{rev}}{T}$$

$$dU = 0 = dq_{rev} + dw_{rev}$$

$$dq_{rev} = -dw_{rev}, \quad dw_{rev} = -PdV$$

$$PV = RT \text{ for an ideal gas}$$

$$dw_{rev} = -RT\frac{dV}{V}$$

$$dq_{rev} = RT\frac{dV}{V}$$

$$dS_1 = R\frac{dV}{V}$$

$$\Delta S_1 = R\int_{V_1}^{V_2}\frac{dV}{V}$$

$$\Delta S_1 = R\,ln\left(\frac{V_2}{V_1}\right)$$

$$P_1V_1 = P_2V_2 = RT \text{ and } \frac{V_2}{V_1} = \frac{P_1}{P_2}$$

$$\Delta S_1 = -R\,ln\left(\frac{P_2}{P_1}\right)$$

Step 2: Reversible, isobaric process

$$dS = \frac{dq_{rev}}{T}$$

$$dq_{rev} = dH \quad \text{when } P \text{ is constant.}$$

$$dH = C_P dT$$

$$dq_{rev} = C_P dT$$

$$dS_2 = \frac{C_P}{T} dT$$

$$\Delta S_2 = \int_{T_1}^{T_2} \frac{C_P}{T} dT$$

The entropy change of the whole process from State 1 to State 2 is then,

$$\Delta S = \Delta S_1 + \Delta S_2$$

$$\Delta S = \int_{T_1}^{T_2} \frac{C_P}{T} dT - R \ln\left(\frac{P_2}{P_1}\right)$$

**Lost Work and Entropy Production**

When we look at the equation for work done by the system,

$$w = -(P_o - \Delta P)V$$

it is clear that the system will do more work as $\Delta P$ becomes smaller, *i.e.*, the pressure difference between the system and the surroundings becomes smaller. The system will do the maximum work when $\Delta P \to 0$, *i.e.*, when the process is reversible.

$$w_{max} = w_{rev} = -P_o V$$

$$\Delta U = q + w$$

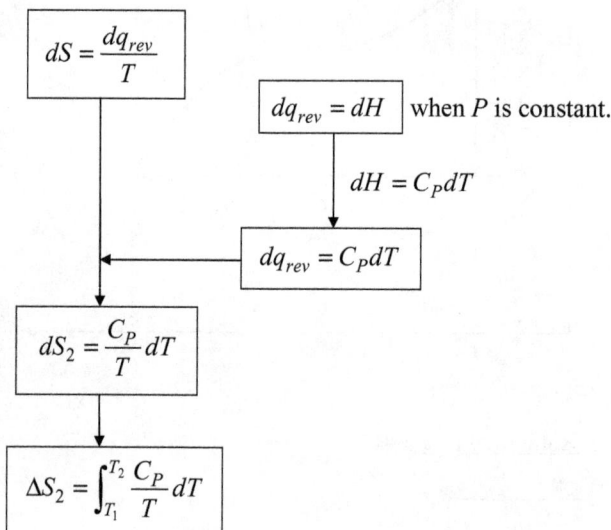

$$q_{max} = q_{rev} = \Delta U + P_o V$$

From the First Law,

$$\Delta U = q + w$$

$$\Delta U = q_{rev} + w_{rev}$$
$$\Delta U = q_{act} + w_{act}$$

$$q_{act} - q_{rev} = w_{rev} - w_{act}$$ Subscript "act" = "actual"

$$q_{rev} = q_{act} - (w_{rev} - w_{act})$$

About $(w_{rev} - w_{act})$

$(w_{rev} - w_{act})$ = The work that could have been done had the process been reversible. – The work that was actually done.

→ The difference is the amount of work lost by the system due to the process being irreversible, and thus is called ***lost work*** ($w_{los}$).
In other words, the system does less amount of work in an irreversible process than in the reversible process by the amount of the lost work ( $w_{los} = w_{rev} - w_{act}$ ).

The work lost due to the process being irreversible is permanent; that is, the amount of the lost work, $w_{los}$, is *permanently lost* so that it can never be recovered.

- $w_{los} < 0$ for irreversible processes: the higher the degree of irreversibility, the larger the absolute amount of the lost work ($|w_{los}|$).
- $w_{los} = 0$ for the reversible process, *i.e.*, no lost work.

Let us consider a system which gains heat $q$ and does work $w$. The diagram below is the schematic representations of the First Law of Thermodynamics to show the position of the lost work:

$$\Delta U = q + w \quad \text{or} \quad w = -q + \Delta U$$

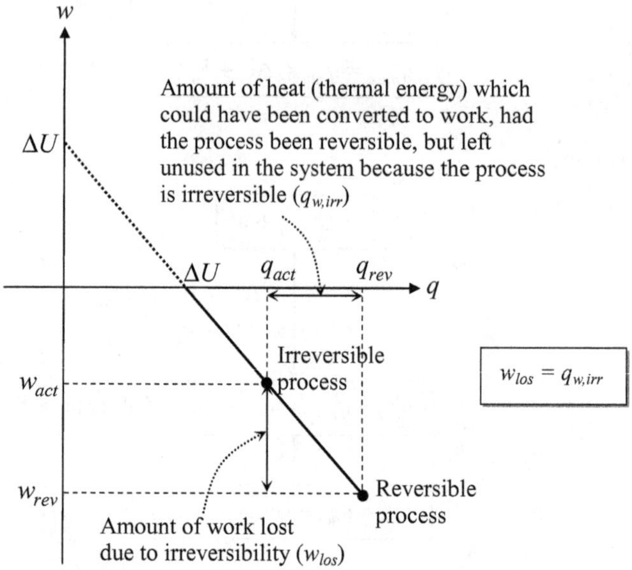


More about the lost work $\left(w_{los} = w_{rev} - w_{act}\right)$

When heat $q$ is supplied to the system, it adds to the molecular motional energy (*i.e.*, kinetic energy: translational, rotational and vibrational) of the system. Then a portion of the motional energy will be expended in doing work $w$, which is the energy transfer due to organized motion of particles (atoms, molecules, etc.). The system can do the maximum amount of work when the process goes in the reversible manner ($w_{rev}$). If the process is irreversible, the amount of work done by the system ($w_{act}$) is less than $w_{rev}$ by the amount of $w_{los}$ as seen in the above figure. This difference is termed the lost work, as the system has lost the chance to do this amount of work due to the process being irreversible. In other words, the motional energy of the system equivalent to $w_{los}$ has lost its chance to conduct work and thus remained in the system as part of the motional energy.

$$\Delta U = q_{rev} + w_{rev} = q_{act} + w_{act}$$

$$\underline{w_{rev} - w_{act}} = -\underline{(q_{rev} - q_{act})}$$

The amount of work lost due to irreversibility ( $w_{los}$ )

The amount of heat which has lost the chance to do work of $w_{los}$ , and thus remains in the system as motional energy.

$$q_{rev} = q_{act} - \left(w_{rev} - w_{act}\right)$$

$$w_{los} = \left(w_{rev} - w_{act}\right)$$

$$q_{rev} = q - w_{los} \quad \text{where } q = q_{act}$$

We know

$$\Delta S_{sys} = \frac{q_{rev}}{T}, \quad \Delta S_{sur} = -\frac{q}{T}$$

$$\Delta S_{tot} = \Delta S_{sys} + \Delta S_{sur}$$

$$\Delta S_{tot} = \frac{q_{rev} - q}{T}$$

$$\Delta S_{sys} = \frac{q}{T} - \frac{w_{los}}{T} \qquad \Delta S_{tot} = -\frac{w_{los}}{T}$$

Since $\frac{w_{los}}{T} \le 0$

$$\Delta S_{sys} \ge \frac{q}{T} \qquad \Delta S_{tot} \ge 0$$

For infinitesimal change

$$dS_{sys} \ge \frac{\delta q}{T} \qquad dS_{tot} \ge 0$$

*Equality* : when the process is reversible.
*Inequality* : when the process is irreversible.

It is now clearly seen that the total entropy change ($\Delta S_{tot}$), which is the sum of the entropy change of the system ($\Delta S_{sys}$) and the entropy change of the surroundings ($\Delta S_{sur}$), is positive, except a special case which is the process being reversible. Thus, entropy does not play the zero sum game, but the sum of the entropy, *i.e.*, the total entropy increases through irreversible or spontaneous or natural processes. The increase in entropy due to irreversibility is called ***entropy production*** or ***entropy generation***.

$$\Delta S_{tot} = -\frac{w_{los}}{T} \ge 0 \quad \text{Entropy production}$$

$$q_{rev} = q - w_{los}$$

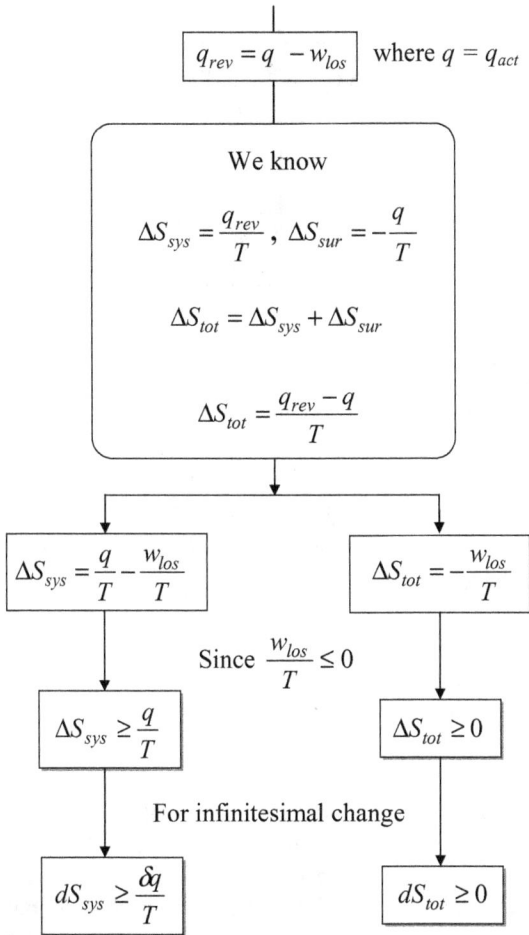

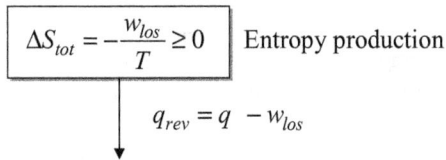

↓

$$\Delta S_{tot} = \frac{q_{rev} - q}{T}$$

↓ Put $q_{irr} = q_{rev} - q$

$$\Delta S_{tot} = \frac{q_{w,irr}}{T} \geq 0$$

In the above, $q_{w,irr} \left(= q_{rev} - q\right)$ is the thermal energy that could have been converted to work (- $w_{los}$), had the process been reversible, but could not, because the process was irreversible. Recall that work $w$ is the energy transfer due to organized motion of molecules (or atoms, etc), while heat $q$ is the transfer of thermal energy; *i.e.*, the transfer of energy via thermal motion (chaotic, random, disorganized motion of molecules or atoms). Thus the above results may be viewed as mechanical energy (- $w_{los}$) having been degraded to thermal energy ( $q_{irr}$ ) due to irreversibility.

More about $\Delta S_{sys} = \frac{q}{T} + (-\frac{w_{los}}{T}) = \frac{q}{T} + \frac{q_{w,irr}}{T}$

**$q$**: Thermal energy transferred from the surroundings to the system which has been degraded by doing so.

**$q_{w,irr}$**: A portion of the motional (thermal) energy of the system which would have been expended by doing work of $w_{los}$ had the process been reversible, but has not been expended due to the process being irreversible, and left in the system.

↓

Thus $q/T$ is the entropy change of the system due to thermal energy transferred from the surroundings. As the thermal energy of the surroundings was reduced by $q$, the entropy of the surroundings has been decreased by the same amount, $-q/T$.

Thus $q_{w,irr}/T$ is the entropy change of the system due to the thermal energy degraded inside the system.

↓

| Entropy change of the system: $\Delta S_{sys}$ | = | Entropy transferred from the surroundings: $\frac{q}{T}$ | + | Entropy produced in the system: $\frac{q_{w,irr}}{T}$ |
|---|---|---|---|---|

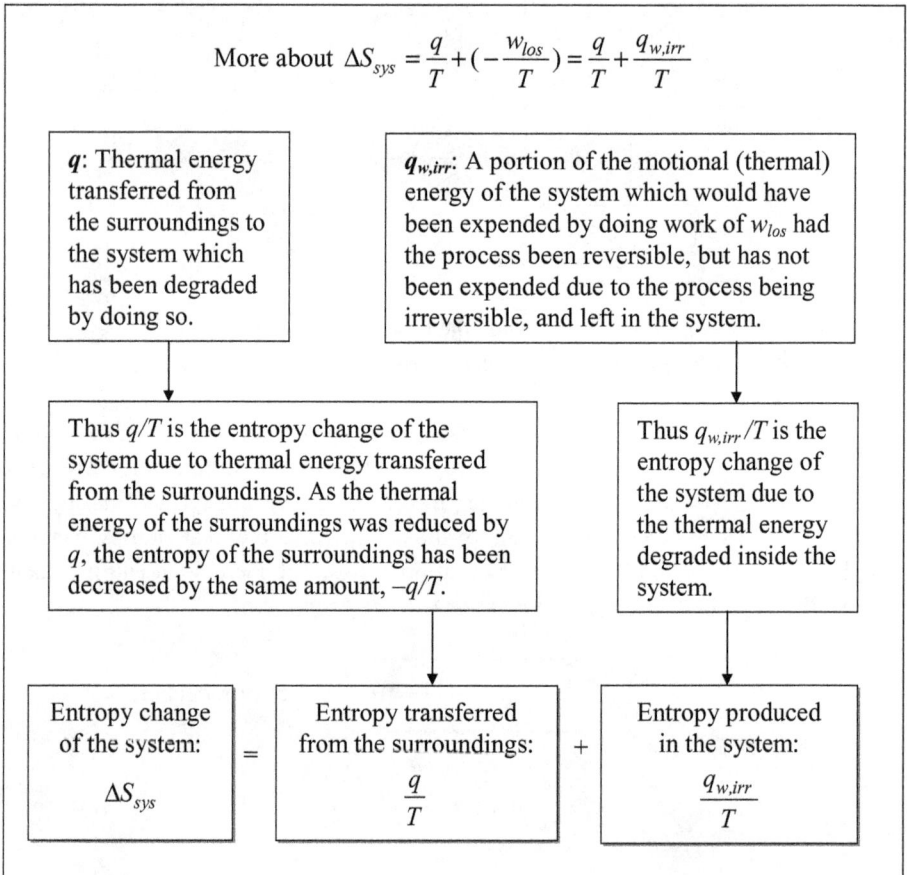

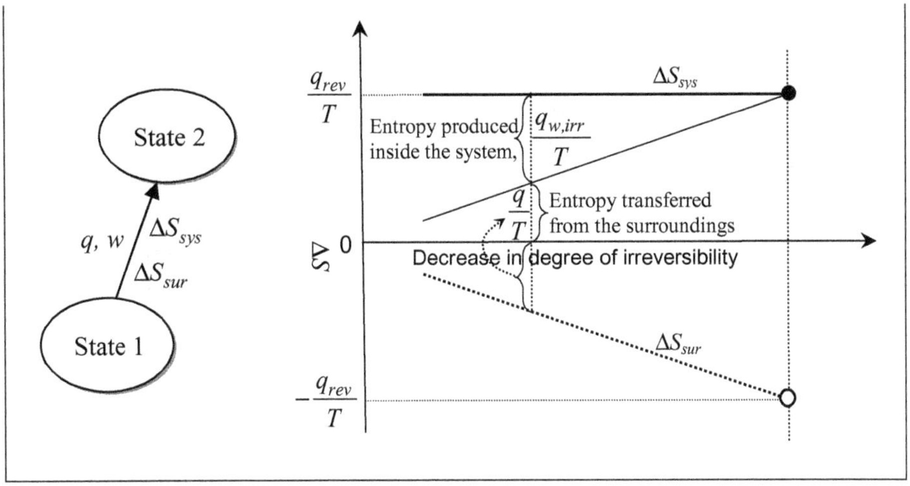


> *Example 3.26*
>
> A cylinder (system) contains an ideal gas at $P_1$ and $V_1$. Calculate the entropy change of the system for the following three different processes:
>
> (1) The gas is expanded reversibly and isothermally until the volume is doubled ($V_2 = 2V_1$).
>
> (2) The gas is expanded isothermally against the pressure which is suddenly decreased to $P_2$ ($=\frac{1}{2}P_1$)
>
> (3) The gas is expanded adiabatically against no pressure (vacuum) until the volume is doubled.

In the reversible, isothermal expansion, the pressure outside cylinder which resists the expansion is always balanced against the pressure inside the cylinder. The graphical representations of three processes are given below:

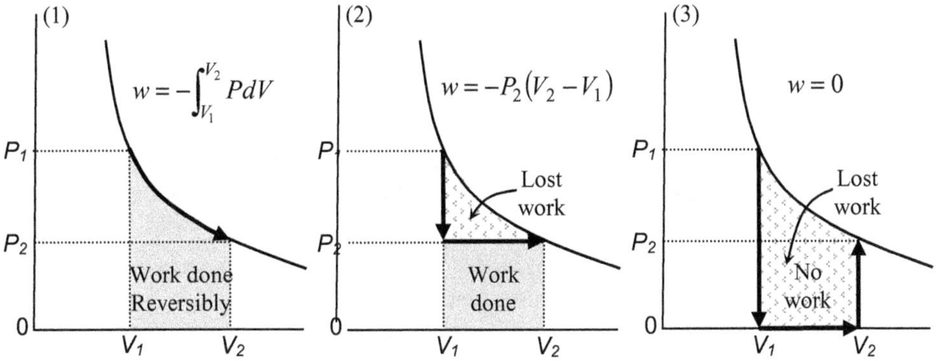

- Process 1 and Process 2: Isothermal process → $T$: constant
- Process 3: Adiabatic process

$$\left.\begin{array}{ll}\text{Adiabatic} & : q = 0 \\ \text{Vacuum} & : w = 0\end{array}\right\} \Delta U = q + w = 0$$

But, the internal energy ($U$) of the ideal gas is a function of temperature alone: $(\Delta U = 0) \rightarrow (T_{intial} = T_{final}) \rightarrow T$: constant

- For all three processes, therefore, even though the paths they take are different, the final states are all identical, and thus the entropy change of the system must be the same.
- We may find the solution of the problem by taking Process 1,

$$w_{rev} = -\int_{V_1}^{V_2} PdV$$

↓ $V_2 = 2V_1$, $PV = RT$

$$w_{rev} = -\int_{V_1}^{2V_1} \frac{RT}{V} dV$$

↓

$$w_{rev} = -RT\,ln\,2$$

↓ $\Delta U = q_{rev} + w_{rev}$
But $\Delta U = 0$ (ideal gas)

$$q_{rev} = RT\,ln\,2$$

↓ $\Delta S = \dfrac{q_{rev}}{T}$

$$\Delta S = R\,ln\,2$$

In the above analysis, a question arises: it has been seen that the entropy change of the system is given by

$$\Delta S = \frac{q_{rev}}{T}$$

For the case 3) above is an adiabatic expansion, that is, $q = 0$. Why the entropy change is not equal to zero, but has a finite value of $\Delta S = R \ln 2$? The answer to this question was already given when the Carnot cycle was discussed and also the energy level change along with the volume change was discussed in Example 3.13. The following shows a practical application of what was discussed earlier.

The free expansion of a gas into vacuum under an adiabatic condition is certainly a spontaneous process so that the entropy must increase. Remember that $q_{rev}$ in the above formula is not the actual amount of heat transferred (in the present case, $q = 0$), but the heat to be transferred if the process were reversible. To answer the above question, first, let us reverse the process so that the expanded gas is compressed reversibly to its original volume ($V_2(=2V_1) \rightarrow V_1$). The work to be done to the system to compress the gas reversibly is,

$$dw_{rev} = -PdV$$

↓ $PV = RT$, $T =$ constant (ideal gas)

$$dw_{rev} = -RT\frac{dV}{V}$$

↓ Integration

$$w_{rev} = -\int_{V_2}^{V_1} RT\frac{dV}{V} = -RT\,ln\left(\frac{V_1}{V_2}\right) = RT\,ln\,2$$

↓ $\Delta U = q_{rev} + w_{rev} = 0$ for ideal gas
Ideal gas $\rightarrow U = f(T)$ only,
but $T =$ constant. Hence $\Delta U = 0$
Therefore $q_{rev} = -w_{rev}$

$$q_{rev} = -RT\,ln\,2$$

↓ $\Delta S = \frac{q_{rev}}{T}$

$$\Delta S = -R\,ln\,2 \quad \text{for compression}$$

↓

$$\Delta S = R\,ln\,2 \quad \text{for expansion}$$

*Exercises*

3.2 Calculate the work done by the reversible, isothermal expansion of 3 moles of an ideal gas from 100 liters initial volume to 300 liters final volume.

3.3 Tin (Sn) transforms from gray to white tin at 286*K*. The heat of transformation ($\Delta H_t$) has been measured as 2.1 kJ mol$^{-1}$.

$$Sn(gray, 268\ K) \rightarrow Sn(white, 268\ K)$$

(1) Calculate the entropy change of the system (Tin).
(2) Calculate the entropy change of the surroundings.
(3) Calculate the total entropy change of the universe (system + surroundings).

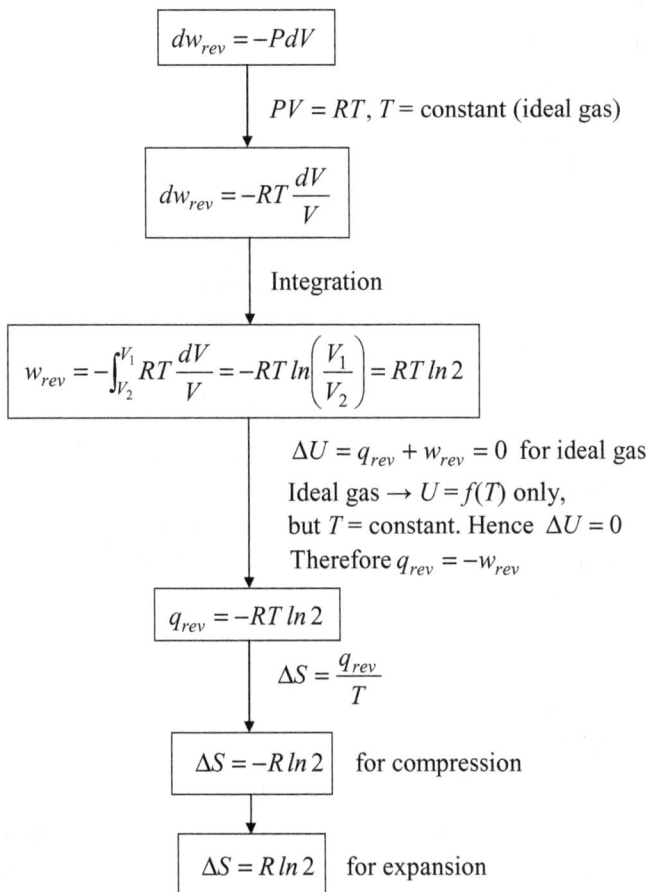

3.3 Suppose that the following reaction takes place at 298 K:

$$Fe_2O_3 + 2Al = 2Fe + Al_2O_3$$

The temperatures of both the system and the surroundings are maintained at 298 K. Calculate the entropy change of the system associated with the reaction. The following data are given:

| | Al | Fe | $Fe_2O_3$ | $Al_2O_3$ |
|---|---|---|---|---|
| $S^o_{298}$ (J mol$^{-1}$K$^{-1}$) | 28.3 | 27.2 | 87.5 | 51.1 |

3.4 One mole of metal block at 1000$K$ is placed in a hot reservoir at 1200$K$. The metal block eventually attains the temperature of the reservoir. Calculate the total entropy change of both the system (the metal block) and the surroundings (the reservoir). The heat capacity of the metal is given as

$$C_P = 23 + 6.3\text{x}10^{-3}T,\ \text{J mol}^{-1}\text{K}^{-1}$$

3.5 Liquid metal can be supercooled to temperatures considerably below their normal solidification temperatures. Solidification of such liquids takes place spontaneously, *i.e.*, irreversibly. Now one mole of silver supercooled to 940°C is allowed to solidify at the same temperature. Calculate the entropy change of the system (silver). The following data are given:

$C_{P(l)} = 30.5$ J mol$^{-1}$K$^{-1}$
$C_{P(s)} = 21.3 + 8.54\times10^{-3}T + 1.51\times10^{5}T^{-2}$, J mol$^{-1}$K$^{-1}$
$\Delta H^o_f = 11{,}090$ J mol$^{-1}$ (Heat of fusion at $T_m = 961$°C)

Calculate the entropy change of the surroundings.
Does the process proceed spontaneously?

3.6 Two blocks of the same metal with equal mass, but at different temperatures, one at 100°C and the other at 200°C, are brought into contact until they come to the same temperature. Assuming these two blocks are isolated from the surroundings, calculate the total entropy change. The heat capacity of the metal is 24 J K$^{-1}$mol$^{-1}$.

## 3.6 Statements of The Second Law of Thermodynamics

The science of thermodynamics deals with energy and energy transfer. The first law is a statement of the conservation of energy in energy transfer. The nature of energy and energy transfer cannot be fully explained by the first law alone. All of the discussions given in this chapter are in fact on some other natural phenomena, and the governing law behind the discussions is termed the ***Second Law of Thermodynamics***. The following is the collection of statements which cannot be explained with the first law:

- Whenever energy is transferred, the grade of energy cannot be conserved; some energy must be permanently reduced to a lower grade.

- It is impossible to construct a device that will operate in a cycle and produce no effect other than the raising of a weight and the exchange of heat with a single reservoir. (Kelvin-Planck)
- It is impossible to construct a device that operates in a cycle and produces no effect other than transfer of heat from a cooler body to a hotter body. (Clausius)
- Heat cannot pass spontaneously (unaided) from a region of lower temperature to a region of higher temperature. (Clausius)
- It is impossible to construct a heat engine which produces no other effects than the extraction of heat from a single source and the production of an equivalent amount of work.
- The maximum efficiency of a heat engine depends only on the temperatures between which it operates and is independent of the nature of the cycle process.
- It is impossible for a device operating in a cyclic manner to completely covert heat into work.
- It is impossible by a cyclic process to take heat from a reservoir and to convert into work without simultaneously transferring heat from a hot to a cold reservoir. (Kelvin)
- Heat absorbed at any one temperature cannot be completely transformed into work without leaving some changes in the system or its surroundings.
- No process is possible in which the sole result is the absorption of heat from a reservoir and its complete conversion into work.

Various statements in the above deals with a common phenomenon in nature and affirm the existence of a state function termed "entropy."

- Entropy, like energy, is a fundamental thermodynamic property.
- Entropy is an index of the capacity to do work.
- Change in entropy measures the degree of irreversibility of a process, or the capacity for spontaneous change of a process.
- Every physical or chemical process in nature takes place in such a way as to increase the sum of entropies of all the bodies taking any part in the process.

To sum up, the second law of thermodynamics can be formulated in terms of entropy as:

"*The entropy of an isolated system must increase or in the limit remains constant.*"

The mathematical expression of the above is,

$$\boxed{\Delta S \geq 0} \quad \text{for an isolated system.}$$

The whole of this chapter have dealt with interpretations and applications of the above inequality equation in a number of different perspectives.

# Chapter 4

# Free Energy Functions

## 4.1 Energy Functions

There are basically two different types of states; an equilibrium state and a non-equilibrium state. When a system and its surroundings are in equilibrium, nothing would happen no matter how long time is given. The universe (the system + the surroundings) is at rest. At a non-equilibrium state, however, the system and the surroundings will interact and both will move toward the equilibrium position.

When a chemical reaction takes place in a closed system, the quantities of components in the system will change as some are consumed and some others are formed. Eventually the change will come to an end and the compositions will remain unchanged as long as the system remains undisturbed. By what mechanism is a process driven toward the equilibrium state?

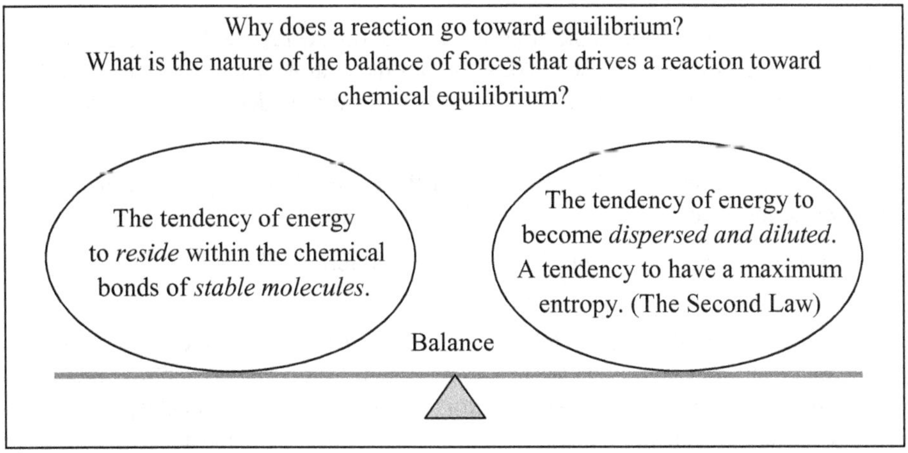

Why does a reaction go toward equilibrium?
What is the nature of the balance of forces that drives a reaction toward chemical equilibrium?
The tendency of energy to *reside* within the chemical bonds of *stable molecules*.
The tendency of energy to become *dispersed and diluted*. A tendency to have a maximum entropy. (The Second Law)
Balance


In the previous chapter, we have seen that the position of equilibrium is at the condition at which the total entropy of both the system and the surroundings is maximum, provided that the total energy is constant. A disadvantage of finding the equilibrium position through the maximum entropy is that we have to determine the entropy (or entropy change) for both the system and the surroundings. It would be much more convenient if we could define a function which determines the position of equilibrium with thermodynamic information of the system alone without necessity of a direct knowledge of changes taking place in the surroundings. Successful energy functions must include terms which represent the above two competing tendencies.

An attempt which we can immediately think of is to make use of a proper utilization of the first and second laws of thermodynamics. Let us combine the equations which represent the two laws.

First Law: $dU = \delta q + \delta w$

Second Law: $dS = \dfrac{dq_{rev}}{T} > \dfrac{\delta q}{T}$

For a reversible process

$$dU = TdS + dw_{rev}$$

For $PV$ work only (no other work) $\quad dw_{rev} = -PdV$

$$dU = TdS - PdV$$

This is an equation of state which relates the internal energy ($U$) to the entropy ($S$) and the volume ($V$). Note that $U$ is treated as the dependent variable, while $S$ and $V$ are the independent variables. From this equation we may deduce the position of equilibrium:

- The equation itself represents the equilibrium state (or the reversible process), provided that the system is closed in such a way that no material exchange is allowed with the surroundings, but energy can be exchanged though the modes of heat and work (*PV* work only).
- If the system is at constant $S$ and $V$, then at the equilibrium $dU = 0$, *i.e.*, the internal energy is at minimum.
- If the system is at constant $U$ and $V$, then at the equilibrium $dS = 0$, *i.e.*, the entropy is at maximum.

The last equation itself is useful and informative for thermodynamic considerations, but from a practical point of view it is not so convenient to use, as the control variables in practice are usually pressure ($P$), temperature ($T$) and volume ($V$), not entropy ($S$). In this regard we need to find functions which are more informative and convenient to use. There are two such functions which are commonly used in thermodynamic considerations and applications. These are ***Helmholtz free energy function*** (or simply ***Helmholtz energy***) and ***Gibbs free energy function*** (or ***Gibbs energy***).

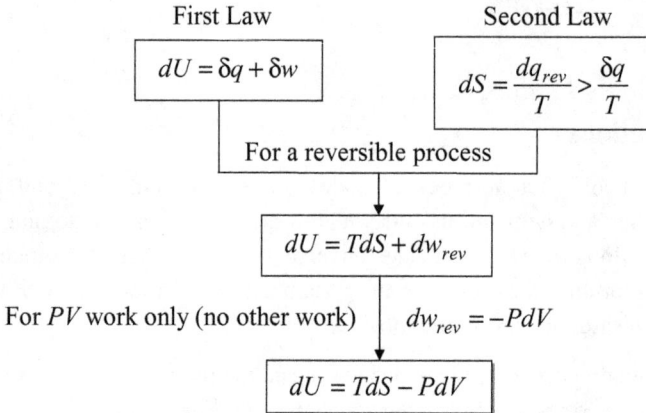

## 4.2 Helmholtz Energy

We start with the equation which combines the first and second laws of thermodynamics.

$$dU = TdS + dw_{rev}$$

$$\downarrow$$

$$dU - TdS = dw_{rev}$$

This equation tells us that, at equilibrium (or in the reversible process), the infinitesimal change of internal energy ($dU$) is balanced with the infinitesimal change of entropy ($dS$) to result in the maximum amount of work to be done on the system ($dw_{rev} = dw_{max}$). It may be useful to define a function which represents the left hand side of the above equation. We now define a new function called ***Helmholtz free energy function*** or **Helmholtz energy** ($A$).

$$A \equiv U - TS$$

↓ Differentiating

$$dA = dU - TdS - SdT$$

↓ At constant T

$$dA = dU - TdS$$

Now the balance between the change in internal energy and the change related to the change of entropy is represented by the change of Helmholtz energy.
When we combine two equations discussed above,

$$dU - TdS = dw_{rev} \qquad dA = dU - TdS$$

$$\downarrow$$

$$dA = dw_{rev}$$

↓ $w_{rev}$ = $PV$ work + all other additional work($w_{add}$)

$$dA = -PdV + dw_{add}$$

↓ At constant $V$ (no $PV$ work)

$$dA = dw_{add}$$

*Additional work* includes electrical work, surface tension work *etc.* which do not involve volume change

↓ No additional work

$$dA = 0$$

It is clear from the last equation that *a system at equilibrium under constant temperature and volume has the minimum value of Helmholtz energy*. Therefore, the Helmholtz energy offers criteria for thermodynamic equilibrium at constant temperature and volume. It is noted that all terms involved in the expression of Helmholtz energy ($dA = dU - TdS$) are for the system only. No term is directly related to the surroundings.

We may discuss Helmholtz energy in somewhat different way. We start with the total entropy change of the universe (the system + the surroundings).

$$dS_{tot} = dS_{sys} + dS_{sur} \geq 0$$

$$dS_{sur} = -\frac{\delta q}{T}$$

↓ At constant $V$, $\delta q = dU$

$$dS_{sur} = -\frac{dU}{T}$$

Put $dS_{sys} = dS$

$$dS_{tot} = dS - \frac{dU}{T} \geq 0$$

↓

$$-TdS_{tot} = dU - TdS \leq 0$$

↓ $dA = dU - TdS$

$$dA = -TdS_{tot} \leq 0$$

We can draw a number of meaningful conclusions from the above equation:

- Helmholtz energy always decreases as the process proceeds ($dA < 0$).
- The more irreversible the process is (*i.e.*, the larger the entropy production, $dS_{tot}$, is), the more rapid decrease Helmholtz energy undergoes (the more negative $dA$ is).

- Helmholtz energy becomes minimum at equilibrium, or stays constant in the reversible process ($dA = 0$).

## 4.3 Gibbs Energy

Helmholtz energy is useful for a process under a constant volume as well as at a constant temperature, for instance, when the process runs in a closed, fixed-volume reactor. However, we are often more interested in systems at constant pressure and temperature rather than at constant volume and temperature. A free energy function, which is based on the constant pressure and temperature, would thus be convenient to use for a process under a constant pressure.

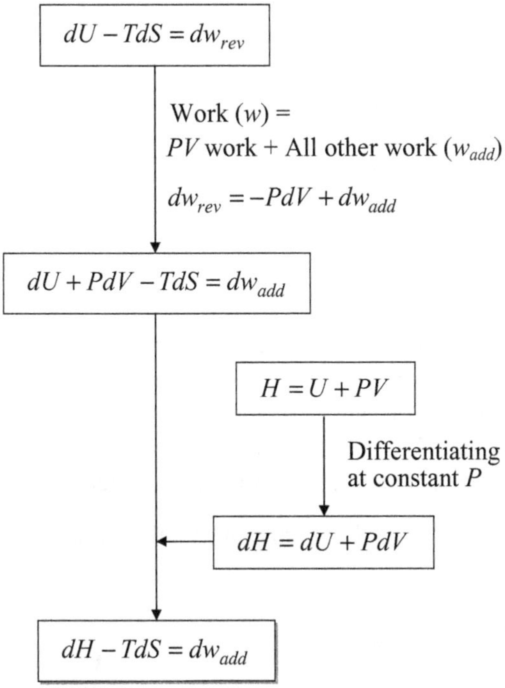
$dU - TdS = dw_{rev}$
Work ($w$) =
$PV$ work + All other work ($w_{add}$)
$dw_{rev} = -PdV + dw_{add}$
$dU + PdV - TdS = dw_{add}$
$H = U + PV$
Differentiating
at constant $P$
$dH = dU + PdV$
$dH - TdS = dw_{add}$

We now define a new function called ***Gibbs free energy function*** or simply ***Gibbs energy*** ($G$).

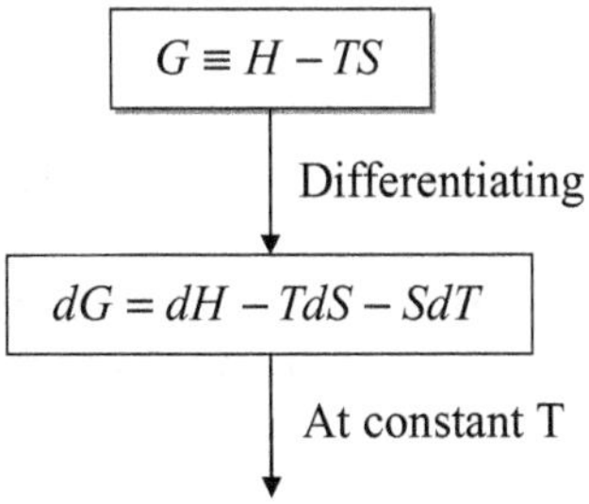
$G \equiv H - TS$
Differentiating
$dG = dH - TdS - SdT$
At constant T

$$dG = dH - TdS$$

Now the balance between the change in enthalpy (*dH*) and the change related to the change of entropy (*TdS*) is represented by the change of Gibbs energy (*dG*).

When we combine two equations discussed above,

$$dH - TdS = dw_{add}$$

$$dG = dH - TdS$$

$$dG = dw_{add}$$

No additional work

$$dG = 0$$

It is clear from the last equation that *a system at equilibrium under constant pressure and constant temperature has the minimum value of Gibbs energy.* Therefore, the Gibbs energy offers criteria for thermodynamic equilibrium at constant temperature and pressure. It is noted that all terms involved in the expression of Gibbs energy ($dG = dH - TdS$) are for the system only. No term is directly related to the surroundings.

The following is the graphical presentation of the last two equations:

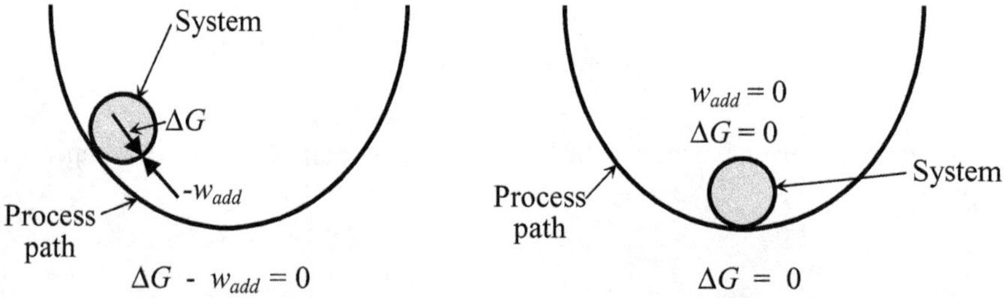


Equilibrium with additional work. Equilibrium without additional work.

We may discuss Gibbs energy in somewhat different way, similar to what we did for Helmholtz energy. Again we start with the total entropy change of the universe (the system + the surroundings).

$$dS_{tot} = dS_{sys} + dS_{sur} \geq 0$$

Put $dS_{sys} = dS$

$$dS_{sur} = -\frac{\delta q}{T}$$

At constant $P$,
$\delta q = dH$

$$dS_{sur} = -\frac{dH}{T}$$

$$dS_{tot} = dS - \frac{dH}{T} \geq 0$$

$$-TdS_{tot} = dH - TdS \leq 0$$

$dG = dH - TdS$

$$dG = -TdS_{tot} \leq 0$$

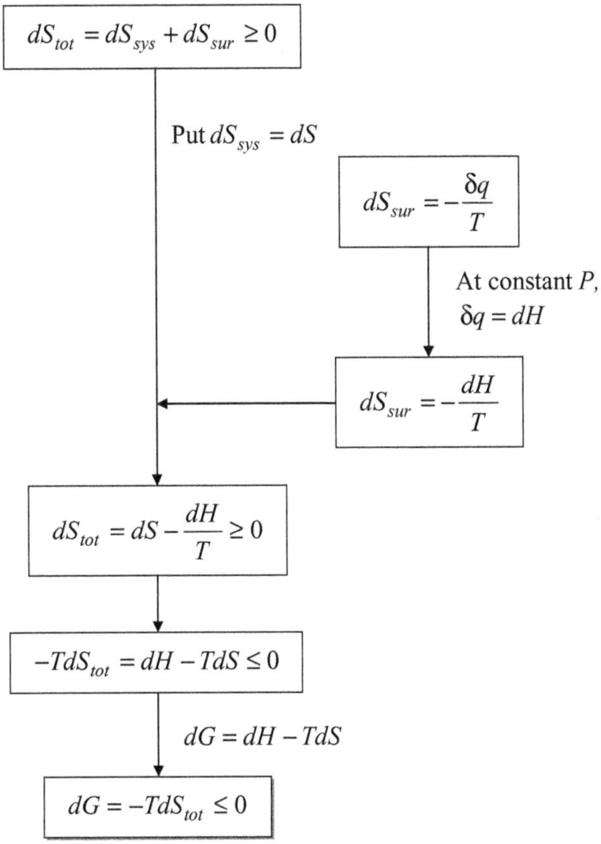

We can draw a number of meaningful conclusions from the above equation:

• Gibbs energy always decreases as the process proceeds ($dG < 0$).

• The more irreversible the process is (*i.e.*, the larger the entropy production, $dS_{tot}$, is), the more rapid decrease Gibbs energy undergoes (the more negative $dG$ is).

• Gibbs energy becomes minimum at equilibrium, or stays constant in the reversible process ($dG = 0$).

---

*Example 4.1*

Is the following statement true?
"Change in Gibbs energy is a measure of work other than $PV$ work done on the system in a reversible process at constant $P$ and $T$."

---

$$dG = dw_{add}$$

For a finite change

$$\Delta G = w_{add}$$

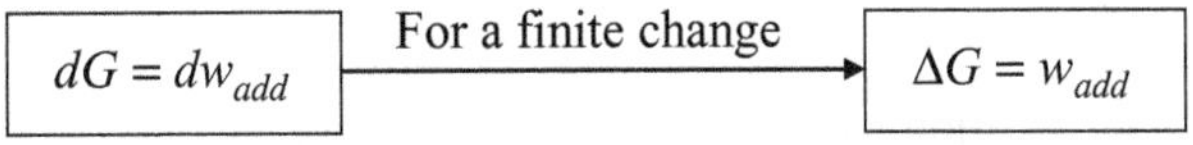

Additional work can be any work other than $PV$ work.

*Example 4.2*

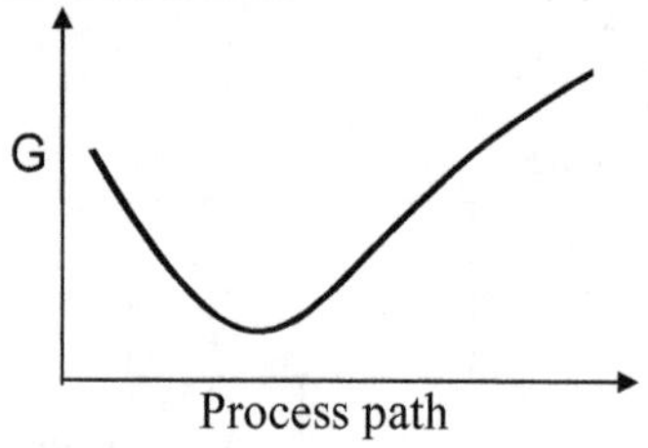


The diagram schematically shows the relationship between the Gibbs energy of the system and the reaction path. Identify the position representing equilibrium of the system at constant temperature and pressure.

Recall that $dG \leq 0$. At equilibrium $dG = 0$; *i.e.*, the minimum point of the curve. A system always moves toward equilibrium and never away from the equilibrium state unless an external agency (the surroundings) does work on the system.

*Exercises*

4.1 Tin transforms from gray to white tin at 286 K and constant pressure of 1 atm. The following thermodynamic data are available:

$C_P = 21.6 + 18.2\times10^{-3}T$, J mol$^{-1}$K$^{-1}$ for both gray and white tin
Heat of transformation ($\Delta H_t^o$) = 2.1 J mol$^{-1}$ at 286 K

(1) Calculate Gibbs energy change ($\Delta G$) of the transformation at 286 K and 1 atm.
(2) Calculate $\Delta G$ of the transformation at 293$K$ and 1atm.
(3) Is the transformation at 293K spontaneous?

## 4.4 Effect of Pressure on Gibbs Energy Change

From the definitions of the Gibbs energy and enthalpy,

$$G = H - TS = U + PV - TS$$

↓

$$dG = dU + PdV + VdP - TdS - SdT$$

↓

$$dU = \delta q + \delta w = dq_{rev} + dw_{rev}$$
$$dq_{rev} = TdS$$
$$dw_{rev} = -PdV \text{ for } PV \text{ work only.}$$

↓

$$dG = VdP - SdT$$

This is an important equation as it tells us how Gibbs energy and thus the equilibrium position vary with pressure and temperature prevailing in the system.

At constant temperature,

$$dG = VdP \quad \text{or} \quad \boxed{\left(\frac{\partial G}{\partial P}\right)_T = V}$$

For an isothermal change from state 1 to state 2,

$$\boxed{\Delta G = G_2 - G_1 = \int_1^2 VdP}$$

If the variation of $V$ with $P$ is known for the substance of interest, this equation can be integrated. For a simple case of one mole of an ideal gas, $PV = RT$. Thus

$$\boxed{\Delta G = RT\,ln\left(\frac{P_2}{P_1}\right)}$$

In an isothermal process for an ideal gas,

$$\boxed{dA = dw_{rev} = -PdV} \qquad \boxed{dG = VdP - \cancel{SdT}} \; \text{T = constant}$$

Integrating

$$\boxed{\Delta A = RT\,ln\left(\frac{V_1}{V_2}\right)} \qquad \boxed{\Delta G = RT\,ln\left(\frac{P_2}{P_1}\right)}$$

Ideal gas

$P_1V_1 - P_2V_2$

$$\boxed{\Delta A = \Delta G = RT\,ln\left(\frac{V_1}{V_2}\right) = RT\,ln\left(\frac{P_2}{P_1}\right)}$$

*Exercises*

4.2 One mole of an ideal gas is compressed isothermally at $298K$ to twice its original pressure. Calculate the change in the Gibbs energy.

## 4.5 Effect of Temperature on Gibbs Energy Change

$$dG = VdP - SdT$$

At constant pressure

$$dG = -SdT$$

$$\left(\frac{\partial G}{\partial T}\right)_P = -S$$

$$\frac{d}{dT}\left(\frac{G}{T}\right) = \frac{1}{T}\frac{dG}{dT} - \frac{G}{T^2}$$

$$\frac{d}{dT}\left(\frac{G}{T}\right) = \frac{1}{T}(-S) - \frac{G}{T^2} = -\frac{TS+G}{T^2}$$

$$G = H - TS$$

$$\left[\frac{\partial}{\partial T}\left(\frac{G}{T}\right)\right]_P = -\frac{H}{T^2}$$

or

$$\left[\frac{\partial\left(\frac{G}{T}\right)}{\partial\left(\frac{1}{T}\right)}\right]_P = H$$

The equations given above in the shaded boxes are called the ***Gibbs-Helmholtz equations***. These equations permit us to calculate the change in enthalpy $\Delta H$ and entropy $\Delta S$ from knowledge of $\Delta G$ at a number of different temperatures. They relate the temperature dependence of Gibbs energy (and thus the position of equilibrium) to the enthalpy and entropy changes.

*Example 4.3*

Internal energy ($U$), enthalpy ($H$), entropy ($S$), Helmholtz energy ($A$) and Gibbs energy ($G$) are functions of state. Each of these can be expressed as a function of two state variables. Prove the following relationships:

$$\left(\frac{\partial G}{\partial T}\right)_P = -S \qquad \left(\frac{\partial G}{\partial P}\right)_T = V \qquad \left(\frac{\partial H}{\partial P}\right)_S = V \qquad \left(\frac{\partial H}{\partial S}\right)_P = T$$

$$dG = VdP - SdT$$

At a constant pressure
$dP = 0$

$$\left(\frac{\partial G}{\partial T}\right)_P = -S$$

$$dG = VdP - SdT$$

At a constant temperature
$dT = 0$

$$\left(\frac{\partial G}{\partial P}\right)_T = V$$

$$H = U + PV$$

$$dH = dU + PdV + VdP$$

$$dU = \delta q + \delta w = dq_{rev} + dw_{rev}$$
$$dq_{rev} = TdS$$
$dw_{rev} = -PdV$ for $PV$ work only.
$$dU = TdS - PdV$$

$$dH = TdS + VdP$$

Constant pressure
$dP = 0$

$$\left(\frac{\partial H}{\partial S}\right)_P = T$$

Constant entropy
$dS = 0$

$$\left(\frac{\partial H}{\partial P}\right)_S = V$$

*Exercises*

4.3 The following equation shows the temperature-dependence of the Gibbs energy change of a reaction :

$$\Delta G = -1{,}750{,}000 - 15.7T \log T + 370T, \ \text{J mol}^{-1}$$

(1) Calculate $\Delta S$ for the reaction at 500 K.
(2) Calculate $\Delta H$ for the reaction at 500 K
(3) Will the reaction take place spontaneously at 500 K?

## 4.6 Some Useful Equations

Based on the relationships between several variables which have been developed until now, we are able to derive several more useful relationships.

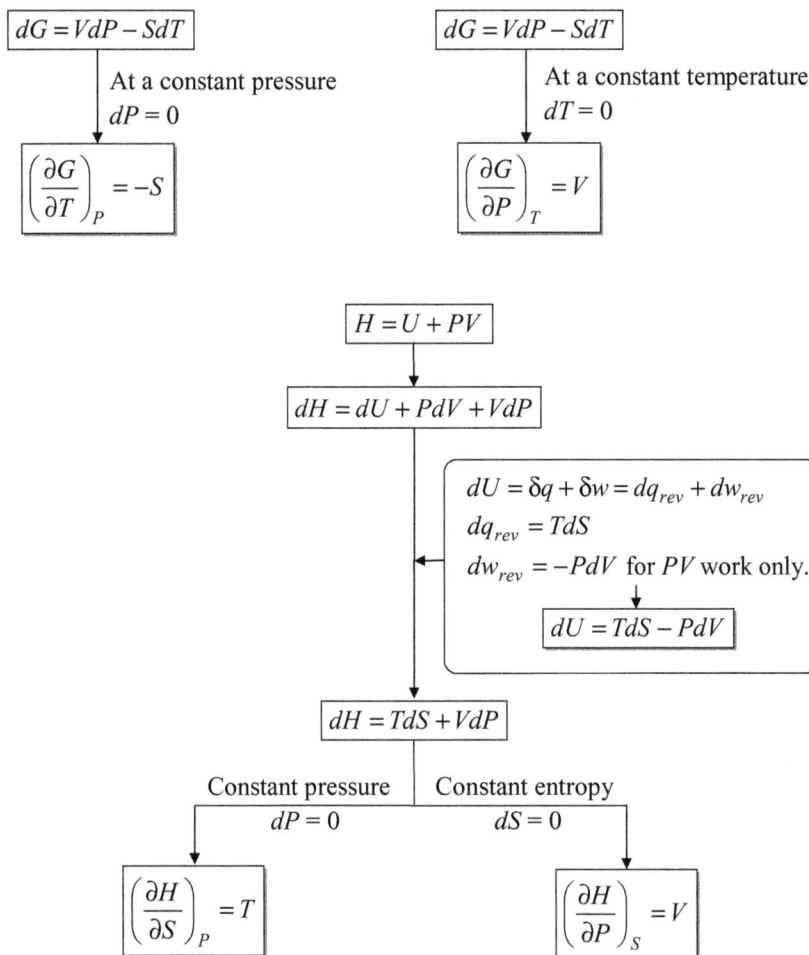

$$dU = TdS - PdV$$

Constant volume: $dV = 0$

$$\left(\frac{\partial U}{\partial S}\right)_V = T$$

Constant entropy: $dS = 0$

$$\left(\frac{\partial U}{\partial V}\right)_S = -P$$

Knowing

$$\frac{\partial}{\partial z}\left(\frac{\partial x}{\partial y}\right)_z = \frac{\partial}{\partial y}\left(\frac{\partial x}{\partial z}\right)_y = \frac{\partial^2 x}{\partial y \partial z}$$

$$\frac{\partial}{\partial V}\left(\frac{\partial U}{\partial S}\right)_V = \left(\frac{\partial T}{\partial V}\right)_S \qquad \frac{\partial}{\partial S}\left(\frac{\partial U}{\partial V}\right)_S = -\left(\frac{\partial P}{\partial S}\right)_V$$

$$\left(\frac{\partial T}{\partial V}\right)_S = -\left(\frac{\partial P}{\partial S}\right)_V$$

Similarly,

$$dH = TdS + VdP \longrightarrow \left(\frac{\partial T}{\partial P}\right)_S = \left(\frac{\partial V}{\partial S}\right)_P$$

$$dA = -PdV - SdT \longrightarrow \left(\frac{\partial S}{\partial V}\right)_T = \left(\frac{\partial P}{\partial T}\right)_V$$

$$dG = VdP - SdT \longrightarrow \left(\frac{\partial V}{\partial T}\right)_P = -\left(\frac{\partial S}{\partial P}\right)_T$$

*An example of the use of these equations*

$(\partial S/\partial V)_T$ and $(\partial S/\partial P)_T$ are difficult to obtain in direct measurements through experiment, but may be calculated from a knowledge of the variation of $P$ with $T$ at constant $V$, $(\partial P/\partial T)_V$, and the variation of $V$ with $T$ at constant $P$, $(\partial V/\partial T)_P$.

These equations are applicable under reversible conditions, and very useful in manipulation of thermodynamic quantities. These equations are known as ***Maxwell equations***.

*Example 4.4*

Is the following statement true?
"The internal energy of an ideal gas at constant temperature is independent of the volume of the gas."

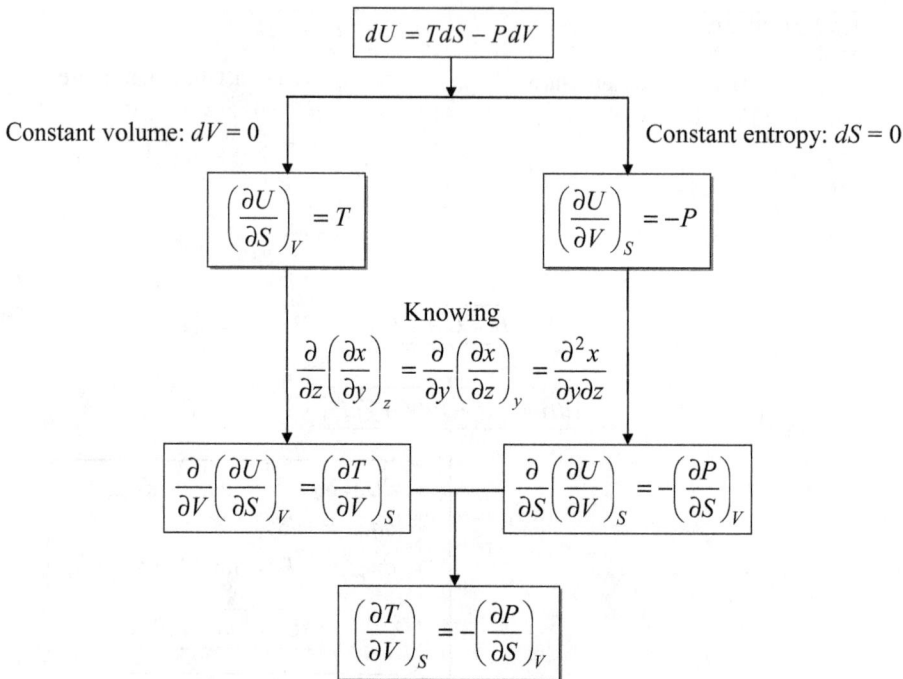

For a closed system,

$$dU = TdS - PdV$$

Differentiating with respect to $V$ at constant $T$,

$$\left(\frac{\partial U}{\partial V}\right)_T = T\left(\frac{\partial S}{\partial V}\right)_T - P$$

From Maxwell equations

$$\left(\frac{\partial S}{\partial V}\right)_T = \left(\frac{\partial P}{\partial T}\right)_V$$

$$\left(\frac{\partial U}{\partial V}\right)_T = T\left(\frac{\partial P}{\partial T}\right)_V - P$$

$PV = RT$ for an ideal gas

$$\left(\frac{\partial P}{\partial T}\right)_V = \frac{R}{V} = \frac{P}{T}$$

$$\left(\frac{\partial U}{\partial V}\right)_T = 0$$

Thus, the statement is true; *i.e.*, the internal energy of an ideal gas is independent of the volume of the gas at constant temperature. In a similar way, one may prove the following statement: *The enthalpy of an ideal gas is independent of the pressure of the gas.*

*Example 4.5*

The volume thermal expansion coefficient $\alpha$ of a substance is defined as,

$$\alpha = \frac{1}{V}\left(\frac{\partial V}{\partial T}\right)_P$$

and the compressibility $\beta$ is defined as

$$\beta = -\frac{1}{V}\left(\frac{\partial V}{\partial P}\right)_T$$

Prove the following relationship:

$$C_P = C_V + \frac{\alpha^2 VT}{\beta}$$

Recall the following two equations:

$$C_P - C_V = \left[P + \left(\frac{\partial U}{\partial V}\right)_T\right]\left(\frac{\partial V}{\partial T}\right)_P \qquad \left(\frac{\partial U}{\partial V}\right)_T = T\left(\frac{\partial P}{\partial T}\right)_V - P$$

Combination of these two equations yields

$$C_P - C_V = T\left(\frac{\partial P}{\partial T}\right)_V\left(\frac{\partial V}{\partial T}\right)_P$$

$$\alpha = \frac{1}{V}\left(\frac{\partial V}{\partial T}\right)_P$$

$$C_P - C_V = \alpha T V\left(\frac{\partial P}{\partial T}\right)_V$$

Consider $V = f(P,T)$ and thus

$$dV = \left(\frac{\partial V}{\partial P}\right)_T dP + \left(\frac{\partial V}{\partial T}\right)_P dT$$

Dividing by $dT$ and holding $V$ constant yields

$$0 = \left(\frac{\partial V}{\partial P}\right)_T\left(\frac{\partial P}{\partial T}\right)_V + \left(\frac{\partial V}{\partial T}\right)_P$$

$$\left(\frac{\partial P}{\partial T}\right)_V = -\frac{\left(\frac{\partial V}{\partial T}\right)_P}{\left(\frac{\partial V}{\partial P}\right)_T} = \frac{\alpha}{\beta}$$

$$C_P = C_V + \frac{\alpha^2 V T}{\beta}$$

This equation shows that $C_V$ can be obtained from $C_P$, compressibility ($\beta$) and thermal expansion coefficient ($\alpha$) for a substance. For solids, $C_V$ is generally more difficult to measure experimentally and the equation offers a way to overcome this difficulty.

# Chapter 5

# The Third Law of Thermodynamics

## 5.1 Statements of the Third Law

- As the temperature decreases, the change in Gibbs energy of a reaction approaches asymptotically its enthalpy change (found experimentally by Richards).
- The above statement can be justified by the basic equation which relates Gibbs energy, enthalpy and entropy:

$$\boxed{dG = dH - TdS} \xrightarrow{T \to 0\,K} \boxed{dG = dH}$$

- The entropy increment accompanying any physical or chemical transformation approaches zero as the absolute temperature approaches zero (Nernst theorem).

If $\Delta G$ and $\Delta H$ for a reaction are plotted as a function of temperature, results are like those shown in the following figure:

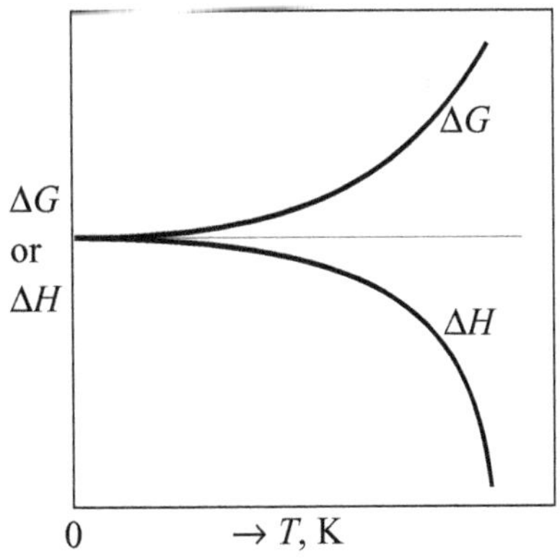


The plots suggest that as the temperature decreases, $\Delta G$ and $\Delta H$ approach equality, and $\Delta G$ approaches $\Delta H$ in magnitude with decreasing T at a much faster rate than $T$ approaches zero.
As the temperature approaches zero,

$$\boxed{\left(\frac{\partial \Delta G}{\partial T}\right)_P = \left(\frac{\partial \Delta H}{\partial T}\right)_P = 0}$$

$$\boxed{\left(\frac{\partial \Delta G}{\partial T}\right)_P = -\Delta S = 0}$$

- The absolute value of the entropy of a pure solid or a pure liquid approaches zero at 0 K (Plank).

$$\boxed{\lim_{T\to 0} S = 0}$$

- If the entropy of each element in some crystalline state is taken as zero at the absolute zero of temperature, every substance has finite positive entropy, but at the absolute zero of temperature the entropy may become zero in the case of perfect crystalline substances (Lewis and Randall).
- The above statement encompasses the following suggestions:
  - Solutions at $0K$ have positive entropy equal to the entropy of mixing (Plank).
  - Supercooled liquids (*e.g.*, glasses) even when composed of a single element, probably retain positive entropies even when the temperature tends to absolute zero (Lewis and Gibson).

Considering the different statements in the above, ***the third law of thermodynamics*** may be stated as "*the entropy of perfect crystalline elements and compounds may be taken to approach zero as the temperature approaches absolute zero*."

The third law is different from the first two laws in that:

- The third law does not give rise to a new thermodynamic function. (the first law $\rightarrow U$, the second law $\rightarrow S$)
- The third law applies only at $T \rightarrow 0$, whereas the first two laws hold at all temperatures.
- The first and second laws do not require information in microscopic level, but are valid for all macroscopic systems, but the third law, by contrast, is understandable only on a molecular basis.

## 5.2 Corollaries to the Third Law

A number of thermodynamic relationships among property variables can readily be derived.

(1) As the temperature decreases, the change in Gibbs energy for a reaction among crystalline substances approaches its enthalpy change.

$$\boxed{\Delta G = \Delta H - T\Delta S}$$

$$\downarrow \quad (T \rightarrow 0\ K) \rightarrow (\Delta S \rightarrow 0)$$

$$\boxed{\Delta G = \Delta H}$$

(2) The entropy of a perfect substance is zero at $0K$ regardless of the pressure.

$$\left(\frac{\partial S}{\partial P}\right)_{T=0\text{K}} = 0$$

(3) The entropy is independent of the volume at 0*K*.

$$\left(\frac{\partial S}{\partial V}\right)_{T=0\text{K}} = 0$$

(4) The thermal expansion coefficient is zero at 0 K. From Maxwell's equations,

$$\left(\frac{\partial S}{\partial P}\right)_T = -\left(\frac{\partial V}{\partial T}\right)_P = 0, \text{ when } T = 0\text{ K} \quad \Big\} \quad \alpha = 0 \text{ at } 0\text{ K}$$

$$\left(\frac{\partial V}{\partial T}\right)_P \rightarrow \alpha V$$

(5) For $\Delta C_P$ in a chemical transformation,

$$\left(\frac{\partial \Delta G}{\partial T}\right)_P = \left(\frac{\partial \Delta H}{\partial T}\right)_P \text{ at } T = 0\text{ K}$$

$$\left(\frac{\partial \Delta G}{\partial T}\right)_P \rightarrow -\Delta S = 0 \qquad \left(\frac{\partial \Delta H}{\partial T}\right)_P \rightarrow \Delta C_P = 0$$

(6) Heat capacity becomes zero at $T = 0$ K.

$$dS = \frac{C_P}{T}dT$$

Integrating from 0 to $T$ K

$$S - S_0 = \int_0^T \frac{C_P}{T}dT$$

The left hand side of the equation will remain finite down to absolute zero, where it will banish. The right hand side must behave in the same way. The integral shows that the heat capacity must decrease to zero at least as rapidly as $T$. Otherwise the integral would diverge because of the lower limit of $T$ is zero, T = 0 K. Therefore,

$$\lim_{T \to 0} C_P = 0$$

In a similar way, we can come to the conclusion that,

$$\lim_{T \to 0} C_V = 0$$

*Example 5.1*

Prove the following statements:
The third law does not give rise to a new thermodynamic function.
(the first law $\rightarrow U$, and the second law $\rightarrow S$)
The third law applies only at $T \rightarrow 0$, whereas the first two laws hold at all temperatures.

What the third law *does* say is that the entropy change $\Delta S$ for any isothermal thermodynamic process becomes zero in the limit of 0 K.

*Example 5.2*

Is the following true?
For the reaction

$$2Ag(s) + PbSO_4 = Pb(s) + Ag_2SO_4(s)$$

(Sum of the entropy of the products)-(Sum of the entropy of the reactants) = 0 at 0 K.

$$\Delta S_{reaction} = \sum (S_{products}) - \sum (S_{reactants})$$

According to the third law, $\Delta S_{reaction} \rightarrow 0$ as $T \rightarrow 0$. In the limit at $T = 0K$, $\Delta S_{reaction} = 0$

*Example 5.3*

Consider the phase change of metallic tin:

$$Sn(s, \text{gray}) = Sn(s, \text{white})$$

The two phases are in equilibrium at 13.2°C and 1 atm. The entropy change $\Delta S$ for the transformation of gray to white tin is 7.82 J mol$^{-1}$K$^{-1}$. Find the entropy change for the transformation at 0 K.

According to the third law, in the limit of absolute zero there is no entropy difference between gray and white tin. The value of the entropy difference decreases from 7.82 J mol$^{-1}$K$^{-1}$ to zero as the temperature decreases from 286.4 K to 0 K.

*Exercises*

5.1 Sulphur has two solid allotropes: Monoclinic sulphur can readily be supercooled to very low temperatures, completely bypassing the phase transformation at 368.5 K. The temperature dependence of the heat capacities of both allotropes can be determined experimentally. It has been found that

$$\Delta S_{rhom} = \int_0^{368.5} C_{P(rhom)} d\,ln\,T = 36.86\,\text{J mol}^{-1}\text{K}^{-1}$$

$$\Delta S_{mono} = \int_0^{368.5} C_{P(mono)} d\,ln\,T = 37.82\,\text{J mol}^{-1}\,\text{K}^{-1}$$

$$\Delta H_t = 402\ \text{J mol}^{-1}$$

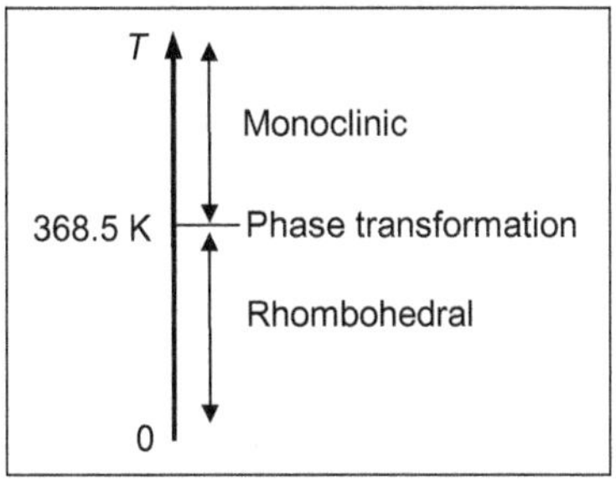


Calculate the entropy change at 0 K.

5.2 Calculate the entropy change for the following dissociation reaction at 0 K:

$$2\text{CuO}(s) = \text{Cu}_2\text{O}(s) + \tfrac{1}{2}\text{O}_2(g)$$

The following data are given:

J mol$^{-1}$ K$^{-1}$

| | $S_{\text{CuO}}$ | $S_{Cu_2O}$ | $S_{O_2}$ |
|---|---|---|---|
| 298 K | 42.64 | 93.10 | 205.04 |

For the dissociation reaction at 298 K and 1 atm.,

$$\Delta H^{\circ} = 140{,}120\ \text{J}$$

$$\Delta G^{\circ} = 107{,}150\ \text{J}$$

## 5.3 Absolute Entropies

As a consequence of the third law, it is possible to evaluate the absolute entropy of a substance.

We now know the following relationships:

$$dS = \frac{dq_{rev}}{T} \xrightarrow{\text{Integrating}} S_T - S_0 = \int_0^T \frac{dq_{rev}}{T} \xrightarrow{S_0 = 0} S_T = \int_0^T \frac{dq_{rev}}{T}$$

Assuming there is no phase change between 0 and $T$ K, the last equation can be further developed:

$$S_T = \int_0^T \frac{dq_{rev}}{T}$$

For $V$ = constant

$$dq_{rev} = dU = C_V dT$$

$$S_T = \int_0^T \frac{C_V}{T} dT$$

For $P$ = constant

$$dq_{rev} = dH = C_P dT$$

$$S_T = \int_0^T \frac{C_P}{T} dT$$

The last two equations enable us to evaluate the absolute entropy at the temperature of interest, provided that there is no phase transformation up to the temperature. If the substance experiences phase change(s) such as melting, the entropy change(s) during the transformation(s) need to be included in the evaluation. For example, the entropy of a substance at temperature $T$ and at a constant pressure can be found by solving the following equation:

$$S = \int_0^{T_{tr}} \frac{C_P}{T} dT + \frac{\Delta H_{tr}}{T_{tr}} + \int_{T_{tr}}^{T} \frac{C_P'}{T} dT$$

where $T_{tr}$ is the temperature of transformation, $C_P$ and $C_P'$ are heat capacities before and after transformation, and $\Delta H_{tr}$ is the enthalpy change of transformation.

*Example 5.4*

At low temperatures, especially near absolute zero, data on heat capacities are lacking for many substances. This lack of data is overcome by making extrapolations to lower temperatures. In this regard the following relationships for heat capacity $C_P$ have proved useful at low temperatures

| | |
|---|---|
| $C_P = aT^3$ | : Most nonmagnetic, nonmetallic crystals |
| $C_P = bT^2$ | : Layer lattice crystals, like graphite and boron nitride, and surface heat capacity |
| $C_P = \gamma T + aT^3$ | : Metals |
| $C_P = jT^{3/2} + aT^3$ | : Ferromagnetic crystals below the magnetic transition temperature |
| $C_P = mT^3$ | : Antiferromagnetic crystals below the magnetic transition temperature |

For metals $\gamma$ can be neglected at low temperatures (but not at $T < 1$ K). Express entropy for metals at low temperatures as a function of heat capacity.

$$S_T = \int_0^T \frac{C_P}{T} dT = \int_0^T (\gamma + aT^2) dT$$

$C_P = \gamma T + aT^3$ for metals
$\gamma = 0$ at low $T$

$$S_T = \frac{aT^3}{3} = \frac{C_P}{3}$$

## 5.4 Molecular Interpretation

In statistical thermodynamics, the entropy is given by Boltzmann equation:

$$S = k \ln W$$

where $W$ is the number of microstates the energy of the system can disperse to.

At 0 K the state of the system in true thermodynamic equilibrium corresponds to a single lowest quantum state (*i.e.*, single microstate). This is the perfectly ordered state of a crystal, for example, with all the molecules in the same lowest energy level. It means that there is only one microstate available at $T = 0$ K, *i.e.*, $W = 1$, and hence $S = 0$.

*Example 5.5*

Solid carbon monoxide (CO) has shown residual entropy of about 4 J $K^{-1}$ at 0 K. The molecular arrangement in the crystal of this material is supposed to be perfect by aligning all in one direction at 0 K:

CO CO CO CO CO …… CO CO

However, at low temperatures many of the molecules are arranged in both forward and reverse positions (CO and OC), and as the crystal is cooled, it may have this disorder frozen because the molecules no longer have the activation energy necessary to rotate and align themselves. Some disorder is therefore retained at very low temperatures.

CO CO OC CO OC …… CO OC

By use of information given above, estimate the range of the residual entropy of CO at $T = 0$ K.

There are two states for each molecule, *i.e.*, forward (CO) and reverse (OC) positions. For $N_0$ (Avogadro number) molecules,

$$W = 2^{N_o}$$

Then, the maximum residual entropy is estimated to be:

$$S = k\,ln\,W = k\,ln\,2^{N_o} = kN_o\,ln\,2 = R\,ln\,2 = 5.76\ \mathrm{J\,K^{-1}mol^{-1}}$$

Therefore the range of the residual entropy of CO at 0 K is

$$0 \leq S \leq 5.76\,\mathrm{J\,K^{-1}mol^{-1}}$$

# Chapter 6

# Enthalpy and Gibbs Energy Changes

## 6.1 Standard States

Absolute values of many thermodynamic properties cannot be obtained. This difficulty is overcome by choosing a ***reference*** or ***standard state*** so that properties can be given in terms of the difference between the state of interest and the reference or standard state.

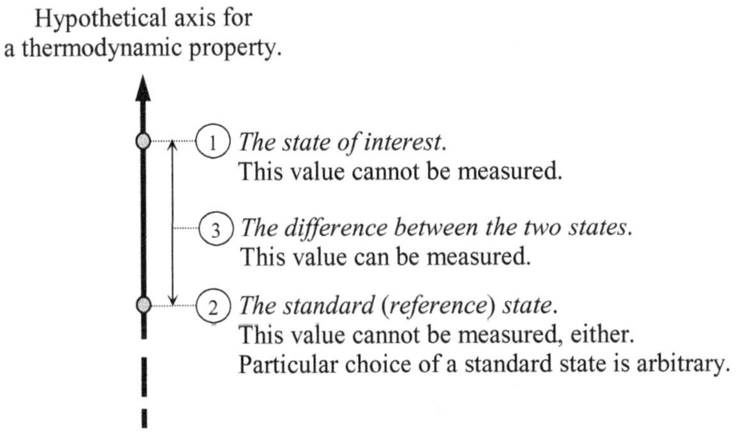

Hypothetical axis for
a thermodynamic property.
1 The state of interest.
This value cannot be measured.
3 The difference between the two states.
This value can be measured.
2 The standard (reference) state.
This value cannot be measured, either.
Particular choice of a standard state is arbitrary.


The standard states for solids, liquids and gases which are most commonly used from the point of view of convenience are,

| States | Standard States |
|---|---|
| Solid | The most stable, pure substance at 1 atm. pressure and the temperature specified. |
| Liquid | The most stable, pure substance at 1 atm. pressure and the temperature specified. |
| Gas | Ideal behavior at 1 atm. pressure of the gas of interest and the temperature specified |

*Example 6.1*

Data on the enthalpy change of titanium (Ti) at 298 K is given as follows:

| | $\Delta H$ |
|---|---|
| α-Ti | 0 |
| β-Ti | 3,350 J $mol^{-1}$ |

Which is incorrectly stated in the following?

(1) α-Ti is the stable form at 298 K.
(2) α-Ti is the standard state at 298 K.
(3) The enthalpy of β-Ti is 3,350 J $mol^{-1}$.
(4) The enthalpy difference between α- and β-Ti, *i.e.*, $H_{\beta\text{-Ti}} - H_{\alpha\text{-Ti}}$, is 3,350 J $mol^{-1}$.

The incorrectly stated one is 3). The value of 3,350 J $mol^{-1}$ is not the enthalpy itself, but it just indicates that the enthalpy of β-Ti is larger than the enthalpy of α-Ti by 3,350 J $mol^{-1}$.

*Example 6.2*

Discuss the validity of the following statements:

(1) Unless otherwise specified, the standard state of an element *i* is customarily chosen to be at a pressure of 1 atm. and in the most stable structure of that element at the temperature at which it is investigated.
(2) However it is possible to choose, as a standard state, one that does not correspond to the most stable form of the species under consideration.
(3) The standard state may also correspond to a virtual state, one that cannot be physically obtained but that can be theoretically defined and for which properties of interest can be calculated.

(1) This statement describes the general definition of the standard state.
(2) This statement is also true. For example, it may be convenient to choose as the standard state of $H_2O$ at 298 K that of the gas instead of the liquid, or one may choose at 298 K the *fcc* structure of iron (austenite) rather than the *bcc* structure (ferrite).
(3) This statement is also correct. One may choose any state as the standard state, if the state chosen is convenient to use for a particular purpose. As will be discussed intensively in the later part (Chapter 10.6), the infinitely dilute solution can be taken as the standard state, in particular to deal with dilute solutions.

## 6.2 Heat of Formation

We shall recall that the enthalpy change for a process (*e.g.*, a chemical reaction), $\Delta H$, is equal to the value of the heat absorbed or evolved when the process (*e.g.*, reaction) takes place at a constant pressure:

$$\boxed{\Delta H = q_P}$$

We shall also recall that it is not possible to measure the absolute value of a thermodynamic property such as the enthalpy of a substance. Nevertheless, let us consider a hypothetical system of which absolute enthalpies of substances at constant temperature and pressure are assumed to be known, say,

| Species | A | $B_2$ | AB | $AB_2$ |
|---|---|---|---|---|
| $H$, J mol$^{-1}$ | 20 | 30 | 30 | 40 |

Then the enthalpy change for the reaction

$A + B_2 = AB_2$ $\quad : \Delta H_1 = H_{AB_2} - (H_A + H_{B_2}) = 40 - (20 + 30) = -10 \text{ J mol}^{-1}$

Similarly,

$A + 1/2B_2 = AB$ $\quad : \Delta H_1 = H_{AB} - (H_A + \frac{1}{2} H_{B_2}) = 30 - (20 + 30/2) = -5 \text{ J mol}^{-1}$

Those enthalpy values listed in the table cannot be real, as it is not possible to measure the absolute value of enthalpy. However, the enthalpy changes of the reactions discussed above can be obtained by measuring heats evolved or absorbed from the reactions. As we are more interested in *enthalpy changes* rather than absolute enthalpies, a new term called ***enthalpy of formation*** **($\Delta H_f$)** or ***heat of formation*** is defined.

The enthalpy of formation is defined as the enthalpy change for the reaction in which one mole of the substance is formed from the *elements* at the temperature of interest. Referring the calculations done above, we assume the following:

| Species | A | $B_2$ | AB | $AB_2$ |
|---|---|---|---|---|
| $\Delta H_f$, J mol$^{-1}$ | 0 | 0 | -5 | -10 |

These are set to *zero* as they themselves are *elements*.

If all elements and species are in their standard states, then we use the symbol $\Delta H_f^o$ and call it ***standard enthalpy of formation***. Let us now check the validity of the concept of the enthalpy of formation by using an example. Consider the following reaction:

$$AB + 1/2\, B_2 = AB_2 \qquad \Delta H_3$$

- By using hypothetical absolute values of $H$ in the above table,

$$\Delta H_3 = 40 - (30 + 30/2) = -5 \text{ J mol}^{-1}$$

- By using enthalpy of formation defined above,

$$\Delta H_3 = -10 - (-5 + 0) = -5 \text{ J mol}^{-1}$$

Note that assigning a value of zero to $\Delta H_f^o$ for each element in its most stable form at the standard state does not affect our calculations in any way. Thus, the enthalpies of formation ($\Delta H_f$) or the standard enthalpies of formation ($\Delta H_f^o$) which are experimentally measurable are of great practical value and give an easy way of determining the enthalpy change accompanying a process or a reaction.

*Example 6.3*

Standard sources of thermodynamic data list the heat or enthalpy of formation at a reference temperature. Most commonly the room temperature (298 K or 298.15 K to be more precise) is chosen as the reference temperature. A few examples are listed in the table given below:

298 K, kJ $mol^{-1}$

| | Ca($s$)* | $H_2$($g$)* | H($g$) | Sn($s$,$w$)* | Sn($s$,$g$) | CaO($s$) | $CaCO_3$($s$) |
|---|---|---|---|---|---|---|---|
| $\Delta H_f^o$ | 0 | 0 | 0 | 0 | 2.51 | -643.3 | -1,207.1 |

"*" represents the standard state.

Find an element or species the standard enthalpy of formation ($\Delta H_f^o$) of which is incorrectly specified. Is the following reaction exothermic or endothermic?

$$Sn(s, \text{white}) \rightarrow Sn(s, \text{gray})$$

(1) As $H_2(g)$ is specified as the standard state, $\Delta H_f^o$ of H($g$) is not zero. It is experimentally found to be 218 kJ $mol^{-1}$.
(2) Sn($s$,white) $\rightarrow$ Sn($s$,gray) : $\Delta H = 2.51 - 0 = 2.51$ kJ $mol^{-1} > 0$ : endothermic reaction.

*Example 6.4*

Are the following statements all true?

(1) The standard enthalpy of formation is defined as the heat change that is resulted in when one mole of a compound is formed from its elements at 1 atm. pressure.

(2) Although the standard state does not specify a temperature, it is customary that we always use $\Delta H_f^o$ values which are measured at 25°C (298.15 K to be exact).

Both statements are correct. Once $\Delta H_f^o$ is known at 25°C, then $\Delta H_f$ at other temperatures can be calculated using information on the heat capacity.

## 6.3 Heat of Reaction

The ***enthalpy of reaction*** or ***heat of reaction*** ($\Delta H_r$) is defined as the difference between the enthalpies of the products and the enthalpies of the reactants.

$$\Delta H_r = \sum H_{\text{products}} - \sum H_{\text{reactants}}$$

Consider an example of the combustion of methane:

$$CH_4 + 2O_2 = CO_2 + 2H_2O \qquad \Delta H_r$$

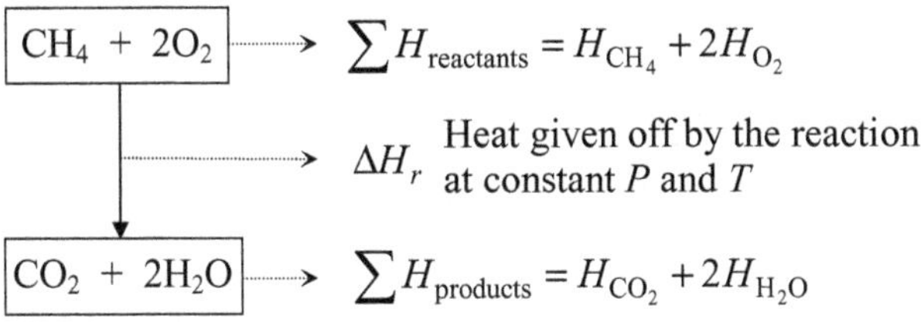

Thus

$$\Delta H_r = \left(H_{CO_2} + 2H_{H_2O}\right) - \left(H_{CH_4} + 2H_{O_2}\right)$$

→ However, these values of absolute enthalpies are not measurable.

The enthalpy of formation, which we have discussed in the previous section, offers an easy way to overcome this difficulty. We now introduce ***Hess's Law***, which states "*the enthalpy change for a chemical reaction is the same whether it takes place in one or several stages*".

Recall that enthalpy is a state function, and hence the enthalpy change depends on the initial and final states only. Hess's law is in fact basically the same as stating that the enthalpy is a state function. Consider the combustion of methane again.



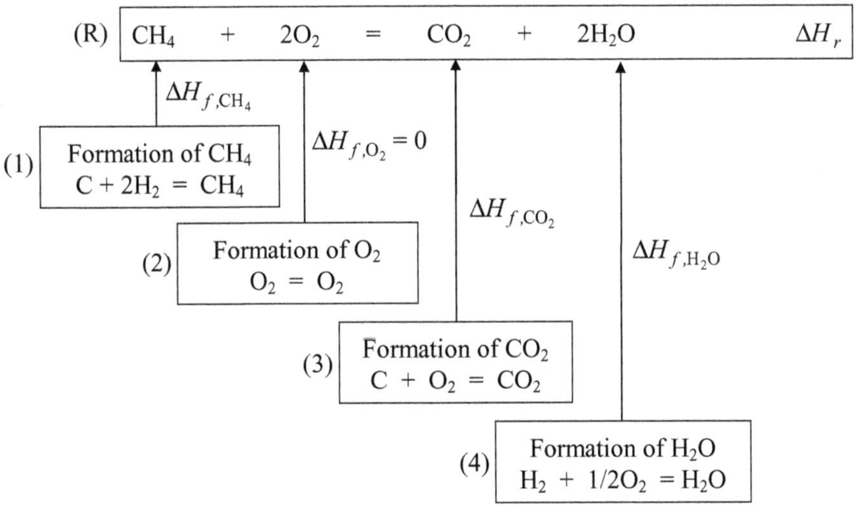

Looking at the above analysis, the following equation holds:

$$(R) = \{(3) + 2\times(4)\} - \{(1) + 2\times(2)\}$$

Then,

$$\Delta H_r = (\Delta H_{f,CO_2} + 2\Delta H_{f,H_2O}) - (\Delta H_{f,CH_4})$$

Now it is in order to generalize our discussion. Consider the following reaction:

$$a\text{A} + b\text{B} = c\text{C} + d\text{D} \qquad \Delta H_r$$

The heat of reaction is given by

$$\Delta H_r = (c\Delta H_{f,\mathrm{C}} + d\Delta H_{f,\mathrm{D}}) - (a\Delta H_{f,\mathrm{A}} + b\Delta H_{f,\mathrm{B}})$$

This equation applies to a system undergoing a chemical reaction at constant pressure and temperature.

- If $\Delta H_r > 0$, the reaction takes place with an absorption of heat from surroundings (*endothermic*).
- If $\Delta H_r < 0$, the reaction takes place with an evolution of heat to the surroundings (*exothermic*).

If temperature is not constant, but changes, recall that

$$\left(\frac{\partial H}{\partial T}\right)_P = C_P \text{ or } dH = C_P dT \text{ at constant } P.$$

For a system undergoing a temperature change from $T_1$ to $T_2$,

$$\Delta H = \int_{T_1}^{T_2} C_P dT$$

Consider a general reaction which occurs at a constant pressure:

$$a\mathrm{A} + b\mathrm{B} = c\mathrm{C} + d\mathrm{D}$$

Enthalpy change of reactants ($a$ moles of A and $b$ moles of B) undergoing temperature change from $T_1$ to $T_2$.

$$\Delta H_{\mathrm{react}} = H_{T_2,\mathrm{react}} - H_{T_1,\mathrm{react}} = \int_{T_1}^{T_2} (aC_{P,\mathrm{A}} + bC_{P,\mathrm{B}})dT$$

Enthalpy change of products ($c$ moles of C and $d$ moles of D) undergoing temperature change from $T_1$ to $T_2$.

$$\Delta H_{\mathrm{prod}} = H_{T_2,\mathrm{prod}} - H_{T_1,\mathrm{prod}} = \int_{T_1}^{T_2} (cC_{P,\mathrm{C}} + dC_{P,\mathrm{D}})dT$$

But the enthalpy change of reaction at $T_2$ is

$$\Delta H_{r,T_2} = H_{T_2,\mathrm{prod}} - H_{T_2,\mathrm{react}}$$

$$H_{T_2,\mathrm{prod}} = H_{T_1,\mathrm{prod}} + \int_{T_1}^{T_2} (cC_{P,\mathrm{C}} + dC_{P,\mathrm{D}})dT$$

$$H_{T_2,\mathrm{react}} = H_{T_1,\mathrm{react}} + \int_{T_1}^{T_2} (aC_{P,\mathrm{A}} + bC_{P,\mathrm{B}})dT$$

Thus

$$\boxed{\Delta H_{r,T_2} = \Delta H_{r,T_1} + \int_{T_1}^{T_2} \Delta C_P dT}$$

where $\Delta H_{r,T_1}$ = heat of reaction at $T_1$, and $\Delta C_P = (cC_{P,\mathrm{C}} + dC_{P,\mathrm{D}}) - (aC_{P,\mathrm{A}} + bC_{P,\mathrm{B}})$

This equation is known as ***Kirchhoff's law*** in integral form. It enables us to calculate the heat of reaction at different temperatures by knowing the heat of reaction at one temperature, say, 298$K$, and heat capacities of reactants and products.
In a differential form,

$$\left(\frac{\partial \Delta H}{\partial T}\right)_P = \Delta C_P$$

*Example 6.5*

Consider a general reaction $a\text{A} + b\text{B} = c\text{C} + d\text{D}$.
The following figure shows the change of enthalpy of reactants and the products with temperature. Answer to each of the following questions:

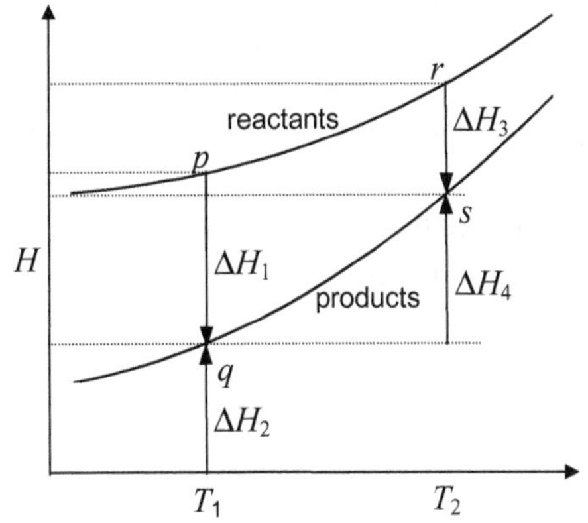


(1) Which point represents the sum of enthalpies of reactants at $T_1$?
(2) Which point represents the sum of enthalpies of the products at $T_2$?
(3) Which one represents the enthalpy change of the products with change in temperature from $T_1$ to $T_2$?
(4) Which one is the heat of reaction at $T_2$?
(5) Is the reaction endothermic or exothermic at $T_2$?

1) $p$ 2) $s$ 3) $\Delta H_4$ $(= s - q)$ 4) $\Delta H_3$ $(= s - r)$ 5) Exothermic ($\Delta H_3 < 0$)

*Example 6.6*

The diagram shows the variation with temperature of enthalpy for a general oxidation reaction:

$$\text{M} + 1/2\text{O}_2 = \text{MO}.$$

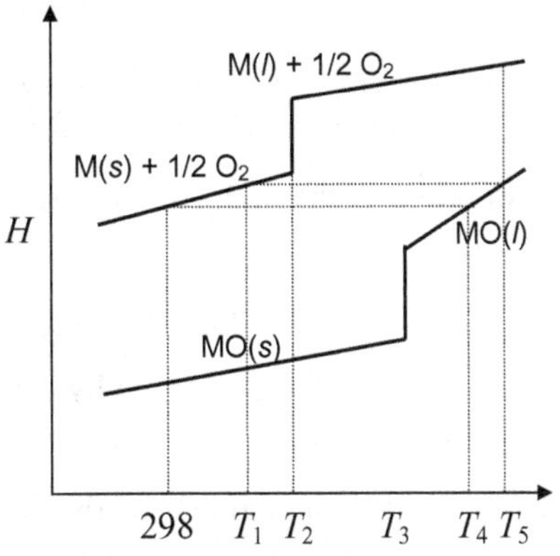


(1) One mole of M($s$) reacts to completion with 1/2 mole of $O_2(g)$ at 298$K$ under an adiabatic condition. Which point represents the temperature of the product MO?
(2) Under the adiabatic condition, it is desired to raise the temperature of the product MO to $T_5$. To what temperature would the reactants need to be preheated?
(3) Which part of the diagram represents the heat of fusion of M?

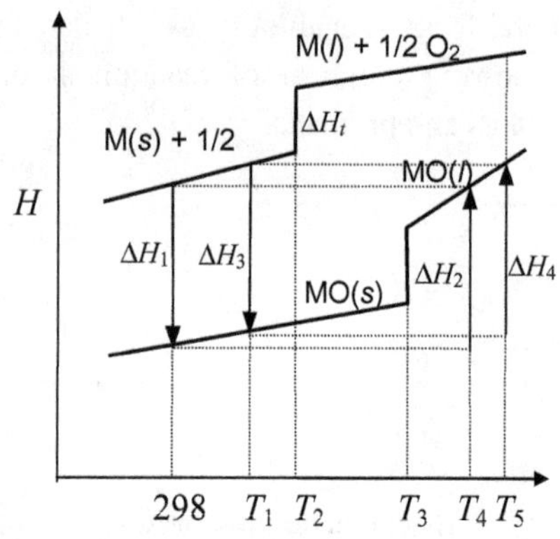


(1) Heat liberated by the reaction = $\Delta H_1$
This heat is used up in raising the temperature of the product MO.
When the temperature of MO is raised to $T_4$, the amount of heat abosorbed by MO is $\Delta H_2$.

$$\Delta H_1 + \Delta H_2 = 0.$$

$\therefore$ Product temperature = $T_4$.

(2) The amount of heat liberated by the reaction = Amount of heat absorbed by the product (adiabatic)
If the reaction takes place at $T_1$, the amount of heat liberated is $\Delta H_3$. This heat is enough to raise the temperature of the product MO from $T_1$ to $T_5$ ($\Delta H_4$). Therefore the reactants need to be preheated to $T_1$.

(3) Fusion is an isothermal phase transformation and thus the enthalpy of M increases without accompanying a temperature change. $\Delta H_t$ represents the heat of fusion of M at $T_2$.

*Exercises*

6.1 The enthalpy change associated with freeze of water at 273 K is -6.0 kJ mol$^{-1}$. The heat capacity ($C_p$) for water is 75.3 J mol$^{-1}$K$^{-1}$ and for ice 37.6 J mol$^{-1}$K$^{-1}$. Calculate the enthalpy change when water freezes at 263 K.

6.2 The enthalpy changes at 298$K$ and 1atm for the hydrogenation and for the combustion of propane are given below :

$$C_3H_6(g) + H_2(g) = C_3H_8(g) \qquad \Delta H_1 = -124 \text{ kJ mol}^{-1}$$
$$C_3H_8(g) + 5O_2(g) = 3CO_2(g) + 4H_2O(l) \qquad \Delta H_2 = -2{,}220 \text{ kJ mol}^{-1}$$

In addition the enthalpy change at 298 K and 1 atm. of the following reaction is known:

$$H_2(g) + 1/2O_2(g) = H_2O(l) \qquad \Delta H_3 = -286 \text{ kJ mol}^{-1}$$

Calculate the heat liberated by the combustion of one mole of propane at 298 K and 1 atm.

6.3 The extraction of zinc by carbothermic reduction of zinc oxide sinter at 1,100°C can be represented by the reaction

$$ZnO(s) + C(s) = Zn(g) + CO(g)$$

Calculate the heat of reaction at 1,100°C. The following data are given:

| | $\Delta H^o_{f,298}$ (kJ mol$^{-1}$) | $T_t$ (K) | $\Delta H_t$ (kJ mol$^{-1}$) | $C_P$ (J mol$^{-1}$K$^{-1}$) |
|---|---|---|---|---|
| ZnO($s$) | -348.1 | | | $49.0 + 5.10\text{x}10^{-3}T - 9.12\text{x}10^{5}T^{-2}$ |
| CO($g$) | -110.5 | | | $28.4 + 4.10\text{x}10^{-3}T - 0.46\text{x}10^{5}T^{-2}$ |
| C($s$) | 0 | | | $17.2 + 4.27\text{x}10^{-3}T - 8.8\text{x}10^{5}T^{-2}$ |
| Zn($s$) | 0 | 693 ($s \rightarrow l$) | 7.28 | $22.4 + 10.0\text{x}10^{-3}T$ |
| Zn($l$) | - | 1,180 ($l \rightarrow g$) | 114.2 | 31.4 |
| Zn($g$) | - | | | 20.8 |

## 6.4 Adiabatic Flame Temperature

Exothermic chemical reactions can be used as energy sources. The more heat evolved per unit mass of fuel, the greater the utility of the fuel as an energy source. The heat released by the combustion raises the temperature of the combustion products. If there is no heat loss to the surroundings and all heat goes into raising the temperature of the products, then the flame temperature will be the highest. This flame temperature is called the ***adiabatic flame temperature***.

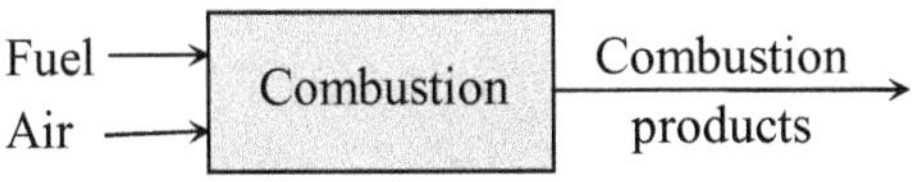


For a steady-flow adiabatic combustion system, the total enthalpy is conserved:

| Sensible heat of fuel | + | Sensible heat of air | + | Heat of combustion | = | Sensible heat products |
|---|---|---|---|---|---|---|

*Example 6.7*

Consider the exothermic reaction as shown in the box below:

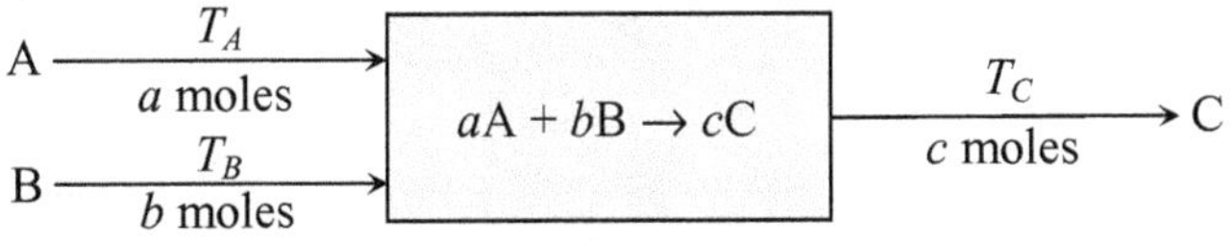

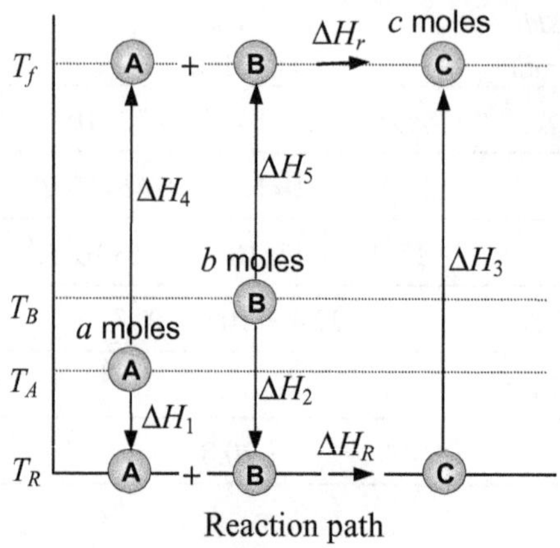


The reaction proceeds to completion and all the heat generated by the reaction is used in raising the temperature of the product. You are asked to calculate the adiabatic flame temperature ($T_f$).
Referring to the diagram given here, answer the following question:
Do the following relationships hold?

(1) $\Delta H_1 + \Delta H_2 + \Delta H_R + \Delta H_3 = 0$
(2) $\Delta H_4 + \Delta H_5 + \Delta H_r = 0$

1) The equation is correct because the total enthalpy is conserved and the enthalpy is a state function so that the enthalpy change is independent of the path the process takes as long as the initial and final states are kept unchanged. The adiabatic flame temperature $T_f$ can be found by solving the equation for $T_f$. The equation physically means;

| Thermal energy released by cooling the reactants A and B to $T_R$ $(\Delta H_1 + \Delta H_2) < 0$ | + | Thermal energy released by chemical reaction at $T_R$ $\Delta H_R < 0$ | + | Thermal energy required to heat the product C from $T_R$ to $T_f$ $\Delta H_3 > 0$ | $= 0$ |
|---|---|---|---|---|---|

2) This equation is also correct. We can construct several different paths between the initial and final states since enthalpy is a state function.

*Exercises*

6.4 A fuel gas containing 22% CO, 13% $CO_2$ and 65% $N_2$ by volume is combusted with the theoretically required amount of air in a furnace to heat a solid burden. The gases enter the furnace at 250°C and the following data are available :

| | $\Delta H^o_{298}$ (J mol$^{-1}$) | $C_P$ (J mol$^{-1}$K$^{-1}$) |
|---|---|---|
| CO(g) | -110,530 | $28.41 + 4.1\text{x}10^{-3}T - 0.46\text{x}10^5T^{-2}$ |
| $CO_2$(g) | -393,510 | $44.14 + 9.04\text{x}10^{-3}T - 8.54\text{x}10^5T^{-2}$ |
| $O_2$(g) | | $29.96 + 4.18\text{x}10^{-3}T - 1.67\text{x}10^5T^{-2}$ |
| $N_2$(g) | | $27.87 + 4.2710^{-3}T$ |

Calculate the adiabatic flame temperature.

## 6.5 Gibbs Energy Changes

The change of the Gibbs energy for a chemical reaction at a specified temperature may be determined from the difference in the Gibbs energy between products and reactants. Let us take an example of a chemical reaction as shown below:

$$aA + bB = cC + dD \qquad \Delta G_r$$

$$(cG_C + dG_D) - (aG_A + bG_B) = \Delta G_r$$

Sum of Gibbs energy of the products

Sum of Gibbs energy of the reactants

Gibbs energy change due to the reaction

If the reactants and products are all in their standard states at the temperature specified,

$$\Delta G_r^o = \left(cG_C^o + dG_D^o\right) - \left(aG_A^o + bG_B^o\right)$$

As seen in the case of enthalpies, absolute values of Gibbs energies are not obtainable and thus we are not able to use the above equation for a practical application to evaluate the change of Gibbs energy due to a reaction. However, this difficulty can be overcome by introducing a term called ***standard Gibbs energy of formation***.

| | | |
|---|---|---|
| Standard Gibbs energy of formation of a compound: $\Delta G_f^o$ | = | Gibbs energy change of the *reaction of formation of the compound from its elements* at their standard states. |

(Examples)

$$C(s) + 1/2O_2(g) = CO(g) \qquad \Delta G_{f,CO}^o$$

$$C(s) + O_2(g) = CO_2(g) \qquad \Delta G_{f,CO_2}^o$$

Elements at their standard states

Compounds at their standard states

Standard Gibbs energy of formation of the compounds

The standard Gibbs energy of formation of a compound is the relative value, relative to the elements to form it and thus the standard Gibbs energy of formation of elements (*e.g.*, C, $O_2$ in the above) is assigned "zero" at all temperatures.

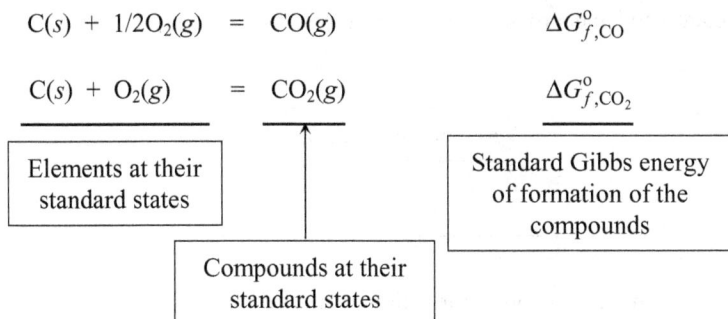

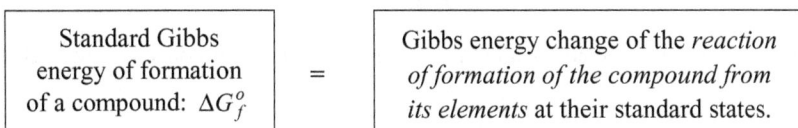

Now we may wish to find the standard Gibbs energy change of a reaction. Let us consider the following reaction:

$$CO(g) + 1/2O_2(g) = CO_2(g) \qquad \Delta G_r^o$$

$$\Delta G_r^o = \Delta G_{f,CO_2}^o - \left(\Delta G_{f,CO}^o + \tfrac{1}{2}\Delta G_{f,O_2}^o\right)$$

zero because it is an element.

Thus the standard Gibbs energy change of the reaction can be calculated from the standard Gibbs energy of formation of species involved in the reaction.

*Reminder*

$\Delta G_r^0$ is the change of Gibbs energy which occurs when all the reactants and products are in their standard states. To deal with systems in which there are some species not in their standard states, but in mixtures (solutions), we shall have to consider ***partial properties*** which will be discussed later.

Recall that Gibbs energy is related to enthalpy and entropy by the following equation:,

$$G = H - TS$$

Applying this equation to a chemical reaction taking place under constant temperature $T$,

$$\boxed{\Delta G_r = \Delta H_r - T\Delta S_r}$$

If all the reactants and the products are in their respective standard states at constant temperature $T$,

$$\boxed{\Delta G_r^o = \Delta H_r^o - T\Delta S_r^o}$$

*Example 6.8*

Consider the following reaction of formation of species $AB_2$:

$$A + 2B = AB_2 \qquad \Delta G_{f,AB_2}^o \qquad T = \text{constant}$$

The following statements are all true. Discuss the application of the statements using the example of the formation of $TiSi_2(s)$.

(1) $\Delta G^o_{f,AB_2}$ is the difference between the free energy of 1 mole of $AB_2$ and the sum of the free energies of 1 mole of A and 2 moles of B, all in their standard states at temperature $T$.
(2) $\Delta G^o_{f,AB_2}$ is the standard free energy of formation of compound $AB_2$ measured with the free energy scale established by setting

$$\Delta G^{\mathrm{o}}_{f,\mathrm{A}} = 0 \quad \text{and} \quad \Delta G^{\mathrm{o}}_{f,\mathrm{B}} = 0.$$

(3) The state of each element (*i.e.*, A or B) for which the above relationships are set, that is, the standard state of each element, is arbitrarily chosen.
(4) The form of element A or B for which the free energy of formation is taken zero must be the same form for which $\Delta H^{\mathrm{o}}_f = 0$ is taken.

(1) $\mathrm{Ti} + 2\mathrm{Si} = \mathrm{TiSi_2}$ $\quad \Delta G^{\mathrm{o}}_{f,\mathrm{TiSi_2}} = G^{\mathrm{o}}_{\mathrm{TiSi_2}} - (G^{\mathrm{o}}_{\mathrm{Ti}} + 2G^{\mathrm{o}}_{\mathrm{Si}})$

(2) $\Delta G^{\mathrm{o}}_{f,\mathrm{TiSi_2}}$ is the standard free energy of formation of $TiSi_2$ measured with the basis of $\Delta G^{\mathrm{o}}_{f,\mathrm{Ti}} = 0$ and $\Delta G^{\mathrm{o}}_{f,\mathrm{Si}} = 0$

(3) The standard state of a species can be arbitrarily chosen

(4) As $G$ is related with $H$ by $G = H - TS$, the standard state should be consistent.

*Example 6.9*

Consider the following reaction:

$$\mathrm{AB} + \mathrm{B} = \mathrm{AB_2} \qquad \Delta G^o_r \qquad T = \text{constant}$$

Check the validity of each of the following statements:

(1) $\Delta G^o_r$ is the difference between the free energy of 1 mole of $AB_2$ and the sum of the free energies of 1 mole of AB and 1 mole of B, all in their standard states at temperature $T$.
(2) The choice of a standard state is arbitrary. If we choose the most stable and pure form of AB at temperature $T$ as its standard state, however, we must choose the most stable and pure forms of all other species (*i.e.*, B and $AB_2$) as their respective standard states.

(1) The standard free energy change of a reaction ($\Delta G^o_r$) is the net free energy change resulted from the reaction. The statement is true. Note that $\Delta G^{\mathrm{o}}_f$ is not the same as $\Delta G^{\mathrm{o}}_r$.
(2) This statement is not true. Each species can take any state as its standard state, but one must be consistent in choosing the standard state of a species throughout the computations

*Exercises*

6.5 Calculate the standard Gibbs energy change of the following reaction at 1000 K and 1 atm.:

$$CO(g) + 1/2\ O_2(g) = CO_2(g) \qquad \Delta G_r^o$$

The following data are available:

$$\Delta G_{f,CO}^o = -112{,}880 - 86.51T,\ \mathrm{J\,mol^{-1}}$$

$$\Delta G_{f,CO_2}^o = -394{,}760 - 0.836T,\ \mathrm{J\,mol^{-1}}$$

6.6 Compute $\Delta G_r^o$ for the reaction

$$CH_4(g) + 2O_2(g) = CO_2(g) + 2H_2O(g)$$

at 298 K. The following data are given:

| | $\Delta H_{f,298}^o$ J mol$^{-1}$ | $S_{f,298}^o$ J mol$^{-1}$K$^{-1}$ |
|---|---|---|
| $CH_4(g)$ | -74,850 | 186.2 |
| $CO_2(g)$ | -393,510 | 213.7 |
| $H_2O(g)$ | -241,810 | 188.7 |
| $O_2(g)$ | 0 | 205.0 |

# Chapter 7

# Behavior of Gases

## 7.1 Ideal Gases

An experimentally found relationship is that,

$$\frac{PV}{RT} \rightarrow 1 \quad \text{as} \quad P \rightarrow 0 \quad \text{for all gases.}$$

$$PV = RT$$

This equation relates the state variables of the system. This equation is called the ***ideal gas equation*** or ***perfect gas equation***. A gas, which obeys this relationship over a range of states of interest, is said to ***behave ideally*** in the range. A gas, which obeys this relationship in all possible states, is called an ***ideal gas*** or a ***perfect gas***.

Recall the following equation:

$$dG - VdP - SdT$$

For an isothermal process

$$dG = VdP$$

$PV = RT$

$$dG = \frac{RT}{P} dP$$

Integrating this equation from the standard state to the state of interest at a constant temperature,

$$G = G^o + RT \ln\left(\frac{P}{P^o}\right)$$

Where $G^{o}$ is the molar Gibbs energy of an ideal gas at its standard state and $G$ is the molar Gibbs energy of an ideal gas at the state of interest, and $P^{o}$ and $P$ are the pressures at the standard state and at the state of interest, respectively.

For gases the standard state is at the pressure of 1 *atm.*, and thus,

$$\boxed{G = G^{o} + RT\ln P}$$

We now consider a mixture of ideal gases to see how the mixing influences the Gibbs energy of the mixture.

**Initial state: Before mixing**

| $n_A$ moles of A($g$) at $T$ and $P$ | $n_B$ moles of B($g$) at $T$ and $P$ |
|---|---|

$$G_A^{i} = G_A^{o} + RT\ln P \qquad G_B^{i} = G_B^{o} + RT\ln P$$

"$i$" stands for "initial."

Total free energy at the initial state $G^{i}$ is then given by

$$\boxed{G^{i} = n_A G_A^{i} + n_B G_B^{i}}$$

$$\boxed{G^{i} = (n_A G_A^{o} + n_B G_B^{o}) + nRT\ln P} \quad \text{where } n = n_A + n_B$$

**Final state: After mixing**

"$f$" stands for "final."

Mixture of $n_A$ moles of A($g$) and $n_B$ moles of B($g$) at $T$ and $P$ → $P_A = N_A P$, $P_B = N_B P$, where $N_i = \dfrac{n_i}{n}$

$$G_A^{f} = G_A^{o} + RT\ln P_A \qquad G_B^{f} = G_B^{o} + RT\ln P_B$$

Total free energy at the final state $G^{f}$ is

$$\boxed{G^{f} = n_A G_A^{f} + n_B G_B^{f}}$$

$$\boxed{G^{f} = (n_A G_A^{o} + n_B G_B^{o}) + n_A RT\ln P_A + n_B RT\ln P_B}$$

$$\boxed{G^{f} = (n_A G_A^{o} + n_B G_B^{o}) + RT(n_A \ln N_A + n_B \ln N_B) + nRT\ln P}$$

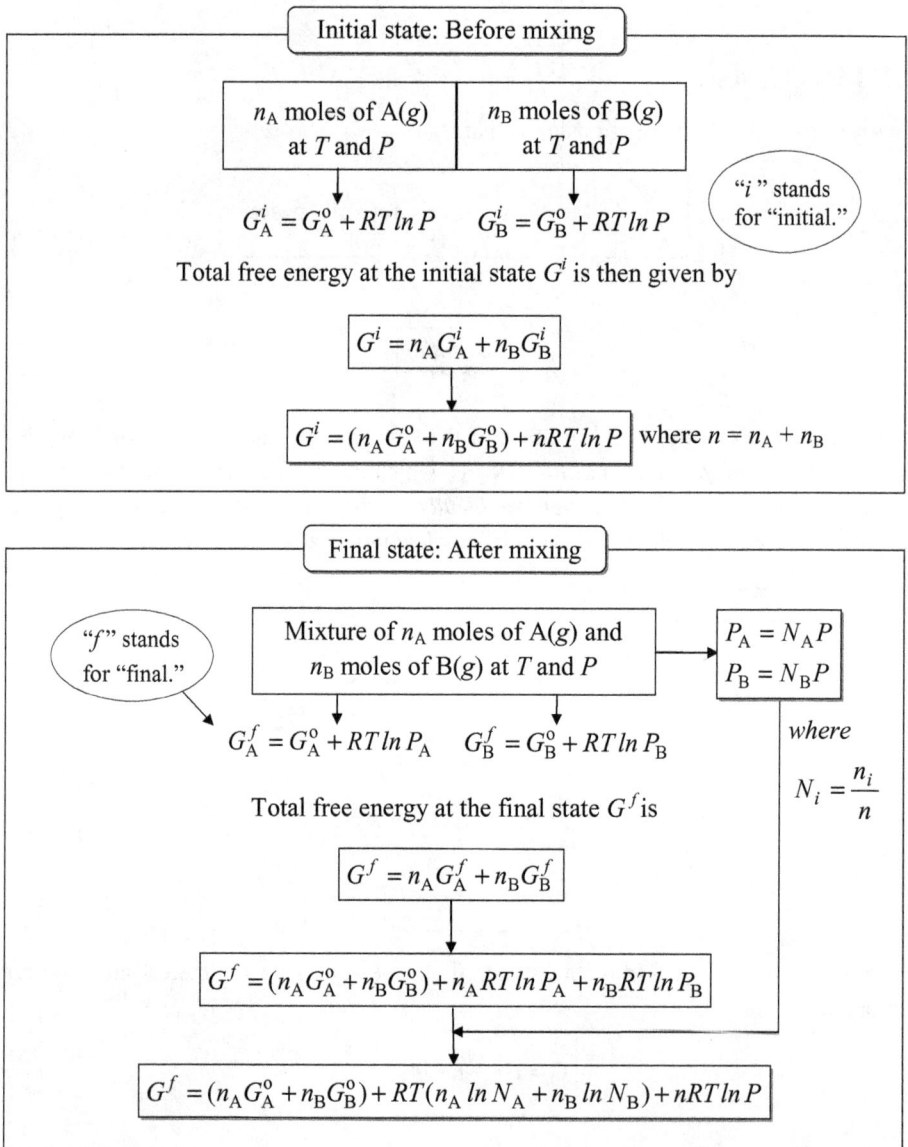

Thus, the change in Gibbs energy on mixing ($G^m$) is given by the difference between $G^f$ and $G^o$.

$$G^m = G^f - G^i$$

↓ From the equations developed above

$$G^m = RT(n_A \ln N_A + n_B \ln N_B)$$

↓ For one mole of the mixture, $G^M = \dfrac{G^m}{n}$,

$$G^M = RT(N_A \ln N_A + N_B \ln N_B)$$

*Summary*

| | |
|---|---|
| $G^M = RT(N_A \ln N_A + N_B \ln N_B)$ | Gibbs energy change for the formation of one mole of mixture or solution. |
| $G_A = G_A^o + RT \ln P_A$ | Gibbs energy of one mole of A after mixing or in the solution |
| $G_B = G_B^o + RT \ln P_B$ | Gibbs energy of one mole of B after mixing or in the solution |

$G_A$ and $G_B$ (later we will be using the symbol of $\overline{G}_i$) are called ***partial molar Gibbs energies*** of A and B, respectively. These are also called ***chemical potentials*** ($\mu_i$). $G^M$ is called ***relative molar Gibbs energy*** or ***Gibbs energy of mixing*** of the solution. A fuller discussion on these properties will be given later (Chapter *8.2*). Note that $G^M < 0$ because $N_i < 0$, and thus mixing is a spontaneous process.

The entropy of mixing ($S^M$) and the enthalpy of mixing ($H^M$) can be found from their relationship with the Gibbs energy:

$$S = -\left(\frac{\partial G}{\partial T}\right)_P$$

↓ $G^M = RT(N_A \ln N_A + N_B \ln N_B)$

$$S^M = -R(N_A \ln N_A + N_B \ln N_B)$$

$$G = H - TS$$

↓

$$H^M = 0$$

→ For ideal gases, no heat is evolved, or absorbed in the mixing process, because there is no interaction between atoms or molecules.

Note that $S^M > 0$, *i.e*, the entropy increases when two ideal gases are mixed, and thus the mixing is spontaneous (no change in the entropy of the surroundings as $H^M = 0$).

*Exercises*

7.1 $n_A$ moles of gas A and $n_B$ moles of gas B are mixed at a constant temperature and pressure. Calculate the ratio of A and B which minimizes the Gibbs energy of the mixture. Assume that both A and B are ideal gases.

7.2 A $1 m^3$ cylinder contains $H_2(g)$ at 298 K and 1 atm., and is connected to another cylinder which contains 3 $m^3$ $O_2(g)$ at 298 K and 0.8 atm. When the valve is opened, the gases diffuse into each other and form a homogeneous mixture under isothermal conditions. Calculate the Gibbs energy of mixing, $G^m$, for the process. Assume the gases behave ideally.

## 7.2 Real Gases and Fugacity

In the previous discussion we have developed the following equation for an ideal gas:

$$G = G^o + RT \ln P$$

If a gas deviates from ideality, this equation ceases to apply. However, it is desired to preserve this simple form of expression as much as possible for non-ideal, real gases. The above equation shows that the Gibbs energy $G$ is a linear function of the logarithm of the pressure of an ideal gas. Now, let us introduce a function which, when used in place of the real pressure, ensures linearity between $G$ and the logarithm of this function in any state of any gas. This function is called ***fugacity*** ($f$), a sort of corrected pressure.

$$\boxed{G = G^o + RT \ln f}$$

How do we evaluate the fugacity $f$? We start with the familiar equation for the Gibbs energy which is expressed as a function of the pressure and temperature:

$$\boxed{dG = VdP - SdT}$$

↓ Keep the temperature constant,

$$\boxed{dG = VdP}$$

In order to integrate the last equation, we need to express the volume ($V$) as a function of the pressure ($P$). Since the gas on hand is not ideal, we cannot use the $PV = RT$

relationship. We now develop an expression which relates $V$ as a function of $P$ for a real gas.

- Volume of real gas : $V_{real}$
- Volume of ideal gas : $V_{ideal} = RT/P$
- The difference ($\alpha$) : $\alpha = V_{ideal} - V_{real} = RT/P - V_{real}$
- Rearrangement yields : $V = RT/P - \alpha$ where $V = V_{real}$

$$dG = VdP$$

↓ $V = RT/P - \alpha$

$$dG = \left(\frac{RT}{P} - \alpha\right)dP$$

↓ $dG = RT\, d\, ln\, f$

$$RT\, d\, ln\, f = \left(\frac{RT}{P} - \alpha\right)dP$$

↓ Integration from $P = 0$ to a state of interest,

$$\int_{P=0}^{P=P} RT\, d\, ln\, f = \int_{P=0}^{P=P} \left(\frac{RT}{P} - \alpha\right)dP$$

↓ Rearrangement yields

$$\left[RT\, ln\left(\frac{f}{P}\right)\right]_{P=0}^{P=P} = -\int_{P=0}^{P=P} \alpha dP$$

↓ Knowing $f / P \rightarrow 1$ as $P \rightarrow 0$.

$$RT\, ln\left(\frac{f}{P}\right) = -\int_{P=0}^{P=P} \alpha dP$$

This equation enables us to evaluate the fugacity at any pressure and temperature, provided that data on $PVT$ (*i.e.*, $\alpha$) for the gas of interest are available.

*Graphical Method*

(1) Plot the deviation ($\alpha$) from ideality of the gas against $P$.
(2) Evaluate the area between the integration limit.

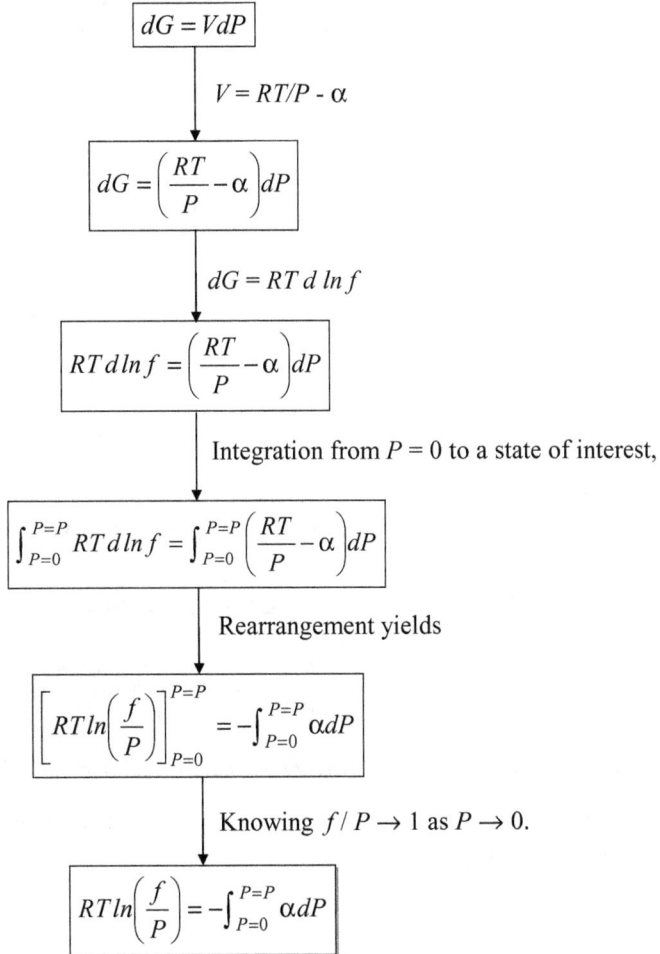

*Analytical Method*

(1) Express $\alpha$ as a function of $P$.
(2) Evaluate the integral analytically.

We now discuss more on the analytical method. Rearrangement of the above equation yields

$$ln\left(\frac{f}{P}\right) = -\int_{P=0}^{P=P}\left(\frac{\alpha}{RT}\right)dP$$

$\alpha = RT/P - V$

We introduce a term called the ***compressibility factor*** $Z$ defined as

$$Z = \frac{PV}{RT}$$

$$ln\left(\frac{f}{P}\right) = \int_{P=0}^{P=P}\frac{Z-1}{P}dP$$

$Z$ is 1 for ideal gases, but for real gases, it is a function of the state of the system, *e.g.*, $Z = f(P,T)$. Some equations of state for non-ideal gases are given below:

$$\frac{PV}{RT} = 1 + B_2P + B_3P^2 + \cdots$$

$$\frac{PV}{RT} = 1 + \frac{C_2}{V} + \frac{C_3}{V^2} + \cdots$$

where $B_i$'s and $C_i$'s are called ***virial coefficients*** and depend only on the temperature.

When a state is at the low pressure or density, the first two terms in the state equations are sufficient to represent the state:

$$\frac{PV}{RT} = 1 + B_2P \qquad \frac{PV}{RT} = 1 + \frac{C_2}{V}$$

Thus it is now possible to evaluate analytically the integral in the previous equation.

*Example 7.1*

A real gas obeys the following equation of state:

$$PV = RT + BP$$

where $B$ is independent of pressure and is a function of temperature only. Choose an incorrect one from the following relationships:

$$\frac{f}{P} = exp\left(\frac{BP}{RT}\right) \qquad \frac{f}{P} = 1 + \frac{BP}{RT} \qquad \frac{f}{P} = ln\left(\frac{BP}{RT}\right) \qquad \frac{f}{P} = Z$$

$$ln\left(\frac{f}{P}\right) = \int_{P=0}^{P=P}\left(\frac{Z-1}{P}\right)dP$$

$$PV = RT + BP$$

$$\frac{PV}{RT} = 1 + \frac{BP}{RT}$$

$$ln\left(\frac{f}{P}\right) = \frac{BP}{RT}$$

$$Z = \frac{PV}{RT}$$

$$Z = 1 + \frac{BP}{RT}$$

$$\frac{f}{P} = exp\left(\frac{BP}{RT}\right)$$

Expanding the exponential term, and taking the first two terms,

$$exp\left(\frac{BP}{RT}\right) = 1 + \frac{BP}{RT} + \frac{1}{2}\left(\frac{BP}{RT}\right)^2 + \cdots$$

$$\frac{f}{P} = 1 + \frac{BP}{RT}$$

$$\frac{f}{P} = Z$$

---

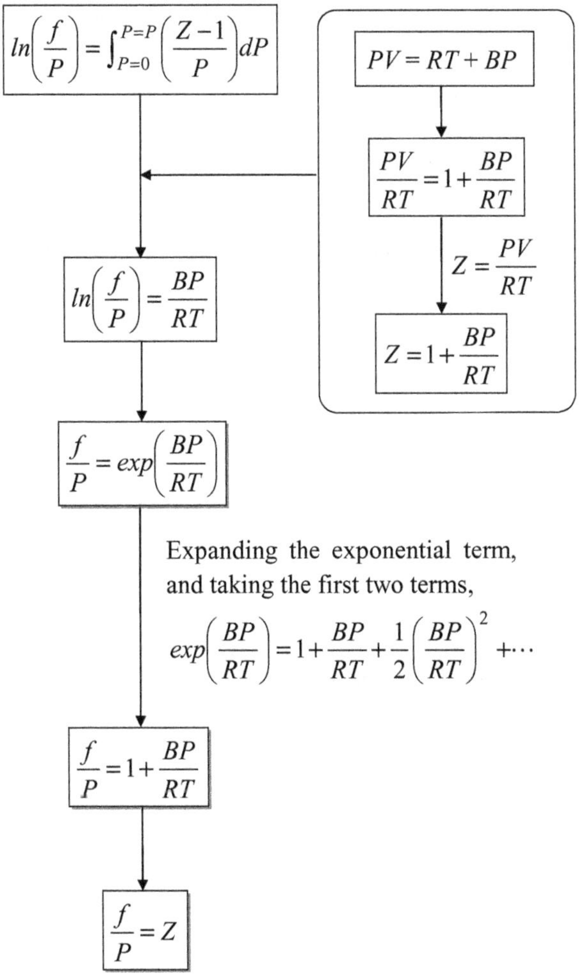

**Example 7.2**

The most powerful equation of state that describes the behavior of real gases rather well is the **van der Waals equation**:

$$\left(P + \frac{a}{V^2}\right)(V - b) = RT$$

where $a$ and $b$ are constants which are characteristic of each gas. This equation was derived by taking into considerations the properties of real gases:

Difference in physical properties between ideal gases and real gases

| | Ideal gases | Real gases |
|---|---|---|
| Particle volume | Volumeless | A finite volume. |
| Particle interaction | No interaction | Interaction |

Which constant in the above equation, $a$ or $b$, is related to the fact that the particles of a real gas occupy a finite volume? Which constant in the equation is related to the fact that interactions occur among the particles of a real gas?

The constant $b$ is related to the correction for the finite volume of the particles in a real gas, and the constant $a$ is related to the correction for particle-particle interactions.

*Exercises*

7.3 The virial equation for hydrogen gas at 298 K in the pressure range 0 to 1,000 atm. is given below:

$$PV = RT(1 + 6.4\text{x}10^{-4}P)$$

(1) Calculate the fugacity of hydrogen at 100 atm. and 298 K.
(2) Calculate the Gibbs energy change associated with the compression of 1 mole of hydrogen at 298 K from 1 atm. to 100 atm.

# Chapter 8

# Thermodynamic Functions of Mixing

## 8.1 Activity

We have seen in the previous section that fugacity, as a sort of modified pressure, plays a key role in dealing with non-ideal, real gases. We extend the concept of fugacity to condensed phases, *i.e.*, solids and liquids. Let us begin our discussion with the vaporization process of substance A in a condensed phase (liquid or solid).

$$\mathrm{A}(c) = \mathrm{A}(g) \qquad \Delta G_{\mathrm{A}}$$

Where "*c*" is for the condensed phase, and $\Delta G_A$ is the molar Gibbs energy change for the vaporization process.

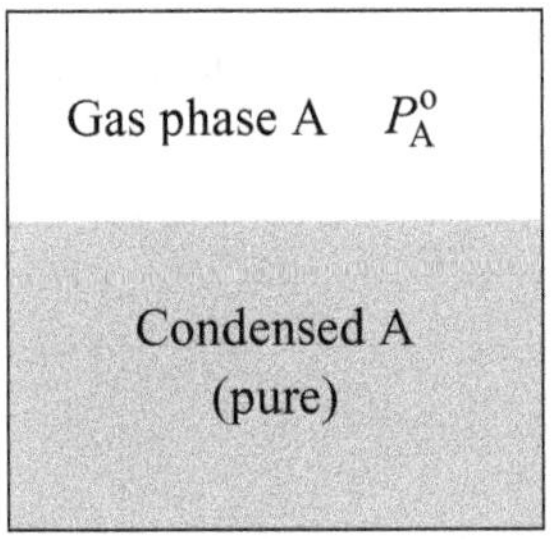


At equilibrium the pressure of A in the gas phase is the saturation vapor pressure of A at temperature *T*, and thus

$$\boxed{\Delta G_{\mathrm{A}} = G^{*}_{\mathrm{A}(g)} - G^{\mathrm{o}}_{\mathrm{A}(c)} = 0}$$

Molar Gibbs energy of gaseous A at $P = P_{\mathrm{A}}^{\mathrm{o}}$ — Molar Gibbs energy of condensed A at the pure state.

Next, consider the vaporization of A not in the pure state, but in the state of solution.

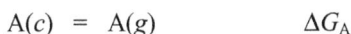

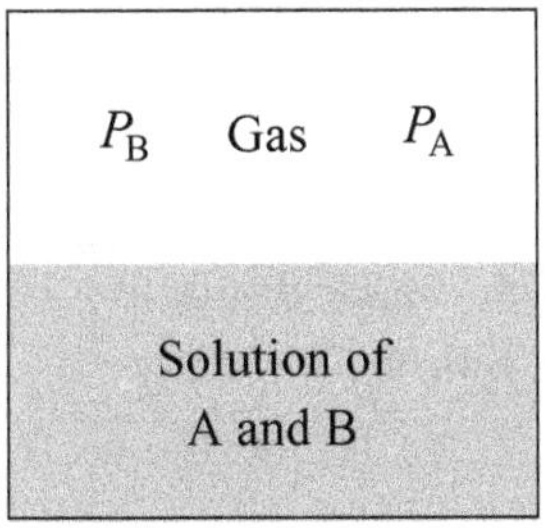


The vapor pressure of A in the gas phase is $P_{\mathrm{A}}$ at temperature *T*, and at equilibrium,

$$\boxed{\Delta G_{\mathrm{A}} = G_{\mathrm{A}(c)} - G_{\mathrm{A}(g)} = 0}$$

Molar Gibbs energy of A in the solution — Molar Gibbs energy of A in the gas phase

Now we consider the Gibbs energy change due to the change in state of A from the pure condensed state to the state of solution (liquid or solid).

Pure, $G^{o}_{A(c)}$ → A in solution, $G_{A(c)}$

Gibbs energy change ( $\Delta G_{A(pure \to solution)}$ )

$$\Delta G_{A(pure \to solution)} = G_{A(c)} - G^{o}_{A(c)}$$

But we know that $G_{A(c)} = G_{A(g)}$ and $G^{o}_{A(c)} = G^{*}_{A(g)}$ at equilibrium.

$$\Delta G_{A(pure \to solution)} = G_{A(c)} - G^{o}_{A(c)} = G_{A(g)} - G^{*}_{A(g)}$$

Recall

$$dG = RT\, d\ln f$$

Integration from the pure state to the state of solution yields

$$G_{A(g)} - G^{*}_{A(g)} = RT\, ln\left(\frac{f_A}{f^{*}_A}\right)$$

$$G_{A(c)} = G^{o}_{A(c)} + RT\, ln\left(\frac{f_A}{f^{*}_A}\right)$$

Note that this equation relates the Gibbs energy change in the condensed phase to the fugacity of the vapor phase.

For substances which have rather low vapor pressures,

$$f \cong P$$

$$G_{A(c)} = G^{o}_{A(c)} + RT\, ln\left(\frac{P_A}{P^{o}_A}\right)$$

Now we define the fugacity of a substance in condensed phase:

| Fugacity of a substance in the condensed phase, *i.e.*, solid or liquid. | = | Fugacity of the vapor that is in equilibrium with the substance. |
|---|---|---|

We introduce a function called ***activity***, $a$, which is defined as

$$a_i \equiv \frac{f_i}{f_i^{\mathrm{o}}}$$

where $i$ is the $i$-th component in the solution.

In most cases of chemical reaction systems, particularly in material systems, vapor pressures are not high so that

$$f \approx P$$

$$a_i = \frac{P_i}{P_i^{\mathrm{o}}}$$

Using the newly introduced function of activity, we have

$$G_{\mathrm{A}(c)} = G^{\mathrm{o}}_{\mathrm{A}(c)} + RT\,ln\left(\frac{P_{\mathrm{A}}}{P^{\mathrm{o}}_{\mathrm{A}}}\right)$$

$$G_{\mathrm{A}(c)} = G^{\mathrm{o}}_{\mathrm{A}(c)} + RT\,ln\,a_{\mathrm{A}}$$

In the equations, the term $G_{\mathrm{A}(c)}$ is the molar Gibbs energy of A in the solution with the concentration which exerts the vapor pressure of $P_{\mathrm{A}}$. This term is called ***partial molar Gibbs energy*** of A and denoted using the symbol $\overline{G}_{\mathrm{A}}$ . It is also called ***chemical potential*** of A and expressed by symbol $\mu_{\mathrm{A}}$.

Thus for $i$-th component in a solution,

$$\overline{G}_i = \mu_i = G_i^{\mathrm{o}} + RT\,ln\,a_i$$

This equation is true for any solutions: gas, liquid or solid solutions; the only approximation made is the use of pressure in place of fugacity.

For a gaseous phase, $P_i^{\mathrm{o}}$ is the pressure of species $i$ at the standard state, *i.e.*, $P_i^{\mathrm{o}}$ = 1 atm. Thus

$$a_i = P_i$$

Note that the activity of a gaseous species is numerically the same as the pressure expressed in atm. unit of the species, and hence $P_i$ in the above equation must be considered as $P_i/1$. If other units of pressure such as Pascal is to be used, we have to return back to the definition of the activity, *i.e.*, $a_i = P_i / P_i^{\circ}$.

We next examine the change in vapor pressure with the change in composition in a solution.

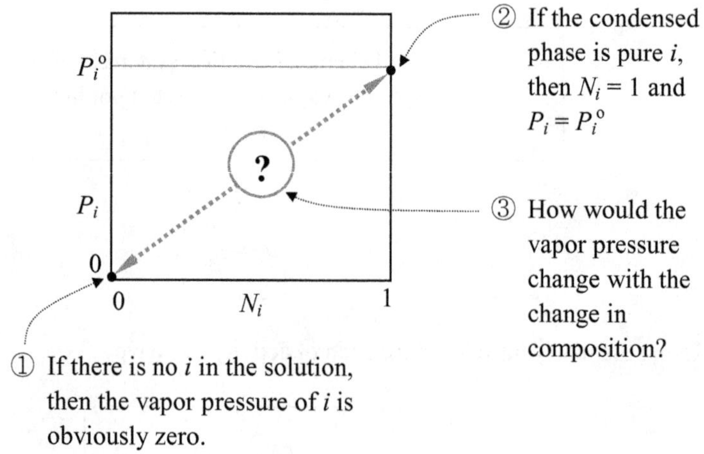

② If the condensed phase is pure $i$, then $N_i = 1$ and $P_i = P_i^{\circ}$
$P_i^{\circ}$
?
$P_i$
③ How would the vapor pressure change with the change in composition?
0
0
$N_i$
1
① If there is no $i$ in the solution, then the vapor pressure of $i$ is obviously zero.

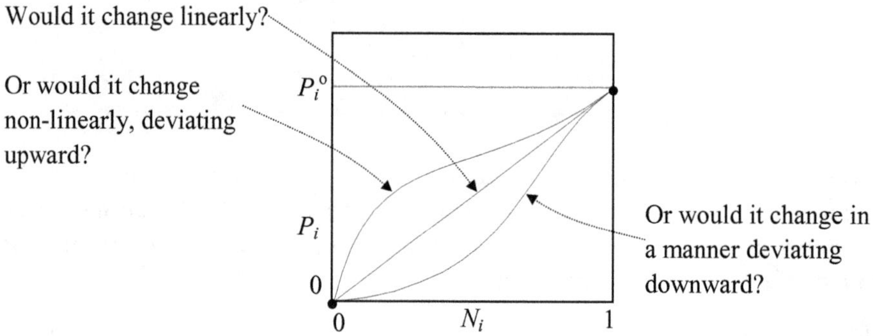

Would it change linearly?
Or would it change non-linearly, deviating upward?
$P_i^{\circ}$
$P_i$
Or would it change in a manner deviating downward?
0
0
$N_i$
1

The answer is "It depends on how species $i$ interacts with other species in the solution." Atoms (or molecules) in the solution interact with their neighboring atoms (or molecules).

Consider an $i$-$j$ binary solution.

- If $i$-$i$, $i$-$j$ and $j$-$j$ interactions are all identical, the vapor pressure will be linearly proportional to the concentration.
- If the $i$-$j$ interaction is different from either the $i$-$i$ or $j$-$j$ interaction, however, the vapor pressure will deviate from linearity.
  - If the $i$-$j$ attraction is weaker than the $i$-$i$ attraction, then $i$ will become freer by having $j$ as its neighbor, and hence more active in the solution and easier to vaporize (upward deviation).

- If the *i-j* attraction is stronger than the *i-i* attraction, on the other hand, then *i* will become more bound by having *j* as its neighbor, and hence less active and more difficult to vaporize (downward deviation).

From the definition of activity, we can convert $P_i$ - $N_i$ relationships to $a_i$ - $N_i$ relationships:

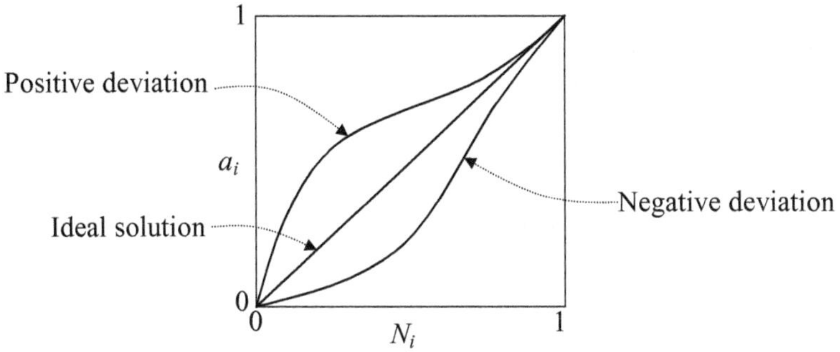


Activity may be regarded as an ***active*** or ***effective concentration***, and can be related to the actual concentration by the equation

$$\boxed{a_i = \gamma_i N_i}$$

where $\gamma_i$ is called ***activity coefficient***.

The value of the activity coefficient is the barometer of the extent of deviation from the ideal behavior and the determination of $\gamma$ is of prime importance in chemical thermodynamics. The variation of the activity coefficient with temperature and composition is generally determined experimentally.

*Example 8.1*

Is the following statement true?
"When activity is defined by

$$a_i = \frac{P_i}{P_i^{\mathrm{o}}}$$

$P_i^{\mathrm{o}}$ is the saturation vapor pressure of species *i* which is in equilibrium with pure *i* at the temperature of interest *irrespective of* the standard state chosen."

The statement is incorrect. $P_i^{\mathrm{o}}$ is the pressure of species of *i* at the standard state chosen. If *i* is a component in the gaseous phase, $P_i^{\mathrm{o}}$ = 1 atm., provided that all pressures are expressed in the unit of atm. If *i* is a component in the liquid or solid solution, $P_i^{\mathrm{o}}$ is the vapor pressure which is in equilibrium with species *i* at the standard state chosen.

Therefore only if the pure state is chosen as the standard state for the species $i$, the statement given is true. If other state is chosen as the standard state, $P_i^o$ must be the vapor pressure at the chosen standard state.

*Example 8.2*

The activity of a species is unity when it is at the standard state. This is true, even though the choice of the standard state is arbitrary. Is this statement true?

$$a_i = \frac{f_i}{f_i^o} \quad \text{or} \quad a_i = \frac{P_i}{P_i^o}$$

$f_i = f_i^o$ or $P_i = P_i^o$ at the standard state.

$$a_i = 1$$

irrespective of
the choice of the standard state.

*Example 8.3*

Is the following statement true?
Although the partial molar Gibbs energy ($\overline{G}_i$) or the chemical potential ($\mu_i$) defined by the equation given below pertains to the individual components of the system, it is a property of the system as a whole. The value of the partial molar Gibbs energy depends not only the nature of the particular substance in question, but also on the nature and relative amounts of the other components present as well.

$$\overline{G}_i = \mu_i = G_i^o + RT\ln a_i$$

In the equation, $G_i^o$ is independent of the system, but dependent only on the standard state chosen for the component $i$. However the activity $a_i$ is dependent on the system, because its value is affected by the nature of interaction with the other components in the system.

*Example 8.4*

Is the following statement true?
The value of the partial molar Gibbs energy or the chemical potential of a component in the solution is independent of the choice of the standard state for the component.

The statement is true.

$$\overline{G}_i = \mu_i = G_i^{o} + RT \ln a_i$$

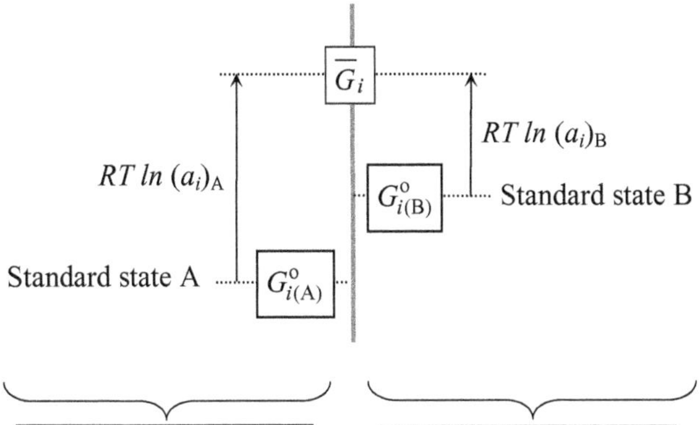


$$\overline{G}_i = G^{o}_{i(A)} + RT \ln (a_i)_A \qquad \overline{G}_i = G^{o}_{i(B)} + RT \ln (a_i)_B$$

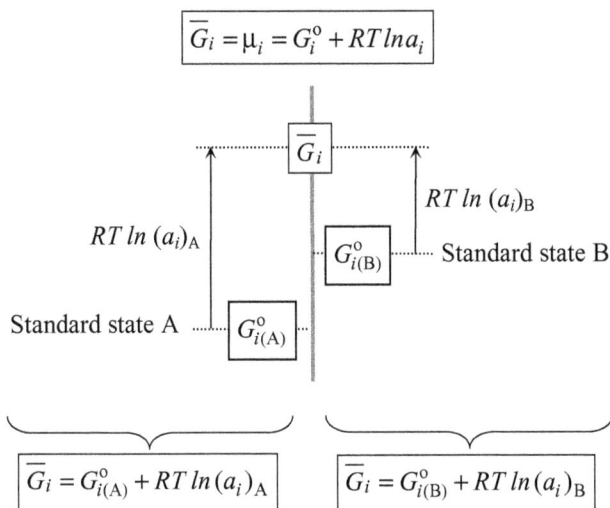

In the above, $(a_i)_A$ and $(a_i)_B$ are the activity of $i$ with respect to the standard state of A and the standard state of B, respectively.

*Exercises*

8.1 The vapor pressure of pure solid silver and solid silver-palladium alloys are given in the following:

For silver,

$$\log P = \frac{-13{,}700}{T} + 8.73 \qquad \text{(torr)}$$

For the solid silver-palladium alloy at $N_{Ag} = 0.8$,

$$\log P = \frac{-13{,}800}{T} + 8.65 \qquad \text{(torr)}$$

1) Calculate the activity of silver in the alloy at 1150 K. Pure solid silver is taken as the standard state for silver in the alloy.

2) Calculate the activity coefficient of silver in the alloy at 1150 K.

8.2 In A-B binary solutions at 600 K, the vapor pressures of A at different compositions are

atm.

| $N_A$ | 1.0 | 0.9 | 0.6 | 0.4 | 0.3 | 0.2 | 0.1 |
|---|---|---|---|---|---|---|---|
| $P_A \times 10^4$ | 4.9 | 4.2 | 2.0 | 0.9 | 0.6 | 0.4 | 0.2 |

Assuming the Gibbs energy of pure A at temperature 600 K is set to zero, calculate the chemical potential of A in the solution of $N_A = 0.6$.

8.3 The activity coefficient of zinc in liquid zinc-copper alloys in the temperature range 1070 – 1300 K can be expressed as follows :

$$RT\,ln\,\gamma_{Zn} = -31{,}600 N^{o}_{Cu} \qquad \text{where } R = 8.314 \text{ J mol}^{-1}\text{K}^{-1}.$$

Calculate the vapor pressure of Zn over a Cu-Zn binary solution of $N_{zn} = 0.3$ at 1280 K. The vapor pressure of pure liquid zinc is given by

$$logP = \frac{-6{,}620}{T} - 2.26 logT + 12.34 \qquad \text{(torr)}$$

## 8.2 Partial Properties

We consider mixtures of substances, *i.e.*, solutions, and assume that the substances in the solution do not react with each other. Consider substance A:

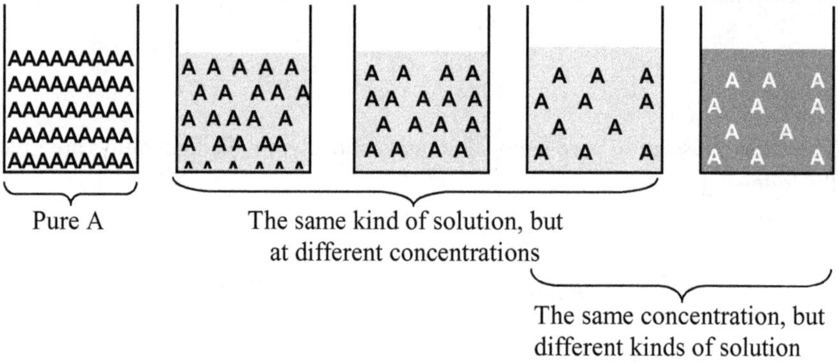


Now we may ask several questions:

- Would A behave the same in all different solutions (different states) above?
- If not, how would A behave in the solution if the concentration is different, or if the kind of solution is different?
- How can thermodynamic properties of A in a solution be systematically expressed?

In general, thermodynamic properties of the components in a solution vary with composition because the environment of each type of atom or molecule changes as the composition changes. The change in interaction force between neighboring atoms or molecules with the change in composition results in the variation of the thermodynamic properties of the solution. The thermodynamic properties that a component has in the solution are called ***partial properties***.

In many processes, we seldom deal with pure material, but with mixtures, *i.e.*, gas mixtures, or liquid or solid solutions, and thus we are concerned with the thermodynamic properties of a component in the solution - ***partial molar properties***:

*Partial molar volume* ($\overline{V}$)
*Partial molar energy* ($\overline{U}$)
*Partial molar enthalpy* ($\overline{H}$)
*Partial molar entropy* ($\overline{S}$)
*Partial molar Gibbs energy* ($\overline{G}$)

Let us begin our discussion with volume, since it is perhaps easier to visualize. The results obtained will then be extended to other properties. We will consider a binary (two components) system initially, and then generalize the results for multi-component systems. The molar volume of a pure substance can be measured with ease. When two different substances form a solution, however, the volume of the solution may not be the same as the algebraic sum of the volumes of the two substances. For example, when we mix water and ethanol as shown below:

| *Water* | | *Ethanol* | | *Water + Ethanol* |
|---|---|---|---|---|
| 100 ml | + | 100 ml | = | ? ml |
| 25°C | | 25°C | | 25°C |

The volume after mixing is not 200 ml, but about 190 ml. Where has the other 10 ml gone? The solution has suffered from shrinking because ethanol and water pack together more tightly than does each species by itself. This is due to the nature of the hydrogen bonding involved in the structure of the liquid.

As seen in this example, the volume of a solution is not, in general, simply the sum of the volumes of the individual components. Consider a binary solution containing $n_A$ moles of A and $n_B$ moles of B. The volume of the solution is $V$.

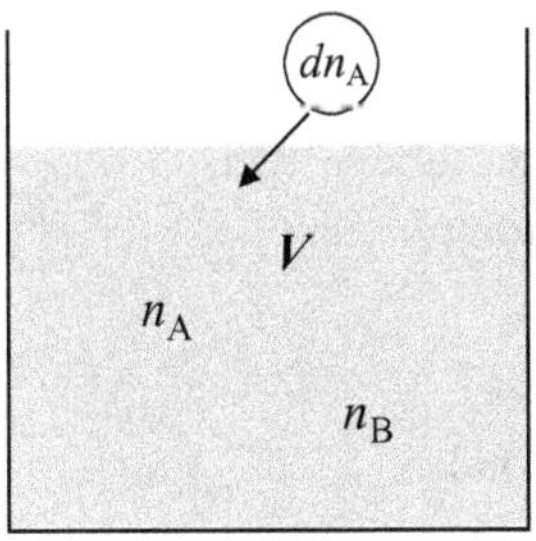


We now add a very small amount of A, $dn_A$, so small that thc addition of this extra amount does not change the concentration of the solution to any appreciable extent. The resulting increase in the volume of the solution at constant temperature and pressure is $dV$. this $dV$ can then be regarded as the volume occupied by $dn_A$ moles of A in the solution of the particular composition. In other words, $dn_A$ moles of A acts in the solution as if it had the volume of $dV$.

Alternatively we may say that one mole of A in the solution acts as if it has the volume of

$$\boxed{\frac{dV}{dn_A}}$$

This volume is called ***partial molar volume*** of A and denoted by symbol. $\overline{V}_A$. This partial volume is dependent on the composition of the solution. If the composition changes, this value also changes. As $T$, $P$ and $n_B$ are kept constant during the process, we may define the partial molar volume as

$$\bar{V}_A = \left(\frac{\partial V}{\partial n_A}\right)_{P,T,n_B}$$

The contribution of A to the total volume is then $n_A\bar{V}_A$ and the contribution of B to the total volume is $n_B\bar{V}_B$. Thus

$$V = \bar{V}_A n_A + \bar{V}_B n_B$$

Volume is a state function with variables of $T$ and $P$ for a closed system, or with variables $T$, $P$, $n_A$, and $n_B$ for an open system.

As $V = V(T,P,n_A,n_B)$, the complete differential at constant $T$ and $P$ yields

$$dV = \left(\frac{\partial V}{\partial n_A}\right)_{P,T,n_B} dn_A + \left(\frac{\partial V}{\partial n_B}\right)_{P,T,n_A} dn_B$$

↓

$$dV = \bar{V}_A dn_A + \bar{V}_B dn_B$$

Recall that

$$V = \bar{V}_A n_A + \bar{V}_B n_B$$

Differentiation of this equation yields

$$dV = \bar{V}_A dn_A + \bar{V}_B dn_B + n_A d\bar{V}_A + n_B d\bar{V}_B$$

↓ Comparison of the above two differential equations yields

$$n_A d\bar{V}_A + n_B d\bar{V}_B = 0$$

↓ Dividing by the total number of moles, $n$,

$$N_A d\bar{V}_A + N_B d\bar{V}_B = 0$$

It is also readily seen that

$$V = \bar{V}_A N_A + \bar{V}_B N_B$$

where $V$ is the ***molar volume of the solution*** or ***integral molar volume***.

Similar equations may be developed for other thermodynamic properties:

| | Enthalpy | Entropy | Gibbs function |
|---|---|---|---|
| Definition of partial properties | $\overline{H}_A = \left(\frac{\partial H}{\partial n_A}\right)_{P,T,n_B}$ | $\overline{S}_A = \left(\frac{\partial S}{\partial n_A}\right)_{P,T,n_B}$ | $\overline{G}_A = \left(\frac{\partial G}{\partial n_A}\right)_{P,T,n_B}$ |
| Integral molar properties | $H = \overline{H}_A N_A + \overline{H}_B N_B$ | $S = \overline{S}_A N_A + \overline{S}_B N_B$ | $G = \overline{G}_A N_A + \overline{G}_B N_B$ |
| Relationships of partial properties | $N_A d\overline{H}_A + N_B d\overline{H}_B = 0$ | $N_A d\overline{S}_A + N_B d\overline{S}_B = 0$ | $N_A d\overline{G}_A + N_B d\overline{G}_B = 0$ |

This equation is particularly important and called ***Gibbs-Duhem equation***.

Generalization for the multi-component systems:

$$\overline{Y}_A = \left(\frac{\partial Y}{\partial n_i}\right)_{P,T,n_j}$$

$$\sum N_i d\overline{Y}_i = 0$$

$$Y = \sum \overline{Y}_i N_i$$

where $Y = V, U, H, S, A$ or $G$.

These equations are of value for solution thermodynamics. For example, if a partial molar property of one component in a binary solution has been determined, then the partial molar property of the other component is fixed by the following relation:

$$N_A d\overline{Y}_A + N_B d\overline{Y}_B = 0$$

All the general thermodynamic relations can readily be expressed in terms of the partial molar properties:

Examples: $G = H - TS \longrightarrow \overline{G}_i = \overline{H}_i + T\overline{S}_i$

$dG = VdP - SdT \longrightarrow d\overline{G}_i = \overline{V}_i dP - \overline{S}_i dT$

Frequently employed is the graphical method of determining the partial molar properties from the data on the integral molar quantities.

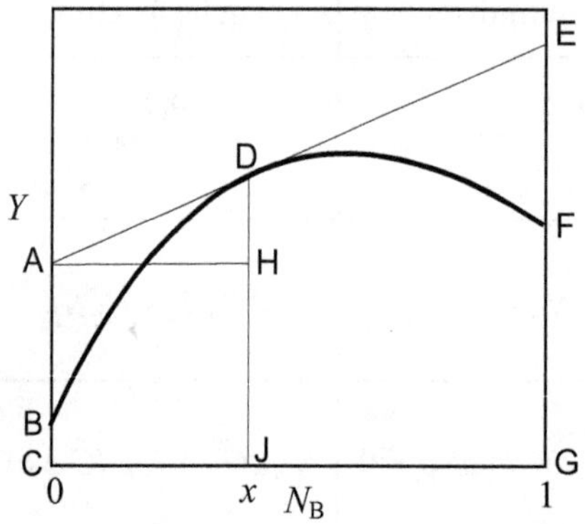


In an A-B two component system, a molar property of the solution or the integral molar property, $Y$, at constant $P$ and $T$ is plotted against the mole fraction of B, $N_B$, in the figure.

$$\boldsymbol{Y} = n_A \overline{Y}_A + n_B \overline{Y}_B$$

Dividing by $n$,

$$Y = N_A \overline{Y}_A + N_B \overline{Y}_B$$

Differentiation of the above equation yields,

$$dY = \overline{Y}_A dN_A + \overline{Y}_B dN_B + N_A d\overline{Y}_A + N_B d\overline{Y}_B$$

$N_A d\overline{Y}_A + N_B d\overline{Y}_B = 0$ (Gibbs-Dehem equation)

$N_A + N_B = 1$

$dN_A + dN_B = 0$

$$\overline{Y}_A = Y + N_B \left( \frac{dY}{dN_A} \right) \rightarrow \overline{Y}_A = Y - N_B \left( \frac{dY}{dN_B} \right)$$

At $N_B = x$ in the figure,

$x = CJ = AH$

$$\frac{dY}{dN_B} = \text{Slope of the tangent} = \frac{DH}{AH}$$

$Y = DJ$

$$\overline{Y}_A = DJ - AH \frac{DH}{AH} = HJ = AC$$

Similarly, $\overline{Y}_B = EG$

This method is known as ***the method of intercepts*** and is of proven value in determining partial molar quantities.

Suppose that $n_A$ moles of pure A and $n_B$ moles of pure B are mixed to form a binary solution at constant $T$ and $P$. The total Gibbs energy change, $G^M$, associated with the mixing may be obtained in the following manner:

$$\bar{G}_i = G_i^o + RT\ln a_i$$

Rearrangement yields

$$\bar{G}_i - G_i^o = RT\ln a_i$$

($\bar{G}_i - G_i^o$) is the change in molar Gibbs energy of $i$ due to the change in state from the standard state to the state of solution of a particular composition. This is called the ***partial molar Gibbs energy of mixing*** or ***relative partial molar Gibbs energy*** of $i$ and is **designated**

$$G_i^M = \bar{G}_i - G_i^o$$

$$\bar{G}_i - G_i^o = RT\ln a_i$$

$$G_i^M = RT\ln a_i$$

Gibbs energy *before* mixing

$$n_A G_A^o + n_B G_B^o$$

Gibbs energy *after* mixing

$$n_A \bar{G}_A + n_B \bar{G}_B$$

The Gibbs energy change *due to mixing* is then,

$$\boldsymbol{G}^M = (n_A \bar{G}_A + n_B \bar{G}_B) - (n_A G_A^o + n_B G_B^o)$$

$$= n_A(\bar{G}_A - G_A^o) + n_B(\bar{G}_B - G_B^o)$$

$$= n_A G_A^M + n_B G_B^M$$

$$= RT(n_A \ln a_A + n_B \ln a_B)$$

$$G^M = RT(N_A \ln a_A + N_B \ln a_B)$$

for one mole of the solution.

This is called the ***relative integral molar Gibbs energy*** or the ***molar Gibbs energy of mixing***.

The method of tangential intercepts which we have applied for determination of partial molar quantities from the integral molar quantities can also apply to the relative properties:

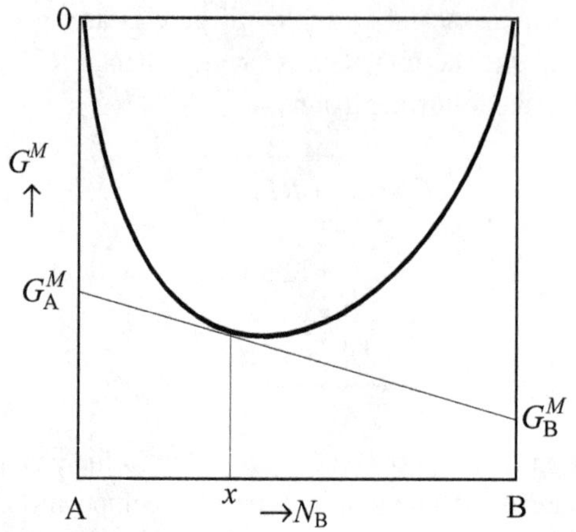


For a solution of specified composition, the ***relative partial molar entropy*** can be determined from the data on activities at different temperatures.

$$\left(\frac{\partial G}{\partial T}\right)_P = -S$$

↓

$$\left(\frac{\partial G_i^M}{\partial T}\right)_P = -S_i^M$$

↓ $G_i^M = RT ln a_i$

$$S_i^M = -R\, ln\, a_i - RT\left(\frac{\partial(ln\, a_i)}{\partial T}\right)_P$$

Note that, if data on the activity at different temperatures are available, the differential in the equation can be evaluated, and hence the relative partial molar entropy can be determined. The ***relative partial molar enthalpy*** is then,

$$\Delta G_i^M = \Delta H_i^M - T\Delta S_i^M$$

↓

$$\Delta H_i^M = \Delta G_i^M + T\Delta S_i^M$$

*Exercises*

8.4 A container having three compartments contains 1 mole of gas A, 2 moles of gas B and 3 moles of gas C, respectively, at the same temperature and pressure (298 K and 1 atm.). The partitions are lifted and the gases are allowed to mix.
Calculate the change in the integral molar Gibbs energy $G^M$. Assume the gases behave ideally.

8.5 The volume of a dilute solution of KCl of molality $m$ ($m$ moles of KCl in 1 kg of water) is given by the equation

$$V = 1{,}003 + 27.15m + 1.744m^2 \quad (\text{cm}^3)$$

1) Calculate the partial molar volume of KCl ($\overline{V}_{\text{KCl}}$) at $m = 0.5$.
2) Calculate the partial molar volume of KCl at the infinitely dilute solution.

8.6 The activity coefficient of zinc in liquid zinc-copper alloys in the temperature range of 1070 to 1300 K can be expresses as follows:

$$RT\, ln\, \gamma_{\text{Zn}} = -31{,}600 N_{\text{Cu}}^2$$

where $R = 8.314$ J mol$^{-1}$K$^{-1}$

Calculate the activity of Cu in Cu-Zn binary solution of $N_{\text{Cu}} = 0.7$ at 1300 K.

# Chapter 9

# Behavior of Solutions

## 9.1 Ideal Solutions

We consider the vaporization of substance A:

Pure A

Vapor

$P_A^*$   A   $\mu_{A(g)}^*$

A   $\mu_{A(l)}^*$

Liquid

A in solution

$P_A$   Vapor   $\mu_{A(g)}$
$P_B$          $\mu_{B(g)}$
$P_C$   A, B, C   $\mu_{C(g)}$

$\mu_{A(l)}$
A, B, C   $\mu_{B(l)}$
Liquid   $\mu_{C(l)}$

At equilibrium,

$$\mu_{A(l)}^* = \mu_{A(g)}^*$$

$$\mu_{A(l)}^* = \mu_{A(g)}^0 + RT \ln P_A^*$$

$$\mu_{A(l)} = \mu_{A(g)}$$

$$\mu_{A(l)} = \mu_{A(g)}^0 + RT \ln P_A$$

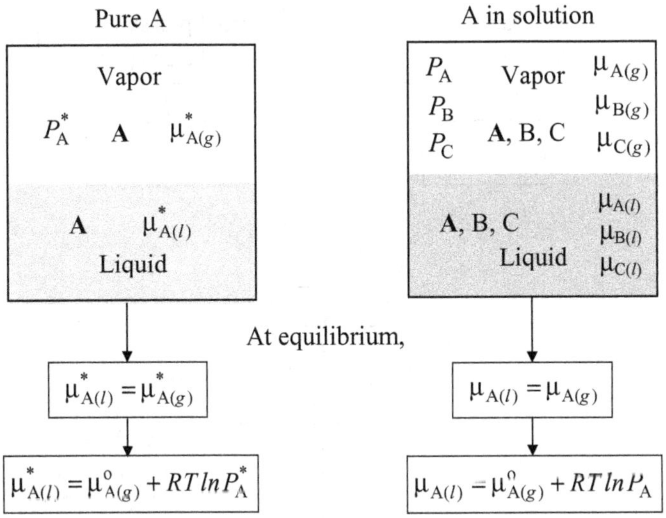

Combination of the above equations yields,

$$\mu_{A(l)} = \mu_{A(l)}^* + RT \ln \left( \frac{P_A}{P_A^*} \right)$$

In a series of experiments on liquid mixtures, the French scientist F. Raoult found that *the ratio of the partial vapor pressure of each component of a solution to its vapor pressure as a pure liquid is approximately equal to the mole fraction of the component in the solution.*

$$\text{Raoult's Law:} \frac{P_A}{P_A^*} = N_A$$

The above statement (or the equation) is known as ***Raoult's Law***. Solutions that obey the equation throughout the composition range are called ***ideal solutions***.

Raoult's law may be understood in molecular interpretation:

*Molecular Interpretation of Raoult's Law*

Suppose that we have a solution containing component A in it. The dynamic situation at the surface of the solution will be simultaneous occurrence of the vaporization of A molecules at the surface into the vapor phase above the solution, and the condensation of A molecules in the vapor phase back to the solution.
At equilibrium,

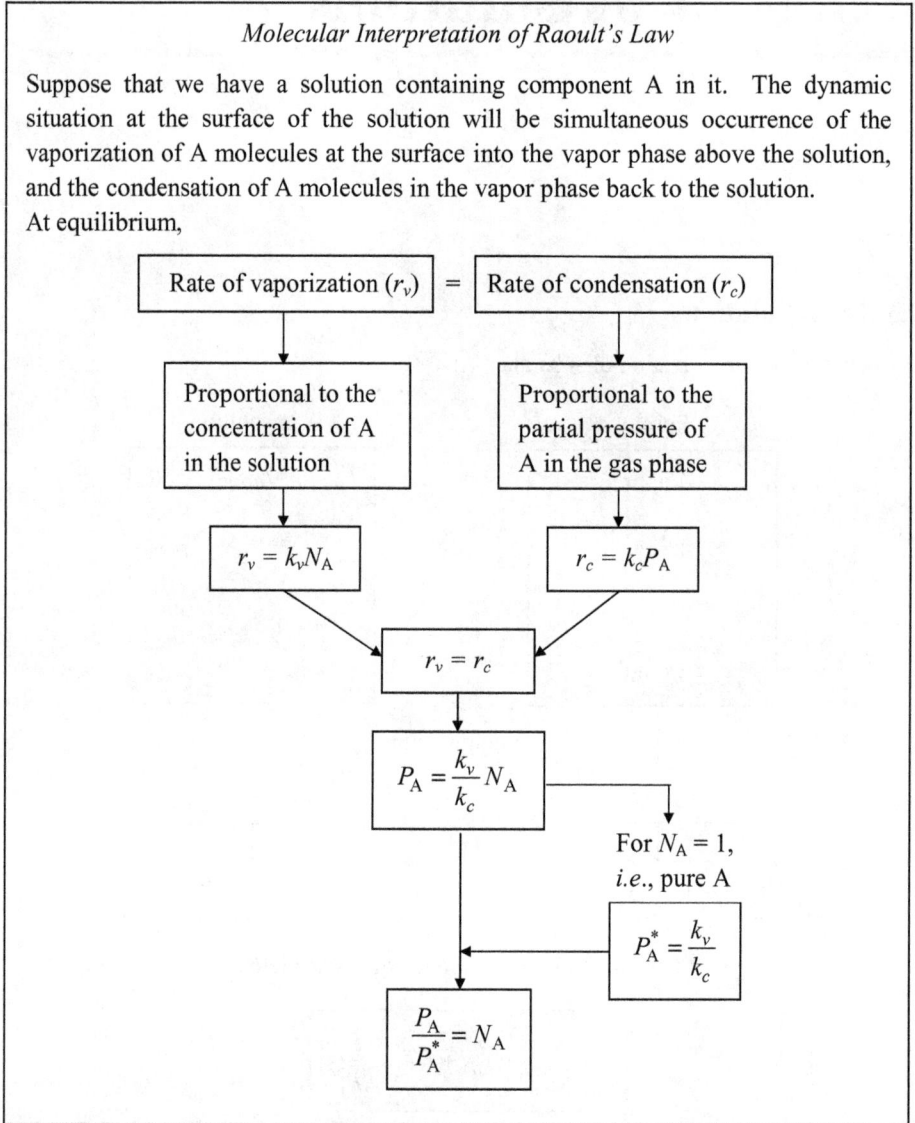

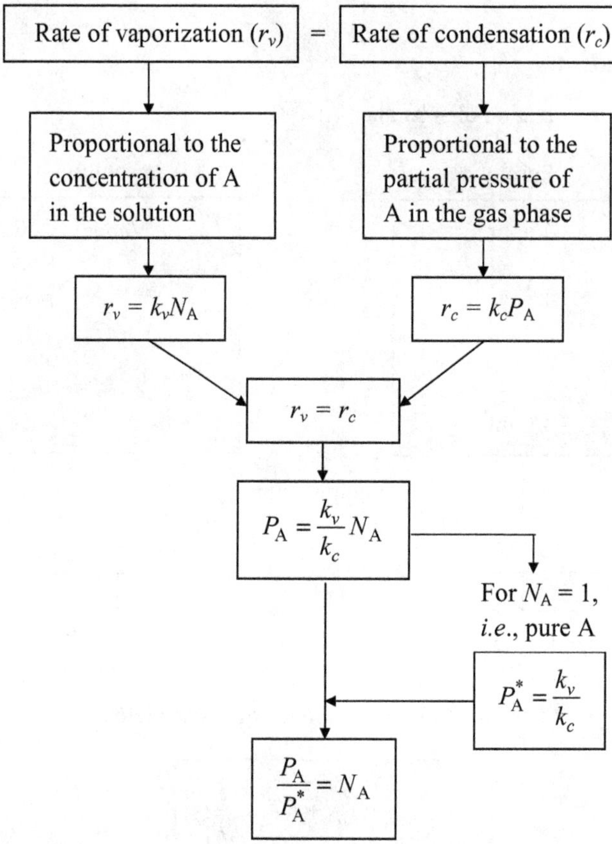
Rate of vaporization ($r_v$)
=
Rate of condensation ($r_c$)
Proportional to the concentration of A in the solution
Proportional to the partial pressure of A in the gas phase
$r_v = k_v N_A$
$r_c = k_c P_A$
$r_v = r_c$
$P_A = \frac{k_v}{k_c} N_A$
For $N_A = 1$, *i.e.*, pure A
$P_A^* = \frac{k_v}{k_c}$
$\frac{P_A}{P_A^*} = N_A$

For an ideal solution which obeys the Raoult's law, we can then derive the following relationship:

$$\mu_{A(l)} = \mu^*_{A(l)} + RT ln N_A$$

Recall that $\mu^*_{A(l)}$ is the chemical potential of pure A at the temperature of interest, and $\mu^o_{A(l)}$ is the chemical potential of pure A at the pressure of $1 atm$. Therefore they are not necessarily the same, but the effect on the chemical potential of the pressure of the system is so small that

$$\mu^*_{A(l)} = \mu^o_{A(l)}$$

Recall the definition of activity:

$$a_A = \frac{P_A}{P^*_A}$$

From the Raoult's law, thus,

$$\boxed{a_A = N_A}$$

That is, the activity (effective concentration) of a component in an ideal solution is the same as the chemical concentration (mole fraction) of the component. On a molecular level, a solution is defined as ideal when the intermolecular potentials are the same between all components in the solution.

*Example 9.1*

Prove the following statement:
When components are mixed to form an ideal solution at constant temperature, no heat is absorbed or released.

$$\boxed{\left[\frac{\partial\left(\frac{G^M_A}{T}\right)}{\partial\left(\frac{1}{T}\right)}\right]_P = H^M_A}$$

$$G^M_A = RT\,ln\,a_A = RT\,ln\,N_A$$

$$\boxed{H^M_A = 0}$$

*Example 9.2*

Prove the following statement:
If species A behaves ideally in the whole composition range of A-B binary solutions, the species B also behaves ideally.

Recall the Gibbs-Duhem equation:

$$N_A dG_A + N_B dG_B = 0$$

↓

$$N_A dG_A^M + N_B dG_B^M = 0$$

↓ $G_i^M = RT\,ln\,a_i$

$$N_A d\,ln\,a_A + N_B d\,ln\,a_B = 0$$

One form of Gibbs-Duhem equation.

↓ ← Ideal behavior of A:
$a_A = N_A$, $d ln a_A = d ln N_A$
$= dN_A/N_A$
$= - dN_B/N_A$

$$-dN_B + N_B d\,ln\,a_B = 0$$

↓

$$d ln a_B = \frac{dN_B}{N_B} = d ln N_B$$

↓ Integration yields

$$ln a_B == ln N_B + \text{constant}$$

↓ But, as $N_B \rightarrow 1$, $a_B \rightarrow 1$.
Thus, constant = 0.

$$a_B = N_B$$

*Exercises*

9.1 Liquids A and B are completely miscible and forms an ideal solution. The vapor pressures of two liquids are $2\times10^{-3}$ atm. and $5\times10^{-3}$ atm. at temperature T, respectively. Calculate the mole fraction of A in liquid phase when the vapor pressure is $4\times10^{-3}$ atm.

9.2 An ideal solution is made of 79 mol% of A, 20 mol% of B and 1 mol% of C at 298 K and 1 atm.

1) Calculate the relative partial molar Gibbs energy of A, $G_A^M$ .
2) Calculate the relative integral molar Gibbs energy of the solution, $G^M$.
3) Calculate the relative partial molar entropy of A, $S_A^M$ .

9.3 A liquid gold-copper alloy contains 45 mol% of copper and behaves ideally at 1320 K. Calculate the amount of heat absorbed in the system when 1g of solid copper is dissolved isothermally at this temperature in a large bath of the alloy of this composition. The following data given:

$$C_{P,\mathrm{Cu}(s)} = 22.64 + 6.28\times10^{-3}T,\ \mathrm{J\ mol^{-1}K^{-1}}$$
$$C_{P,\mathrm{Cu}(l)} = 31.38\ \mathrm{J\ mol^{-1}K^{-1}}$$
$$\Delta H^o_{f,Cu} = 12{,}980\ \mathrm{J\ mol^{-1}}\ \text{(Heat of fusion of Cu)}$$
$$T_{m,\mathrm{Cu}} = 1{,}083°\mathrm{C}\ \text{(melting point of Cu)}$$
$$M_{\mathrm{cu}} = 63.5\ \text{(atomic weight of Cu)}$$

Calculate the change in entropy of the system in the above process.

9.4 A solution is composed of benzene (B) and toluene (T). The Raoult's law holds for both benzene and toluene. The equilibrium vapor pressures of benzene and toluene are 102.4 kPa and 39.0 kPa, respectively, at 81°C. Calculate the mole fraction of benzene in the vapor which is in equilibrium with the $N_B = 0.5$ solution.

## 9.2 Non-ideal Solutions and Excess Properties

Solutions may be classified into "ideal solutions" and "non-ideal (real) solutions." The non-ideal solutions may be further divided into "solutions with a positive deviation from the ideal behavior" and "solutions with a negative deviation from the ideal behavior." The above statements can be summarized by the following diagram:

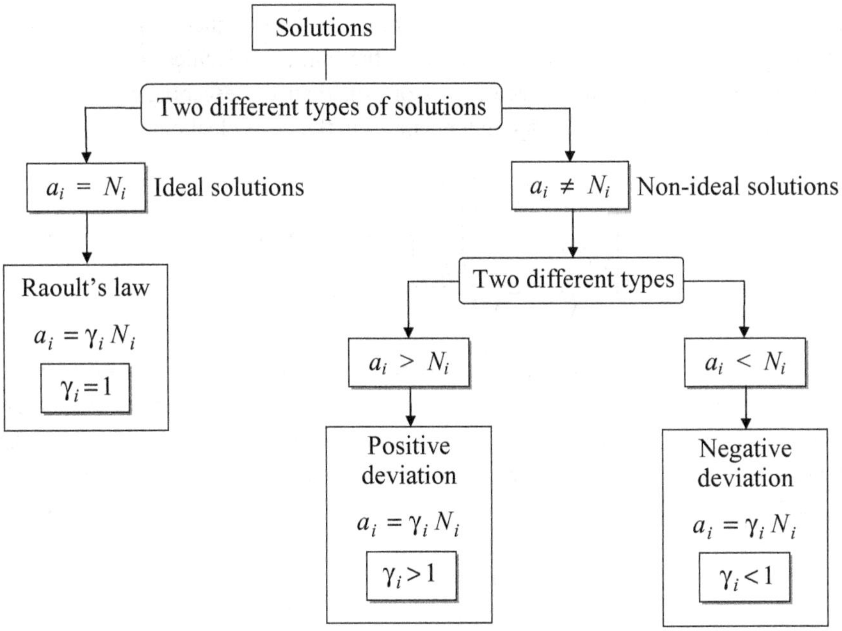


We can also represent the above classifications by a composition-activity diagram:

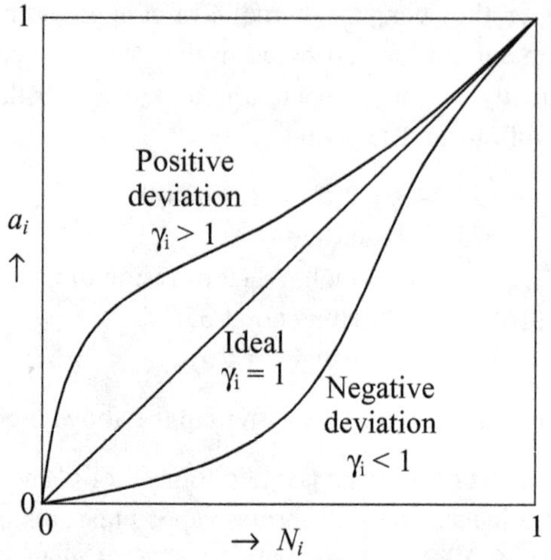


Knowledge of the variation of the value of $\gamma$ with temperature and composition is of prime importance in solution thermodynamics. Now a question arises as to how to express the extent of deviation of this value from the ideal behavior. As we have discussed in the preceding section, in ideal solutions attractive forces between the unlike molecules in solution are the same as those between the like molecules. Therefore the escaping tendency of the component $i$ in an ideal solution is the same as that in its pure state. From the Raoult's law, for ideal solutions

$$P_i = N_i P_i^*$$

The positive deviation is characterized by vapor pressures higher than those calculated for the ideal solution. If the attraction between the unlike molecules ($i$-$j$) is weaker than the mutual attraction of like molecules ($i$-$i$ or $j$-$j$), then the escaping tendencies of the molecules are higher than the escaping tendencies in the individual pure states.

$$\boxed{\underbrace{\left(\frac{P_i}{P_i^*}\right)}_{\gamma_i N_i} > \underbrace{\left(\frac{P_i}{P_i^*}\right)_{ideal}}_{N_i}} \longrightarrow \boxed{\gamma_i > 1}$$

The negative deviation, on the other hand, is characterized by vapor pressures lower than those calculated for the ideal solution. If the attraction between the unlike molecules ($i$-$j$) is stronger than the mutual attraction of the like molecules ($i$-$i$ or $j$-$j$), then the escaping tendencies of the molecules in the solution are lower than escaping tendencies in the individual pure states. Thus

$$\boxed{\left(\frac{P_i}{P_i^*}\right) < \left(\frac{P_i}{P_i^*}\right)_{ideal}} \longrightarrow \boxed{\gamma_i < 1}$$

We have seen that the properties of non-ideal or real solutions differ from those of ideal solutions.

Another convenient method of expressing deviation from the ideal behavior is by the use of ***excess properties***. We may divide thermodynamic mixing properties into two parts:

| | | | | |
|---|---|---|---|---|
| A thermodynamic mixing property $Y^M$ | = | The property which would be exhibited if the solution were ideal. $Y^{M,\,ideal}$ | + | An *excess contribution* which expresses the deviation from ideal $Y^E$ |

$$Y^M = Y^{M,ideal} + Y^E$$

$$Y^E = Y^M - Y^{M,ideal}$$

$Y^E$ is called the *excess property*.

When we apply the above equation to the Gibbs energy, then the ***excess Gibbs energy*** is,

$$G_i^E = G_i^M - G_i^{M,ideal}$$

$$G_i^M = RT\,ln\,a_i = RT\,ln(\gamma_i N_i)$$

and

$$G_i^{M,ideal} = RT\,ln\,N_i$$

$$G_i^E = RT\,ln\,\gamma_i$$

This is called the ***excess partial molar Gibbs energy*** of *i*. This is the deviation of Gibbs energy of component *i* in the real solution from the ideal behavior.

In a similar way we may obtain the ***excess partial molar entropy***, $S_i^E$ , and ***excess partial molar enthalpy***, $H_i^E$ :

$$S_i^E = S_i^M - S_i^{M,ideal}$$

$$S_i^M = -R\,ln\,a_i - RT\left(\frac{\partial\,ln\,a_i}{\partial T}\right)_P \qquad S_i^{M,ideal} = -R\,ln\,N_i$$

$$S_i^E = -R\,ln\,\gamma_i - RT\left(\frac{\partial\,ln\,a_i}{\partial T}\right)_P$$

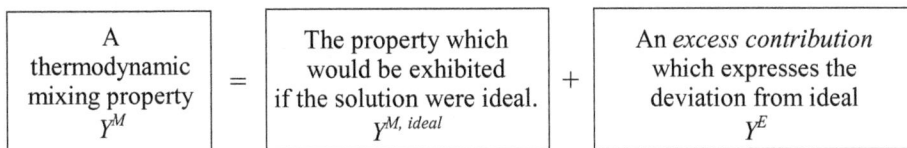

$$\left(\frac{\partial \ln a_i}{\partial T}\right)_P = \left(\frac{\partial \ln \gamma_i N_i}{\partial T}\right)_P$$

$$= \left(\frac{\partial \left(\ln \gamma_i + \ln N_i\right)}{\partial T}\right)_P$$

$N_i \neq f(T)$

$$= \left(\frac{\partial \ln \gamma_i}{\partial T}\right)_P$$

$$S_i^E = -R \ln \gamma_i - RT\left(\frac{\partial \ln \gamma_i}{\partial T}\right)_P$$

$$H_i^E = H_i^M - H_i^{M,ideal}$$

$H_i^{M,ideal} = 0$

$$H_i^E = H_i^M$$

$$H_i^E = -RT^2\left(\frac{\partial \ln a_i}{\partial T}\right)_P$$

$$H_i^E = -RT^2\left(\frac{\partial \ln \gamma_i}{\partial T}\right)_P$$

The ***excess integral molar Gibbs energy*** or the ***excess molar Gibbs energy of solution*** is given by

$$G^E = \sum\left(N_i G_i^E\right)$$

$G_i^E = RT \ln \gamma_i$

$$G^E = RT\sum\left(N_i \ln \gamma_i\right)$$

The general thermodynamic relationships for partial molar quantities are also valid for the excess quantities.

As for the temperature dependence of the activity coefficient, as the temperature increases, in general the extent of deviation from ideal behavior of a non-ideal solution decreases. In other words, the solution moves towards the ideal solution as the temperature increases.

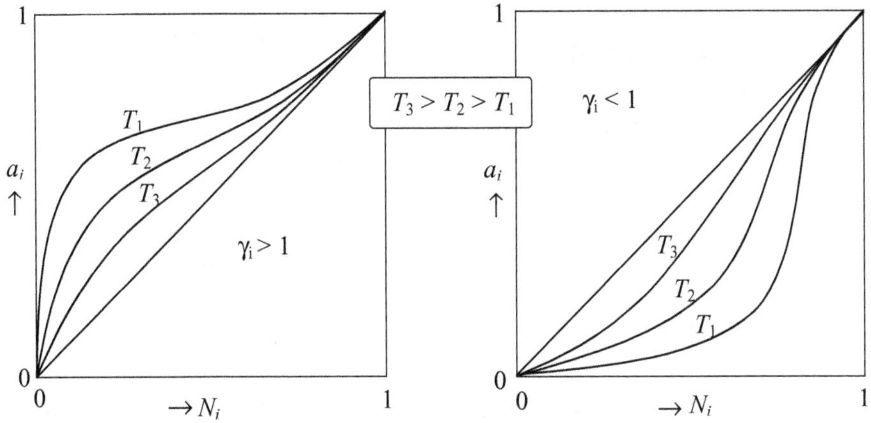


Thus in a solution,

- If $\gamma_i > 1$, then $\gamma_i$ decreases as $T$ increases.
- If $\gamma_i < 1$, then $\gamma_i$ increases as $T$ increases.

The activity coefficient of a particular component in the solution is a measure of the interaction that occurs between atoms or molecules in solution. If $\gamma_i > 1$, then $G_i^E > 0$ (recall $G_i^E = RT\, ln\gamma_i$).

*Example 9.3*

Prove that, if a component in a solution exhibits positive deviation from the ideal, *i.e.*, $\gamma_i > 1$, then the solution process is endothermic.

Knowing that,

$$\left[\frac{\partial}{\partial T}\left(\frac{G}{T}\right)\right]_P = -\frac{H}{T^2}$$

↓

$$\left[\frac{\partial}{\partial T}\left(\frac{G_i^E}{T}\right)\right]_P = -\frac{H_i^E}{T^2}$$

↓ $G_i^E = RT\, ln\,\gamma_i$

$$\left[\frac{\partial(R\ln\gamma_i)}{\partial T}\right]_P = -\frac{H_i^E}{T^2}$$

If $\gamma_i > 1$, $\gamma_i$ decreases with increase in $T$.

$$\left[\frac{\partial(R\ln\gamma_i)}{\partial T}\right]_P < 0$$

$$H_i^E > 0$$

Endothermic,

If $\gamma_i < 1$, $\gamma_i$ increases with increase in $T$.

$$\left[\frac{\partial(R\ln\gamma_i)}{\partial T}\right]_P < 0$$

$$H_i^E < 0$$

Exothermic,

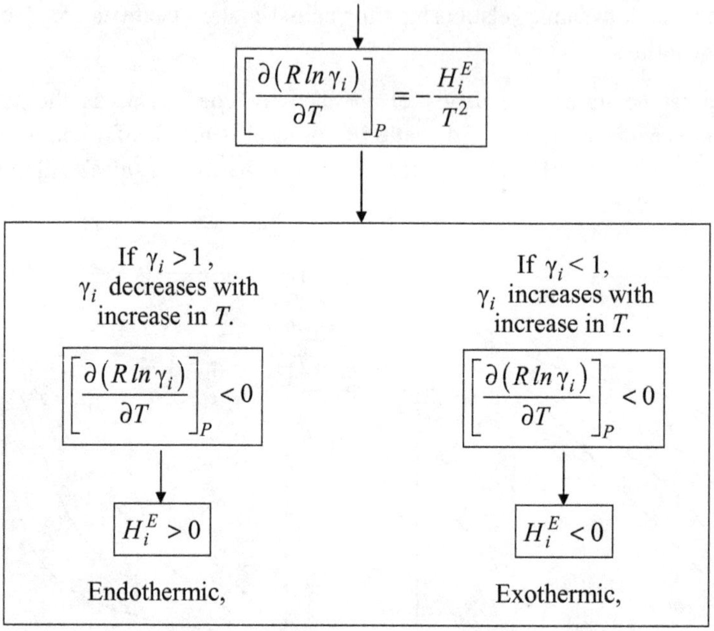

If the solution process in an *i-j* binary system is endothermic, the *i-i* and *j-j* attractions are greater than the *i-j* attraction. *i* atoms attempt to have only *i* atoms as nearest neighbors – a tendency towards ***phase separation*** or ***clustering***. If the solution process is exothermic, the *i-i* and *j-j* attractions are weaker than the *i-j* attraction. In this case, *i* atoms attempt to have only *j* atoms as nearest neighbors, and *j* atoms attempt to have only *i* atoms as nearest neighbors - a tendency towards ***compound formation***.

*Exercises*

9.5 The excess integral molar Gibbs energy of the Ga-P binary solution containing up to 50 mol% P is

$$G^E = (-7.53T - 2{,}500)N_P N_{Ga},\ \text{J mol}^{-1}$$

Calculate the amount of heat associated with the formation of one mole of solution containing 20 mol% P.

9.6 The integral molar enthalpy and entropy of the Cd-Zn liquid alloy at 432°C are described by the following empirical equations:

$$H^M = 6{,}700N_{Cd}N_{Zn} - 1{,}500N_{Zn}\ln N_{Zn},\quad \text{J mol}^{-1}$$
$$S^M = -8.4(N_{Cd}\ln N_{Cd} + N_{Zn}\ln N_{Zn})\quad \text{J mol}^{-1}\text{K}^{-1}$$

Calculate the excess partial molar Gibbs energy of cadmium, $G_{Cd}^E$ at $N_{cd} = 0.5$.

## 9.3 Dilute Solutions

An ideal solution which obeys the Raoult's law is represented by the equation:

$$P_i = P_i^* N_i$$

Components in non-ideal, real solutions may also show the relationship similar to the above:

$$\boxed{P_i = K_i N_i}$$

where $K_i$ is the proportionality constant which is not the same as $P_i^*$, the vapor pressure of pure component *i*.

The above relationship is known as ***Henry's Law***. Henry's Law holds only when component *i* is dilute in the solution. The relationship of the above two equations are graphically represented in the figure here. Component *i* is said to follow Henry's Law in the concentration range indicated by the circle in the figure where the tangential line and the partial pressure curve coincide.

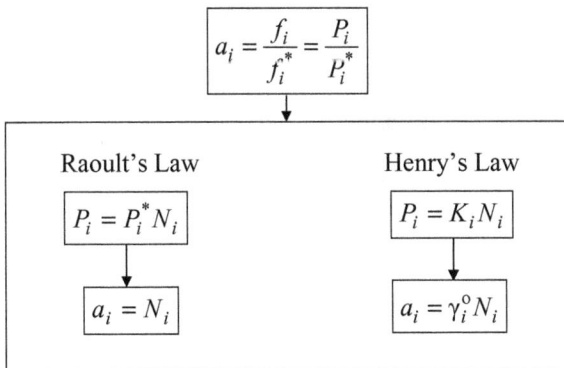

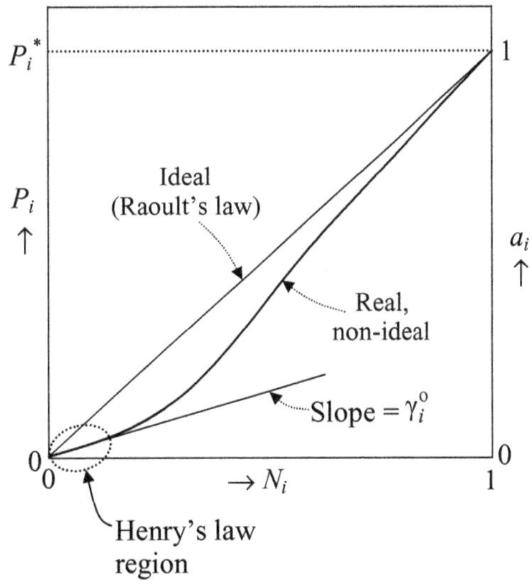


Summary of Raoult's law and Henry's law

$$\boxed{a_i = \frac{f_i}{f_i^*} = \frac{P_i}{P_i^*}}$$

↓

| Raoult's Law | Henry's Law |
| --- | --- |
| $P_i = P_i^* N_i$ | $P_i = K_i N_i$ |
| ↓ | ↓ |
| $a_i = N_i$ | $a_i = \gamma_i^o N_i$ |

In the above, $\gamma_i^o = \dfrac{K_i}{P_i^*}$ and called ***Henrian activity coefficient***.

In the figure above, the right-hand ordinate is for the activity of *i* according to its definition ($a_i = P_i / P_i^*$) and the slope of the tangential line is $\gamma_i^o$.

> *Example 9.4*
>
> Explain Henry's law from the molecular point of view.

In an *i-j* binary solution in which *i* is sufficiently dilute in *j*, the neighboring species of *i* will be essentially *j*. The intermolecular interaction the species *i* experiences will the *i-j* interaction for virtually all *i*'s, as they see only species *j* in the concentration range at which Henry's law applies. This is the reason why the partial pressure (which reflects the escaping tendency from the solution) is linearly proportional to the concentration. But, as the *i-j* interaction is different from the *i-i* or *j-j* interaction, the proportionality constant is also different from that for Raoult's law.

> *Example 9.5*
>
> ***Colligative properties*** are properties of liquid solutions that depend mainly on the number of solute molecules in a given solvent rather than the properties (*e.g.* size or mass) of the molecules. Colligative properties include lowering of vapor pressure, elevation of boiling point, depression of freezing point and osmotic pressure.
> Discuss the principles by which the boiling point is elevated and the freezing point is depressed, and develop equations by which the elevation and depression can be quantitatively evaluated.

<Principles>

We first make two assumptions:

(1) The solute is not volatile so that there is no vapor pressure of the solute in the vapor phase.

(2) The solute does not form solid solutions with the solvent when the solvent freezes.

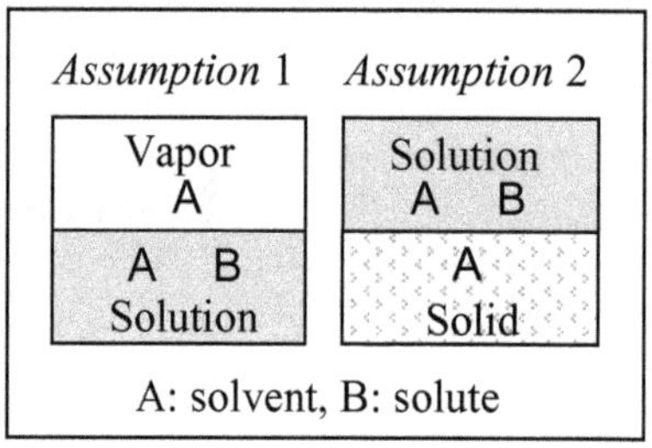


Consider the following relationships:

$$G_A^o = \mu_A^o = H_A^o - TS_A^o$$

$$\overline{G}_A = G_A^o + RT\,ln\,a_A$$

$$\mu_A = \mu_A^o + RT\,ln\,a_A$$

And we know that,

1) Entropy of solid ( $S_{A(s)}^o$ ) < Entropy of liquid ( $S_{A(l)}^o$ ) < Entropy of vapor ( $S_{A(g)}^o$ )

2) Chemical potential of A in solution ( $\mu_A$ ) < Chemical potential of pure A ( $\mu_A^o$ ), since $RT\,ln\,a_A < 0$.

3) In the above, the activity term ($a_A$) can be replaced with the mole fraction ($N_A$), as the solute B is so dilute that the solvent A is nearly pure and thus it behaves ideally.

$$\boxed{\mu_A = \mu_A^o + RT\,ln\,N_A}$$

When we make a graphical representation of all the above statements, it may look like the following figure:

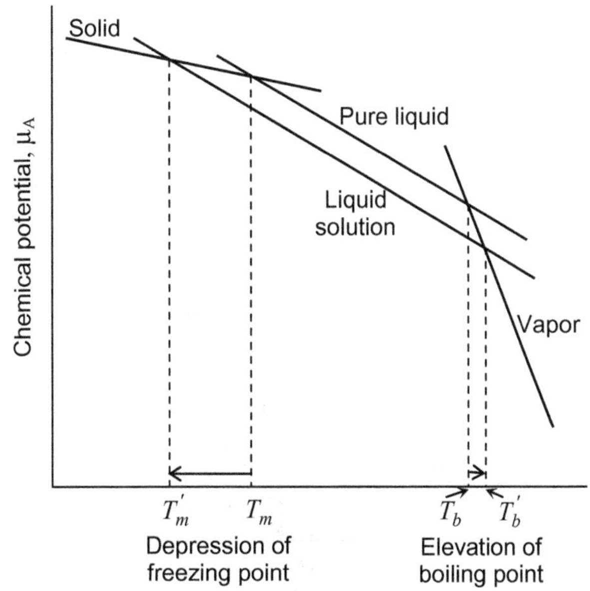


The lowering of the chemical potential, which results in the depression of the freezing point and the elevation of the boiling point, is not because of the energy associated with the interaction between the solvent and solute molecules (enthalpy). This conclusion is evidenced by the fact that even in ideal solutions the lowering occurs.
The lowing is due to the increase in the entropy by mixing ( $G_A^M = \mu_A^M = H_A^M - TS_A^M$ ).

<Quantitative analysis>

<u>Elevation of the boiling point</u>

The boiling point is the temperature at which the vapor pressure of the liquid equals the environmental pressure surrounding the liquid which is normally taken to be 1 *atm*. At equilibrium the chemical potential of a component (say, the solvent) is the same throughput the system (say, in the liquid and the vapor).

$$\begin{array}{|c|}
\hline
\text{Vapor} \\
\text{A}\;\;\mu_{A(g)} \\
\hline
\text{B}\;\;\text{A}\;\;\mu_{A(l)} \\
\text{Solution} \\
\hline
\end{array}$$

$$\mu_{A(g)} = \mu_{A(g)}^{o} + RT\,ln\,P_{A} \qquad \xrightarrow{P_{A}=1\,atm} \qquad \mu_{A(g)} = \mu_{A(g)}^{o}$$

$$\mu_{A(l)} = \mu_{A(l)}^{o} + RT\,ln\,N_{A}$$

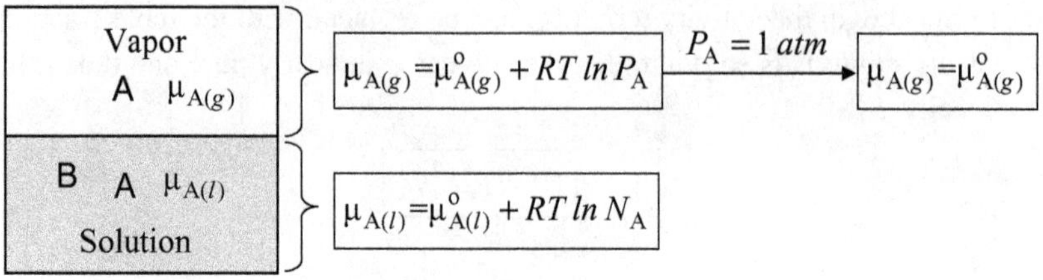

At equilibrium,

$$\mu_{A(g)} = \mu_{A(l)}$$

$$\Delta G_{vap}^{o} = \mu_{A(g)}^{o} - \mu_{A(l)}^{o}$$

$$\Delta G_{vap}^{o} = RT\,ln\,N_{A}$$

$$\Delta G_{vap}^{o} = \Delta H_{vap}^{o} - T\Delta S_{vap}^{o}$$

$$ln\,N_{A} = \frac{\Delta H_{vap}^{o}}{RT} - \frac{\Delta S_{vap}^{o}}{R}$$

For pure A (melting point $T_m$)

$$N_A = 1 \;\;\rightarrow\;\; ln\,N_A = 0$$

$$\frac{\Delta S_{vap}^{o}}{R} = \frac{\Delta H_{vap}^{o}}{RT_m}$$

$$ln\,N_{A} = \frac{\Delta H_{vap}^{o}}{R}\left(\frac{1}{T} - \frac{1}{T_m}\right)$$

As $N_B \ll 1$,

$$ln\,N_A = ln\left(1 - N_B\right) \cong -N_B$$

$$N_B = \frac{\Delta H_{vap}^{o}}{R}\left(\frac{1}{T_m} - \frac{1}{T}\right)$$

As $N_B > 0$ and , $\Delta H_{vap}^{o} > 0$

$$T > T_m$$

The above analysis proves the elevation of the boiling point by adding a solute to form a dilute solution, and also provides the equation which enables us to find the boiling point of the solution.

In a similar way, we may derive an equation for the depression of the freezing point as shown below:

$$\boxed{N_B = \frac{\Delta H^o_{fus}}{R}\left(\frac{1}{T} - \frac{1}{T_f}\right)}$$

*Exercises*

9.7 The table below shows the vapor pressures of A exerted by A-B alloys at 1273 K:

| $N_A$ | 0.1 | 0.2 | 0.3 | 0.4 | 0.5 | 0.6 | 0.7 | 0.8 | 0.9 | 1.0 |
|---|---|---|---|---|---|---|---|---|---|---|
| $P_a \times 10^6$,atm. | 0.25 | 0.5 | 0.75 | 1.2 | 2.0 | 2.85 | 3.75 | 4.8 | 5.4 | 6.0 |

Up to what mole fraction does the solute A obey the Henry's law?

9.8 The activity of carbon, $a_C$, in liquid Fe-C alloys is given by the following equation:

$$log\, a_C = log\left(\frac{N_C}{1-2N_C}\right) + \frac{1180}{T} - 0.87 + \left(0.72 + \frac{3400}{T}\right)\left(\frac{N_C}{1-N_C}\right)$$

where the standard state of carbon is pure graphite.

1) Find Henrian activity coefficient, $\gamma^o_C$, of liquid Fe-C solutions at 1,600°C.
2) Calculate $H^M_C$ in the composition range over which carbon obeys Henry's law.

## 9.4 Gibbs-Duhem Equation

We have previously developed the following general form of the Gibbs-Duhem equation:

$$\boxed{\sum N_i d\overline{Y}_i = 0}$$

↓ Applying to A-B binary solution for partial molar Gibbs energies,

$$\boxed{N_A d\overline{G}_A + N_B d\overline{G}_B = 0}$$

↓ $G^M_i = \overline{G}_i - G^o_i$

$dG^M_i = d\overline{G}_i$ at constant $T$

$$\boxed{N_A dG^M_A + N_B dG^M_B = 0}$$

↓ $G^M_i = RT\, ln a_i$

$$N_A d\,ln a_A + N_B d\,ln a_B = 0$$

↓ $a_i = \gamma_i N_i$

$N_A + N_B = 1 \rightarrow dN_A + dN_B = 0$

$$N_A d\,ln\gamma_A + N_B d\,ln\gamma_B = 0$$

All these equations are called Gibbs-Duhem equation, and tell us that

- The thermodynamic properties of one component in a solution cannot vary independently, but they are interrelated to those of the other components. This is true for multi-component solutions as well.

The last two equations are particularly useful in determining activities. If the activity of one component is known over a range of compositions at a temperature, then it is possible to determine the activity of the other component by employing one of the two equations.

Suppose that the activity of component B is known over a range of compositions.

$$N_A d\,ln a_A + N_B d\,ln a_B = 0$$

↓ Rearrangement and integration

$$\int_{a_A=1}^{a_A=a_A} d\,ln a_A = \int_{a_A=1}^{a_A=a_A} -\frac{N_B}{N_A} d\,ln a_B$$

↓ $a_A = a_A \rightarrow N_A = N_A$

$a_A = 1 \rightarrow N_A = 1$

$$\int_{a_A=1}^{N_A=N_A} d\,ln a_A = \int_{N_A=1}^{N_A=N_A} -\frac{N_B}{N_A} d\,ln a_B$$

↓

$$ln a_A \Big|_{N_A=N_A} = \int_{N_A=1}^{N_A=N_A} -\frac{N_B}{N_A} d\,ln a_B$$

In the above equation, $ln\, a_A \big|_{N_A=N_A}$ is the activity of A in logarithm at the concentration of $N_A$. This equation can be solved in principle by graphical integration. If we plot the data of the activity of A at a number of different concentrations of B by putting $(-\,ln\, a_B)$ as abscissa and $(N_B / N_A)$ as ordinate, then we can obtain the integral value of the right side of the above equation by measuring the area under the curve. The following figure explains how to integrate the above equation graphically:

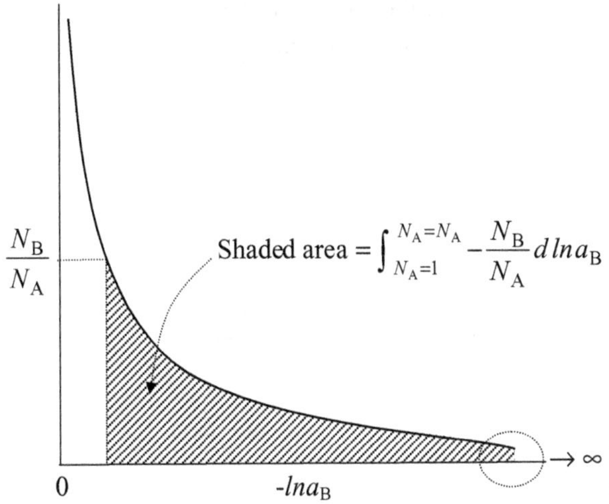


Thus it is difficult to accurately evaluate the area under the curve.

Now, let us consider the other form of the Gibbs-Duhem equation which is expressed with activity coefficients in place of activities:

$$N_A d\,ln\gamma_A + N_B d\,ln\gamma_B = 0$$

$$\downarrow$$

$$d\,ln\gamma_A = -\frac{N_B}{N_A} d\,ln\gamma_B$$

$$\downarrow$$

$$\int_{\gamma_A=1}^{\gamma_A=\gamma_A} d\,ln\gamma_A = \int_{\gamma_A=1}^{\gamma_A=\gamma_A} -\frac{N_B}{N_A} d\,ln\gamma_B$$

$$\downarrow \quad \gamma_A = \gamma_A \rightarrow N_A = N_A, \quad \gamma_A = 1 \rightarrow N_A = 1$$

$$\int_{\gamma_A=1}^{N_A=N_A} d\,ln\gamma_A = \int_{N_A=1}^{N_A=N_A} -\frac{N_B}{N_A} d\,ln\gamma_B$$

$$\downarrow$$

$$ln\gamma_A \Big|_{N_A=N_B} = \int_{N_A=1}^{N_A=N_A} -\frac{N_B}{N_A} d\,ln\gamma_B$$

This equation can be solved by graphical integration. If we plot the data of the activity coefficient of A at a number of different concentrations of B by putting ($-ln\gamma_A$) as abscissa and ($N_B/N_A$) as ordinate, Then we can obtain the integral value of the right side of the above equation by measuring the area under the curve. The following figure explains how to integrate the above equation graphically:

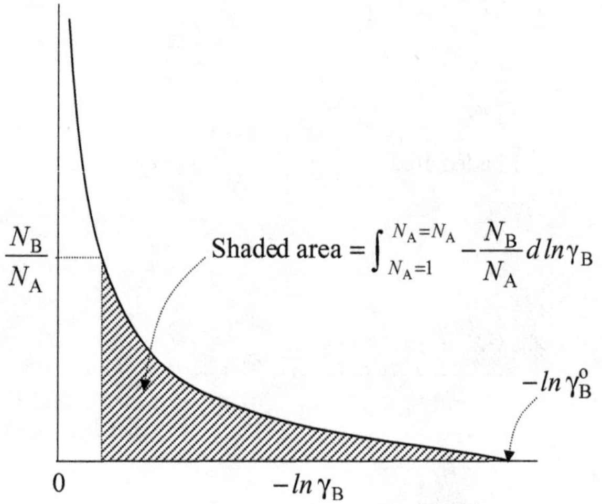


Darken and Gurry have introduced a new function called the ***alpha function***, $\alpha$:

where $i$ = A or B.

$$\alpha_i = \frac{ln\gamma_i}{(1-N_i)^2}$$

$$d\,ln\gamma_B = d\left(\alpha_B N_A^2\right) = N_A^2 d\alpha_B + 2\alpha_B N_A dN_A$$

$$ln\gamma_A = \int_{N_A=1}^{N_A} -\frac{N_B}{N_A} d\,ln\gamma_B$$

$$ln\gamma_A = -\int_{N_A=1}^{N_A} N_A N_B d\alpha_B - \int_{N_A=1}^{N_A} 2\alpha_B N_B dN_A$$

Integration by parts of the first term above yields,

$$\int_{N_A=1}^{N_A} N_A N_B d\alpha_B = \left(\alpha_B N_A N_B\right)\Big|_{N_A=1}^{N_A} - \int_{N_A=1}^{N_A} \alpha_B (1-2N_A) dN_A$$

$$ln\gamma_A = -\alpha_B N_A N_B - \int_{N_A=1}^{N_A} \alpha_B dN_A$$

The above equation enables us to calculate the activity coefficient of A at the composition of $N_A$ by graphically integrating $\alpha_B$ values over the range from $N_A = 1$ to $N_A = N_A$.

The activity coefficient of A at infinite dilution of A, *i.e.*, the Henrian activity coefficient, $\gamma_A^o$ can be found by applying $N_A \rightarrow 0$ to the above equation:

$$ln\,\gamma_A = -\alpha_B N_A N_B - \int_{N_A=1}^{N_A} \alpha_B dN_A$$

$$\downarrow N_A \rightarrow 0$$

$$ln\,\gamma_A^o = -\int_{N_A=1}^{0} \alpha_B dN_A$$

*Exercises*

9.9 The activity of zinc in liquid cadmium-zinc alloys at 708*K* is related to the alloy composition by the following equation:

$$ln\,\gamma_{Zn} = 0.87N_{Cd}^2 - 0.3N_{Cd}^3$$

Calculate the activity of cadmium at $N_{Cd} = 0.1$.

9.10 The relative integral molar enthalpy or the molar heat of mixing of liquid Sn-Bi solutions at 330°C is represented by

$$H^M = 400N_{Sn}N_{Bi}\ \text{J mol}^{-1}$$

Determine $H_{Sn}^M$ at $N_{Sn} = 0.3$.

## 9.5 Solution Models

The activity of component *i* in a solution is related to the concentration of *i* in the solution and determination of the variation of $\gamma_i$ with the temperature ($T$) and composition ($N_i$) is of critical importance in chemical thermodynamics.

Once $\gamma_i$ is known

↓

$a_i$.is determined by

$$a_i = \gamma_i N_i$$

↓

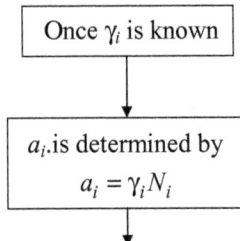

↓

Then Gibbs energy of mixing can be evaluated by

$$G_i^M = RT\ln a_i$$

↓ ↓

And then entropy of mixing

$$S_i^M = -\left(\frac{\partial G_i^M}{\partial T}\right)_P$$

Also enthalpy of mixing

$$H_i^M = \left[\frac{\partial\left(\frac{G_i^M}{T}\right)}{\partial\left(\frac{1}{T}\right)}\right]_P$$

As we can see from the above, once the activity coefficient is known, other important thermodynamic terms of a solution are determined accordingly. The activity coefficient reflects the nature of interaction of particles (atoms or molecules) in the solution.

Several formalisms have been developed to relate the activity coefficient of a component in the solution to the temperature and composition of the solution. The simplest model for a solution is obviously the ideal solution discussed previously. The behavior of an ideal solution can be determined by knowing the composition and temperature of the solution. As the ideal solution model contains no adjustable thermodynamic parameter, it fails to describe the differences that may exist in two different solutions of the same composition at the same temperature.

We now discuss the following three models one after another:

- Regular solution model
- Margules formalism
- Darken's quadratic formalism

***Regular Solutions***

The solution model that is the most conveniently used for a number of different solution may be so-called the ***regular solution model***. The model was developed with some basic assumptions:

(1) Particles (atoms or molecules) are distributed at random in the solution (*i.e.*, ideal mixing, and thus $S^E = 0$).

(2) Heat absorbed or released for forming the solution is non-zero (*i.e.*, the enthalpy change for formation of the solution is not zero: $H^E (= H^M) \neq 0$).

Based on the above assumptions, we are able to develop useful thermodynamic relationships.

For the A-B binary solution,

Entropy of Mixing
Random Mixing
(Ideal entropy)

$$S_A^M = S_A^{M,ideal} = -R\,ln\,N_A$$
$$S_B^M = S_B^{M,ideal} = -R\,ln\,N_B$$

$$G_i^M = H_i^M - TS_i^M$$
$$G_i^M = RT\,ln\,a_i$$

Enthalpy of Mixing
Non-zero heat of mixing
(Non-ideal enthalpy)

$$H_A^M = -RT\,ln\,\gamma_A \neq H_A^{M,ideal}(=0)$$
$$H_B^M = -RT\,ln\,\gamma_B \neq H_B^{M,ideal}(=0)$$

$$G_A^E = RT\,ln\,\gamma_A = H_A^M = H_A^E$$
$$G_B^E = RT\,ln\,\gamma_B = H_B^M = H_B^E$$

The relationships given by the last two equations are the unique feature of the regular solution. A statistical-mechanical analysis based on the assumptions behind the regular solution model leads to the following relationships for A-B binary solution (Derivation follows.):

$$ln\,\gamma_A = \frac{\Omega}{RT}(1-N_A)^2 \qquad ln\,\gamma_B = \frac{\Omega}{RT}(1-N_B)^2$$

Where $\Omega$ is called the ***interaction parameter***, and is independent of composition and also temperature.

For the Gibbs energy of mixing of the solution ($G^M$),

$$G^M = RT\left(N_A\,ln\,a_A + N_B\,ln\,a_B\right)$$

$$a_i = \gamma_i N_i$$

$$G^M = RT\left(N_A\,ln\,N_A + N_B\,ln\,N_B\right) + RT\left(N_A\,ln\,\gamma_A + N_B\,ln\,\gamma_B\right)$$

$$RT\,ln\,\gamma_A = \Omega N_B^2 \qquad RT\,ln\,\gamma_B = \Omega N_A^2$$

$$G^M = RT\left(N_A\,ln\,N_A + N_B\,ln\,N_B\right) + \Omega N_A N_B$$

The last term in the above equation is the excess Gibbs energy of mixing of the solution ($G^E$), which is indicative of deviation from ideal behavior. Note that it is also the same as $H^M$.

$$G^E = \Omega N_A N_B$$

$$\left(\frac{\partial G}{\partial T}\right)_P = -S$$

$$\left(\frac{\partial G^E}{\partial T}\right)_P = -S^E = 0$$

Random mixing,
*i.e.*, ideal entropy.

$$\left(\frac{\partial \Omega}{\partial T}\right)_P = 0$$

The interaction parameter, $\Omega$, is thus proved to be independent of temperature.
For regular solutions, the activity at one temperature is known, the activity at other temperature can be determined:

$$RT\,ln\gamma_A = \Omega N_B^2$$

$T_1$: $$RT_1 ln(\gamma_A)_{T_1} = \Omega N_B^2$$

$T_2$: $$RT_2 ln(\gamma_A)_{T_2} = \Omega N_B^2$$

$$\frac{ln(\gamma_A)_{T_1}}{ln(\gamma_A)_{T_2}} = \frac{T_2}{T_1}$$

The statistical approach to the regular solution model is now in order. Suppose the system which consists of $n_A$ particles (atoms or molecules) of A and $n_B$ particles of B and total number of particles is $n$.

$$n = n_A + n_B$$

The model assumes that the particles are distributed at random in the solution and thus the entropy of mixing is the same as that of an ideal solution. It is also assumed that each particle interacts only with the nearest neighbors (the number of nearest neighbors is called the ***coordination number***, $Z$.) Now we count the number of different pairs ($n_{AA}$ = the number of AA pairs, $n_{BB}$ = the number of BB pairs, $n_{AB}$ = the number of AB pairs).

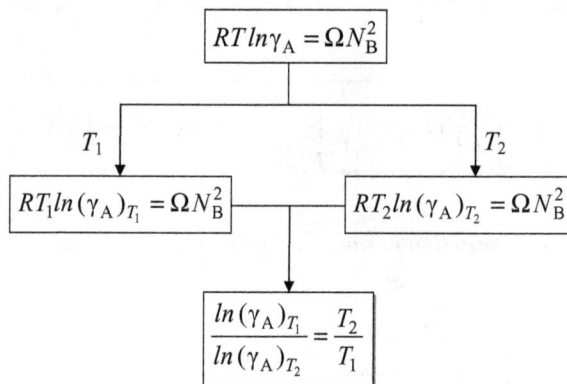

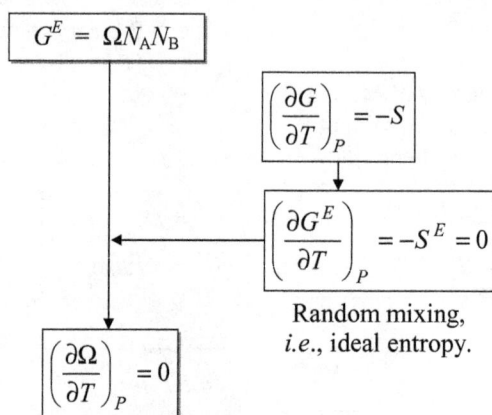

(Symbols to be used)

$n^{o}_{AA}, n^{o}_{BB}$ : The number of AA and BB pairs, respectively, before mixing

$u_{AA}, u_{BB}, u_{AB}$ : Bonding energies of AA, BB and AB pairs, respectively.

$U^{o}, U, U^{M}$ : Energies of the system, before, after, and net change, respectively.

• *Before mixing*

<u>The number of AA pairs in pure A</u>

$$n^{o}_{AA} = \tfrac{1}{2} n_A Z$$

*(Each A has Z numbers of nearest neighbors making $n_A Z$ number of pairs, but divided by 2 to avoid double-counting.)*

<u>The number of BB pairs in pure B</u>:

$$n^{o}_{BB} = \tfrac{1}{2} n_B Z$$

<u>Total pairing energies</u>:

$$U^{o} = n^{o}_{AA} u_{AA} + n^{o}_{BB} u_{BB} = \tfrac{1}{2} Z \left( n_A u_{AA} + n_B u_{BB} \right)$$

• *After mixing*

<u>The number of AA pairs in the solution</u>:

$$n_{AA} = \tfrac{1}{2} Z n_A \left( n_A / n \right)$$

*(Each A has Z numbers of nearest neighbors making $\tfrac{1}{2} n_A Z$ numbers of pairs, but the probability of a nearest neighbor being A is $n_A / n$ )*

<u>The number of BB pairs in the solution</u>:

$$n_{BB} = \tfrac{1}{2} Z n_B \left( n_B / n \right)$$

<u>The number of AB pairs in the solution</u>:

$$n_{AB} = Z n_A \left( n_B / n \right)$$

<u>Total pairing energies</u>:

$$U = n_{AA} u_{AA} + n_{BB} u_{BB} + n_{AB} u_{AB}$$
$$= \tfrac{1}{2} Z \left( n_A^2 / n \right) u_{AA} + \tfrac{1}{2} Z \left( n_B^2 / n \right) u_{BB} + Z \left( n_A n_B / n \right) u_{AB}$$

Then the energy of mixing of the system will be,

$$U^M = U - U^{\circ}$$

$$U^M = \tfrac{1}{2}Z\left(n_A^2/n\right)u_{AA} + \tfrac{1}{2}Z\left(n_B^2/n\right)u_{BB} + Z\left(n_A n_B/n\right)u_{AB} - \tfrac{1}{2}Z\left(n_A u_{AA} + n_B u_{BB}\right)$$

$$N_A = n_A/n \qquad N_B = n_B/n$$

$$U^M = nZ\left[\frac{\left(N_A^2 - N_A\right)u_{AA} + \left(N_B^2 - N_B\right)u_{BB}}{2}\right] + nZN_A N_B u_{AB}$$

$$U^M = nZN_A N_B\left[u_{AB} - \tfrac{1}{2}\left(u_{AA} + u_{BB}\right)\right]$$

Define $\Omega$

$$\Omega = nZ\left[u_{AB} - \tfrac{1}{2}\left(u_{AA} + u_{BB}\right)\right]$$

$$U^M = \Omega N_A N_B$$

Note that this last equation is the same as that for the excess enthalpy.

$$H^E = H^M = G^E = \Omega N_A N_B$$

Recall that the partial properties (*i.e.*, the properties of individual components in a solution) can be determined from the integral properties (*i.e.*, the properties of the solution) by the following relationship:

$$\overline{Y}_A = Y + N_B\left(\frac{dY}{dN_A}\right)$$

Applying this equation to the excess enthalpy,

$$H_{\mathrm{A}}^{E} = H^{E} + N_{\mathrm{B}}\left(\frac{dH^{E}}{dN_{\mathrm{A}}}\right)$$

$$\downarrow \quad H^{E} = \Omega N_{\mathrm{A}} N_{\mathrm{B}}$$

$$H_{\mathrm{A}}^{E} = \Omega N_{\mathrm{B}}^{2}$$

$$\downarrow \quad H_{\mathrm{A}}^{E} = G_{\mathrm{A}}^{E} = RT\, ln\, \gamma_{\mathrm{A}}$$

$$ln\, \gamma_{\mathrm{A}} = \frac{\Omega}{RT} N_{\mathrm{B}}^{2}$$

$$\downarrow$$

$$ln\, \gamma_{\mathrm{B}} = \frac{\Omega}{RT} N_{\mathrm{A}}^{2}$$

The last two equations prove the same equations introduced earlier without proof.

The physical significance of the interaction parameter, $\Omega = nZ\left[u_{\mathrm{AB}} - \frac{1}{2}\left(u_{\mathrm{AA}} + u_{\mathrm{BB}}\right)\right]$:

- It is the difference between the bonding energy of AB pairs and the arithmetic mean of the bonding energies of AA and BB pairs.
- If the AB bonding is stronger (*i.e.*, more negative bonding energy) than the AA and BB bondings, then $\Omega$ is negative, and the excess functions are negative, and thus the deviation from ideality is negative.
- If the AB bonding is weaker (*i.e.*, less negative bonding energy) than the AA and BB bondings, then $\Omega$ is positive, and the excess functions are positive, and thus the deviation from ideality is positive.
- If the AB bonding energy is the same as the mean of the bonding energies of AA and BB pairs, *i.e.*, if $\Omega$ is equal to zero, then the solution behaves ideally.

***Margules Formalism***

Margules suggested a power series formula for expressing the activity-composition variation of a binary solution:

$$ln\, \gamma_{\mathrm{A}} = \alpha_1 N_{\mathrm{B}} + \frac{1}{2}\alpha_2 N_{\mathrm{B}}^{2} + \frac{1}{3}\alpha_3 N_{\mathrm{B}}^{3} + \cdots$$

$$ln\,\gamma_B = \beta_1 N_A + \frac{1}{2}\beta_2 N_A^2 + \frac{1}{3}\beta_3 N_A^3 + \cdots$$

Applying the Gibbs-Duhem equation with ignoring coefficients $\alpha_i$'s and $\beta_i$'s higher than $i = 3$, we can obtain

$$\alpha_1 = \beta_1 = 0, \quad \beta_2 = \alpha_2 + \alpha_3, \quad \beta_3 = -\alpha_3$$

Then

$$ln\,\gamma_A = \frac{1}{2}\alpha_2 N_B^2 + \frac{1}{3}\alpha_3 N_B^3$$

$$ln\,\gamma_B = \frac{1}{2}(\alpha_2 + \alpha_3) N_A^2 - \frac{1}{3}\alpha_3 N_A^3$$

or

$$ln\gamma_A = N_B^2\left[A - 2N_A(A - B)\right]$$

$$ln\gamma_B = N_A^2\left[A - 2N_B(A - B)\right]$$

where, $A = 1/2\ \alpha_2 + 1/3\ \alpha_3$ and $B = 1/2\ \alpha_2 + 1/6\ \alpha_3$

These equations are known as ***three suffix Margules equations***.

### *Darken's Quadratic Formalism*

Darken suggested that, for binary metallic solutions, the entire concentration range can be divided into three distinct regions: two terminal regions and an interconnecting central region. It is seen in the following figure that the logarithm of the activity coefficient of component A is linearly dependent on the square of the concentration of component B.

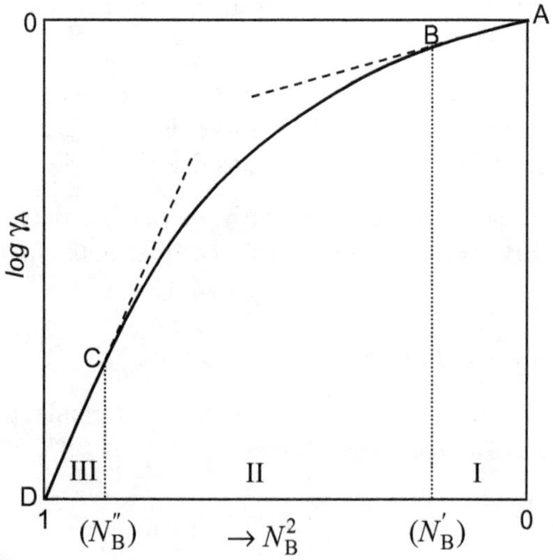


Referring to the figure here,

<Region I (AB) where component A is concentrated.>

$$\log \gamma_A = \alpha_I N_B^2$$

↓ Gibbs-Dehem equation
$N_A d \log \gamma_A + N_B d \log N_B = 0$

$$d \log \gamma_B = 2\alpha_I N_A dN_A$$

↓ Integrating from $N_A = 1$
($N_B = 0$ or $\gamma_B = \gamma_B^o$) to $N_A$.

$$\int_{\gamma_B^o}^{\gamma_B} d \log \gamma_B = 2\alpha_I \int_1^{N_A} N_A dN_A$$

↓

$$\log\left(\frac{\gamma_B}{\gamma_B^o}\right) = \alpha_I \left(N_A^2 - 1\right)$$

<Region III (CD) where component B is concentrated.>

In a similar way to the above,

$$\log\left(\frac{\gamma_A}{\gamma_A^o}\right) = \alpha_{III} \left(N_B^2 - 1\right)$$

*Exercises*

9.11 Zinc and cadmium liquid alloys conform to regular solution behavior. The following table shows the relative integral molar enthalpies (or the molar heats of mixing) at various zinc concentrations at 723 K.

| $N_{Zn}$ | 0.06 | 0.09 | 0.15 | 0.37 | 0.61 | 0.76 | 0.86 | 0.95 |
|---|---|---|---|---|---|---|---|---|
| $H^M$ (J mol$^{-1}$) | 493 | 714 | 1,126 | 1,985 | 2,030 | 1,585 | 1,039 | 423 |

(1) Calculate the average interaction parameter $\Omega$.
(2) Calculate the activity of Zn in the solution containing $N_{Zn} = 0.3$.

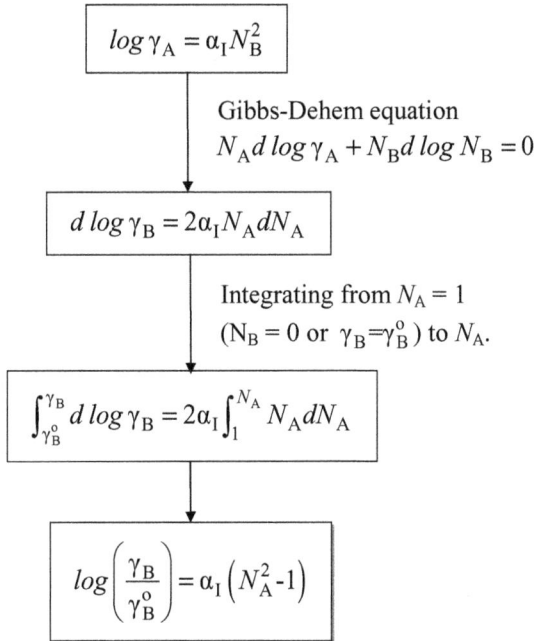

9.12 In the A-B binary solution the activity coefficients are given by the three-suffix Margules equations:

$$ln\gamma_A = \frac{1}{2}\alpha_2 N_B^2 + \frac{1}{3}\alpha_3 N_B^3$$

$$ln\gamma_B = \frac{1}{2}(\alpha_2 + \alpha_3) N_A^2 - \frac{1}{3}\alpha_3 N_A^3$$

It was found that the Henrian activity coefficients of A and B at the temperature $T$, $\gamma_A^o$ and $\gamma_B^o$ were 0.75 and 0.54, respectively. Determine $\alpha_2$.

# Chapter 10

# Reaction Equilibria

## 10.1 Equilibrium Constants

Consider a general chemical reaction occurring at a constant temperature and pressure:

$$a\text{A} + b\text{B} + \cdots = m\text{M} + n\text{N} + \cdots$$

where $a$, $b$, ..., $m$, $n$, ... are stoichiometric coefficients indicating the number of moles of respective species A, B, ..., M, N, ....

Let $\Delta G_r$ denote the Gibbs energy change associated with the above reaction.

$$\Delta G_r = \Sigma G_{products} - \Sigma G_{reactants}$$

$$\Sigma G_{reactants} = a\bar{G}_\text{A} + b\bar{G}_\text{B} + \cdots$$

$$\Sigma G_{products} = m\bar{G}_\text{M} + n\bar{G}_\text{N} + \cdots$$

and for any species in the solution,

$$\bar{G}_i = G_i^\text{o} + RT\,ln\,a_i$$

$$\Delta G_r = \Delta G_r^\text{o} + RT\,ln\left(\frac{a_\text{M}^m\, a_\text{N}^n \cdots}{a_\text{A}^a\, a_\text{B}^b \cdots}\right)$$

where

$$\Delta G_r^\text{o} = (mG_\text{M}^\text{o} + nG_\text{N}^\text{o} + ..) - (aG_\text{A}^\text{o} + bG_\text{B}^\text{o} + ..)$$

This is the Gibbs energy change when all reactants and products are at their respective standard states - ***Standard Gibbs energy change of reaction.***

We now define the ***activity quotient***, $Q$ as

$$Q = \frac{a_\text{M}^m\, a_\text{N}^n \cdots}{a_\text{A}^a\, a_\text{B}^b \cdots}$$

$$\Delta G_r = \Delta G_r^\text{o} + RT\,lnQ$$

$\Delta G_r = 0$ at equilibrium.

$$0 = \Delta G_r^{\circ} + RT \ln Q_{eq}$$ where $Q_{eq} = Q$ at equilibrium.

$$Q_{eq} = exp\left(-\frac{\Delta G_r^{\circ}}{RT}\right)$$

Let $K = Q_{eq}$

$$K = exp\left(-\frac{\Delta G_r^{\circ}}{RT}\right)$$ $K$ is called ***equilibrium constant.***

Note that,

- The equilibrium constant ($K$) of a reaction varies with temperature, but at a given temperature it is constant, independent of concentrations of individual species involved in the reaction.
- Only if the activities of individual reactants and products of the reaction satisfy the activity quotient ($Q$) being equal to the value of the equilibrium constant ($K$), then the reaction is in equilibrium. Otherwise the reaction is not in equilibrium, but still under progress.

At equilibrium,

$$Q_{eq} = K$$

$$Q_{eq} = \left(\frac{a_M^m \, a_N^n \cdots}{a_A^a \, a_B^b \cdots}\right)_{eq} \qquad K = exp\left(-\frac{\Delta G_r^{\circ}}{RT}\right)$$

$$\left(\frac{a_M^m \, a_N^n \cdots}{a_A^a \, a_B^b \cdots}\right)_{eq} = K = exp\left(-\frac{\Delta G_r^{\circ}}{RT}\right)$$

Note that,

- The values of $a_i$'s for $K$ are the ones when the reaction has arrived at the equilibrium state, and thus $\Delta G_r$ becomes zero when these values are plugged into the equation for $\Delta G_r$.
- Since $a_i$'s are dimensionless, both $K$ and $Q$ are also dimensionless.
- The larger the value of $K$ is, the more the formation of products is favored, and vice versa.

The following equation may be obtained by proper combination of the equations for $\Delta G_r$, $Q$ and $K$:

$$\Delta G_r = RT\ln\left(\frac{Q}{K}\right)$$

Note that,

- If $Q/K < 1 \rightarrow$ the reaction is spontaneous from left to right as written
- If $Q/K = 1 \rightarrow$ the reaction is at equilibrium
- If $Q/K > 1 \rightarrow$ the reaction is spontaneous from right to left as written

---

**Example 10.1**

The figure given below depicts some reaction paths that the following general reaction may take:

$$aA + bB = mM + nN, \qquad \Delta G_r$$

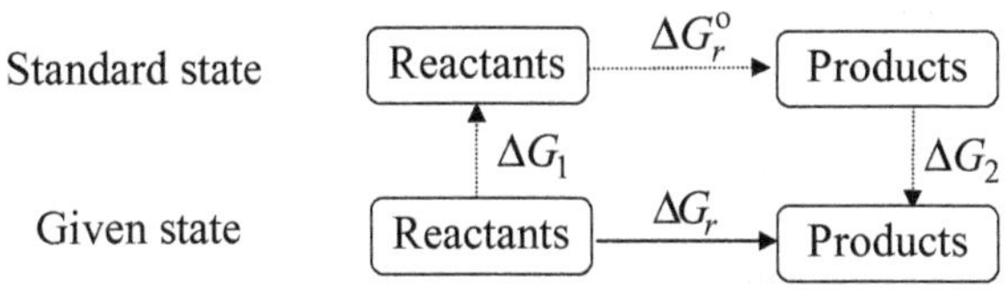

Prove the following relationships:

$$\Delta G_1 = -RT\ln\left(a_A^a a_B^b\right) \qquad \Delta G_2 = -RT\ln\left(a_M^m a_N^n\right) \qquad \Delta G_r = \Delta G_r^o + RT\ln\left(\frac{a_M^m a_N^n}{a_A^a a_B^b}\right)$$

---

**For reactants**

$$\Delta G_1 = a(G_A^o - \overline{G}_A) + b(G_B^o - \overline{G}_B)$$

$$\overline{G}_i = G_i^o + RT\ln a_i$$

$$\Delta G_1 = -RT\ln(a_A^a a_B^b)$$

**For products**

$$\Delta G_2 = m(\overline{G}_M - G_M^o) + n(\overline{G}_N - G_N^o)$$

$$\overline{G}_i = G_i^o + RT\ln a_i$$

$$\Delta G_2 = RT\ln(a_M^m a_N^n)$$

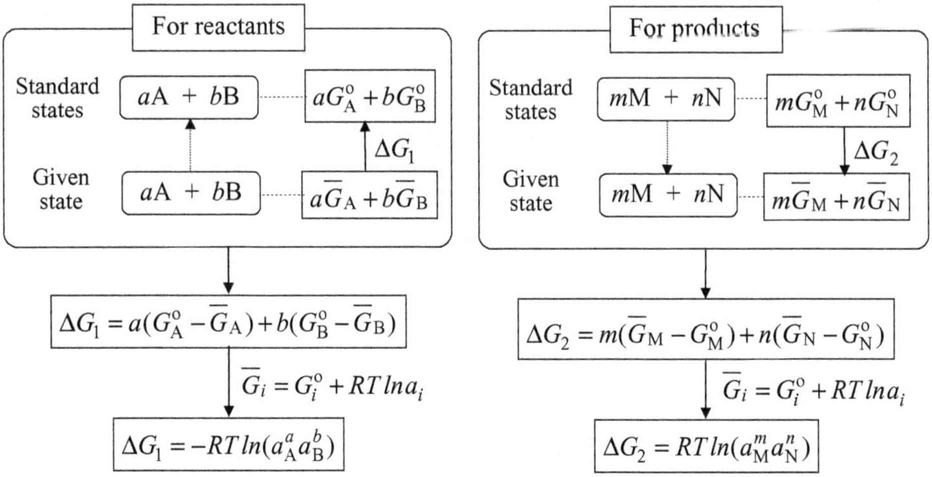

The Gibbs energy change of reaction at the state of interest is then

$$\Delta G_r = \Delta G_1 + \Delta G_r^{\circ} + \Delta G_2$$

$$\downarrow$$

$$\Delta G_r = \Delta G_r^{\circ} + RT\,ln\left(\frac{a_{\mathrm{M}}^{m} a_{\mathrm{N}}^{n}}{a_{\mathrm{A}}^{a} a_{\mathrm{B}}^{b}}\right)$$

This example clearly shows the distinction between $\Delta G_r$ and $\Delta G_r^{\circ}$.

$\Delta G_r^{\circ}$ : Gibbs energy change of the reaction when all reactants and products are at their respective *standard* states

Difference $\longleftrightarrow$ $RT\,ln\left(\frac{a_{\mathrm{M}}^{m} a_{\mathrm{N}}^{n}}{a_{\mathrm{A}}^{a} a_{\mathrm{B}}^{b}}\right)$

$\Delta G_r$ : Gibbs energy change of the reaction when all reactants and products are at their respective *arbitrarily given* states

*Example 10.2*

Is the following statement true or false?
"Since the equilibrium constant *K* does not have a unit, one does not have to concern with units in the formulation of *K*."

In principle, the answer is "true", but we must be careful in evaluating the numerical value of *K*, when, in particular, a gaseous species is involved in the reaction. Consider the following simple reaction involving a gas species:

$$2\mathrm{A}(c) \ + \ \mathrm{B}(g) \ = \ \mathrm{M}(c)$$

where $c$ = condensed phase, and $g$ = gas phase.

$$K = \frac{a_M}{a_{\mathrm{A}}^2 a_{\mathrm{B}}}$$

$$\downarrow \quad a_{\mathrm{B}} = \frac{f_{\mathrm{B}}}{f_{\mathrm{B}}^{\circ}} \cong \frac{P_{\mathrm{B}}}{P_{\mathrm{B}}^{\circ}}$$

$$K = \frac{a_{\mathrm{M}}}{a_{\mathrm{A}}^2\left(\frac{P_{\mathrm{B}}}{P_{\mathrm{B}}^{\circ}}\right)}$$

where $P_{\mathrm{B}}^{\circ}$ = the pressure at the standard state of the gas (= the gas at 1 atm. pressure, and behaves ideally).

Then the above equation becomes,

$$K = \frac{a_M}{a_A^2\left(\frac{P_B}{1}\right)}$$

$$\downarrow$$

$$K = \frac{a_M}{a_A^2 P_B}$$

It is customary to express activities of gaseous species by the pressure symbol *P*, but we must always keep in mind that $P_i$ in the formulation of *K* is in fact $P_i / P_i^o$.

*Exercises*

10.1 Consider the following reaction at 873 K:

$$2SO_2(g) + O_2(g) = 2SO_3(g)$$

(1) Calculate the standard Gibbs energy change of reaction ($\Delta G_r^o$) at 873 K. The following data are given:

$$\Delta G^o_{f,SO_2} = -361{,}670 + 72.68T\,\text{J}$$

$$\Delta G^o_{f,SO_3} = -457{,}900 + 163.34T\,\text{J}$$

(2) Calculate the equilibrium constant *K* of the reaction at 873 K.

(3) A gas mixture includes $SO_2$, $SO_3$ and $O_2$. Partial pressures of these species are given below:

$$P_{SO_2} = 0.1\,\text{atm.},\quad P_{SO_3} = 0.01\,\text{atm.},\quad P_{O_2} = 0.21\,\text{atm.}$$

Calculate the activity quotient *Q*.

(4) Calculate the free energy change of the reaction, $\Delta G_r$.

(5) Is this reaction spontaneous from left to right as written under the conditions given above?

(6) If the partial pressures are changed as indicated below, calculate the Gibbs Energy change of the reaction:

$$P_{SO_2} = 0.01\,\text{atm.},\quad P_{SO_3} = 0.1\,\text{atm.},\quad P_{O_2} = 0.21\,\text{atm.}$$

(7) Is this reaction spontaneous from left to right as written?

(8) Calculate the partial pressure of $SO_3$ which would be in equilibrium at 873 K with

$$P_{SO_2} = 0.01\,\text{atm.},\quad P_{O_2} = 0.21\,\text{atm.}$$

10.2 Excess amount of pure carbon is reacted with $H_2O(g)$ at 1,000°C.

$$C(s) + H_2O(g) = CO(g) + H_2(g)$$

The gas phase does not contain any other species and is maintained at 1atm. of the total pressure. Calculate the equilibrium partial pressure of $H_2O(g)$. The following data are given:

$$C(s)+\tfrac{1}{2}O_2(g)=CO(g) \qquad \Delta G_1^o=-112{,}880-86.51T,\ \ \mathrm{J\,mol^{-1}}$$

$$H_2(g)+\tfrac{1}{2}O_2(g)=H_2O(g) \qquad \Delta G_2^o=-247{,}390+55.85T,\ \ \mathrm{J\,mol^{-1}}$$

## 10.2 Criteria of Reaction Equilibrium

Prior to deriving conditions for equilibrium of a system, mathematics on the ***extremum principle*** is revisited in the following:

Consider a function $g$ of a variable $x$,

$$g = g(x)$$

An extreme value of $g$ can be found by differentiating the equation, and the extreme values occur at the points on the curve at which the derivative is zero.

$$\boxed{\frac{dg}{dx}=0}$$

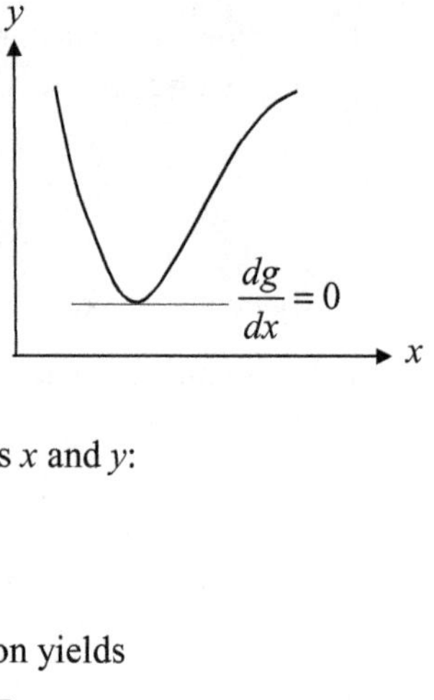


If the function $g$ is dependent on two independent variables $x$ and $y$:

$$\boxed{g = g(x,y)}$$

Differentiation yields

$$\boxed{dg=\left(\frac{\partial g}{\partial x}\right)_y dx+\left(\frac{\partial g}{\partial y}\right)_x dy}$$

At the extremum, $dg = 0$, but dx $\neq 0$, and $dy \neq 0$. Thus,

$$\boxed{\left(\frac{\partial g}{\partial x}\right)_y=0 \quad \text{and} \quad \left(\frac{\partial g}{\partial y}\right)_x=0}$$

These two equations determine the extremum of the function $g$.

In general, for a function $g$,

$$g = g(x_1, x_2, \cdots)$$

Differentiation yields

$$dg = \left(\frac{\partial g}{\partial x_1}\right)dx_1 + \left(\frac{\partial g}{\partial x_2}\right)dx_2 + \cdots$$

At the extremum,

$$\left(\frac{\partial g}{\partial x_1}\right) = 0 \quad \left(\frac{\partial g}{\partial x_2}\right) = 0 \quad \cdots$$

Now, suppose that we have a function $g = g(x,y)$, in which $y$ is *not* independent of $x$, but related to $x$ as $y = y(x)$. The extremum with this constraint will differ from the unconstrained absolute extremum, as shown in the following figure:

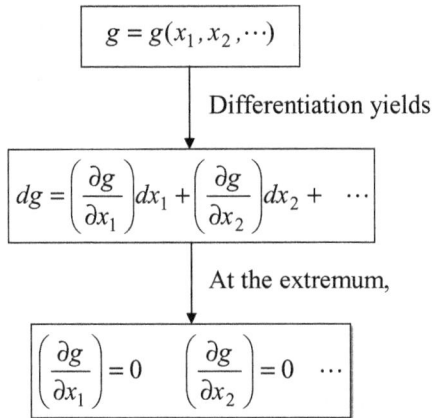

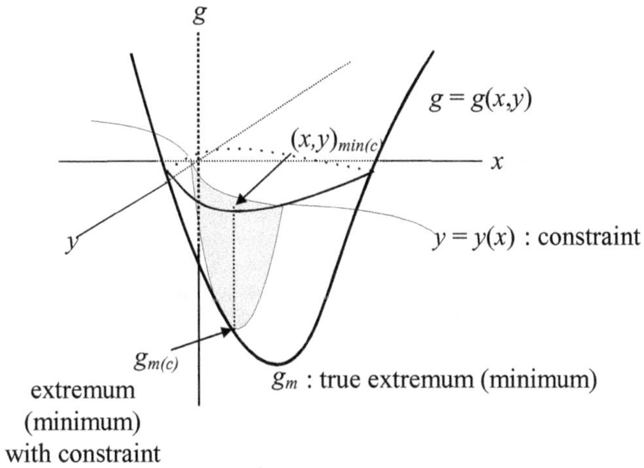


As an example, consider a function $g$ as given below:

$$g = x^2 + y^2 - 2x - 6y + 14$$

Differentiation yields

$$dg = 2xdx + 2ydy - 2dx - 6dy = (2x - 2)dx + (2y - 6)dy$$

<u>Finding extremum (minimum in this case)</u>

**Without a constraint**

$2x - 2 = 0$ and $2y - 6 = 0$,
Thus $x = 1$, and $y = 3$.

**With a constraint of $y = 2x$**

Differentiation yields

$$dy = 2dx$$

Then,

$$\begin{aligned} dg &= 2xdx + 2(2x)(2dx) - 2dx - 6(2dx) \\ &= (10x - 14)dx \end{aligned}$$

$$10x - 14 = 0$$

$$x = 1.4$$

and $y = 2.8$.

The above example clearly shows the difference between the absolute extremum and the constrained extremum

Now we apply the above extremum principle to minimization of Gibbs energy change of a reacting system. At equilibrium a system has the minimum value of the Gibbs energy. Suppose we have a system which consists of three phases, namely gas, liquid and solid solutions, with a number of different species (1, 2, 3, ...):

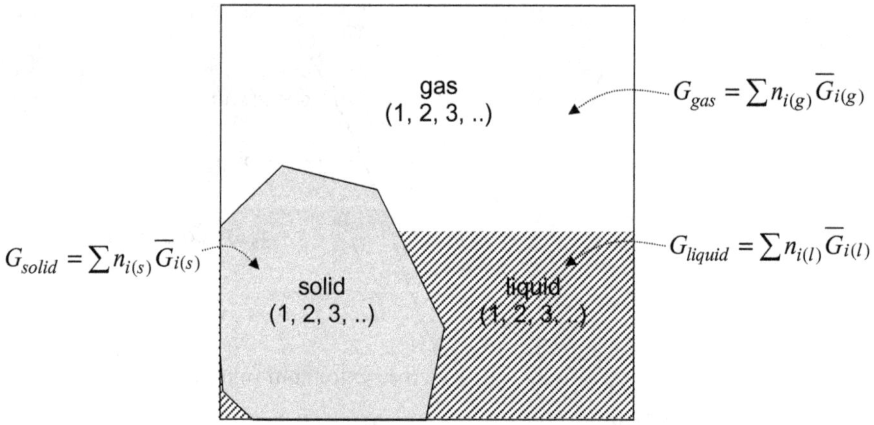


The total Gibbs free energy of the system ($G_{system}$)

$$G_{system} = G_{gas} + G_{liquid} + G_{solid}$$

↓

$$G_{system} = \sum n_{i(g)} \overline{G}_{i(g)} + \sum n_{i(l)} \overline{G}_{i(l)} + \sum n_{i(s)} \overline{G}_{i(s)}$$

↓ $\overline{G}_i = G_i^o + RT \ln a_i$

$$G_{system} = \left(\sum n_{i(g)} G^o_{i(g)} + \sum n_{i(l)} G^o_{i(l)} + \sum n_{i(s)} G^o_{i(s)}\right) + RT\left(\sum n_{i(g)} \ln a_{i(g)} + \sum n_{i(l)} \ln a_{i(l)} + \sum n_{i(s)} \ln a_{i(s)}\right)$$

At equilibrium, the total Gibbs free energy of the system, $G_{system}$, is at minimum: *i.e.*,

$$\left(\frac{\partial G_{system}}{\partial n_{i(k)}}\right)_{n_j} = 0$$

where $i$: component 1, 2, ...
$k$: phase $g$, $l$, $s$

The above equation determines the equilibrium state of the system under consideration, but is subject to two fundamental constraints:

(1) The atom balance relations must be satisfied (mass conservation).
(2) $n_i$ and $a_i$ are to have non-negative values.

Solving simultaneously the above set of equations under the two constraints, we can find, at least in principle, the distribution of each component between phases ($n_{i(k)}$) when the system arrives at equilibrium. The following diagram shows a two dimensional graphical interpretation of the equilibrium conditions discussed above:

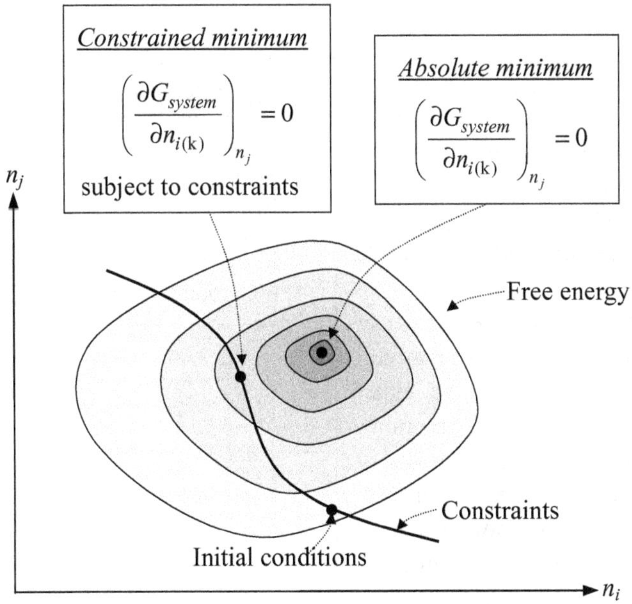

*Example 10.3*

In a reactor CO($g$) and $O_2$($g$) were introduced and allowed to react to form $CO_2$($g$) at constant temperature $T$ and 1 *atm*. pressure. At equilibrium it was found that all three gaseous species coexisted together. Find the equilibrium conditions of the system by utilizing the fact that the Gibbs energy of the system is minimum at equilibrium.

Initial: CO $O_2$ → Reactions → Equilibrium: CO $O_2$ $CO_2$

Total free energy at equilibrium, $G_{\text{total}}$,

$$G_{total} = n_{CO}\overline{G}_{CO} + n_{O_2}\overline{G}_{O_2} + n_{CO_2}\overline{G}_{CO_2}$$

↓ Differentiation

$$dG_{total} = (n_{CO}d\overline{G}_{CO} + n_{O_2}d\overline{G}_{O_2} + n_{CO_2}d\overline{G}_{CO_2}) + (\overline{G}_{CO}dn_{CO} + \overline{G}_{O_2}dn_{O_2} + \overline{G}_{CO_2}dn_{CO_2})$$

↓ Gibbs-Duhem equation

$$n_{CO}d\overline{G}_{CO} + n_{O_2}d\overline{G}_{O_2} + n_{CO_2}d\overline{G}_{CO_2} = 0$$

$$dG_{total} = \overline{G}_{CO}dn_{CO} + \overline{G}_{O_2}dn_{O_2} + \overline{G}_{CO_2}dn_{CO_2}$$

↓ At equilibrium

$$dG_{total} = 0 = \overline{G}_{CO}dn_{CO} + \overline{G}_{O_2}dn_{O_2} + \overline{G}_{CO_2}dn_{CO_2}$$

Since chemical reactions are allowed to occur in the system, the number of moles of the components are not conserved. Even if the process proceeds in a closed system which does not allow matter to cross the system boundary, the number of moles of each of the components may still vary: *i.e.*, a component may be either consumed or produced by reactions. However, the number of gram atoms of each of the elements in the system must be conserved as atoms cannot be created or destroyed. This fact imposes constraints to the above equation.

<u>*Atom balance*</u>

Oxygen : $m_O = n_{CO} + 2n_{CO_2} + 2n_{O_2}$

Carbon : $m_C = n_{CO} + n_{CO_2}$

↓

<u>*Atom conservation*</u>

$dm_O = 0 = dn_{CO} + 2dn_{CO_2} + 2dn_{O_2}$

$dm_C = 0 = dn_{CO} + dn_{CO_2}$

↓

$$dG_{total} = 0 = (\overline{G}_{CO_2} - \overline{G}_{CO} - \tfrac{1}{2}\overline{G}_{O_2})dn_{CO_2}$$

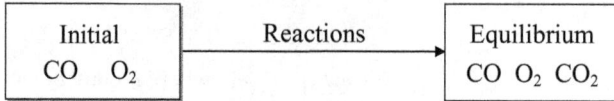

As $dn_{CO_2} \neq 0$

$$\overline{G}_{CO_2} - \overline{G}_{CO} - \tfrac{1}{2}\overline{G}_{O_2} = 0$$

$$\overline{G}_{CO_2} - (\overline{G}_{CO} + \tfrac{1}{2}\overline{G}_{O_2}) = 0$$

This equation is significant. Notice that the left-hand side of the equation is in fact the change in the Gibbs energy for the reaction

$$CO + \tfrac{1}{2}O_2 = CO_2$$

This Gibbs energy change is zero as shown in the above equation, and thus there is no driving force for the reaction to proceed in either direction. This means that the above reaction is in equilibrium.

Using the relationship

$$\overline{G}_i = G_i^{o} + RT\, ln\, P_i$$

$$G^{o}_{CO_2} - \left(G^{o}_{CO} + \tfrac{1}{2}G^{o}_{O_2}\right) + RT\, ln\left(\frac{P_{CO_2}}{P_{CO}P_{O_2}^{\frac{1}{2}}}\right) = 0$$

$$\frac{P_{CO_2}}{P_{CO}P_{O_2}^{\frac{1}{2}}} = \exp\left(-\frac{\Delta G^{o}}{RT}\right)$$

where

$$\Delta G^{o} = G^{o}_{CO_2} - (G^{o}_{CO} + \tfrac{1}{2}G^{o}_{O_2})$$

The above equation determines the relationship between the partial pressures of CO, $CO_2$ and $O_2$ at equilibrium. The analysis given above is the basis of the concept of *equilibrium constant* discussed in the prior section.

*Example 10.4*

Pure FeO(*s*) is reduced to Fe(*s*) by CO(*g*) in a reactor at constant temperature *T* and the total pressure of 1 atm. At equilibrium, Fe(*s*) and FeO(*s*) coexist with CO(*g*) and $CO_2$(*g*). Relate the equilibrium ratio of $CO_2$/CO to the standard Gibbs energies of formation of species existing in the system and to the temperature *T*.

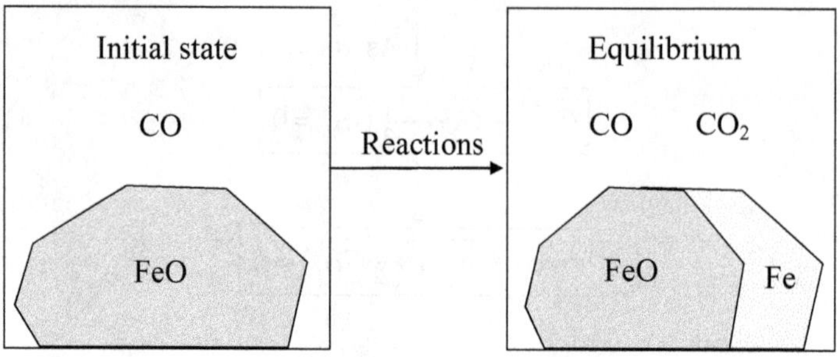


Total Gibbs energy of the system

$$G_{total} = n_{\text{Fe}}\overline{G}_{\text{Fe}} + n_{\text{FeO}}\overline{G}_{\text{FeO}} + n_{\text{CO}}\overline{G}_{\text{CO}} + n_{\text{CO}_2}\overline{G}_{\text{CO}_2}$$

↓ Differentiation

$$dG_{total} = n_{\text{Fe}}d\overline{G}_{\text{Fe}} + n_{\text{FeO}}d\overline{G}_{\text{FeO}} + n_{\text{CO}}d\overline{G}_{\text{CO}} + n_{\text{CO}_2}d\overline{G}_{\text{CO}_2} + \overline{G}_{\text{Fe}}dn_{\text{Fe}} + \overline{G}_{\text{FeO}}dn_{\text{FeO}} + \overline{G}_{\text{CO}}dn_{\text{CO}} + \overline{G}_{\text{CO}_2}dn_{\text{CO}_2}$$

Pure Fe: $\overline{G}_{\text{Fe}} = G^{\text{o}}_{\text{Fe}} = \text{constant}$

Pure FeO: $\overline{G}_{\text{FeO}} = G^{\text{o}}_{\text{FeO}} = \text{constant}$

Therefore, $d\overline{G}_{\text{Fe}} = d\overline{G}_{\text{FeO}} = 0$

Gibbs-Duhem equation for the gas phase

$$n_{\text{CO}}d\overline{G}_{\text{CO}} + n_{\text{CO}_2}d\overline{G}_{\text{CO}_2} = 0$$

↓

$$dG_{total} = \overline{G}_{\text{Fe}}dn_{\text{Fe}} + \overline{G}_{\text{FeO}}dn_{\text{FeO}} + \overline{G}_{\text{CO}}dn_{\text{CO}} + \overline{G}_{\text{CO}_2}dn_{\text{CO}_2}$$

*Atom balance*

$$m_{\text{Fe}} = n_{\text{Fe}} + n_{\text{FeO}}$$
$$m_{\text{C}} = n_{\text{CO}} + n_{\text{CO}_2}$$
$$m_{\text{O}} = n_{\text{CO}} + n_{\text{FeO}} + 2n_{\text{CO}_2}$$

*Atom conservation*

$$dm_{\text{Fe}} = 0 = dn_{\text{Fe}} + dn_{\text{FeO}}$$
$$dm_{\text{C}} = 0 = dn_{\text{CO}} + dn_{\text{CO}_2}$$
$$dm_{\text{O}} = 0 = dn_{\text{CO}} + dn_{\text{FeO}} + 2dn_{\text{CO}_2}$$

↓

$$dG_{total} = (\overline{G}_{\text{Fe}} - \overline{G}_{\text{FeO}} - \overline{G}_{\text{CO}} + \overline{G}_{\text{CO}_2})dn_{\text{CO}_2}$$

↓ At equilibrium

$$dG_{total} = 0 = (\overline{G}_{\text{Fe}} - \overline{G}_{\text{FeO}} - \overline{G}_{\text{CO}} + \overline{G}_{\text{CO}_2})dn_{\text{CO}_2}$$

↓

This equation shows the equilibrium criterion with the atom balance constraints of the reaction system in which Fe and FeO coexist with CO and $CO_2$ in the gas phase.

Since $dn_{CO_2} \neq 0$,

$$\overline{G}_{Fe} - \overline{G}_{FeO} - \overline{G}_{CO} + \overline{G}_{CO_2} = 0$$

Rearrangement

$$\overline{G}_{Fe} + \overline{G}_{CO_2} - (\overline{G}_{FeO} + \overline{G}_{CO}) = 0$$

This equation shows that the Gibbs energy change associated with the following reaction is zero:

FeO + CO = Fe + $CO_2$

$\overline{G}_i = G_i^o + RT \ln a_i$

$$(G_{Fe}^o + G_{CO_2}^o) - (G_{FeO}^o + G_{CO}^o) + RTln\left(\frac{a_{Fe}P_{CO_2}}{a_{FeO}P_{CO}}\right) = 0$$

Rearrangement

$$\frac{a_{Fe}P_{CO_2}}{a_{FeO}P_{CO}} = exp\left(-\frac{\Delta G_r^o}{RT}\right)$$

$\Delta G_r^o = \left(G_{Fe}^o + G_{CO_2}^o\right) - \left(G_{FeO}^o + G_{CO}^o\right)$

Pure Fe and FeO : $a_{Fe} = 1$, $a_{FeO} = 1$

$$\frac{P_{CO_2}}{P_{CO}} = exp\left(-\frac{\Delta G_r^o}{RT}\right)$$

*Example 10.5*

In a steel refining process, molten steel eventually comes into equilibrium with slag and gas phases coexisting in the furnace. The furnace can be considered as a closed system. The species identified in each phase are given in the following:

Metal phase : Fe, Si, Mn, C, O, N, S, Al
Slag phase : CaO, $SiO_2$, MnO, FeO, CaS, $Al_2O_3$
Gas phase : $N_2$, CO

Determine thermodynamic relationships between these species by using the *free energy minimization method*, *i.e.*, the fact that Gibbs energy is the minimum at equilibrium.

**Gas**: $N_2$, CO

$$G_{gas} = n_{N_2}\overline{G}_{N_2} + n_{CO}\overline{G}_{CO}$$

**Slag**: CaO, $SiO_2$, MnO, FeO, CaS, $Al_2O_3$

$$G_{slag} = n_{CaO}\overline{G}_{CaO} + n_{SiO_2}\overline{G}_{SiO_2} + n_{MnO}\overline{G}_{MnO} + n_{FeO}\overline{G}_{FeO} + n_{CaS}\overline{G}_{CaS} + n_{Al_2O_3}\overline{G}_{Al_2O_3}$$

**Metal**: Fe, Si, Mn, C, O, N, S, Al

$$G_{metal} = n_{Fe}\overline{G}_{Fe} + n_{Si}\overline{G}_{Si} + n_{Mn}\overline{G}_{Mn} + n_{C}\overline{G}_{C} + n_{O}\overline{G}_{O} + n_{N}\overline{G}_{N} + n_{S}\overline{G}_{S} + n_{Al}\overline{G}_{Al}$$

$$G_{total} = G_{gas} + G_{slag} + G_{metal}$$

↓ Differentiation

$$dG_{total} = dG_{gas} + dG_{slag} + dG_{metal}$$

↓ Substituting $G_{gas}$, $G_{slag}$, and $G_{metal}$ | Applying the Gibbs-Duhem equation,

$$dG_{total} = \overline{G}_{N_2}dn_{N_2} + \overline{G}_{CO}dn_{CO} + \overline{G}_{CaO}dn_{CaO} + \overline{G}_{SiO_2}dn_{SiO_2} + \overline{G}_{MnO}dn_{MnO} + \overline{G}_{FeO}dn_{FeO} + \overline{G}_{CaS}dn_{CaS} + \overline{G}_{Al_2O_3}dn_{Al_2O_3} + \overline{G}_{Fe}dn_{Fe} + \overline{G}_{Si}dn_{Si} + \overline{G}_{Mn}dn_{Mn} + \overline{G}_{C}dn_{C} + \overline{G}_{O}dn_{O} + \overline{G}_{N}dn_{N} + \overline{G}_{S}dn_{S} + \overline{G}_{Al}dn_{Al}$$

*Atom balances*

$$dm_N = 0 = 2dn_{N_2} + dn_N$$
$$dm_C = 0 = dn_{CO} + dn_C$$
$$dm_O = 0 = dn_{CO} + dn_{CaO} + 2dn_{SiO_2} + dn_{MnO} + dn_{FeO} + 3dn_{Al_2O_3} + dn_O$$
$$dm_S = 0 = dn_{CaS} + dn_S$$
$$dm_{Fe} = 0 = dn_{FeO} + dn_{Fe}$$
$$dm_{Ca} = 0 = dn_{CaO} + dn_{CaS}$$
$$dm_{Mn} = 0 = dn_{MnO} + dn_{Mn}$$
$$dm_{Si} = 0 = dn_{SiO_2} + dn_{Si}$$
$$dm_{Al} = 0 = 2dn_{Al_2O_3} + dn_{Al}$$

↓

$$dG_{total} = (\overline{G}_N - \tfrac{1}{2}\overline{G}_{N_2})dn_N + (\overline{G}_{CO} - \overline{G}_C - \overline{G}_O)\,dn_{CO} + (\overline{G}_{CaS} + \overline{G}_O - \overline{G}_{CaO} - \overline{G}_S)dn_{CaS} + (\overline{G}_{SiO_2} - \overline{G}_{Si} - \overline{2G_O})\,dn_{SiO_2} + (\overline{G}_{MnO} - \overline{G}_{Mn} - \overline{G}_O)\,dn_{MnO} + (\overline{G}_{FeO} - \overline{G}_{Fe} - \overline{G}_O)dn_{FeO} + (\overline{G}_{Al_2O_3} - 2\overline{G}_{Al} - 3\overline{G}_O)\,dn_{Al_2O_3}$$

↓ At equilibrium, $dG_{total} = 0$ and $dn_i$'s $\neq 0$

$$
\begin{aligned}
&0 = \overline{G}_{N} - \tfrac{1}{2}\overline{G}_{N_2}\\
&0 = \overline{G}_{CO} - (\overline{G}_{C} + \overline{G}_{O})\\
&0 = \overline{G}_{CaS} + \overline{G}_{O} - (\overline{G}_{CaO} + \overline{G}_{S})\\
&0 = \overline{G}_{MnO} - (\overline{G}_{Mn} + \overline{G}_{O})\\
&0 = \overline{G}_{SiO_2} - (\overline{G}_{Si} + 2\overline{G}_{O})\\
&0 = \overline{G}_{FeO} - (\overline{G}_{Fe} + \overline{G}_{O})\\
&0 = \overline{G}_{Al_2O_3} - (2\overline{G}_{Al} + 3\overline{G}_{O})
\end{aligned}
$$

Note that each of these reactions indicates that the Gibbs energy change of the corresponding reaction must be zero: *i.e.*, the reaction is in equilibrium.

$$
\begin{aligned}
&1/2\ N_2 = N\\
&C + O = CO\\
&CaO + S = CaS + O\\
&Mn + O = MnO\\
&Si + 2O = SiO_2\\
&Fe + O = FeO\\
&2Al + 3O = Al_2O_3
\end{aligned}
$$

The above analysis results in an important conclusion:
"If a system is in equilibrium, all subsystems in the system must also be in equilibrium."

Combining each of the above equations with the relationship of $\overline{G}_i = G_i^{o} + RT\ln a_i$

$$\frac{a_N}{P_{N_2}^{\frac{1}{2}}} = exp\left(-\frac{\Delta G_N^{o}}{RT}\right) \qquad \text{where} \quad \Delta G_N^{o} = G_N^{o} - \tfrac{1}{2}G_{N_2}^{o}$$

$$\frac{P_{CO}}{a_C a_O} = exp\left(-\frac{\Delta G_{CO}^{o}}{RT}\right) \qquad \text{where} \quad \Delta G_{CO}^{o} = G_{CO}^{o} - G_C^{o} - G_O^{o}$$

$$\frac{a_{CaS} a_O}{a_{CaO} a_S} = exp\left(-\frac{\Delta G_{CaS}^{o}}{RT}\right) \qquad \text{where} \quad \Delta G_{CaS}^{o} = G_{CaS}^{o} + G_O^{o} - G_{CaO}^{o} - G_S^{o}$$

$$\frac{a_{MnO}}{a_{Mn} a_O} = exp\left(-\frac{\Delta G_{MnO}^{o}}{RT}\right) \qquad \text{where} \quad \Delta G_{MnO}^{o} = G_{MnO}^{o} - G_{Mn}^{o} - G_O^{o}$$

$$\frac{a_{SiO_2}}{a_{Si} a_O^2} = exp\left(-\frac{\Delta G_{SiO_2}^{o}}{RT}\right) \qquad \text{where} \quad \Delta G_{SiO_2}^{o} = G_{SiO_2}^{o} - G_{Si}^{o} - 2G_O^{o}$$

$$\frac{a_{FeO}}{a_{Fe} a_O} = exp\left(-\frac{\Delta G_{FeO}^{o}}{RT}\right) \qquad \text{where} \quad \Delta G_{FeO}^{o} = G_{FeO}^{o} - G_{Fe}^{o} - G_O^{o}$$

$$\frac{a_{Al_2O_3}}{a_{Al}^2 a_O^3} = exp\left(-\frac{\Delta G_{Al_2O_3}^{o}}{RT}\right) \qquad \text{where} \quad \Delta G_{Al_2O_3}^{o} = G_{Al_2O_3}^{o} - 2G_{Al}^{o} - 3G_O^{o}$$

- If the system is at equilibrium, all of the above equations must be satisfied simultaneously.
- Each of the equations is the condition for equilibrium of the corresponding reaction which occurs in the system.
- Notice that the number of equations, *i.e.*, the number of independent reactions ($r$) is related with the number of components ($c$) and the number of elements ($e$) by the equation

$$r = c - e$$

For instance, for the system under discussion,
$c$ = 16 ($N_2$, CO, CaO, $SiO_2$, MnO, FeO, CaS, $Al_2O_3$, Fe, Si, Mn, C, O, N, S, Al)
$e$ = 9 (N, C, O, Ca, Si, Mn, Fe, S, Al)
$r$ = 16 - 9 = 7

*Exercises*

10.3 A system consisting of ZnO($s$), C($s$), CO($g$), Zn($g$), $CO_2$($g$) and $O_2$($g$) is at equilibrium at 1200 K and 1 atm. Using the free energy minimisation method, calculate the equilibrium partial pressures for gaseous species in the system. The following data are given:

$$\text{Zn}(g)+\tfrac{1}{2}\text{O}_2(g)=\text{ZnO}(s) \quad \Delta G^o = -460{,}240+198.32T, \text{J mol}^{-1}$$
$$\text{C}(s)+\tfrac{1}{2}\text{O}_2(g)=\text{CO}(g) \quad \Delta G^o = -112{,}880-86.51T, \text{J mol}^{-1}$$
$$\text{C}(s)+\text{O}_2(g)=\text{CO}_2(g) \quad \Delta G^o = -394{,}760-0.836T, \text{J mol}^{-1}$$

10.4 Excess $K_2CO_3$ and C are heated together in an initially evacuated vessel to a temperature of 1400 K and allowed to reach equilibrium. No liquid or solid phases are formed. The gas phase contains K, CO and $CO_2$. Calculate the partial pressures of these species at 1440 K. Use the free energy minimization method. The following data are given:

| Species | K | CO | $CO_2$ | $K_2CO_3$ |
|---|---|---|---|---|
| $\Delta G_f^o$ (J mol$^{-1}$) | 0 | -235,100 | -396,300 | -708,700 |

10.5 A reducing gas mixture consisting of 24 mol% CO, 4 mol% $CO_2$, 60 mol% $H_2$ and 12 mol% $H_2O$ is passed through a packed bed of wustite (FeO) pellets held at 1100 K. The pressure is maintained constant at 2.8 atm. Assuming that the gas phase and the solids are in equilibrium, calculate the composition of the exit gas phase using the free energy minimisation method. The following data are given:

$$\text{C}(s)+\tfrac{1}{2}\text{O}_2(g)=\text{CO}(g) \quad \Delta G^o = -112{,}880-86.51T, \text{J mol}^{-1}$$
$$\text{C}(s)+\text{O}_2(g)=\text{CO}_2(g) \quad \Delta G^o = -394{,}760-0.836T, \text{J mol}^{-1}$$
$$\text{Fe}(s)+\tfrac{1}{2}\text{O}_2(g)=\text{FeO}(s) \quad \Delta G^o = -264{,}000+64.59T, \text{J mol}^{-1}$$
$$\text{H}_2(g)+\tfrac{1}{2}\text{O}_2(g)=\text{H}_2\text{O}(g) \quad \Delta G^o = -247{,}39+55.85T, \text{J mol}^{-1}$$

## 10.3 Effect of Temperature on Equilibrium Constant

How will the position of equilibrium change when the temperature is altered? We can answer this question by using the thermodynamic relations we have developed so far.

Recall that,

$$\left[\frac{\partial\left(\dfrac{G}{T}\right)}{\partial\left(\dfrac{1}{T}\right)}\right]_P = H$$

↓

$$\left[\frac{\partial\left(\dfrac{\Delta G_r^{\circ}}{T}\right)}{\partial\left(\dfrac{1}{T}\right)}\right]_P = \Delta H_r^{\circ}$$

↓ $K = exp\left(-\dfrac{\Delta G_r^{\circ}}{RT}\right)$ or $\Delta G_r^{\circ} = -RT\,ln\,K$

$$\left(\frac{\partial\, ln\, K}{\partial(1/T)}\right)_P = -\frac{\Delta H_r^{\circ}}{R}$$

↓

$$\left(\frac{\partial\, ln\, K}{\partial T}\right)_P = \frac{\Delta H_r^{\circ}}{RT^2}$$

The last two equations are basically the same and known as ***the van't Hoff equation***, and it expresses the temperature dependence of the equilibrium constant in terms of the heat (standard enthalpy change) of reaction. This equation tells us that

- If the reaction is endothermic, *i.e.*, $\Delta H_r^{\circ} > 0$, $K$ increases with increasing $T$,
- If the reaction is exothermic, *i.e.*, $\Delta H_r^{\circ} < 0$, $K$ decreases with increasing $T$.

If $\Delta H_r^{\circ}$ is independent of $T$ (*), the integration of the van't Hoff equation yields

$$ln\left(\frac{K_2}{K_1}\right) = \frac{\Delta H_r^{\circ}}{R}\left(\frac{1}{T_1} - \frac{1}{T_2}\right)$$

This equation enables us to determine the equilibrium constant

---

* This is generally the case when the range of temperatures involved is not appreciable, and in the absence of any phase changes in the participating species.

From the van't Hoff equation the temperature dependence of the equilibrium constant may be examined by plotting $ln\,K$ versus $1/T$. The following figure is an example of the plot:

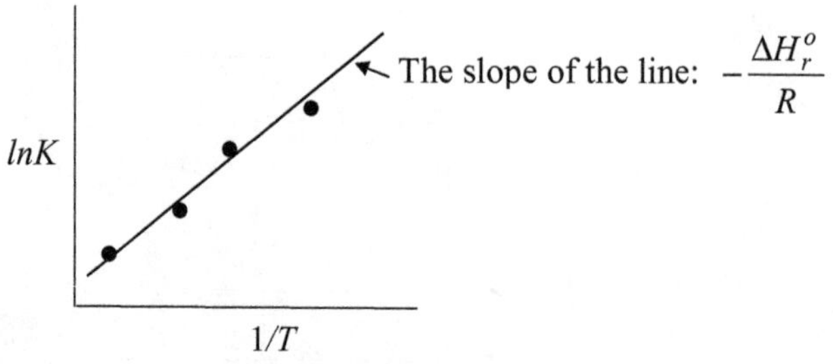


From the value of the slope the heat of reaction can be estimated as indicated in the figure. The above example shows a positive slope, and thus the reaction is exothermic, *i.e.*, $\Delta H_r^\circ < 0$.

*Exercises*

10.6 The equilibrium constant for the ammonia synthesis reaction

$$\tfrac{1}{2}N_2(g)+\tfrac{3}{2}H_2(g)=NH_3(g)$$

is 775 at 25°C based on 1 atm. ideal gas standard states. The enthalpy change associated with the reaction, or simply the heat of reaction, $\Delta H_r^\circ$, is -45.9 kJ. Assuming that $\Delta H_r^\circ$ is independent of temperature, estimate the equilibrium constant for the reaction at 45°C.

10.7 Equilibrium of the system containing MnO(*s*), $Mn_3O_4$(*s*) and $O_2$(*g*) was examined. Both MnO and $Mn_3O_4$ were pure and stable states. It was found that equilibrium partial pressures of oxygen were

$$P_{O_2} = 1.4\times10^{-6} \text{ atm. at } 1{,}000°\text{C}$$

$$P_{O_2} = 2.8\times10^{-5} \text{ atm. at } 1{,}100°\text{C}$$

Calculate the heat of reaction $\Delta H_r^\circ$ of the reaction. Assume that $\Delta H_r^\circ$ is independent of *T*.

$$3MnO(s)+\tfrac{1}{2}O_2(g)=Mn_3O_4(s)$$

## 10.4 Effect of Pressure on Equilibrium Constant

The equilibrium constant depends on the value of the standard Gibbs energy change of a reaction:

$$K = exp\left(-\frac{\Delta G_r^o}{RT}\right)$$

Recall that in the equation $\Delta G_r^o$ is the Gibbs energy change when all the reactants and products are at their respective standard states in which $P_{total}$ = 1 atm. Therefore $K$ is independent of pressure. This conclusion does not necessarily mean that the equilibrium composition is independent of pressure. Consider the reaction

$$X_2(g) = 2X(g) \qquad K = \frac{P_X^2}{P_{X_2}}$$

$K$ is independent of pressure. This does not mean that the individual terms in the above expression for $K$ are independent of the total pressure. What is independent of the total pressure of the system is the ratio of $P_X^2 / P_{X_2}$, *i.e.*, the equilibrium constant $K$, but not individual partial pressures. The individual partial pressures at equilibrium change as the total pressure of the system changes. Recall that $P_i = N_i P_{total}$. Substitution of this equation into the equation for $K$ yields

$$K = \frac{P_X^2}{P_{X_2}} = \left(\frac{N_X^2}{N_{X_2}}\right) P_{total}$$

If the total pressure, $P_{total}$, changes, values of the individual mole fractions should change in such a way that the ratio cancels the change of $P_{total}$.

Let $K_C = \left(\dfrac{N_X^2}{N_{X_2}}\right)$

$$K = K_C P_{total}$$

Independent of pressure ($K$) Dependent on pressure ($P_{total}$)

For a general reaction in which all species are gaseous:

$$aA(g) + bB(g) = mM(g) + nN(g)$$

$$K = \frac{P_M^m P_N^n}{P_A^a P_B^b} = \left(\frac{N_M^m N_N^n}{N_A^a N_B^b}\right) P^{(m+n-a-b)}$$

$$\text{Let } K_C = \frac{N_M^m N_N^n}{N_A^a N_B^b}$$

$$K = K_C P^{(m+n-a-b)}$$

$K_C$ is the equilibrium constant expressed in concentrations, and note that it can be independent of the total pressure only if $m+n-a-b = 0$ *i.e.*, there is no net change in the number of moles by the reaction.

*Example 10.6*

Consider the reaction

$$A(g) = 2B(g) \qquad K = \frac{P_B^2}{P_A}$$

1) If the total pressure in the reactor is increased by injecting an inert gas into the reactor, would the equilibrium constant $K$ change?
2) If the total pressure in the reactor is increased by compression, would the equilibrium constant $K$ change? Would the partial pressures of the individual species change?

1) Consider the definition of partial pressure: Partial pressures of perfect gases are the pressures that each species would exert if it were alone in the system. Therefore the presence of another gas has no effect on the equilibrium constant and on the equilibrium molar concentrations (*e.g.*, $mol\ cm^{-3}$) of species so long as the gases are perfect.

2) Recall that $K = exp\left(-\frac{\Delta G_r^o}{RT}\right)$.

$\Delta G_r^o$ is independent of pressure and hence $K$ is independent of the total pressure.

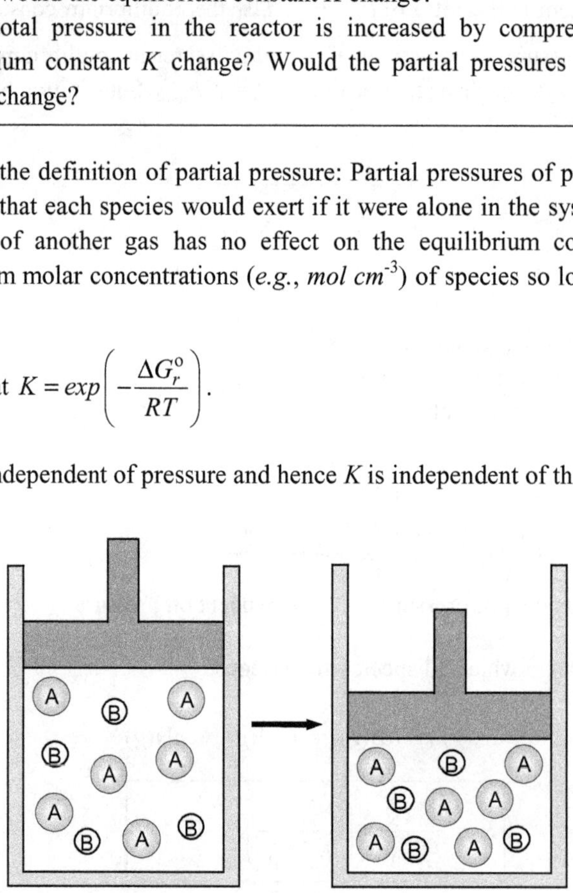

As $K$ is independent of pressure, $P_B^2 / P_A$ (= $K$) should also be independent of the total pressure. However, the compression, or pressure change in general, adjusts the individual partial pressures of the species in such a way that, although the partial pressure of each species changes, their ratio appearing in the equilibrium constant expression remains unchanged.

*Exercises*

10.8 The standard Gibbs energy change of the dissociation reaction

$$N_2O_4(g) = 2NO_2(g)$$

is 4,770 J at 25°C. The system initially contains $N_2O_4$ only.

1) Calculate the equilibrium partial pressure of $NO_2$ at 1 atm. total pressure.
2) Calculate the equilibrium partial pressure of $NO_2$ at 10 atm. total pressure.

10.9 The equilibrium constant $K$ for the reaction

$$2SO_2(g) + O_2(g) = 2SO_3(g)$$

is 110.7 at 600°C. Assuming that the gases behave ideally, calculate the equilibrium constant $K_c$ when the concentrations are expressed in mol per liter. (Gas constant $R = 0.082$ liter atm. $mol^{-1}K^{-1}$)

## 10.5 Le Chatelier's Principle

Consider the general reaction which is in equilibrium:

$$aA + bB = mM + nN$$

The reaction is then subjected to a change in conditions that affect the reaction equilibrium. This perturbation will cause the reaction to proceed toward a new equilibrium. But in which direction? Toward right or left or unaffected?

***Le Chatelier's principle*** provides a convenient way of predicting the direction in which the reaction proceeds toward a new equilibrium state.

| Le Chatelier's Principle |
|---|
| Perturbation of a system at equilibrium will cause the equilibrium position to change in such a way as to tend to remove the perturbation. |

(*Examples*)

- If a reaction is exothermic, the reaction will be promoted by lowering the temperature.
- If a reaction results in a change in volume, then increase in pressure will cause the reaction to proceed in the direction which results in decrease in volume.

Le Chatelier's principle provides a good guide to the effects of pressure, temperature and concentration. For a quantitative analysis, however, more rigorous treatments are required as seen in the previous sections.

*Example 10.7*

Consider the reaction which is in equilibrium

$$C(s) + CO_2(g) = 2CO(g)$$

The gases are assumed to behave ideally.

(1) If the equilibrium is disturbed by adding some additional $CO(g)$ into the reactor, in which direction will the reaction proceed?
(2) If some additional solid carbon is added in the reactor, what would happen?
(3) If the total pressure in the reactor is increased by compression, in which direction will the reaction proceed?
(4) The reaction is endothermic as written. In which direction will an increase in the temperature shift the reaction?

(1) To the left.
(Because the concentration of CO is increased, the reaction proceeds in the direction which results in the consumption of $CO(g)$.)
(2) Unaffected
(The concentration of a solid is independent of the amount of the solid present so that there is no shift in the reaction equilibrium.)
(3) To the left
(An increase in the total pressure will shift the reaction equilibrium towards the side with the smaller number of moles of gas.)
(4) To the right
(An increase in temperature favours the absorption of heat (endothermic) and thus shifts the reaction equilibrium to the right.)

*Exercises*

10.10 Consider the reaction at equilibrium:

$$A(s) + 2B(g) = M(s) + N(g) \quad \Delta H_r^\circ < 0$$

Determine the direction of the reaction for each of the following changes of the thermodynamic conditions:

(1) Increase in temperature
(2) Decrease in pressure
(3) Increase in the concentration of B
(4) Increase in the concentration of N.

## 10.6 Alternative Standard States

**Recapitulation**

$$a_i = \frac{f_i}{f_i^o} \cong \frac{P_i}{P_i^o}$$

- The choice of the standard state is arbitrary, and the activity is always unity at the standard state chosen.
- The activity of a component in a solution is essentially a relative quantity.
- From the definition of activity it follows that the numerical value of the activity of a particular component is dependent on the choice of the standard state.
- There is no fundamental reason for preferring one standard state over another.

Up to now, we have chosen *the pure state as the standard state*. That is, a pure component in its stable state of existence at the specified temperature and 1atm. pressure is chosen as the standard state. This particular choice is also known as the ***Raoultian standard state***.

The Raoultian standard state is quite satisfactory in dealing with many solution systems. But there are some inconvenience and limitations associated with this standard state:

- If the pure component exists in a physical state which is different from that of the solution at the temperature of interest (*e.g.*, pure oxygen is a gas, but it is liquid when dissolved in water.), how can the pressure terms in the definition of activity be determined?
- With the Raoultian standard state, it is found not infrequently that the activity of a solute in a dilute solution is very small.

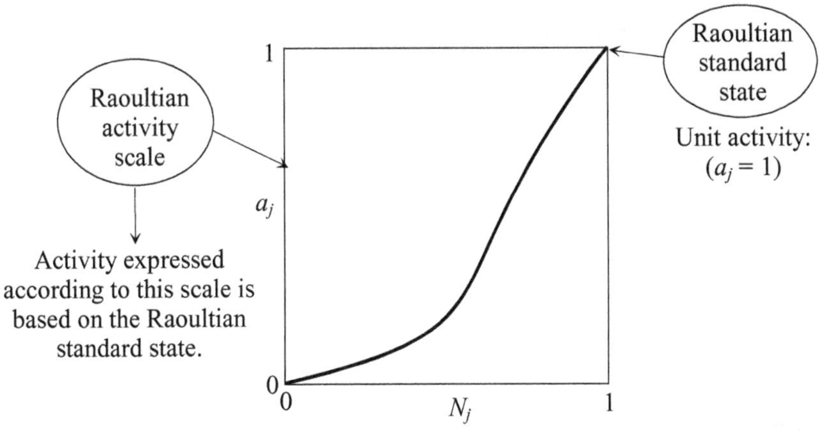

To resolve these problems, we now define a new standard state called the ***Henrian standard state*** which originates from Henry's law.

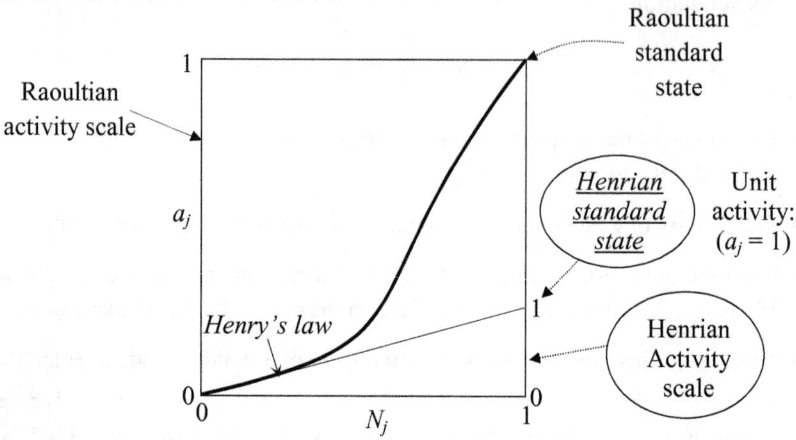


The Henrian standard state where the activity is unity for component $j$ is a hypothetical, non-physical state.

We now have two different activity scales, which means that the activity of a component in a solution can be expressed in either one of the two scales. In other words, there may be two different numerical values of the activity for the same component in the same solution. Analogy may be found in the choice of temperature scales: Here is a picture of the temperature conversion index for Celsius and Fahrenheit. For the same temperature we may get a different numerical value according to the choice of the scale. The choice of the scale is basically according to the matter of convenience. The equation of [°F] = 9/5[°C] + 32 relates the two temperature scales.

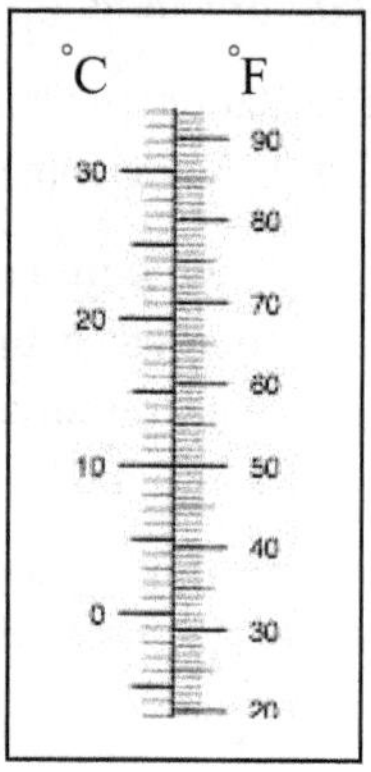


Similarly, the choice of the standard state for thermodynamic analysis is basically according to the matter of convenience.

Suppose that we are interested in the composition marked x in the following figure:

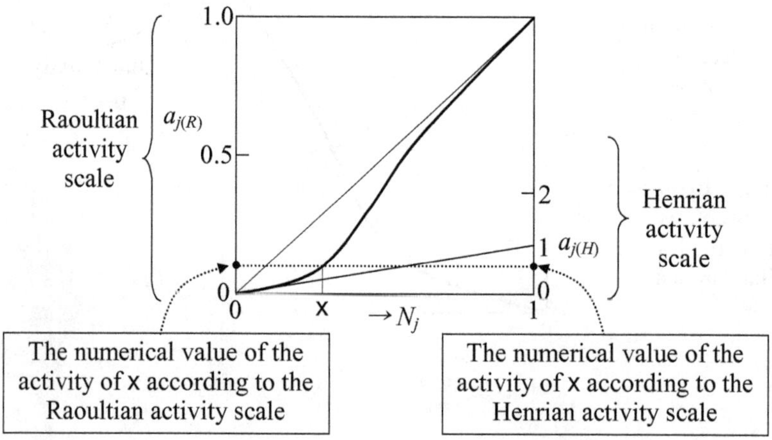

Note that the numerical value of the activity for the composition x depends on the choice of the standard state.

> <u>*Two important points*</u>
>
> 1. If $j$ obeys Henry's law, the value of the activity on the Henrian scale is numerically equal to the mole fraction of $j$.
> 2. The numerical value of the activity of $j$ at the Henrian *standard state* is unity(1) on the Henrian activity scale, but $\gamma_j^{\circ}$ on the Raoultian activity scale.

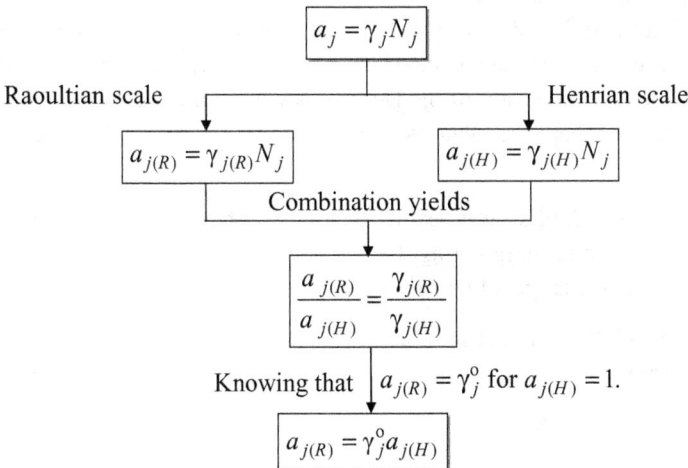

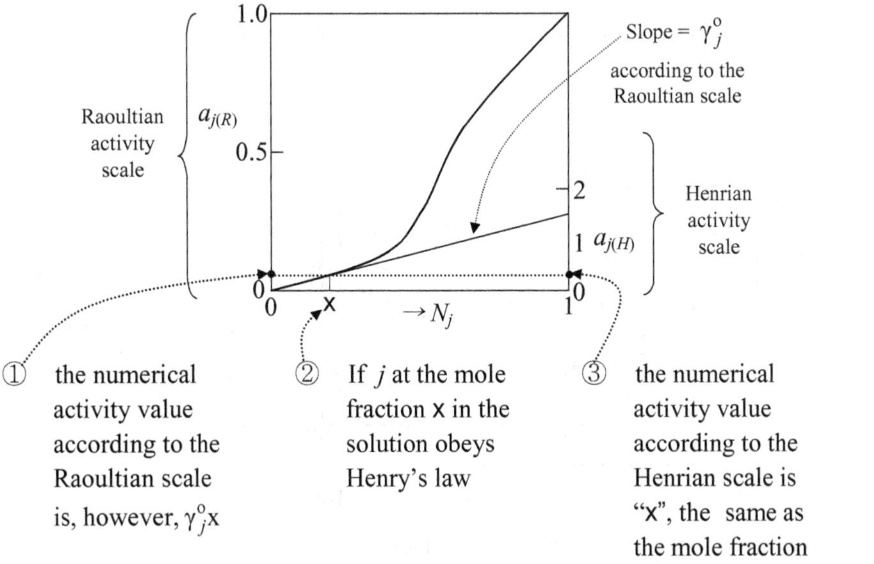

1.0
Slope = $\gamma_j^{\circ}$
according to the Raoultian scale
Raoultian activity scale
$a_{j(R)}$
0.5
2
Henrian activity scale
1 $a_{j(H)}$
0
0
x
$\rightarrow N_j$
1
0
① the numerical activity value according to the Raoultian scale is, however, $\gamma_j^{\circ}$x
② If $j$ at the mole fraction x in the solution obeys Henry's law
③ the numerical activity value according to the Henrian scale is "x", the same as the mole fraction

Based on the principles for the decision on the activity scales (Raoultian and Henrian) we may develop a relationship which relates the activities according to the different standard states.

$$a_j = \gamma_j N_j$$

Raoultian scale: $a_{j(R)} = \gamma_{j(R)} N_j$ Henrian scale: $a_{j(H)} = \gamma_{j(H)} N_j$

Combination yields

$$\frac{a_{j(R)}}{a_{j(H)}} = \frac{\gamma_{j(R)}}{\gamma_{j(H)}}$$

Knowing that $a_{j(R)} = \gamma_j^{\circ}$ for $a_{j(H)} = 1$.

$$a_{j(R)} = \gamma_j^{\circ} a_{j(H)}$$

This equation relates the activity on the Henrian scale to the activity on the Raoultian scale.

The Henrian standard state is sometimes called the ***infinitely dilute solution standard state*** because it is based on the Henry's law and is mostly used for dilute solutions.

The concentration of solutions is frequently given with the unit of weight (mass) percent, *wt%*., in particular for industrial applications. It will thus be of use to develop a new standard state which is based on the wt% expression of concentrations of a solution. We now define a standard state called the ***1 wt% standard state***.

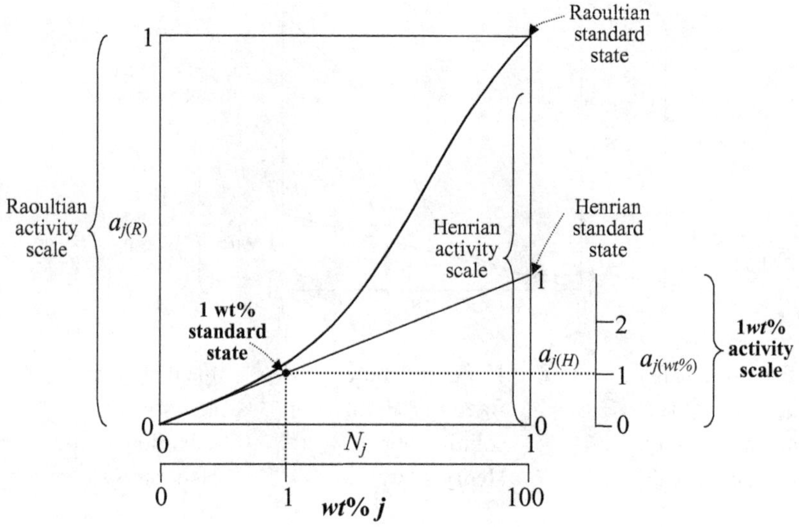


The position of 1 wt% in the scale is a bit exaggerated for better visualization.

Note that the Henrian standard state and 1*wt%* standard state are both based on Henry's law, but these standard states differ from each other in view of the physical state. Remembering that the standard state is the state at which the activity of the component of interest is unity regardless of the choice of the standard state, the Henrian standard state is the state at which the extension of the Henry's law line meets at $N_j = 1$ whereas the 1 wt% standard state is where the Henry's law line meets at $wt\%\ j = 1$ (Refer to the last figure.)

Note that the 1*wt%* standard state is real if the solution obeys Henry's law up to 1 *wt%*, in other words, if the point representing the 1*wt%* standard state lies in the region where the activity curve and the tangent line coincide. Otherwise it is hypothetical.

Note also that in dilute solutions in which $j$ obeys Henry's law the value of $a_{j(wt\%)}$ is numerically the same as ($wt\%\ j$).

**Relationship between mole fraction ($N_j$) and weight percent ($wt\%\,j$) of $i$-$j$ binary solution**

Base: 100g of solution

$$N_j = \frac{n_j}{n_i + n_j}$$

$$n_i = \frac{wt\%i}{M_i} \qquad n_j = \frac{wt\%j}{M_j}$$

↓

$$N_j = \frac{\dfrac{wt\%j}{M_j}}{\dfrac{wt\%i}{M_i} + \dfrac{wt\%j}{M_j}}$$

↓ For dilute solution of $j$ ($wt\%\,j \to 0$)

$$N_j = \frac{\dfrac{wt\%j}{M_j}}{\dfrac{wt\%i}{M_i} + \dfrac{wt\%j}{M_j}}$$

$wt\%i = (100 - wt\%\,j)$, with $wt\%\,j \approx 0$; $\dfrac{wt\%j}{M_j} \to 0$

↓

$$N_j = \frac{(wt\%j)M_i}{100M_j}$$

where $M_i$ and $M_j$ are molecular weights of $i$ and $j$, respectively, and $n_i$ and $n_j$ are the number of moles of $i$ and $j$, respectively.

We may develop a relationship between the activity by the Henrian standard state and the activity by the $1wt\%$ standard state when the component of interest is sufficiently dilute. Referring to the activity scales of the Henrian standard state and the 1wt% standard state, we can get the following relationship:

$$\frac{a_{j(H)}}{a_{j(wt\%)}} = \frac{N_j}{wt\%j}$$

↓

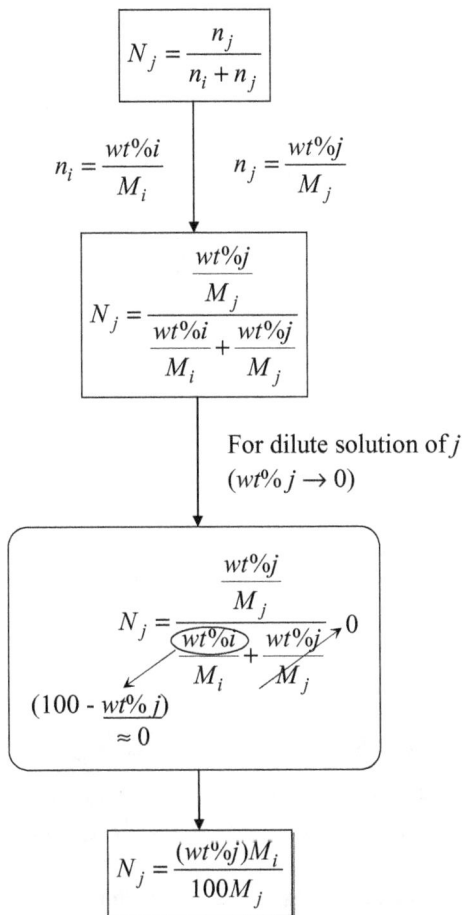

$$N_j = \frac{(wt\%j)M_i}{100M_j} \quad \text{when } j \text{ is dilute.}$$

$$a_{j(wt\%)} = \left(\frac{100M_j}{M_i}\right) a_{j(H)}$$

The following diagram is the summary of the relationships between activities on the different standard states:

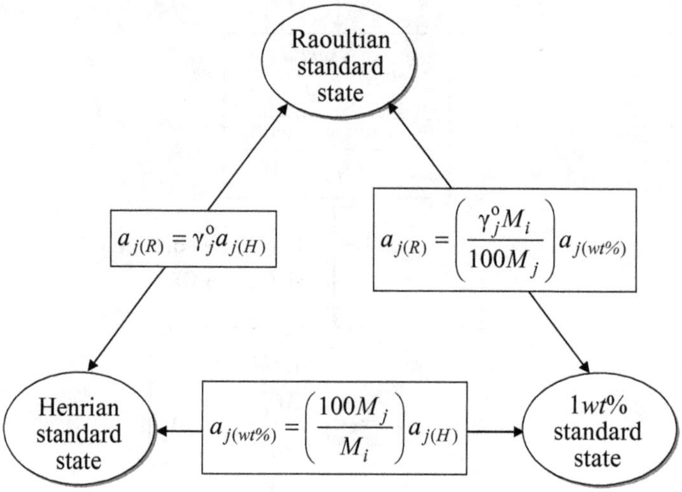


As we can see in the above diagram, the Henrian activity coefficient, $\gamma_j^{o}$, plays a pivotal role in converting the activity from one standard state to another. This is understandable because both new standard states are based on the Henry's law.

Recall that the partial molar Gibbs energy or chemical potential of a component in a solution is independent of the standard state chosen. In other words, it is an absolute thermodynamic property of the component in the solution.

Recall the following equation:,

$$\overline{G}_j = G_j^{o} + RT\, ln\, a_j$$

This equation is applicable to any standard state. The second term in the right side of the equation represents the deviation from the standard state.

- $\overline{G}_j$ is independent of the standard state chosen.
- Therefore the sum $(G_j^{o} + RT\, ln\, a_j)$ must also be independent of the standard state chosen.

The figure below shows the relationships when the above equation is applied to different standard states:

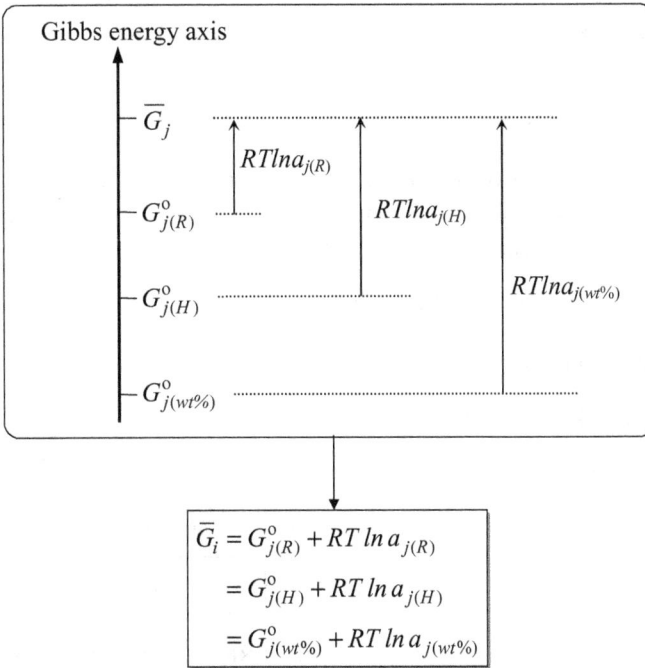

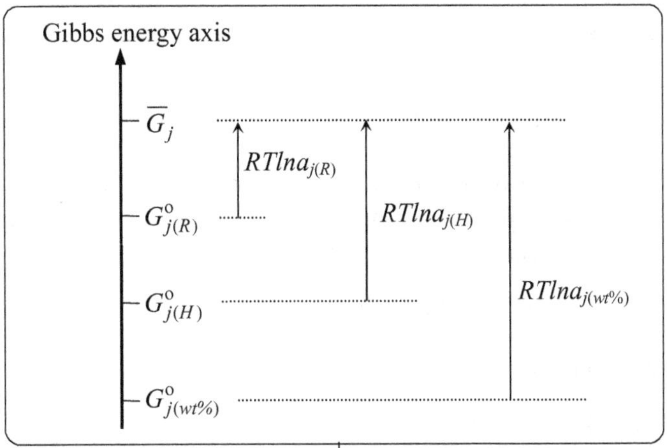


$$\overline{G}_i = G^{o}_{j(R)} + RT\,ln\,a_{j(R)} = G^{o}_{j(H)} + RT\,ln\,a_{j(H)} = G^{o}_{j(wt\%)} + RT\,ln\,a_{j(wt\%)}$$

In thermodynamic considerations of A-B binary solution, if the standard state of B is changed from the Raoultian to the Henrian standard state, the standard molar Gibbs energy should also change accordingly.

$$B_{(\text{Raoultian standard state})} \rightarrow B_{(\text{Henrian standard state})}$$

$$\Delta G^{o}_{B(R\rightarrow H)} = G^{o}_{B(H)} - G^{o}_{B(R)}$$

$$\overline{G}_B = G^{o}_{B(R)} + RT\,ln\,a_{B(R)} = G^{o}_{B(H)} + RT\,ln\,a_{B(H)}$$

$$a_{j(R)} = \gamma^{o}_{j} a_{j(H)}$$

$$\Delta G^{o}_{B(R\rightarrow H)} = RT\,ln\,\gamma^{o}_{B}$$

In a similar way we may find the Gibbs energy change associated with the change in the standard state from the Raoultian to the $1wt\%$ standard state.

$$\Delta G_{B(R\rightarrow wt\%)} = RT\,ln\left(\frac{\gamma^{o}_{B} M_A}{100 M_B}\right)$$

Consider the heterogeneous reaction

$$a\text{A}(l) + b\text{B}(g) = m\text{M}(s)$$

The standard Gibbs energy change at temperature $T$ for the reaction is $\Delta G_r^o$ when Raoultian standard states, i.e., pure liquid A, gaseous B and pure solid M at 1 atm. are used.

$$a\text{A}\,(l, \text{Raoultian}) + b\text{B}(g, \text{Raoultian}) = m\text{M}(s, \text{Raoultian}) \qquad \Delta G_{r,1}^o$$

It is sometimes more convenient to use alternative standard states for the species involved in the reaction. When the standard state of liquid A is changed from the Raoultian to Henrian standard state, the Gibbs energy change of the reaction

$$a\text{A}\,(l, \text{Henrian}) + b\text{B}(g, \text{Raoultian}) = m\text{M}(s, \text{Raoultian}) \qquad \Delta G_{r,2}^o$$

can be obtained as follows :

(1) $a\text{A}\,(l, R) + b\text{B}(g, R) = m\text{M}(s, R)$ $\quad \Delta G_{r,1}^o$

(2) $\text{A}\,(l, R) = \text{A}\,(l, H)$ $\quad \Delta G^o_{\text{A}(R\to H)} = RT\,ln\,\gamma_\text{A}^o$

(3) $a\text{A}\,(l, R) = a\text{A}\,(l, H)$ $\quad a\Delta G^o_{\text{A}(R\to H)} = aRT\,ln\,\gamma_\text{A}^o$

(1) - (3) $a\text{A}\,(l, H) + b\text{B}(g, R) = m\text{M}(s, R)$ $\quad \boxed{\Delta G_{r,2}^o = \Delta G_{r,1}^o - aRT\,ln\,\gamma_\text{A}^o}$

Similarly, if the standard state of A is changed from Raoultian to 1*wt*% standard state,

$$a\text{A}\,(l, 1wt\%) + b\text{B}(g, \text{Raoultian}) = m\text{M}(s, \text{Raoultian}) \qquad \Delta G_{r,3}^o$$

$$\boxed{\Delta G_{r,3}^o = \Delta G_{r,1}^o - aRTln\left(\frac{\gamma_\text{A}^o M_\text{X}}{100 M_\text{A}}\right)}$$

where $M_\text{A}$ and $M_\text{X}$ are molecular weights of A and solvent X in the A-X binary solution

Recall that

$\boxed{a_j = \gamma_j N_j}$ This expression is valid for all standard states.

| Raoultian standard state | Henrian standard state | 1*wt*% standard state |
|---|---|---|
| $a_{j(R)} = \gamma_{j(R)} N_j$ | $a_{j(H)} = \gamma_{j(H)} N_j$ | $a_{j(wt\%)} = \gamma_{j(wt\%)}(wt\%j)$ |

We often use symbol $f$ to denote the activity coefficient for 1*wt*% standard state.

$\boxed{a_{j(wt\%)} = f_j(wt\%j)}$

As both the Henrian and 1wt% standard states are based on the Henry's law, the following relationship holds

$$\boxed{f_j = \gamma_{j(H)}}$$

*Exercises*

10.11 The activity of silicon in a binary Fe-Si liquid alloy containing $N_{si}$ = 0.02 is 0.000022 at 1,600°C relative to the Raoultian standard state. The Henrian activity coefficient $\gamma_{Si}^{o}$ is experimentally determined to be 0.0011.

1) Calculate the activity of silicon in the same alloy, but relative to the Henrian standard state.
2) Calculate the change of the standard molar Gibbs energy of silicon for the change of the standard state from Raoultian to Henrian.
3) Calculate the activity of silicon relative to the 1wt% standard state. The molecular weights of Fe and Si are 55.85 and 28.09, respectively.

10.12 A liquid Fe-Al alloy is in equilibrium at 1,600°C with a $Al_2O_3$-saturated slag and a gas phase containing oxygen. The partial pressure of oxygen in the gas phase is maintained at $5\times10^{-14}$ atm. Calculate the activity of aluminium in the alloy in 1wt% standard state. The following data are given:

$$2\text{Al}(l) + \tfrac{3}{2}\text{O}_2(g) = \text{Al}_2\text{O}_3(s) \qquad \Delta G_r^{o} = -1{,}077{,}500 \text{ J}$$

for all the species at Raoultian standard states.

$$\gamma_{Al}^{o} = 0.029 \qquad M_{Fe} = 55.85 \qquad M_{Al} = 26.98$$

10.13 The residual oxygen present in a copper melt can be removed by equilibrating the melt with a $H_2O$-$H_2$ gas mixture. It is desired to keep the oxygen concentration in the copper melt lower than 0.001 wt% at 1,200°C. Calculate the ratio of $H_2O$ to $H_2$ in the gas mixture which is in equilibrium with 0.001 *wt*% oxygen in the melt. The following data are given:

$$\text{H}_2(g) + \tfrac{1}{2}\text{O}_2(g) = \text{H}_2\text{O}(g) \qquad \Delta G_1^{o} = -247{,}400 + 55.85T, \text{ J}$$

$$\tfrac{1}{2}\text{O}_2(g) = \text{O(wt\%)} \qquad \Delta G_2^{o} = -85{,}350 + 18.54T, \text{ J}$$

$$\log f_O = -0.158\,[\text{wt\%O}]$$

## 10.7 Interaction Coefficients

Much of the discussion thus far has been concerned with binary solutions. Most practical systems, however, are more complex and consist of more than two components. It is now in order to examine thermodynamics of solutions which contain several dilute solutes.

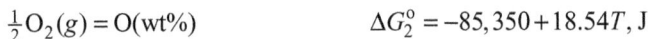

The activity of B in a dilute solution with respect to the Henrian standard state is given by

$$a_{B(H)} = f_B N_B$$

where $f_B$ is the activity coefficient of B on the Henrian activity scale.
First, suppose that we have the A-B binary solution, and let $f_B^B$ denote the activity coefficient of B in the A-B binary solution in which B is the only solute. Then

$$f_B = f_B^B$$

Next, consider the A-B-C ternary solution. Holding the concentration of B constant, we add a small amount of C. C will then influence the interaction between A and B in the solution, and hence alter the activity coefficient of B.

$$f_B = f_B^B f_B^C$$

where $f_B^C$ is the effect of C on the activity coefficient of B.
For the A-B-C-D quaternary solution,

$$f_B = f_B^B f_B^C f_B^D$$

In this relationship it is assumed that addition of D in the solution does not give any influence on $f_B^C$ and vice versa. This condition is generally satisfactory in most practical systems.
For a multicomponent system in general,

$$f_B = f_B^B f_B^C f_B^D \cdots$$

Taking logarithms,

$$\ln f_B = \ln f_B^B + \ln f_B^C + \ln f_B^D + \cdots$$

Since $\ln f_B$ is the function of mole fractions of B, C, D, ..., the Taylor-series expansion yields

$$\ln f_B = N_B \left( \frac{\partial \ln f_B}{\partial N_B} \right)_{N_B=0} + N_C \left( \frac{\partial \ln f_B}{\partial N_C} \right)_{N_C=0} + N_D \left( \frac{\partial \ln f_B}{\partial N_D} \right)_{N_D=0} + \cdots$$

Define
***interaction coefficient*** $\varepsilon_i^j$

$$\varepsilon_i^j = \left( \frac{\partial \ln f_i}{\partial N_j} \right)_{N_j=0}$$

$$\ln f_B = \varepsilon_B^B N_B + \varepsilon_B^C N_C + \varepsilon_B^D N_D + \cdots$$

From a practical point of view it is often more convenient to use *wt*% for the unit of concentrations, which facilitates use of the 1*wt*% standard state. The above equation may be revised as follows:.

$$\log f_{\mathrm{B}} = e_{\mathrm{B}}^{\mathrm{B}}(\mathrm{wt\%B}) + e_{\mathrm{B}}^{\mathrm{C}}(\mathrm{wt\%C}) + e_{\mathrm{B}}^{\mathrm{D}}(\mathrm{wt\%D}) + \cdots$$

where the logarithm is with the base ten.

The last two equations offer an important means in calculating the activity coefficient of a dilute solute in multi-component solutions.

- These equations are valid only for dilute solution because of approximations taken in the Taylor-series expansion:

$$\ln f_{\mathrm{B}} = \ln f_{\mathrm{B}}^{\mathrm{o}} + \left[ N_{\mathrm{B}}\left(\frac{\partial \ln f_{\mathrm{B}}}{\partial N_{\mathrm{B}}}\right)_{N_{\mathrm{B}}=0} + N_{\mathrm{C}}\left(\frac{\partial \ln f_{\mathrm{B}}}{\partial N_{\mathrm{C}}}\right)_{N_{\mathrm{C}}=0} + N_{\mathrm{D}}\left(\frac{\partial \ln f_{\mathrm{B}}}{\partial N_{\mathrm{D}}}\right)_{N_{\mathrm{D}}=0} + \cdots \right]$$
$$+ \left[ \frac{1}{2} N_{\mathrm{B}}^2 \left(\frac{\partial^2 \ln f_{\mathrm{B}}}{\partial N_{\mathrm{B}}^2}\right)_{N_{\mathrm{B}}=0} + N_{\mathrm{B}}N_{\mathrm{C}}\left(\frac{\partial^2 \ln f_{\mathrm{B}}}{\partial N_{\mathrm{B}}\partial N_{\mathrm{C}}}\right)_{N_{\mathrm{B}}=N_{\mathrm{C}}=0} + \cdots \right] + \cdots$$

The above equation was simplified by applying the following approximations:

- The Henrian standard state or 1*wt*% standard state is chosen, and thus, when $N_{\mathrm{B}} \rightarrow 0$, then $f_B = f_{\mathrm{B}}^{\mathrm{o}} = 1$.
- For dilute solutions, the second and higher order terms are negligible.
- It should be noted that, if the solution is not dilute, the second order terms should also be considered.

- The following relations hold between interaction coefficients:

$$\varepsilon_{\mathrm{A}}^{\mathrm{B}} = \varepsilon_{\mathrm{B}}^{\mathrm{A}}, \quad e_{\mathrm{A}}^{\mathrm{B}} = \frac{M_{\mathrm{A}}}{M_{\mathrm{B}}} e_{\mathrm{B}}^{\mathrm{A}}, \quad e_{\mathrm{A}}^{\mathrm{B}} = \frac{M_{\mathrm{S}}}{230.3 M_{\mathrm{B}}} \varepsilon_{\mathrm{A}}^{\mathrm{B}}$$

where, A, B: solutes, and S: solvent.

*Exercises*

10.14 Oxygen dissolved in molten steels is reduced by using deoxidizing elements like Al and Si. In a steel refining process at 1,600°C, aluminium content in the melt was found to be 0.01 wt% by sample analysis. Assuming that oxygen is in equilibrium with Al in the melt and pure $Al_2O_3(s)$ which is the oxidation product, calculate the *wt*% of oxygen dissolved in the melt. The following data are given:

$$2\text{Al}(l)+\tfrac{3}{2}\text{O}_2(g)=\text{Al}_2\text{O}_3(s) \qquad \Delta G_1^\circ=-1{,}682{,}900+323.24T,\ \text{J}$$

$$\text{Al}(l)=\text{Al}(l,\text{wt\%}) \qquad \Delta G^\circ_{2(R\to wt\%)}=-63{,}180-27.91T,\ \text{J}$$

$$\tfrac{1}{2}\text{O}_2(g,1\,\text{atm})=\text{O}(l,\text{wt\%}) \qquad \Delta G^\circ_{3(R\to wt\%)}=-117{,}150-2.89T,\ \text{J}$$

$$e_{\text{Al}}^{\text{Al}}=0.048,\quad e_{\text{Al}}^{\text{O}}=-6.6,\quad e_{\text{O}}^{\text{Al}}=-3.9,\quad e_{\text{O}}^{\text{O}}=-0.20$$

10.15 A liquid copper containing dissolved oxygen and sulfur is in equilibrium with the gaseous phase consisting of $N_2$, $O_2$, SO, $SO_2$ and $SO_3$ at 1,206°C. Calculate the $SO_2$ partial pressure which is in equilibrium with the melt containing 0.02 wt% S and 0.1 wt% O. The following data are available:

$$\tfrac{1}{2}\text{S}_2(g)+\text{O}_2(g)=\text{SO}_2(g) \qquad \Delta G_1^\circ=-361{,}670+72.68T,\ \text{J}$$

$$\tfrac{1}{2}\text{O}_2(g)=\text{O}(l,\text{wt\%}) \qquad \Delta G_2^\circ=-85{,}350+18.54T,\ \text{J}$$

$$\tfrac{1}{2}\text{S}_2(g)=\text{S}(l,\text{wt\%}) \qquad \Delta G_3^\circ=-119{,}660+25.23T,\ \text{J}$$

$$e_{\text{S}}^{\text{O}}=-0.33,\quad e_{\text{S}}^{\text{S}}=-0.19,\quad e_{\text{O}}^{\text{O}}=-0.16,\quad e_{\text{O}}^{\text{S}}=-0.16$$

## 10.8 Ellingham Diagram

The standard Gibbs energy of formation ($\Delta G_f^\circ$) of a compound varies with temperature. The variation with temperature is usually presented by means of a table or some simple equations like

$$\Delta G_f^\circ = A + BT\ln T + CT, \quad \text{or}$$

$$\Delta G_f^\circ = A + BT$$

Ellingham presented the variation of $\Delta G_f^\circ$ with temperature in a graphical form in that it was plotted against temperature. He found that relationships in general were *linear* over temperature ranges in which no change in physical state occurred. The relations could well be represented by means of the simple equation:

$$\boxed{\Delta G_f^\circ = A + BT}$$

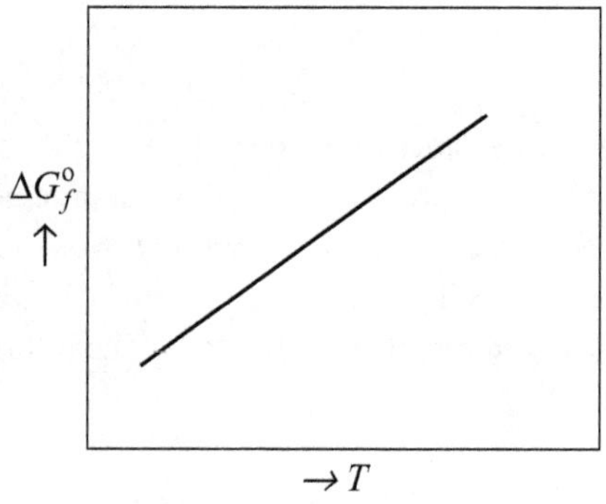

In plotting $\Delta G_f^o$ - $T$ diagrams Ellingham made use of $\Delta \tilde{G}_f^o$ which is the standard Gibbs energy of formation of the compound, *not per mole of the compound, but per mole of the gaseous element (for example oxygen $O_2$)* consumed. For instance,

$$2\text{Cr}(s)+\frac{3}{2}\text{O}_2(g)=\text{Cr}_2\text{O}_3(s) \qquad \Delta G_f^o \text{ per mole of the compound } (\text{Cr}_2\text{O}_3)$$

$$\frac{4}{3}\text{Cr}(s)+\text{O}_2(g)=\frac{2}{3}\text{Cr}_2\text{O}_3(s) \qquad \Delta \tilde{G}_f^o \text{ per mole of the gaseous species } (\text{O}_2)$$

Consider a general reaction of formation:

$$\text{M} + \text{O}_2(g) = \text{MO}_2 \qquad \Delta \tilde{G}_f^o = \text{A} + \text{B}T$$

We may now be able to represent the Gibbs energy change of the reaction versus temperature in a graphical form:

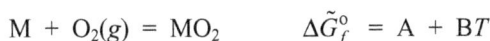

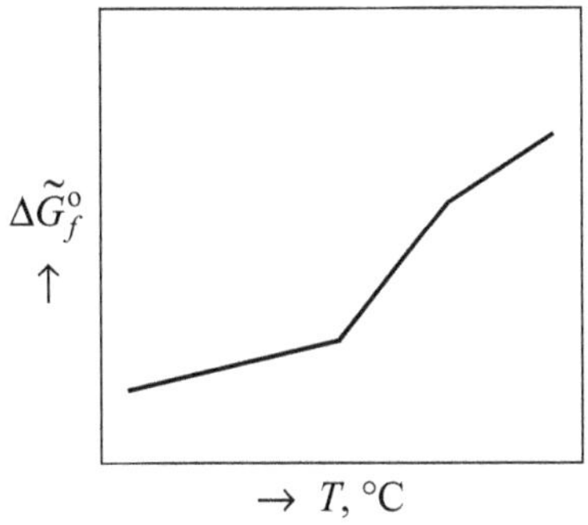


$\Delta \tilde{G}_f^o$

This figure is known as the ***Ellingham diagram*** which graphically shows the change in $\Delta \tilde{G}_f^o$ with $T$

Recall that

$$\boxed{\Delta \tilde{G}_f^o = -RT\, ln\, K}$$

$$K = \frac{a_{\text{MO}_2}}{a_{\text{M}} P_{\text{O}_2}} \qquad a_{\text{M}} = a_{\text{MO}_2} = 1 \text{ when M and MO}_2 \text{ are pure.}$$

$$\boxed{\Delta \tilde{G}_f^o = RT\, ln\, P_{\text{O}_2}}$$

The above equation relates $\Delta \tilde{G}_f^o$ to the oxygen pressure in equilibrium with M($s$) and $MO_2$($s$) at temperature $T$. Each discontinuity point in the diagram indicates the phase change of a species involved in the formation reaction. For instance,

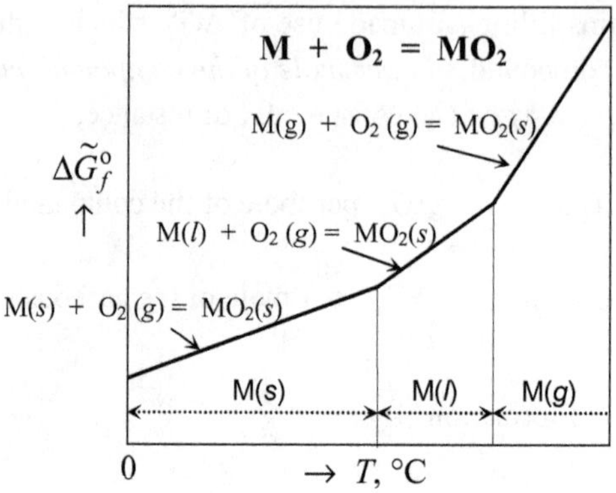


Recall that,

$$\Delta \tilde{G}_f^o = A + BT$$

$$\left(\frac{\partial G}{\partial T}\right)_P = -S \qquad \left[\frac{\partial\left(\frac{G}{T}\right)}{\partial\left(\frac{1}{T}\right)}\right]_P = H$$

$$B = -\Delta \tilde{S}_f^o \qquad A = \Delta \tilde{H}_f^o$$

$$\Delta \tilde{G}_f^o = \Delta \tilde{H}_f^o - T\Delta \tilde{S}_f^o$$

Thus the slope of the line in the Ellingham diagram is $\Delta \tilde{S}_f^o$ and the intercept of the line at $0\ K$ is $\Delta \tilde{H}_f^o$. Simplifying the notations in the diagram, we have

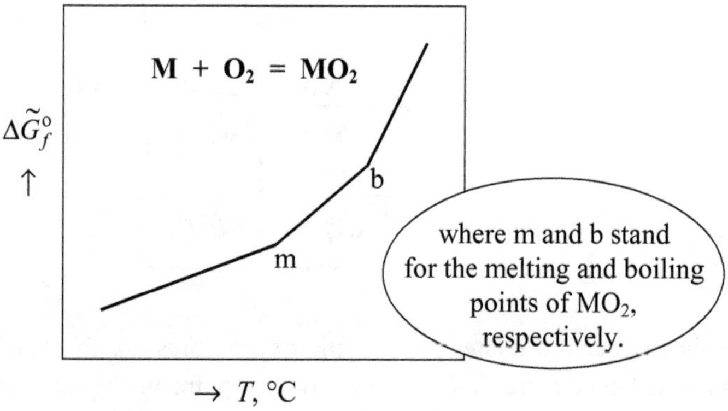

As we may add as many formation reactions as we want, this method of presentation provides a large amount of thermodynamic data and shows the relative stability of compounds for given conditions. The Ellingham diagram for some oxides is given below.

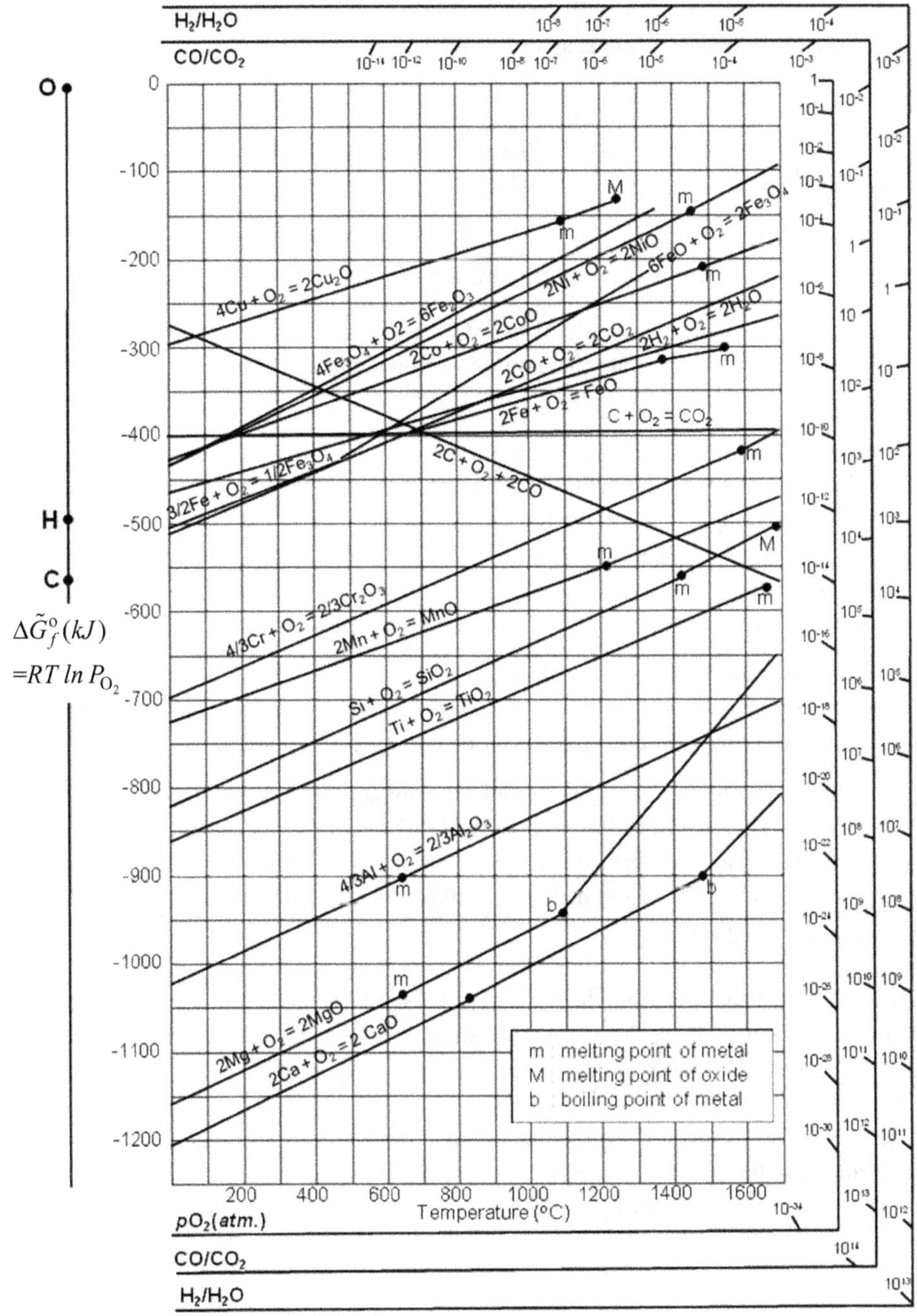

The following are some important information which we can get from the diagram:

- The majority of the lines slope upwards, because the metal (condensed phase: solid or liquid) reacts with a gas (oxygen) to produce another condensed phase, which reduces the entropy.

- The oxidation of solid carbon is an exception. The reaction of

$$C+O_2 = CO_2$$

  is a solid (C) reacting with a mole of gas ($O_2$) to produce a mole of gas ($CO_2$), and so there is little change in entropy and the line is nearly horizontal.

- For the reaction

$$2C+O_2 = 2CO$$

  a solid (C) reacts with one mole of gas ($O_2$) to produce two moles of gas (CO), and hence there is a substantial increase in entropy and the line runs rather sharply downward.

- The position of the line for a given reaction on the Ellingham diagram shows the stability of the oxide as a function of temperature. The lower the position of a metal in the Ellingham diagram, the higher the stability of its oxide. For example, the Ellingham diagram for Al is found to be below Ti and thus $Al_2O_3$ is more stable than $TiO_2$. In other words, as we move down toward the bottom of the diagram, the metals become progressively more reactive and their oxides become harder to reduce

- A metal can reduce the oxides of all other metals whose lines lie above it on the diagram. For example, the $2Mg + O_2 = 2MgO$ line lies below the $Ti + O_2 = TiO_2$ line, and so magnesium can reduce titanium oxide to metallic titanium.

- Since the $2C + O_2 = 2CO$ line is downward-sloping, it cuts across the lines for many other metals. This makes carbon unusually useful as a reducing agent, because as soon as the carbon oxidation line goes below a metal oxidation line, the carbon can then reduce the metal oxide to metal. For less stable oxides, carbon monoxide is often an adequate reducing agent.

It must be noted that the Ellingham diagram is based on the condition that all species involved in reactions are at their respective standard states, and thus the forgoing general discussions are valid only if all the reactants and products involved are in their respective standard states.

In case that, among species involved in a reaction, one or more of them are not in their standard states, the general equation for the Gibbs energy change of the reaction given earlier in this chapter must be applied:

$$\Delta G_r = \Delta G_r^{\mathrm{o}} + RT\,ln\left(\frac{a_{\mathrm{M}}^m\, a_{\mathrm{N}}^n \cdots}{a_{\mathrm{A}}^a\, a_{\mathrm{B}}^b \cdots}\right)$$

> *Example 10.8*
>
> An experiment is devised to confirm if it is possible to use manganese metal (Mn) as a reductant to produce Magnesium metal (Mg) from its oxide (MgO) at 1,200°C. Do you expect that magnesium metal can be produced?

From the Ellingham diagram at 1,200°C,

(1) $2Mg(g) + O_2 = 2MgO(s)$ $\qquad \Delta \tilde{G}^o_{f(1)} = -880$ kJ

(2) $2Mn(s) + O_2 = 2MnO(s)$ $\qquad \Delta \tilde{G}^o_{f(2)} = -550$ kJ

Combining the two reactions,

(3) $2Mn(s) + 2MgO(s) = 2Mg(g) + 2MnO(s)$ $\qquad \Delta G^o_{(3)} = 330$ kJ

or

(4) $Mn(s) + MgO(s) = Mg(g) + MnO(s)$ $\qquad \Delta G^o_{(4)} = 150$ kJ

Reaction (4) gives a positive Gibbs energy change, and thus the reaction will not occur as written if all the species involved in the reaction are in their respective standard states, and magnesium metal cannot be produced by using manganese as a reductant.
However, some or all species are non-standard states, then the situation becomes different.
The Gibbs energy change of reaction (4) under non-standard states is given by

$$\Delta G_{(4)} = \Delta G^o_{(4)} + RT \ln\left(\frac{a_{MnO} P_{Mg}}{a_{Mn} a_{MgO}}\right)$$

↓ Assuming Mn, MnO and MgO are all at their standard states (*i.e.*, pure states)

$$\Delta G_{(4)} = \Delta G^o_{(4)} + RT \ln P_{Mg}$$

↓ $\Delta G^o_{(4)} = 150$ kJ
If $\Delta G_{(4)} < 0$, the reaction proceeds to the right.

$$P_{Mg} < 4 \times 10^{-6} \text{ atm.}$$

The above result tells us that, when the Mg vapor pressure at the reaction site is kept lower than $4 \times 10^{-6}$ atm., manganese metal is able to reduce MgO. This condition can be met by maintaining a flow of inert gas through the system to remove magnesium vapor produced so that the magnesium vapor pressure is kept below the equilibrium value.

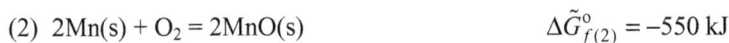

*Example 10.9*

Knowing that

$$\Delta\tilde{G}_f^o = RT\ln P_{O_2}$$

Discuss whether the Ellingham diagram can be converted to a phase diagram which relates stable phases to the oxygen potential and temperature.

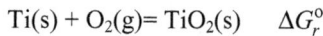

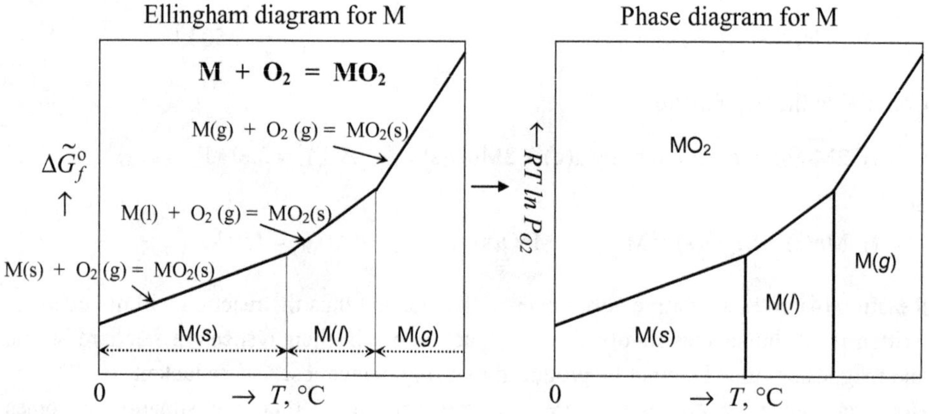


This diagram is sometimes termed "***predominance diagram***."

*Example 10.10*

It is desired to reduce the oxygen content of argon gas less than $10^{-24}$ atm. by passing the gas through a furnace containing metallic titanium sponge. Determine the temperature at which the furnace should be run.

$$Ti(s) + O_2(g) = TiO_2(s) \qquad \Delta G_r^o$$

From the table in Appendix II,

$$\Delta G_r^o = -941{,}000 + 177.6T, \quad J$$

The equilibrium constant of the above reaction:

$$K = exp\left(-\frac{\Delta G_r^o}{RT}\right) = \left[\frac{a_{TiO_2}}{a_{Ti}P_{O_2}}\right]_{eq}$$

$$a_{TiO_2} = 1, \quad a_{Ti} = 1, \quad P_{O_2} = 10^{-24}$$

$$T = 1{,}204°C$$

The more convenient way, though perhaps not a very accurate approach, to solve this problem is to make use of the Ellingham diagram. To find the equilibrium partial pressure of oxygen, the scale on the right side of the diagram labeled "$pO_2$(atm.)" is used together with the pivotal point "O" at the far left vertical axis. The procedure is simply to connect the point "O" to the point representing $pO_2 = 10^{-24}$ atm. with a straight line. The point of intersection between the straight line and the Gibbs energy plot for $TiO_2$ gives the desired temperature.

The oxygen scale (***nomographic scale***) in the Ellingham diagram provides a convenient and rapid means of determining the equilibrium oxygen partial pressure for a metal/oxide system at any temperature. It was first devised by Richardson and Jeffes.

*Example 10.11*

Wustite (FeO) can be reduced to metallic iron (Fe) by a CO-$CO_2$ gas mixture at elevated temperatures. Determine the partial pressures of $O_2$, CO and $CO_2$ in the gas which is in equilibrium with FeO and Fe at 1,000°C and 1 atm. total pressure.

Reactions prevailing in the system:

(1) $FeO + CO = Fe + CO_2$ $\Delta G_1^o$

(2) $Fe + 1/2O_2 = FeO$ $\Delta G_2^o$

$$\Delta G_1^o = \left(\Delta G_{f,Fe}^o + \Delta G_{f,CO_2}^o\right) - \left(\Delta G_{f,FeO}^o + \Delta G_{f,CO}^o\right)$$

From Appendix II

$\Delta G_{f,Fe}^o = 0$

$\Delta G_{f,CO_2}^o = -394800 + 0.836T, \ \text{Jmol}^{-1}$

$\Delta G_{f,FeO}^o = -264000 + 64.6T, \ \text{Jmol}^{-1}$

$\Delta G_{f,CO}^o = -112900 - 86.5T, \ \text{Jmol}^{-1}$

$$\Delta G_1^o = -17900 + 22.7T, \text{J}$$

$T$ = 1273K

$$\Delta G_1^o = 10180 \text{ J}$$

For $\Delta G_2^o$, from Appendix II,

$$\Delta G_2^o = -264000 + 64.6 \text{ J}$$

$T$ = 1273K

$$\Delta G_2^o = -181800 \text{ J}$$

From the equilibrium constant relationships,

$$K_1 = \left(\frac{a_{Fe} P_{CO_2}}{a_{FeO} P_{CO}}\right)_{eq} = exp\left(-\frac{\Delta G_1^o}{RT}\right)$$

$$a_{Fe} = a_{FeO} = 1$$
$$\Delta G_1^o = 10{,}180 \text{ J}$$

$$\frac{P_{CO_2}}{P_{CO}} = 0.38$$

$$K_2 = \left(\frac{a_{FeO}}{a_{Fe} P_{O_2}^{\frac{1}{2}}}\right)_{eq} = exp\left(-\frac{\Delta G_2^o}{RT}\right)$$

$$a_{Fe} = a_{FeO} = 1$$
$$\Delta G_2^o = -181{,}800 \text{ J}$$

$$P_{O_2} = 1.2 \times 10^{-15}\, atm.$$

Knowing that,

$$P_t = 1 = P_{CO} + P_{CO_2} + P_{O_2}$$

$$P_{CO} = 0.72 \text{ atm.} \quad P_{CO_2} = 0.28 \text{ atm.}$$

Simpler and more convenient way of solving this problem is to make use of the Ellingham diagram, similar to what we did earlier for finding equilibrium oxygen partial pressures. To find the equilibrium $CO/CO_2$ ratio, the scale on the right side of the diagram labeled "$CO/CO_2$" is used together with the pivotal point "C" at the far left vertical axis. The procedure is simply to connect the point "C" to the point at which the line of "$2Fe + O_2 = 2FeO$" crosses the 1,200°C temperature line, and extend it to the "$CO/CO_2$" line, and then read off the value from the "$CO/CO_2$"scale. The equilibrium oxygen partial pressure may be found in a similar way, but by using the pivotal point "O" and $pO_2$ scale. The best values we can get from the diagram are,

$$\frac{P_{CO}}{P_{CO_2}} = 3\text{, and } P_{O_2} = 2\times10^{-15}\text{ atm.}$$

Then $P_{CO}$ = 0.75 atm. and $P_{CO2}$ = 0.25 atm. The results are reasonably close to the values obtained by rigorous calculations which we did above.

*Exercises*

10.16 A gas mixture of $H_2$ and $H_2O$ at 1 atm. pressure is in equilibrium with pure iron ($\gamma$ iron) and pure wustite (FeO) at 950°C.

(1) Calculate the $H_2/H_2O$ ratio in the gas mixture using the data of the Gibbs energy of formation of the species involved. Compare the results with what you can estimate from the $H_2/H_2O$ scale in the Ellingham diagram.

(2) In another experiment with iron-nickel alloy (81 *mol*% Fe) and pure wustite at the otherwise same conditions as above, the equilibrium $H_2/H_2O$ ratio in the gas mixture was found to be 1.44. Find the activity and activity coefficient of iron in the alloy.

10.17 A mixture of CoO(*s*) and $Fe_3O_4$(*s*) is to be processed in a furnace at 800°C and 1 *atm.* pressure to produce cobalt. However, the iron must not be reduced, but retained as $Fe_3O_4$(*s*).

(1) In the case that the CO and $CO_2$ gas mixture is used in the process, suggest the range of $CO/CO_2$ ratio which can satisfy the above requirements.

(2) Find the range of $H_2/H_2O$ ratio, if the available gas is the $H_2$-$H_2O$ gas mixture.

## 10.9 Adsorption Equlibria

When the macroscopic thermodynamic properties of a system were considered in the preceding discussions, the contribution of the surface or interface was not included under implicit assumption that the physical size of the surface or interface is negligibly small in comparison with the bulk of the system. However, heterogeneous systems inevitably involve the surface or interface which separates homogeneous phases in contact. In fact, changes to a substantial extent occur at surfaces and interfaces.

In this section, thermodynamics of surface (interface), in particular, of the adsorption of solutes at the surface (interface) is discussed. Capillary effects of the surface (interface) induced by surface tension is discussed separately in the chapter that follows.

> *Surface and Interface*
>
> *Surface*: the term normally used for gas-liquid and gas-solid boundaries
>
> *Interface*: the term used for solid-solid, solid-liquid, and liquid-liquid boundaries

The interface which separates two phases, say $\alpha$ and $\beta$, is in fact three dimensional. In other words, the interface has a finite thickness. For the convenience of thermodynamic analysis, the real three-dimensional interface is replaced by a hypothetical two-dimensional surface which divides the two phases. This two dimensional surface is termed the ***Gibbs dividing surface***, **Σ**. The figures below schematically compare the real interface and the Gibbs dividing surface. Any excess thermodynamic properties due to existence of the interface are assigned to the dividing surface. That is,

- The bulk phases are assumed to continue uniformly right to the dividing surface.
- Thermodynamic properties of the system obtained under the above assumptions are different from the properties of the real system which includes heterogeneity of the interfaces.
- The difference of the properties, *i.e.*, excess thermodynamic properties, which may be positive or negative, are assigned to the Gibbs dividing surface.

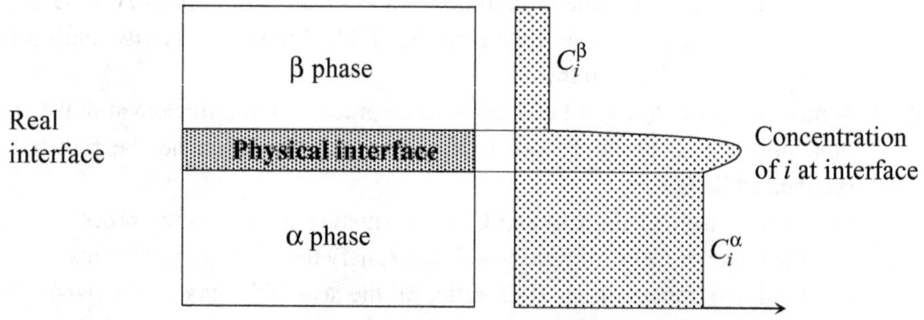


Schematic of real interface    Concentration profile

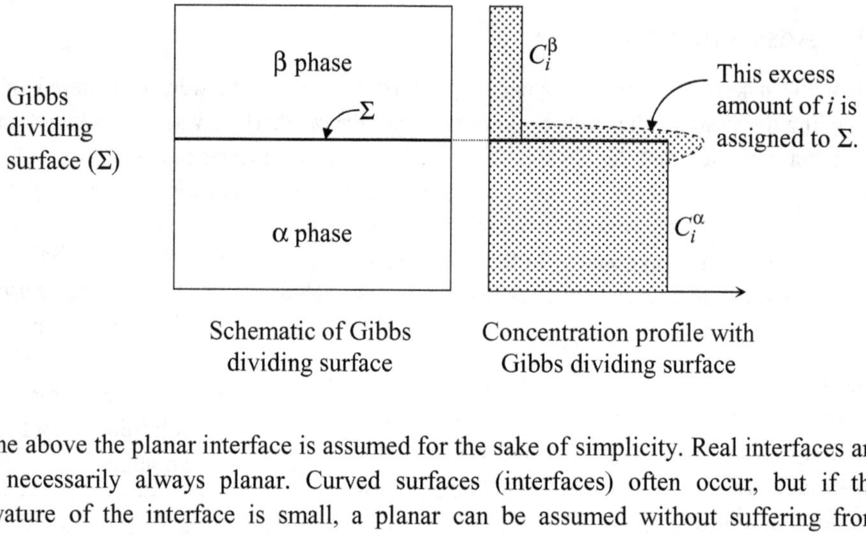


Schematic of Gibbs dividing surface    Concentration profile with Gibbs dividing surface

In the above the planar interface is assumed for the sake of simplicity. Real interfaces are not necessarily always planar. Curved surfaces (interfaces) often occur, but if the curvature of the interface is small, a planar can be assumed without suffering from unacceptable deviation.

Let us now apply the concept of the Gibbs dividing surface to the interface between two multi-component phases.

- Mass balance: $n_i^t = n_i^\alpha + n_i^\beta + n_i^\sigma$ or

  $n_i^\sigma = n_i^t - (n_i^\alpha + n_i^\beta)$

  where $n_i^t$ is the total number of moles, and $n_i^\sigma$ is the number of moles of $i$ assigned to the Gibbs dividing surface Σ)

• Thermodynamic properties:

$$U^\sigma = U - (U^\alpha + U^\beta)$$

$$S^\sigma = S - (S^\alpha + S^\beta)$$

$$G^\sigma = G - (G^\alpha + G^\beta)$$

$$V^\sigma = V - (V^\alpha + V^\beta) = 0 \quad (\Sigma \text{ has no thickness.})$$

Let us define ***surface excess concentration*** of component $i$ ($\Gamma_i$) as

$$\boxed{\Gamma_i = \frac{n_i^\sigma}{A}} \quad \text{where } A \text{ is the area of } \Sigma$$

Referring to the diagrams given above, the choice of the position of the dividing surface appears to be arbitrary. If the position is changed, all related properties such as the excess moles, internal energy, entropy and Gibbs energy vary accordingly. Thermodynamics of the interface thus depends on the position of the dividing surface. In order to overcome this ambiguity, the concept of the ***relative adsorption*** is now introduced.

$$\boxed{n_i^\sigma = n_i^t - (n_i^\alpha + n_i^\beta)}$$

$$n_i^\alpha = C_i^\alpha V^\alpha \text{ and } n_i^\beta = C_i^\beta V^\beta$$

$$\boxed{n_i^\sigma = n_i^t - (C_i^\alpha V^\alpha + C_i^\beta V^\beta)}$$

$$V^\alpha = V - V^\beta$$

$$\boxed{n_i^\sigma = n_i^t - C_i^\alpha V + (C_i^\alpha - C_i^\beta)V^\beta}$$

$$\Gamma_i = \frac{n_i^\sigma}{A}$$

$$\boxed{\Gamma_i = \frac{1}{A}\left(n_i^t - C_i^\alpha V + (C_i^\alpha - C_i^\beta)V^\beta\right)}$$

Applying to component A,

$$\boxed{\Gamma_A = \frac{1}{A}\left(n_A^t - C_A^\alpha V + (C_A^\alpha - C_A^\beta)V^\beta\right)}$$

In the right-hand sides of the last two equations, $V^\beta$ is the only term which depends on the position of the dividing surface Σ. Elimination of $V^\beta$ by combining these two equations yields,

$$\Gamma_i - \Gamma_A\left(\frac{C_i^\alpha - C_i^\beta}{C_A^\alpha - C_A^\beta}\right) = \frac{1}{A}\left[(n_i - C_i^\alpha V) - (n_A - C_A^\alpha V)\left(\frac{C_i^\alpha - C_i^\beta}{C_A^\alpha - C_A^\beta}\right)\right]$$

Note that none of the terms in the right-hand side of the above equation is related to the location of the dividing surface. In other words, the right-hand side of the above equation is independent of the position choice of the dividing surface. Thus the left-hand side must also be independent of the position of the dividing surface. This means that, although individual terms of $\Gamma_i$ and $\Gamma_A$ vary by varying the position of the dividing surface, their combination seen in the left-hand side is independent of the position of the dividing surface. This combination is termed the ***relative adsorption of component i with respect to component* A** ($\Gamma_i^{(A)}$). That is,

$$\Gamma_i^{(A)} = \Gamma_i - \Gamma_A\left(\frac{C_i^\alpha - C_i^\beta}{C_A^\alpha - C_A^\beta}\right)$$

Consider A-B two components

$$\Gamma_B^{(A)} = \Gamma_B - \Gamma_A\left(\frac{C_B^\alpha - C_B^\beta}{C_A^\alpha - C_A^\beta}\right)$$

Since $\Gamma_i^{(A)}$ ($\Gamma_B^{(A)}$ for the A-B two components) is independent of the location of the dividing surface Σ, we have freedom to choose a position which gives a particular value to $\Gamma_A$. We can then conveniently choose Σ which gives a value of $\Gamma_A$ equal to zero. For the relative adsorption becomes,

$$\Gamma_i^{(A)} = \Gamma_i \quad \text{or}$$

$$\Gamma_B^{(A)} = \Gamma_B \quad \text{for the A-B two components}$$

The concept of ($\Gamma_A = 0$) is graphically illustrated below for two different concentration profiles:

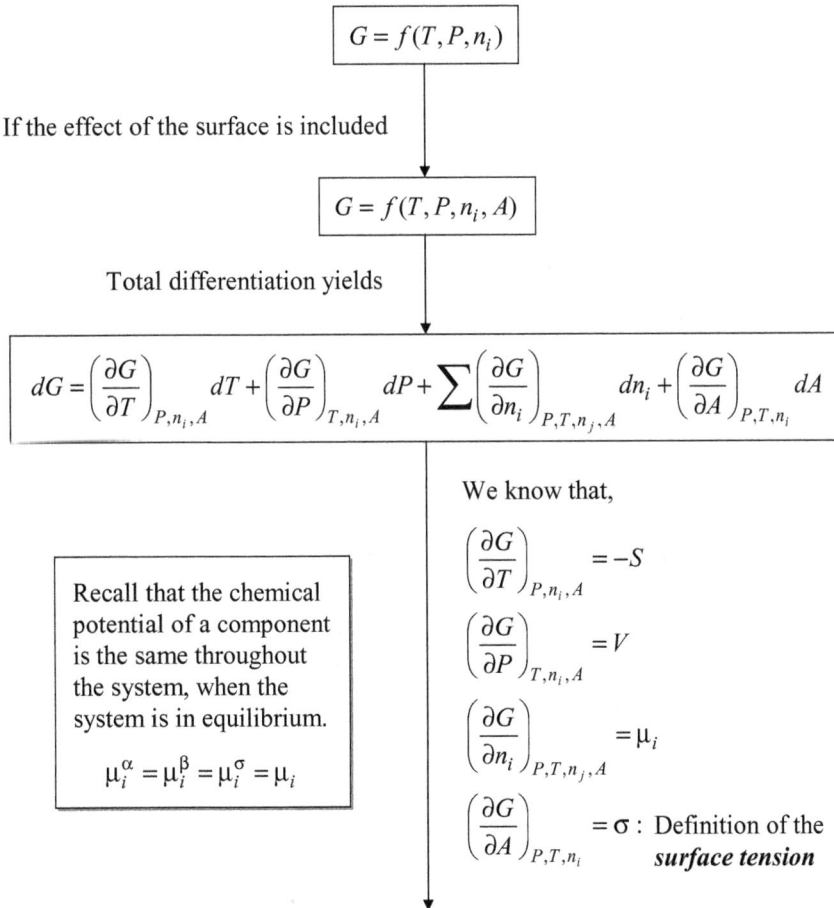

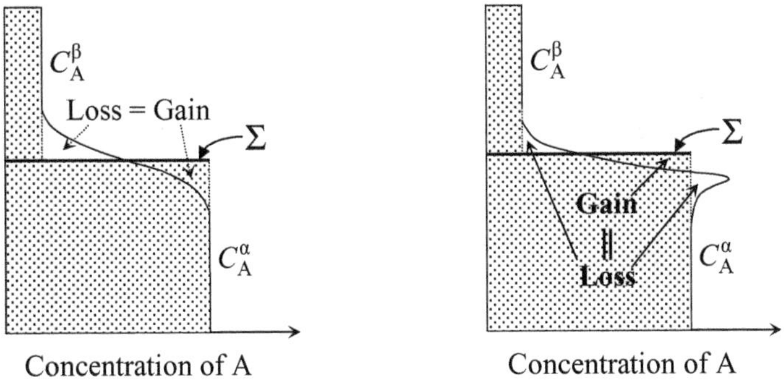


Surface properties including the surface tension are related to the concentration of the surface. We now develop the relationship between surface tension and surface concentration.

Recall the general discussion on the Gibbs energy of a system which the effect of the interface is now included:

$$G = f(T, P, n_i)$$

If the effect of the surface is included

$$G = f(T, P, n_i, A)$$

Total differentiation yields

$$dG = \left(\frac{\partial G}{\partial T}\right)_{P,n_i,A} dT + \left(\frac{\partial G}{\partial P}\right)_{T,n_i,A} dP + \sum \left(\frac{\partial G}{\partial n_i}\right)_{P,T,n_j,A} dn_i + \left(\frac{\partial G}{\partial A}\right)_{P,T,n_i} dA$$

We know that,

$$\left(\frac{\partial G}{\partial T}\right)_{P,n_i,A} = -S$$

$$\left(\frac{\partial G}{\partial P}\right)_{T,n_i,A} = V$$

$$\left(\frac{\partial G}{\partial n_i}\right)_{P,T,n_j,A} = \mu_i$$

$$\left(\frac{\partial G}{\partial A}\right)_{P,T,n_i} = \sigma$$ : Definition of the ***surface tension***

Recall that the chemical potential of a component is the same throughout the system, when the system is in equilibrium.

$$\mu_i^\alpha = \mu_i^\beta = \mu_i^\sigma = \mu_i$$

$$dG = -SdT + VdP + \sum \mu_i dn_i + \sigma dA$$

For α and β phases,

$$dG^{\alpha} = -S^{\alpha}dT + V^{\alpha}dP + \sum \mu_i dn_i^{\alpha}$$

$$dG^{\beta} = -S^{\beta}dT + V^{\beta}dP + \sum \mu_i dn_i^{\beta}$$

Knowing that,

$$G^{\sigma} = G - (G^{\alpha} + G^{\beta})$$

$$dG^{\sigma} = -\left[S - (S^{\alpha} + S^{\beta})\right]dT + \left[V - (V^{\alpha} + V^{\beta})\right]dP + \sum \mu_i \left[dn_i - (dn_i^{\alpha} + dn_i^{\beta})\right] + \sigma dA$$

$$S^{\sigma} = S - (S^{\alpha} + S^{\beta})$$

$$V^{\sigma} = V - (V^{\alpha} + V^{\beta}) = 0$$

$$dn^{\sigma} = dn - (dn^{\alpha} + dn^{\beta})$$

$$dG^{\sigma} = -S^{\sigma}dT + \sum \mu_i dn_i^{\sigma} + \sigma dA$$

At a constant $T$,

$$dG^{\sigma} = \sum \mu_i dn_i^{\sigma} + \sigma dA$$

On the other hand, when we look at the dividing surface separately, the Gibbs energy of the surface ( $G^{\sigma}$ ) at a constant temperature can be determined by,

| Gibbs energy of the surface ( $G^{\sigma}$ ) | = | Sum of the contribution of each component ( $\sum \mu_i n_i^{\sigma}$ ) | + | Contribution by the surface ( $\sigma A$ ) |
|---|---|---|---|---|

That is,

$$G^{\sigma} = \sum \mu_i n_i^{\sigma} + \sigma A$$

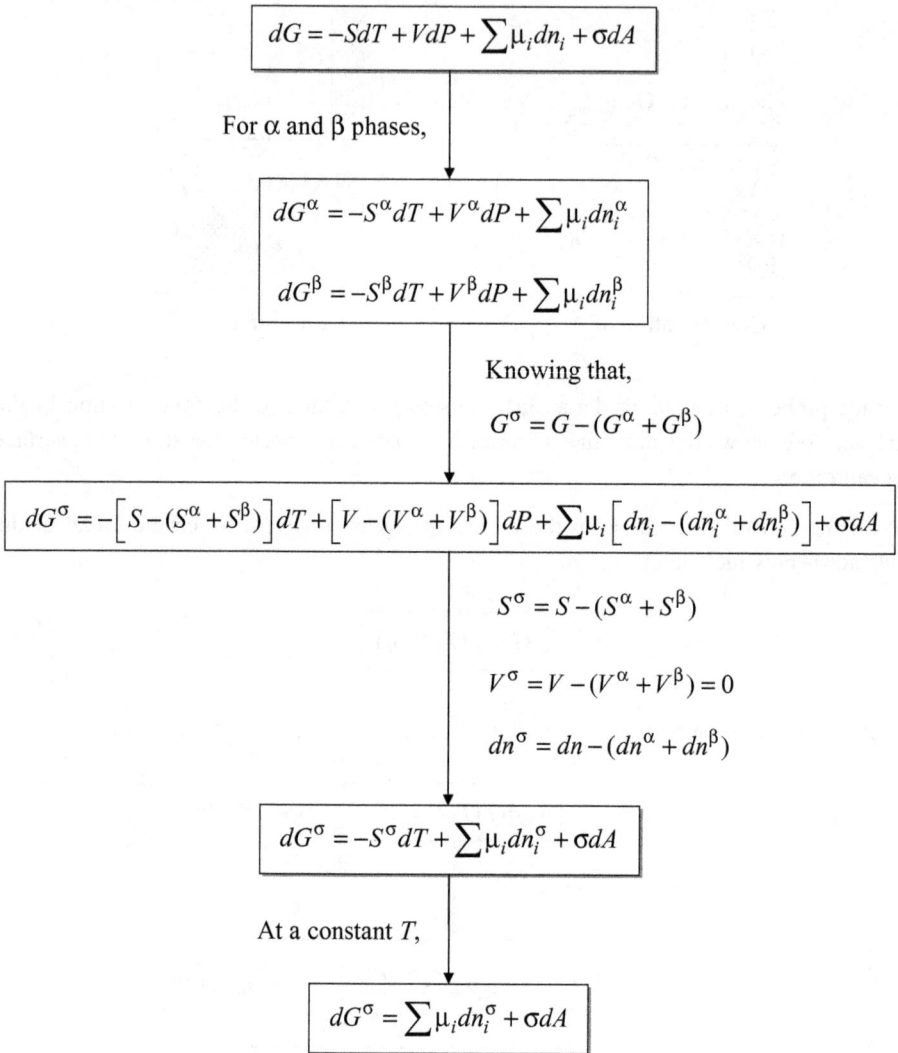

The total differentiation yields,

$$dG^{\sigma} = \sum \mu_i dn_i^{\sigma} + \sum n_i^{\sigma} d\mu_i + \sigma dA + A d\sigma$$

Comparison of the last two equations in the shaded boxes gives us that,

$$\sum n_i^{\sigma} d\mu_i + A d\sigma = 0$$

Recall this is the *Gibbs-Duhem equation* which now includes the surface effect.

Applying the definition of the surface excess concentration,

$$\Gamma_i = \frac{n_i^{\sigma}}{A}$$

$$d\sigma = -\sum \Gamma_i d\mu_i$$

This equation is known as the ***Gibbs adsorption equation*** which relates the effect on the surface tension of the surface excess concentrations of all components in a multi-component system.

Let us apply the Gibbs adsorption equation to an A-B binary system.

$$d\sigma = -(\Gamma_{\mathrm{A}} d\mu_{\mathrm{A}} + \Gamma_{\mathrm{B}} d\mu_{\mathrm{B}})$$

But, from the Gibbs –Duhem equation,

$$C_{\mathrm{A}}^{\alpha} d\mu_{\mathrm{A}} + C_{\mathrm{B}}^{\alpha} d\mu_{\mathrm{B}} = 0$$
$$C_{\mathrm{A}}^{\beta} d\mu_{\mathrm{A}} + C_{\mathrm{B}}^{\beta} d\mu_{\mathrm{B}} = 0$$

Subtraction and rearrangement yield,

$$d\mu_{\mathrm{A}} = -\left(\frac{C_{\mathrm{B}}^{\alpha} - C_{\mathrm{B}}^{\beta}}{C_{\mathrm{A}}^{\alpha} - C_{\mathrm{A}}^{\beta}}\right) d\mu_{\mathrm{B}}$$

$$d\sigma = -\left[\Gamma_{\mathrm{B}} - \left(\frac{C_{\mathrm{B}}^{\alpha} - C_{\mathrm{B}}^{\beta}}{C_{\mathrm{A}}^{\alpha} - C_{\mathrm{A}}^{\beta}}\right) \Gamma_{\mathrm{A}}\right] d\mu_{\mathrm{B}}$$

We already derived the following two equations:

$$\Gamma_B^{(A)} = \Gamma_B - \Gamma_A\left(\frac{C_B^\alpha - C_B^\beta}{C_A^\alpha - C_A^\beta}\right) \text{ and}$$

$$\Gamma_B^{(A)} = \Gamma_B \text{ for relative adsorption,}$$

$$d\sigma = -\Gamma_B d\mu_B \text{ or } \Gamma_B = -\left(\frac{\partial\sigma}{\partial\mu_B}\right)_T$$

Applying the relationship

$$\mu_B = \mu_B^o + RT\ln a_B$$

$$\Gamma_B = -\frac{1}{RT}\left(\frac{\partial\sigma}{\partial \ln a_B}\right)_T$$

This equation enables us to find $\Gamma_B$ or $\Gamma_B^{(A)}$ from the knowledge of the dependence of the surface tension on the activity of the solute B at a given temperature. This equation is called the ***Gibbs adsorption isotherm***.

If a solute exerts a large effect on the surface (interfacial) tension even at a very dilute concentration, the solute is called to be ***surface active***. Those elements in the group VIb in the periodic table (O, S, Se, Te) are strongly surface-active when they exist as solutes in metals.

The following figure shows the graphical representation of the last equation:

As $\ln a_B$ increases, the curve eventually reaches a point after which the slope of the curve stays constant, *that is*,

$$\left(\frac{\partial\sigma}{\partial \ln a_B}\right)_T \text{ is constant.}$$

At this point the surface is saturated with B and the slope is $-RT\Gamma_B^o$ where $\Gamma_B^o$ is the *surface excess at saturation*.

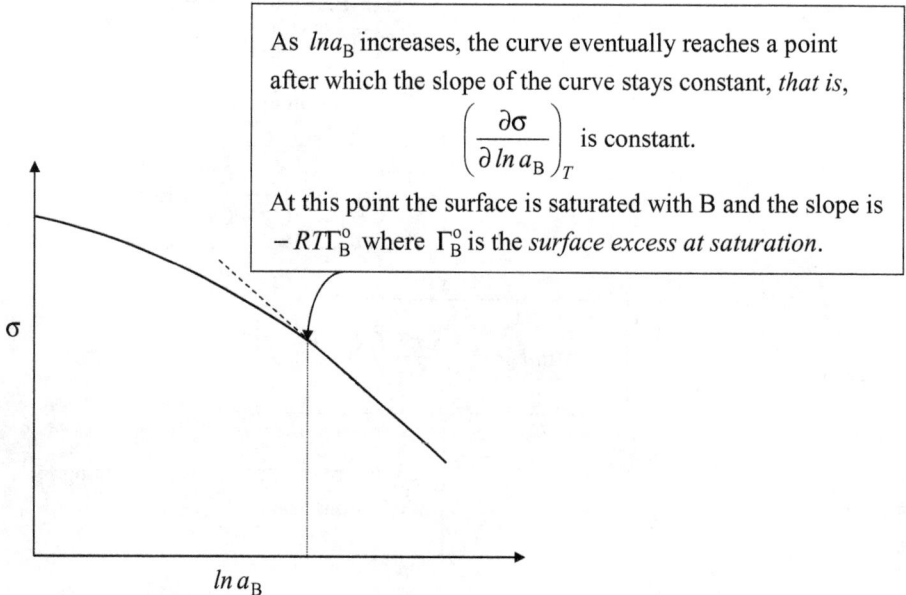

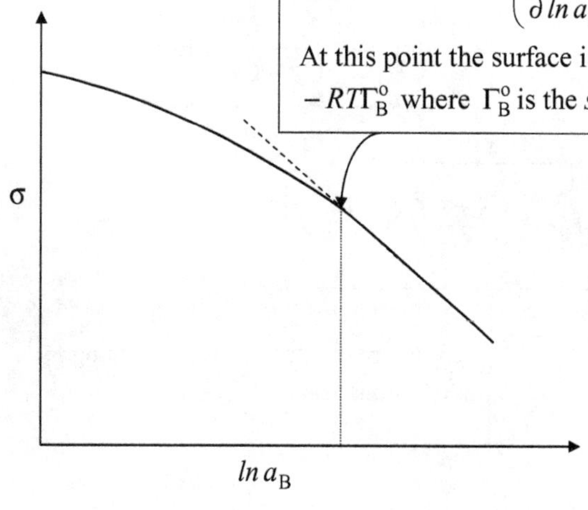

*Example 10.12*

Two immiscible liquids are in contact forming a flat interface. The surface active species S accumulates at the interface, but all others do not. Prove that the interfacial tension decreases as the surface active species accumulates at the interface.

$$\Gamma_{\mathrm{B}} = -\frac{1}{RT}\left(\frac{\partial \sigma}{\partial \ln a_{\mathrm{B}}}\right)_T$$

S: Surface active → low concentration in the bulk → Henry's law applicable →

$a_{\mathrm{S}} = \gamma_{\mathrm{S}}^{\mathrm{o}} N_{\mathrm{S}}$

$$\Gamma_{\mathrm{B}} = -\frac{1}{RT}\left(\frac{\partial \sigma}{\partial \ln N_{\mathrm{B}}}\right)_T$$

$$d\ln N_{\mathrm{B}} = \frac{1}{N_{\mathrm{B}}} dN_{\mathrm{B}}$$

$$\left(\frac{\partial \sigma}{\partial N_{\mathrm{B}}}\right)_T = -\frac{\Gamma_{\mathrm{B}} RT}{N_{\mathrm{B}}} < 0$$

This equation implies that the change in $\sigma$ with increase in $N_{\mathrm{B}}$ is negative, and thus the interfacial tension decreases.

Now it is in order to discuss the adsorption of ***surface active species***. The physical interface between two metals is not usually in the form of a mono-atomic layer. In the case of surface active species such as oxygen, sulphur, selenium and tellurium, in a liquid metal, however, they usually form a mono-layer at the surface. When this is the case, we can consider the following reaction at the surface:

$$\mathrm{B} + v = \mathrm{B}_v$$

where B is a solute, $v$ a vacant site available for adsorption of B, and $\mathrm{B}_v$ the adsorbed B.

For simplicity, a number of assumptions are made:

- Adsorption occurs only to form a mono-layer.
- The surface is uniform and all sites for adsorption are the same.
- There is no interaction among adsorbing species at the surface.

Based on the above assumptions, it can be considered that the surface is a two-dimensional ideal solution consisting of adsorbed species ($\mathrm{B}_v$) and adsorption sites ($v$).

Then the equilibrium constant expression ($K$, termed ***adsorption coefficient***) for the above reaction will yield,

$$K = \frac{\Gamma_B}{a_B(\Gamma_B^o - \Gamma_B)}$$

Define the ***fractional surface coverage*** θ as

$$\theta = \frac{\Gamma_B}{\Gamma_B^o}$$

$$K = \frac{\theta}{a_B(1-\theta)}$$

Rearrangement gives

$$\theta = \frac{K a_B}{1 + K a_B}$$

The last two equations are called the ***Langmuir adsorption isotherm***.

If several different species also adsorb on the surface, the fractional site available for adsorption of a particular species may have to be modifies as,

$$(1-\theta) \rightarrow (1-\sum \theta_i)$$

In the case of a gaseous species adsorbing on a substrate which follows all the assumptions specified in the above, the Langmuir isotherm may be expressed as,

$$K_P = \frac{\theta}{P_B(1-\theta)} \quad \text{or} \quad \theta = \frac{K_P P_B}{1 + K_P P_B}$$

When the gaseous species dissociates on adsorption into two parts (for instance, a diatomic molecule dissociates into two atoms), it will require two adsorption sites. The relevant reaction will be

$$B_2 + 2v = 2B_v$$

Then the equilibrium constant expression will become,

$$K_P = \frac{\theta^2}{P_{B_2}(1-\theta)^2} \quad \text{or} \quad \theta = \frac{\sqrt{K_P P_{B_2}}}{1+\sqrt{K_P P_{B_2}}}$$

It is seen from the square root dependence on the pressure that, for a given pressure of the adsorbing species, the dissociative adsorption requires a higher pressure for the same coverage. This is due to the fact that the dissociative adsorption requires two sites for each adsorbing species (for instance, molecules).

When Belton (1976) combined the Gibbs adsorption isotherm and the Langmuir adsorption isotherm together,

$$\Gamma_B = -\frac{1}{RT}\left(\frac{\partial \sigma}{\partial \ln a_B}\right)_T$$

$$\theta = \frac{Ka_B}{1+Ka_B}$$

$$\theta = \frac{\Gamma_B}{\Gamma_B^0}$$

$$d\sigma = -RT\left(\frac{\Gamma_B^0 K a_B}{1+Ka_B}\right) d\ln a_B$$

Integration of this equation from the clean surface state ($\sigma^o$, $a_B = 0$) to the adsorbed surface state yields

$$\sigma = \sigma^o - RT\Gamma_B^0 \ln(1+Ka_B)$$

This ideal adsorption equation was first deduced empirically by Szyszkowski (1908). Belton first reported that this equation was found to give a close description of the available surface tension data for the Group VIb elements in some liquid metals.

The Langmuir model was further extended by Fowler and Guggenheim to include interaction between adsorbed species (atoms or molecules) by keeping all other assumptions of the Langmuir model still applied. The derivation of the model yields,

$$Ka_B = \frac{\theta}{1-\theta} \exp\left[-2\left(\frac{z\Omega}{kT}\right)\theta\right]$$

where $z$ is the number of the nearest neighbors within the surface layer, and $\Omega$ is the interaction parameter of the regular solution model. Note that,

- $\Omega = 0$ : The equation reduces to the Langmuir isotherm.
- $\Omega > 0$ : Attraction between the adsorbed atoms or molecules
- $\Omega < 0$ : Repulsion between the adsorbed atoms or molecules

The adsorption may form a multilayer instead of a monolayer of the adsorbed atoms. The initial monolayer may act as a substrate for further adsorption on top of it. In this case the thickness of the adsorbed layer may grow indefinitely. The ***BET isotherm*** (after Brunauer, Emmett and Teller) is the most widely used isotherm for the multilayer adsorption.

$$\frac{P}{m_B(P^o - P)} = \frac{1}{m_B^m C}\left[1 + (C-1)\left(\frac{P}{P^o}\right)\right]$$

where $P$ is the vapor pressure of the adsorbing species (B), $P^o$ is the saturation vapor pressure of B, $m_B$ is the total amount of B adsorbed, $m_B^m$ is the amount of B for monolayer coverage, and $C$ is a constant.

The Langmuir isotherm assumes that all adsorption sites are identical. If adsorption sites are different from each other, then the energetically more favorable sites will be preferentially occupied first. To take into account the variation due to non-identical sites, a number of suggestions have been made:

***Temkin isotherm***:

$$\theta = C_1 \, ln(C_2 P)$$

where $C_1$ and $C_2$ are constants.

***Freundlich isotherm***:

$$\theta = C_1 P^{\frac{1}{C_2}}$$

where $C_1$ and $C_2$ are constants.

Chapter 11

# Phase Equilibria

## 11.1 Phase Rule

Suppose that we have one mole of an ideal gas A.

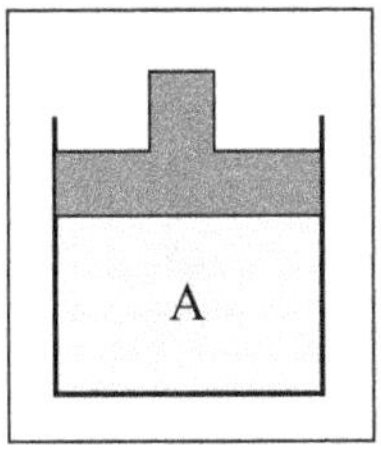


Are we free to vary temperature? *Yes.*
Are we free to vary both temperature and pressure independently? *Yes.*
Can we then vary temperature, pressure and volume all independently? *No.*
From the equation of state or ideal gas law ($PV = nRT$), we know that we can specify only two variables, *i.e.*, $(P,V)$, $(P,T)$ or $(V,T)$, but not all three variables. The third variable is uniquely determined by the equation of state. In other words, the equilibrium condition of the system is fully defined by specifying two variables. We have *two variables* under our control. We then say that the number of ***degrees of freedom*** = 2.

What if the gas phase is the mixture of two ideal gases A and B?

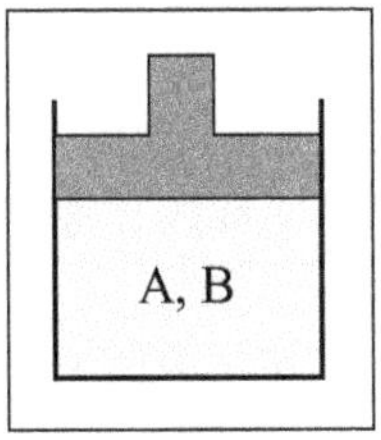


Is the number of degrees of freedom still the same? In other words, can we fully define the thermodynamic state of the system by specifying two variables, say, $T$ and $P$? *No.*

Specifying $T$ and $P$ will determine $V$, but the composition of the gas phase is left undetermined until the concentration of either A or B is fixed. That is, even after fixing $T$ and $P$ we can freely vary the concentration of either A or B (but not both because $N_A + N_B = 1$), and hence we have *one additional degree of freedom.* Thus, the number of degrees of freedom is three in this case.

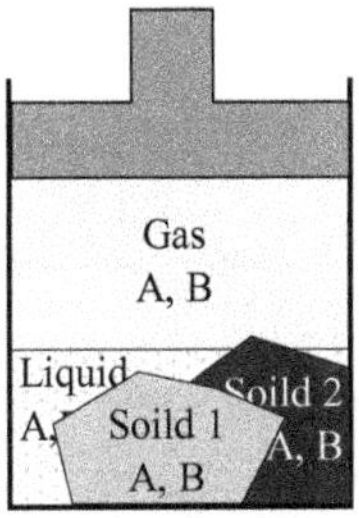


What if a liquid phase coexists with the gas phase?
What if a solid phase exists together as well?
Would it be thermodynamically feasible to have two solid phases together with a liquid phase and a gas phase in equilibrium?
*Are there any systematic methods available to answer these questions?*

J.W. Gibbs developed thermodynamic methods for characterization of equilibrium states of heterogeneous systems involving any number of substances. Before deriving a thermodynamic method for the characterization of equilibrium states of heterogeneous systems, we define two important terms: *i.e.*, *phase* and *component*.

### *Phase*

Phase is defined as a physically distinct, homogeneous and mechanically separable part of a system.

(*Examples*)

- The *ice – water – steam* system at 0°C has three distinct phases: solid(ice), liquid(water) and gas(steam).
- *Vapors* and *gases*, either pure or mixed, constitute a single phase because the component gases are miscible.
- *Solutions* (liquid and solid) are single phases.
- *Immiscible liquids* constitute separate phases (*e.g.*, liquid slag and metal in a furnace)

### *Component*

The number of components of a system is the minimum number of composition variables that must be specified in order to completely define the composition of each phase in the system. The number of components is not necessarily the same as the number of elements, species or compounds present in the system, but given by

$$\boxed{c = s - r}$$

where $c$ = number of components,
$s$ = number of chemically distinct constituents,
$r$ = number of algebraic relationships among the composition variables

(*Examples*)

In the *nitrogen-hydrogen-ammonia* system

- A non-reactive mixture of $N_2(g)$, $H_2(g)$ and $NH_3(g)$ at a low temperature

$$s = 3,\ r = 0 \rightarrow c = 3$$

- A mixture of $N_2(g)$, $H_2(g)$ and $H_2(g)$ at a high temperature where the following equilibrium is established:

$$N_2(g) + 3H_2(g) = 2NH_3(g) \qquad K = \frac{P_{NH_3}^2}{P_{N_2} P_{H_2}^3}$$

Thus, $s = 3,\ r = 1 \rightarrow c = 2$

One less degree of freedom than the previous case is due to the fact that the independent chemical reaction at equilibrium gives rise to a restrictive condition.

- If the above mixture was initially obtained by heating $NH_3(g)$, there exists an additional restrictive condition:

$$P_{H_2} = 3P_{N_2}$$

Thus, $s = 3,\ r = 2 \rightarrow c = 1$

Note from the above analysis that each stoichiometric relation or constraint gives rise to a restrictive condition.

We now consider more on the equality of chemical potentials at equilibrium.

**$\boldsymbol{G}$ for a closed system**

$\boldsymbol{G} = f(T, P)$

↓ Differentiation

$d\boldsymbol{G} = -\boldsymbol{S}dT + \boldsymbol{V}dP$

↓ At constant $T$ and $P$

$d\boldsymbol{G} = 0$

**$\boldsymbol{G}$ for an open system**

$\boldsymbol{G} = f(T, P, n_1, n_2, ...)$

↓ Differentiation

$d\boldsymbol{G} = -\boldsymbol{S}dT + \boldsymbol{V}dP + \mu_1 dn_1 + \mu_1 dn_1 + ...$

↓ At constant $T$ and $P$

$d\boldsymbol{G} = \mu_1 dn_1 + \mu_1 dn_1 + ...$

↓

This equation applies to each phase in the system, as each phase is allowed to exchange components with other phases in the system.

↓

$d\boldsymbol{G}^{\delta} = \mu_1^{\delta} dn_1^{\delta} + \mu_2^{\delta} dn_2^{\delta} + \cdots$

$d\boldsymbol{G}^{\gamma} = \mu_1^{\gamma} dn_1^{\gamma} + \mu_2^{\gamma} dn_2^{\gamma} + \cdots$

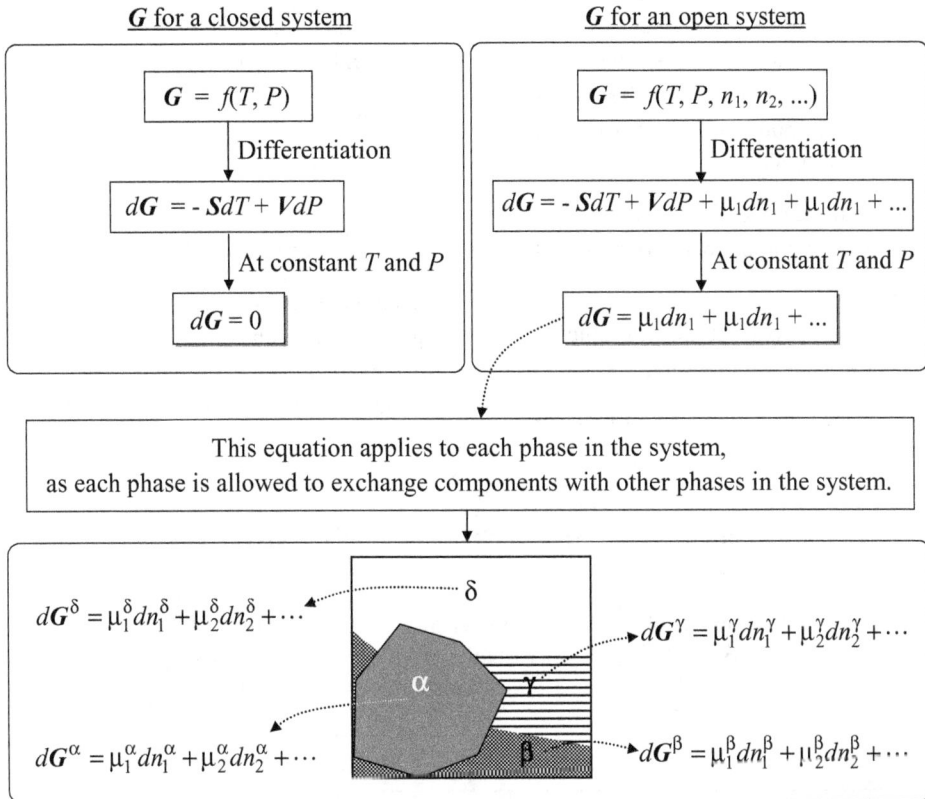

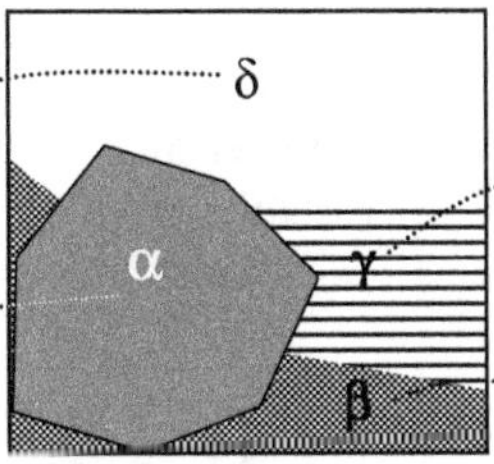


$d\boldsymbol{G}^{\alpha} = \mu_1^{\alpha} dn_1^{\alpha} + \mu_2^{\alpha} dn_2^{\alpha} + \cdots$

$d\boldsymbol{G}^{\beta} = \mu_1^{\beta} dn_1^{\beta} + \mu_2^{\beta} dn_2^{\beta} + \cdots$

For simplicity let us first consider two-component (1 and 2), two-phase (α and β) system.

$$d\boldsymbol{G} = d\boldsymbol{G}^{\alpha} + d\boldsymbol{G}^{\beta} = \mu_1^{\alpha} dn_1^{\alpha} + \mu_2^{\alpha} dn_2^{\alpha} + \mu_1^{\beta} dn_1^{\beta} + \mu_2^{\beta} dn_2^{\beta}$$

↓ From mass balance $dn_1^{\alpha} + dn_1^{\beta} = 0$ and $dn_2^{\alpha} + dn_2^{\beta} = 0$

$$d\boldsymbol{G} = (\mu_1^{\alpha} - \mu_1^{\beta})dn_1^{\alpha} + (\mu_2^{\alpha} - \mu_2^{\beta})dn_2^{\alpha} = 0$$ at equilibrium.

↓ Since $dn_1^{\alpha}$ and $dn_2^{\alpha}$ are arbitrary, and thus they are non-zero. Therefore, $(\mu_1^{\alpha}-\mu_1^{\beta})=0$, and $(\mu_2^{\alpha}-\mu_2^{\beta})=0$

$$\mu_1^{\alpha} = \mu_1^{\beta} \quad \text{and} \quad \mu_2^{\alpha} = \mu_2^{\beta}$$

At equilibrium, therefore, the chemical potential of each component is constant throughout the entire system.

We are now ready to derive an equation known as the ***Gibbs phase rule*** (or simply ***phase rule***) which applies to both homogeneous and heterogeneous systems in equilibrium. Consider a system composed of $c$ components that are distributed between $p$ phases.

*Number of variables*

There are ($c$ - 1) component variables in each phase, because by knowing the concentrations (e.g., mole fractions) of ($c$ - 1) components the last one can be found from $N_1 + N_2 + ... + N_c = 1$

Since there are $p$ phases, the total number of component variables of the system is $p(c - 1)$.

There are two additional variables: temperature and pressure.

Thus the total number of variables is $p(c - 1) + 2$

*Number of restricting equations*

The chemical potential of component $i$ is constant throughout the system:

$$\mu_1^\alpha = \mu_1^\beta = \mu_1^\gamma = \cdots = \mu_1^p$$
$$\mu_2^\alpha = \mu_2^\beta = \mu_2^\gamma = \cdots = \mu_2^p$$
$$\vdots$$
$$\mu_i^\alpha = \mu_i^\beta = \mu_i^\gamma = \cdots = \mu_i^p$$

Hence the number of independent equations for each component is ($p$ - 1). Since there are $c$ components in the system, the total number of restricting equations is $c(p - 1)$.

Therefore, the number of undetermined variables is given by,

| Number of undermined variables | = | Number of variables in the system | − | Number of restricting equations |
|---|---|---|---|---|

Number of undetermined variables $= [p(c - 1) + 2] - [c(p - 1)] = c - p + 2$

The number of undetermined variables is called the ***number of degrees of freedom*** and denoted by the symbol $f$.

$$f = c - p + 2$$

This equation is called the *Gibbs phase rule* or simply the *phase rule*. The significance of the phase rule is that,

- The phase rule offers a simple means of determining the minimum number of intensive variables that have to be specified in order to unambiguously determine the thermodynamic state of the system.

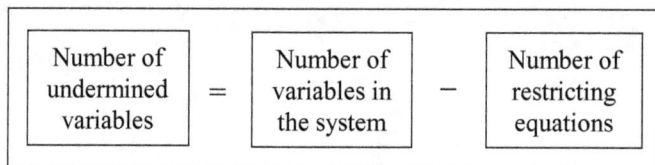

- The application of the phase rule does not require knowledge of the actual constituents of a phase.
- The phase rule applies only to systems which are in equilibrium.

*Example 11.1*

Suppose we have a system which is composed of water and steam in equilibrium.
1) Can we choose at will the equilibrium temperature of the system?
2) We add ice to the system and want to have all three phases in equilibrium. Can we choose at will the equilibrium pressure of the system?

1) $c = 1$ ($H_2O$), and $p = 2$ (gas, liquid). Thus, from the phase rule, $f = 1$.
   The significance of the value $f = 1$ is that we can choose either the equilibrium temperature or equilibrium pressure, but not both.
2) $c = 1$ ($H_2O$), and $p = 3$ (gas, liquid, solid). Thus $f = 0$
   The significance of the value $f = 0$ is that we are not allowed to freely choose either $T$ or $P$. In other words the system can only exist at a unique temperature and a unique pressure. This unique point is called the ***triple point***, which is discussed in more detail in the section to follow.

*Example 11.2*

Consider the substance M which has two allotropes α and β. Is it possible that four phases (solid α, solid β, liquid and vapor) coexist in equilibrium?

$c = 1$ (M), $p = 4$ (α, β, liquid and vapor) $\rightarrow f = -1$
The significance of the negative value of $f$ is that the system is not capable of having all the phases coexisting in equilibrium.

*Example 11.3*

The term *phase* signifies a state of matter that is uniform throughout, not only in chemical composition, but also in physical state. Determine the number of phases in each of the systems specified in the following:
(1) Ice chipped into small pieces
(2) An alloy of two metals which are immiscible
(3) An alloy of two metals which are miscible
(4) A gas mixture composed of $N_2$, $O_2$ and CO
(5) A liquid solution of A and B in equilibrium with their respective pure solids
(6) A liquid solution of A and B in equilibrium with a liquid solution of their oxides

(1) 1, (2) 2, (3) 1, (4) 1, (5) 3, (6) 2

> *Example 11.4*
>
> A system is composed of Fe(*s*), FeO(*s*), $Fe_3O_4$(*s*), CO(*g*) and $CO_2$(*g*).
>
> (1) Determine the number of degrees of freedom of the system.
>
> (2) Find the temperature at which all these five species exist together in equilibrium.
>
> (3) Determine the partial pressures of CO and $CO_2$ at 1 *atm*. total pressure.

(1) The number of independent reactions:

① Fe(*s*) + $CO_2$(*g*) = FeO(*s*) + CO(*g*) $\Delta G_1^o$

② 3FeO(*s*) + $CO_2$(*g*) = $Fe_3O_4$(*s*) + CO(*g*) $\Delta G_2^o$

Therefore,

Number of algebraic relationships among the composition variables r = 2

Number of chemically distinct constituents s = 5 (Fe, FeO, $Fe_3O_4$, CO and $CO_2$).

Thus, Number of components c = s − r = 5 − 2 = 3

Number of phases p = 4 [Fe(*s*), FeO(*s*), $Fe_3O_4$(*s*), gas(CO, $CO_2$)]

Number of degrees of freedom: f = c − p + 2 = 3 − 4 + 2 = 1

(2) For Reaction (1) Fe(*s*) + $CO_2$(*g*) = FeO(*s*) + CO(*g*)

$$K_1 = \left(\frac{P_{CO}}{P_{CO_2}}\right)_{1,eq} = \exp\left(-\frac{\Delta G_1^o}{RT}\right)$$

$$\Delta G_1^o = \left(\Delta G_{f,FeO}^o + \Delta G_{f,CO}^o\right) - \left(\Delta G_{f,CO_2}^o\right)$$

$$\Delta G_{f,FeO}^o = -264000 + 64.6T, \text{ J}$$

$$\Delta G_{f,CO}^o = -112900 - 86.5T, \text{ J}$$

$$\Delta G_{f,CO_2}^o = -394800 + 0.836T, \text{ J}$$

$$\Delta G_1^o = 17{,}900 - 22.74T, \text{ J}$$

$$\left(\frac{P_{CO}}{P_{CO_2}}\right)_{1,eq} = exp\left(-\frac{17900 - 22.7T}{RT}\right)$$

Similarly for Reaction (2) 3FeO(*s*) + $CO_2$(*g*) = $Fe_3O_4$(*s*) + CO(*g*)

$$\left(\frac{P_{CO}}{P_{CO_2}}\right)_{2,eq} = exp\left(-\frac{-29200 + 26.26T}{RT}\right)$$

In order for the two reactions to be in equilibrium together in the same system,

$$\left(\frac{P_{CO}}{P_{CO_2}}\right)_{1,eq} = \left(\frac{P_{CO}}{P_{CO_2}}\right)_{2,eq}$$

↓

$$exp\left(-\frac{17900-22.7T}{RT}\right) = exp\left(-\frac{-29200+26.26TT}{RT}\right)$$

↓

$$T = 960K = 686^{\circ}C$$

(3)

$$K_1 = K_2 = 0.61 \text{ at } 960K$$

↓

$$K_1 = K_2 = \left(\frac{P_{CO}}{P_{CO_2}}\right)$$

↓ $P_{CO} + P_{CO_2} = 1$

$$P_{CO} = 0.38atm., \quad P_{CO_2} = 0.62atm.$$

*Example 11.5*

In a reaction chamber metallic Co and Ni, and their oxides CoO and NiO exist together with oxygen at $P_{O_2}$ and temperature *T*.

(1) Can we change $P_{O_2}$ and/or T at our own discretion without losing any of the solid phases? Assume that no solid solution of either Co-Ni or CoO-NiO forms.

(2) What if solid solutions of Co-Ni and CoO-NiO form?

(1) The number of degrees of freedom is to be checked to see if we have any freedom to assign a value to any variable of the system.
Independent reactions in the system:
① Co(*s*) + 1/2$O_2$(*g*) = CoO(*s*)
② Ni(*s*) + 1/2$O_2$(*g*) = NiO(*s*)
Therefore, we have two restricting relationships: $r = 2$.
And $s = 5$ (Co, Ni, CoO, NiO, $O_2$), then $c = s - r = 5 - 2 = 3$
Number of phases: $p = 5$ [Co(*s*), Ni(*s*), CoO(*s*), NiO(*s*), $O_2$(*g*)]

Number of degrees of freedom: $f = c - p + 2 = 3 - 5 + 2 = 0$.
We have no degree of freedom: *i.e.*, this particular equilibrium condition is invariant and thus the system can exist under a unique set of variables.
The equilibrium oxygen partial pressure and temperature may be found by solving simultaneously the two equilibrium constant expressions:

$$K_{\text{Co/CoO}} = exp\left(-\frac{\Delta G^{o}_{\text{Co/CoO}}}{RT}\right) = \frac{a_{\text{CoO}}}{a_{\text{Co}} P_{\text{O}_2}^{\frac{1}{2}}} = P_{\text{O}_2}^{-\frac{1}{2}}$$

$$K_{\text{Ni/NiO}} = exp\left(-\frac{\Delta G^{o}_{\text{NiNioO}}}{RT}\right) = \frac{a_{\text{NiO}}}{a_{\text{Ni}} P_{\text{O}_2}^{\frac{1}{2}}} = P_{\text{O}_2}^{-\frac{1}{2}}$$

By using the standard Gibbs energy of formation available in the thermodynamic table in the appendix, the solution of the above two simultaneous equations gives $T$ = 123 K and $P_{\text{O}_2} = 10^{-48}$ atm. Such a low equilibrium oxygen partial pressure indicates that the coexistence of the four solid phases in equilibrium is impractical.

One convenient way of finding equilibrium temperature and oxygen partial pressure is to use of the Ellingham diagram. The crossing point of the lines of the above two reactions in the Ellingham diagram is the point at which the two reactions are in equilibrium at the same time under the same oxygen potential.

(2) When solid solutions of Co-Ni and CoO-NiO form, activities of Co, Ni, CoO and NiO are not unity any more.
Number of phases: $p = 3$ (Co-Ni solid solution, CoO-NiO solid solution, $O_2$)
Number of components: $c = 3$ (= 5 – 2), the same as the previous case.
Number of degrees of freedom: $f = c - p + 2 = 3 - 3 + 2 = 2$.
There are two degrees of freedom we can exercise. When we look at the equilibrium constant expressions given above, apparently there are 6 variables, namely,

$$T,\ P_{\text{O}_2},\ a_{\text{Co}},\ a_{\text{Ni}},\ a_{\text{CoO}},\ a_{\text{NiO}}$$

However, activities of Co and Ni in the metallic solution, and CoO and NiO in the oxide solution are respectively interrelated with each other by Gibbs-Duhem equation:

$$N_i d\ ln\ a_i + N_j d\ ln\ a_j = 0$$

And the activity itself is related to the concentration by $a_i = \gamma_i N_i$.
Therefore the number of variables is reduced to 4, but we have two equations which related these variables, and thus fixing two variables suffices in finding the rest of the variables. The choice can be, for instance, the mole fractions of Co and NiO:

$$\boxed{N_{\text{Co}}} \xrightarrow{a_{\text{Co}} = \gamma_{\text{Co}} N_{\text{Co}}} \boxed{a_{\text{Co}}} \xrightarrow{N_{\text{Co}} d\ ln\ a_{\text{Co}} + N_{\text{Ni}} d\ ln\ a_{\text{Ni}} = 0} \boxed{a_{\text{Ni}}}$$

$$\boxed{N_{\text{NiO}}} \xrightarrow{a_{\text{NiO}} = \gamma_{\text{NiO}} N_{\text{NiO}}} \boxed{a_{\text{NiO}}} \xrightarrow{N_{\text{NiO}} d\ ln\ a_{\text{NiO}} + N_{\text{CoO}} d\ ln\ a_{\text{CoO}} = 0} \boxed{a_{\text{CoO}}}$$

Alternatively we may specify the temperature and oxygen partial pressure, and then the compositions of the solutions can be found by reversing the sequence in the above.

*Exercises*

11.1 Shown below is a one-component phase diagram. There exist 4 different phases, namely, $\alpha$, $\beta$, $\gamma$ and $\delta$.

(1) Calculate the number of degrees of freedom in the area A.
(2) Calculate the number of degrees of freedom on the line B.
(3) Calculate the number of degrees of freedom at the point C.
(4) An experimenter has reported that a new phase $\varepsilon$ was found to coexist in equilibrium with $\beta$, $\gamma$ and $\delta$ phases at a particular set of conditions. What is your opinion about this report?

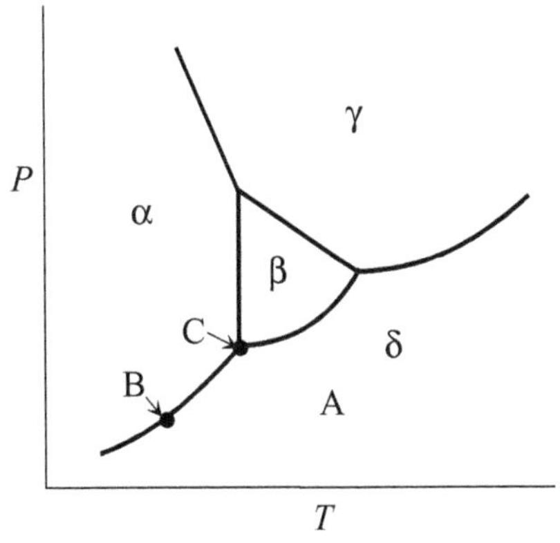


11.2 Consider a system in which the following equilibrium occurs:

$$CaCO_3(s) = CaO(s) + CO_2(g)$$

Calculate the number of degrees of freedom.

11.3 Solid zinc oxide is reduced by solid carbon at a high temperature. The system is in equilibrium and found to contain the following species:

$$ZnO(s),\ C(s),\ Zn(g),\ CO(g),\ CO_2(g)$$

(1) Calculate the number of degrees of freedom.
(2) Calculate the number of degrees of freedom if the system is initially prepared from $ZnO(s)$ and $C(s)$.

11.4 Solid solutions consisting of GaAs and InAs can be produced by equilibrating a gas mixture composed of $H_2$, HCl, InCl, GaCl and $As_4$. The composition of the solid solution is determined by the thermodynamic conditions of the system. In order to produce GaAs-InAs solid solution with a particular composition, how many intensive thermodynamic variables need to be fixed?

## 11.2 Phase Transformations

The intensive properties of a system include temperature, pressure and the chemical potentials (or partial molar Gibbs energies) of the various species present.

- If there is a difference in *temperature*, the transfer of energy occurs as *heat*: Temperature is a measure of the tendency of thermal energy to leave the system.
- If there is a difference in *pressure*, the transfer of energy occurs as *PV work*: Pressure is a measure of the tendency towards mechanical work
- If there is a difference in *chemical potential*, the transfer of energy occurs as *transfer of matter*. Chemical potential is a measure of the tendency of the species to leave the phase, to react or to spread throughout the phase through chemical reaction, diffusion, *etc.*

A substance with a higher chemical potential has a spontaneous tendency to move to a state with lower chemical potential.

Why do solid substances melt upon heating?
Why do liquid substances vaporize upon heating, but not solidify?
Why do phase transitions occur?

The chemical potential provides the key to these questions. Consider a one-component system:

$$\boxed{\left(\frac{\partial G}{\partial T}\right)_P = -S} \quad \text{or} \quad \boxed{\left(\frac{\partial \mu}{\partial T}\right)_P = -S} \qquad (G = \mu)$$

This equation shows that, because entropy is always positive, the chemical potential of a pure substance decreases as the temperature is increased. As $S_{(g)} > S_{(l)} > S_{(s)}$, the slope of the plot of μ versus $T$ is steeper for the vapor than for the liquid, and steeper for the liquid than for the solid.

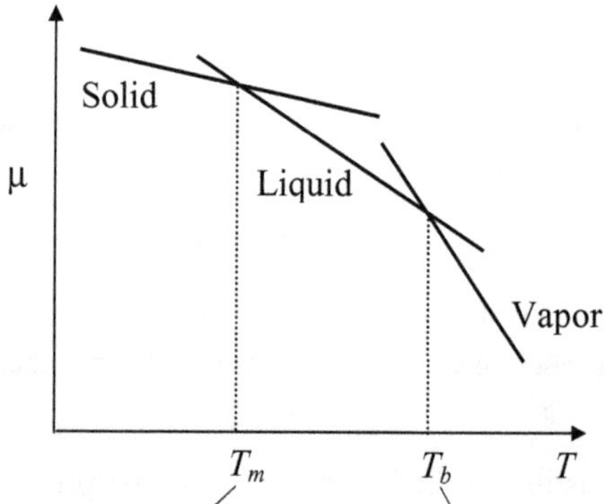


*Below* $T_m$: $\mu_{(s)} < \mu_{(l)}$: solid is stable.
*Above* $T_m$: $\mu_{(s)} > \mu_{(l)}$: liquid is stable.
*At* $T_m$: $\mu_{(s)} = \mu_{(l)}$: both are stable.
***$T_m$* : *melting point***

*Below* $T_b$: $\mu_{(l)} < \mu_{(g)}$, liquid is stable.
*Above* $T_b$: $\mu_{(l)} > \mu_{(g)}$: gas is stable.
*At* $T_b$: $\mu_{(l)} = \mu_{(g)}$: both are stable.
***$T_b$*: *boiling point***

When a component is heated, the supplied heat raises the enthalpy and entropy at the rates of their respective relationship with heat capacity of the component ($C_P$).
We know the following relationships:

$$\left(\frac{\partial H}{\partial T}\right)_P = C_P\,,\ \left(\frac{\partial S}{\partial T}\right)_P = \frac{C_P}{T} \text{ and } \left(\frac{\partial G}{\partial T}\right)_P = -S$$

We also know that,

$$G = H - TS$$

From the above relationships we may draw the following schematic diagram:

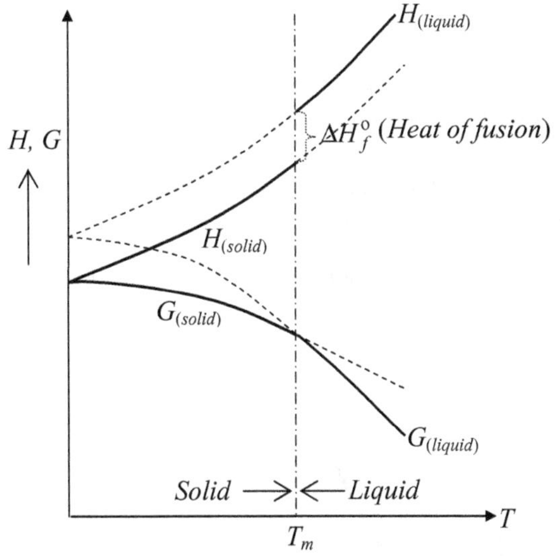


Enthalpy increases with temperature at the rate of the value of $C_p$.

At $T_m$ it increases stepwise due to the ***heat of fusion*** (sometimes called ***latent heat***).

The Gibbs energy curves of solid and liquid cross at the melting point as the curve for liquid is steeper downward in slope due to higher entropy of liquid.

A ***phase transition***, the spontaneous conversion from one phase to another, occurs at a characteristic temperature for a given pressure. This characteristic temperature is called the ***transition temperature***.

**Effect of pressure**

Next, examine the effect of pressure on the chemical potential:

$$\boxed{\mu = VdP - SdT}$$

$\downarrow$ at constant $T$

$$\boxed{\left(\frac{\partial \mu}{\partial P}\right)_T = V}$$

As $V$ is always positive, an increase in pressure increases the chemical potential of any pure substance. For most substances, $V_{(l)} > V_{(s)}$ and hence the increase in pressure increases the chemical potential, but more for the liquid than the solid.

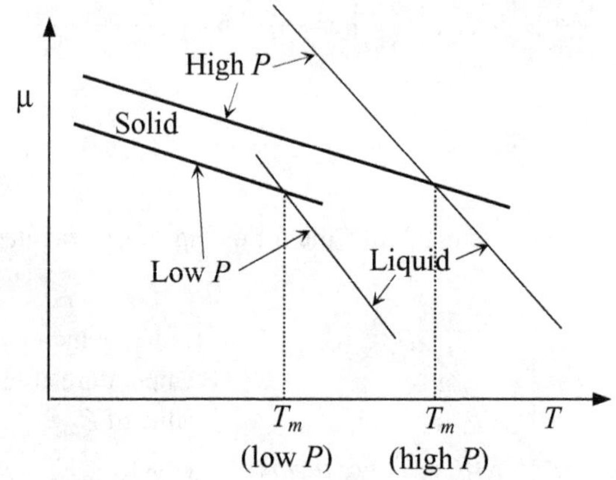


Thus the increase in pressure results in increase in the melting temperature. If $V_{(l)} < V_{(s)}$, however, an increase in pressure effects a decrease in $T_m$. (*e.g.*, $V_{water} < V_{ice}$)

It would be useful to combine the effects of temperature and pressure on phase transition of a substance in a same diagram.

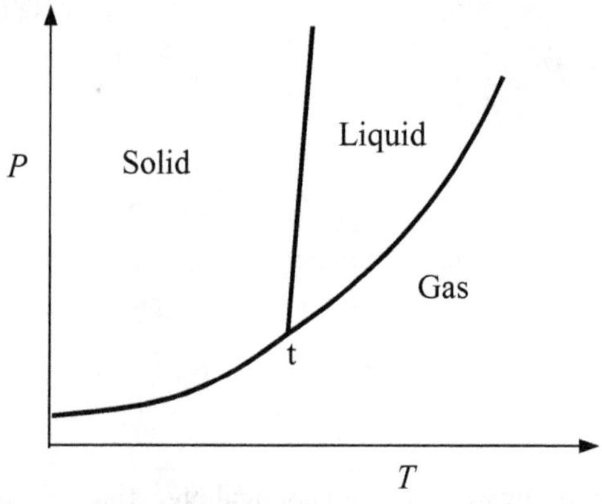


This figure is known as the ***phase diagram*** of a substance. It shows the thermodynamically stable phases at different pressures and temperatures. The lines separating phases are called as ***phase boundaries*** at which two phases coexist in equilibrium. The point "t" is the ***triple point*** at which three phases coexist in equilibrium.

Now, a question arises, "Is there a way to quantitatively describe the phase boundaries in terms of $P$ and $T$?" The phase rule predicts the existence of the phase boundaries, but does not give any clue on the shape (slope) of the boundaries. To answer the above question, we make use of the fact that at equilibrium the chemical potential of a substance is the same in all phases present.

Consider the phases α and β which are in equilibrium.

$$\boxed{\mu^{\alpha} = \mu^{\beta}}$$

↓ Differentiation

$$\boxed{d\mu^{\alpha} = d\mu^{\beta}}$$

Recall that

$$dG = d\mu = VdP - SdT$$

Therefore

$$d\mu^{\alpha} = V^{\alpha}dP - S^{\alpha}dT$$

$$d\mu^{\beta} = V^{\beta}dP - S^{\beta}dT$$

↓

$$\boxed{V^{\alpha}dP - S^{\alpha}dT = V^{\beta}dP - S^{\beta}dT}$$

↓ Rearrangement yields

$$\boxed{\frac{dP}{dT} = \frac{(S^{\beta} - S^{\alpha})}{(V^{\beta} - V^{\alpha})}}$$

$(S^{\beta} - S^{\alpha})$
Entropy change ($\Delta S$) associated with the phase change of $\alpha \rightarrow \beta$

$(V^{\beta} - V^{\alpha})$
Volume change ($\Delta V$) associated with the phase change of $\alpha \rightarrow \beta$

↓

$$\boxed{\frac{dP}{dT} = \frac{\Delta S}{\Delta V}}$$

For a transition from one phase to another <u>under equilibrium conditions</u> at constant temperature and pressure:

$$\Delta G = \Delta H - T\Delta S = 0$$

Therefore

$$\Delta S = \frac{\Delta H}{T}$$

↓

$$\boxed{\frac{dP}{dT} = \frac{\Delta H}{T\Delta V}}$$

This equation is known as the ***Clapeyron equation***. The application of Clapeyron equation is limited to equilibrium involving phases of fixed composition (*e.g.*, one-component system) because μ has been assumed independent of composition:

$$d\mu = VdP - SdT + \cancel{\mu_1 dn_1} + \cancel{\mu_2 dn_2} +$$

For the solid-liquid phase boundary,

$$\boxed{\frac{dP}{dT} = \frac{\Delta H_f}{T\Delta V_f}}$$

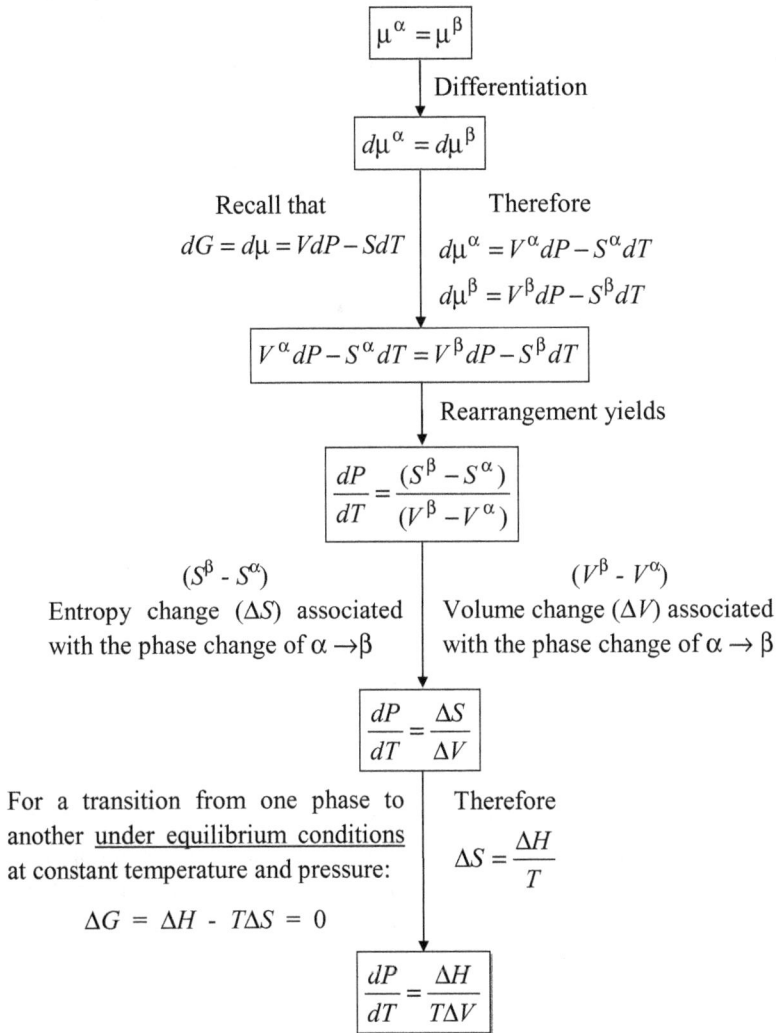

where $\Delta H_f$ is the heat(enthalpy change) of fusion, and it always has a positive value, and $\Delta V_f$ is the volume change upon melting, and greater than zero in most cases and always small. Therefore $dP/dT$ is positive and very large. This indicates that the slope of the solid-liquid phase boundary is positive and steep in most cases.

Integrating the Clapeyron equation assuming that $\Delta H_f$ and $\Delta V_f$ are constant,

$$P_2 - P_1 = \frac{\Delta H_f}{\Delta V_f} ln\left(\frac{T_2}{T_1}\right) \cong \frac{\Delta H_f}{T_1 \Delta V_f}(T_2 - T_1)$$

For the liquid-vapor phase boundary,

$$\frac{dP}{dT} = \frac{\Delta H_v}{T\Delta V_v}$$

$\Delta H_v$ : Heat of vaporization > 0 $\Delta V_v$ : Volume change upon vaporization $= V_{(g)} - V_{(l)} \cong V_{(g)} = RT/P$

$$\frac{d\,ln\,P}{dT} = \frac{\Delta H_v}{RT^2}$$

This equation is known as the ***Clausius-Clapeyron equation***. Integration yields,

$$ln\left(\frac{P_2}{P_1}\right) = -\frac{\Delta H_v}{R}\left(\frac{1}{T_2} - \frac{1}{T_1}\right)$$

Using this equation $\Delta H_v$ can be estimated with knowledge of the equilibrium vapor pressure of a liquid at two different temperatures. For the solid-vapor phase boundary (sublimation), an analogous equation is obtained by replacing $\Delta H_v$ with the heat of sublimation $\Delta H_s$.

*Example 11.6*

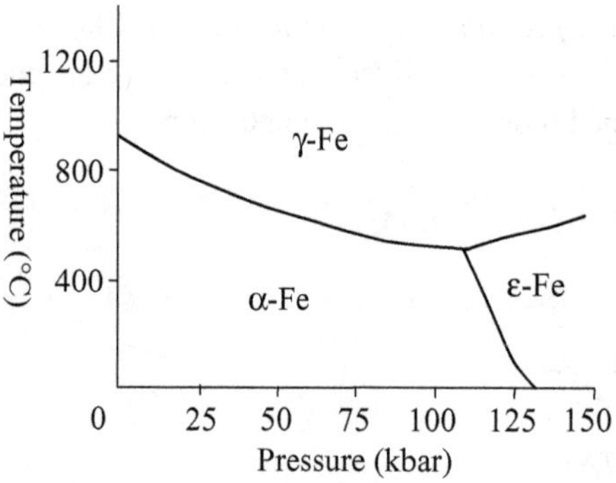


Shown is the equilibrium phase diagram of pure iron. It shows the effect of pressure on the equilibrium temperature. It is seen from the diagram that increasing pressure effects depressing the equilibrium temperature of α- and γ-phases. Suggest a reason for this.
It is known that $\Delta H = H_\gamma - H_\alpha > 0$.

From Clapeyron equation,

$$\frac{dP}{dT} = \frac{\Delta H}{T\Delta V}$$

$\Delta H > 0$ and, γ-Fe(*fcc* structure) has smaller molar volume than α-Fe (bcc structure).

$$\Delta V = V_{\gamma(fcc)} - V_{\alpha(bcc)} < 0$$

$$\frac{dP}{dT} < 0$$

This result tells us that pressure and temperature adversely affect each other: *that is*,

- Increasing temperature has the effect of depressing temperature, or
- Increasing pressure has the effect of decreasing temperature.

*Example 11.7*

Substance A in a condensed phase is in equilibrium with its own vapor at temperature $T$. A is the only species in the gas phase. Now an inert gas is introduced into the system so that the total pressure rises from $P^{\circ}$ to $P'$. Assuming that the inert gas behaves ideally with A and does not dissolve in the condensed phase, choose the correct one from the following:

(1) The vapor pressure of A is independent of the total pressure of the system.
(2) The vapor pressure of A is affected by the total pressure, but the change in the vapor pressure is generally negligibly small.
(3) The vapor pressure of A is affected to a large extent by the total pressure.

From the knowledge that A in the condensed phase is in equilibrium with A in the gaseous phase,

$$dG_{A(c)} = dG_{A(g)}$$

At constant $T$

$$dG = VdP - S\cancel{dT}$$

$$V_{A(c)}dP_{A(c)} = V_{A(g)}dP_{A(g)}$$

$P_{A(c)}$ : Pressure exerted on the condensed phase, *i.e.*, the total pressure of the gas.

$P_{A(g)}$ : Vapor pressure of A in the gas phase, *i.e.*, the partial pressure of A in the gas phase.

$$V_{A(g)} = \frac{RT}{P_{A(g)}}$$

$$V_{A(c)} dP_{A(c)} = \left(\frac{RT}{P_{A(g)}}\right) dP_{A(g)}$$

Integration with the following limits:
Before adding the inert gas

$$P_{A(g)} = P^b_{A(g)}, \quad P_{A(c)} = P^b_{A(g)}$$

After adding the inert gas:

$$P_{A(g)} = P^a_{A(g)}, \quad P_{A(c)} = P$$

$$ln\left(\frac{P^a_{A(g)}}{P^b_{A(g)}}\right) = \frac{V_{A(c)}}{RT}\left(P - P^b_{A(g)}\right) > 0$$

$$P^a_{A(g)} > P^b_{A(g)}$$

However, $V_{A(c)}/RT$ in the equation is small and hence the difference between $P^a_{A(g)}$ and $P^b_{A(g)}$ is negligibly small.

*Exercises*

11.5 A system contains ice at 0°C and 1 atm. Calculate the change in the chemical potential of ice associated with the increase in the pressure from 1 atm. to 2 atm. The density of ice is 0.915 g cm$^{-3}$. Calculate the change in the chemical potential of water associated with the increase in the pressure from 1 atm. to 2 atm. Ice is now in equilibrium with water at 0°C and 1 atm. If the pressure in the system is increase to 2 atm., will ice melt, freeze or remain unchanged?

11.6 Consider the phase transformation

$$\text{HgS}(s, \text{red}) = \text{HgS}(s, \text{black}) \quad \Delta G^\circ_r = 4{,}180 - 5.44T, \text{ J}$$

(1) Determine the enthalpy change of the reaction.
(2) Determine the transition temperature at $P = 1$ *atm*.
(3) Determine the pressure at which both HgS($s$, red) and HgS($s$, black) coexist in equilibrium at 500°C.

$$M_{HgS} = 232.7 \text{ g mol}^{-1}, \quad \rho_{HgS(red)} = 8.1 \text{ g cm}^{-3}, \quad \rho_{HgS(black)} = 7.7 \text{ g cm}^{-3}$$

(4) Calculate $dP/dT$ at 495°C using the Clapeyron equation.

11.7 The vapor pressure of liquid iron is given by the equation

$$\log P_{\mathrm{Fe}} = \frac{-19,710}{T} - 1.27 \log T + 13.27 \qquad \text{(torr)}$$

Calculate the standard heat of vaporization at 1,600°C

11.8 The vapor pressures of solid and liquid zinc are given by

$$\ln P_{\mathrm{Zn}(s)} = \frac{-15,780}{T} - 0.755 \ln T + 19.3 \qquad \text{(atm)}$$

$$\ln P_{\mathrm{Zn}(l)} = \frac{-15,250}{T} - 2.255 \ln T + 21.3 \qquad \text{(atm)}$$

Calculate the temperature at which solid, liquid and gaseous zinc coexist in equilibrium (triple point).

11.9 The equilibrium vapor pressures of solid and liquid $NH_3$ are given by

$$\ln P_{\mathrm{NH_3}(s)} = 23.03 - \frac{3,754}{T} \qquad \text{(torr)}$$

$$\ln P_{\mathrm{NH_3}(l)} = 19.49 - \frac{3,063}{T} \qquad \text{(torr)}$$

Calculate the heat of fusion ($\Delta H_f$) of $NH_3$.

## 11.3 Phase Equilibria and Gibbs Energies

When a liquid solution is cooled slowly, temperature will eventually reach the ***liquidus*** point, and a solid phase will begin to separate from the liquid solution.

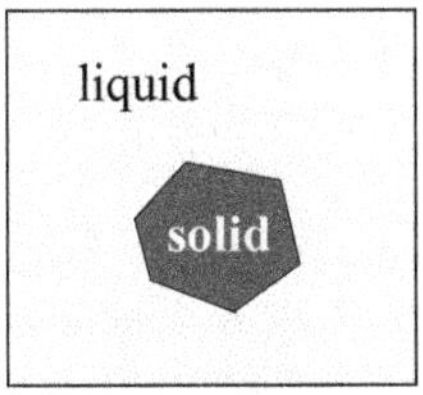


This solid phase precipitated could be a pure component, a solid solution or a compound. Now a question arises as to in which direction the system will proceed with cooling: Precipitating a pure component, a solid solution, a compound or something else? The answer is *to the state at which the Gibbs energy of the system is at minimum under given thermodynamic conditions*. Therefore phase changes can be predicted from thermodynamic information on the Gibbs energy-composition-temperature relationship.

Suppose that we have a binary A-B solution. The Gibbs energy of mixing of the solution ($G^M$) varies with the composition of the solution according to the following relationship which has already been discussed:

$$\boxed{G^M = N_{\mathrm{A}} G_{\mathrm{A}}^M + N_{\mathrm{B}} G_{\mathrm{B}}^M}$$

$$\downarrow \quad G_i^M = RT \ln a_i$$

$$G^M = RT\left(N_A \ln a_A + N_B \ln a_B\right)$$

Suppose that the temperature of the solution is sufficiently higher than the melting points of both A and B:

$$T_1 >> T_{m(A)},\ T_{m(B)}$$

Thus liquid is stable for both A and B. The natural choice of the standard state for A and B will then be the pure liquid A and pure liquid B.
Therefore,

$G^M_{A(l)} = 0$ and $G^M_{B(l)} = 0$ for pure *liquid* A and B, but

$G^M_{A(s)} > 0$ and $G^M_{B(s)} > 0$ for pure *solid* A and B, as they are unstable.

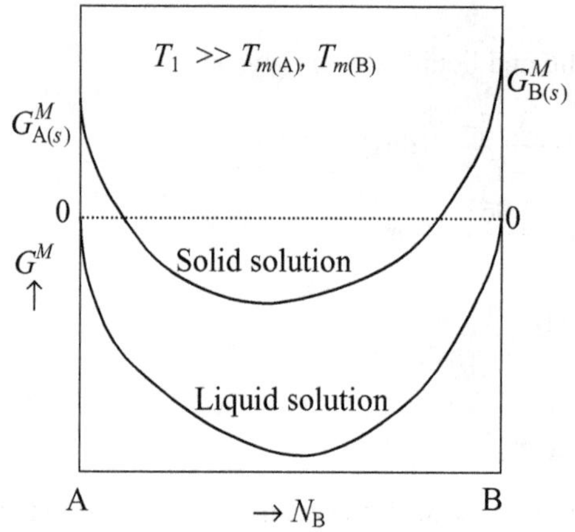


The plot of Gibbs energy of mixing of the solution according to the equation

$$G^M = RT\left(N_A \ln a_A + N_B \ln a_B\right)$$

may look like the figure seen on the left. As the prevailing temperature is higher than the melting points of both components, the liquid solution has lower Gibbs energy of mixing over the entire range of composition, and hence the liquid phase is more stable than the solid phase at $T_1$.

Next, consider the opposite case: The temperature is sufficiently lower than the melting points of both A and B:

$$T_2 << T_{m(A)},\ T_{m(B)}$$

Then the solid phase is more stable for both A and B, and hence the natural choice of the standard state for A and B will be the pure solid A and pure solid B.

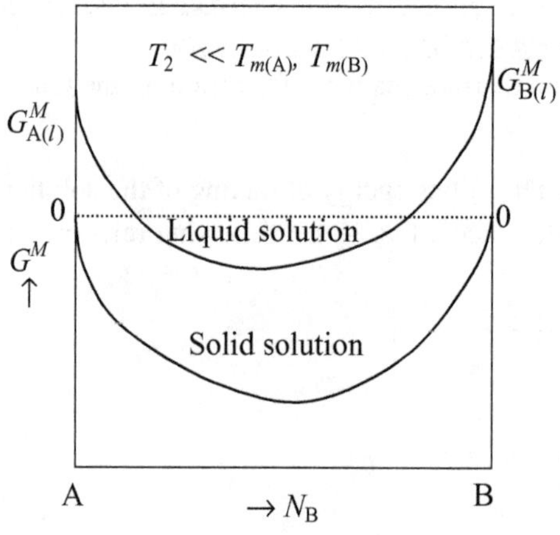


Therefore,

$G^M_{A(s)} = 0$ and $G^M_{B(s)} = 0$

for pure *solid* A and B, but

$G^M_{A(l)} > 0$ and $G^M_{B(l)} > 0$

for pure *liquid* A and B.

As the temperature is much lower than the melting points of both components, the solid solution has lower Gibbs energy of mixing than the liquid solution over the entire range of composition, and thus the solid phase is more stable than the liquid phase at $T_2$.as shown in the figure.

If the change in the Gibbs energy of mixing ($G^M$) of a mixture with composition is convex downward at constant temperature and pressure, why is the solution homogeneous over the entire range of composition? Would there be any possibility of phase separation, for instance, two immiscible liquid phases? To answer this question, consider a mixture of A and B with an average composition of "*a*" shown in the following figure.

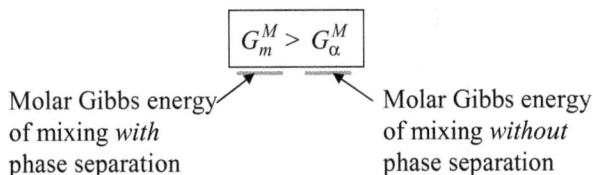

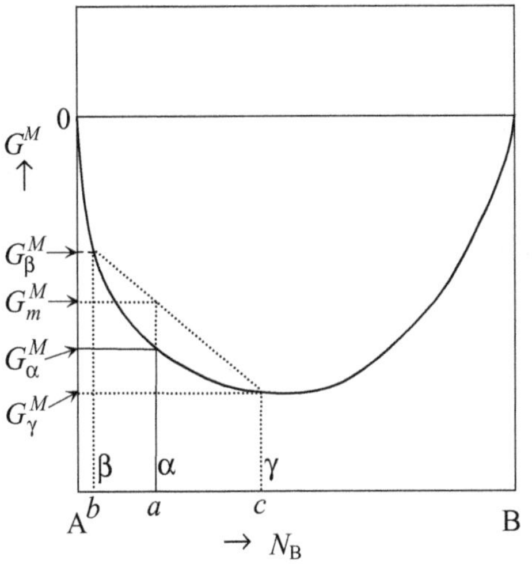


If this mixture forms a homogeneous solution (α), the Gibbs energy of mixing of the solution is given by $G_\alpha^M$.

Is there any way to lower the Gibbs energy of mixing below this value? What if the mixture forms two separate phases, β and γ, with the composition of *b* and *c*, respectively? Then the molar Gibbs energies of mixing of β and γ are $G_\beta^M$ and $G_\gamma^M$, respectively.

As the average composition of the system after phase separation (sum average of β and γ phases) must be the same as the initial mixture (*a*), the fraction of each phase ($N_\beta$ and $N_\gamma$) is given by

$$N_\beta = \frac{ac}{bc} \qquad N_\gamma = \frac{ba}{bc}$$

Then the average free energy of mixing of the mixture of β and γ phases ($G_m^M$) is given by

$$G_m^M = N_\beta G_\beta^M + N_\gamma G_\gamma^M$$

This value of $G_m^M$ is shown in the above figure, and it is clear that

$$\boxed{G_m^M > G_\alpha^M}$$

Molar Gibbs energy of mixing *with* phase separation ($G_m^M$) — Molar Gibbs energy of mixing *without* phase separation ($G_\alpha^M$)

Therefore, we may conclude that, if the change in the Gibbs energy of mixing ($G^M$) of a mixture with composition is convex downward at constant temperature and pressure, the phase separation is energetically not feasible and thus the mixture forms a homogeneous solution in equilibrium over the entire range of composition.

Now we have a question as to the shape of the Gibbs energy curve: Why is the curve of $G^M$ versus $N_i$ convex downward? Always convex downward? Or is it also possible to show other shapes? What are the important factors which determine the shape of the Gibbs energy curve?

Recall that,

$$G^M = H^M - TS^M$$

$$G^M = RT\left(N_A \ln a_A + N_B \ln a_B\right)$$

These two equations are basically the same, but for convenience we will discuss both of these equations individually.

From the first equation, $G^M = H^M - TS^M$ , it is clear that the shape of the $G^M$ curve is determined by how the two terms of $H^M$ and $-TS^M$ vary with the composition of the solution.

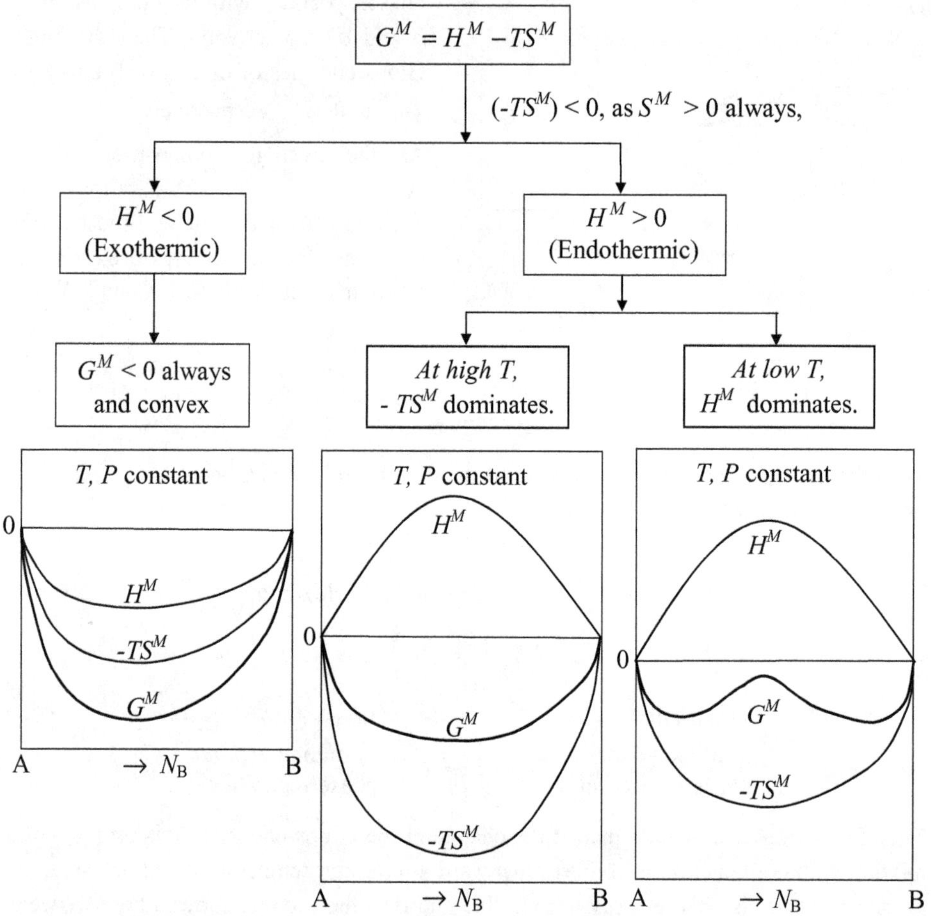


In the case that the Gibbs energy curve versus concentration is not monotonically convex downward as seen in the above at low temperature and $H^M > 0$, the mixture of A and B does not form a homogeneous solution over the entire range of the composition. Instead there will be a phase separation in a certain range of the composition where the Gibbs energy of solution can be lowered by the separation.

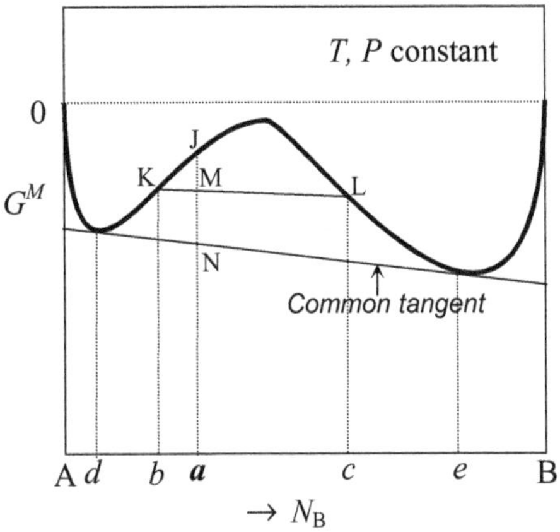


Consider a solution of composition "***a***" at constant temperature and pressure. If the solution does not dissociate, but maintains homogeneity, $G^M$ of the solution is represented by "J" in the figure. It can be seen from the figure that it is possible to lower $G^M$ below J by dissociating into two separate coexisting solutions. For instance, if the solution dissociates into two solutions of compositions "*b*" and "*c*", respectively,

$G^M$ for solution $b$ = K

$G^M$ for solution $c$ = L

↓ Applying the lever rule,

Average $G^M$ of the two solutions = M

↓ Note M < J

Minimum $G^M$ occurs
when the solution dissociates into two solutions of composition "*d*" and "*e*" which are the intercepts of common tangent to the Gibbs energy curve.

↓

Average $G^M$ of the two solutions = N

Fraction of solution $d$ = $\dfrac{ae}{de}$, Fraction of solution $e$ = $\dfrac{da}{de}$

*Summary*

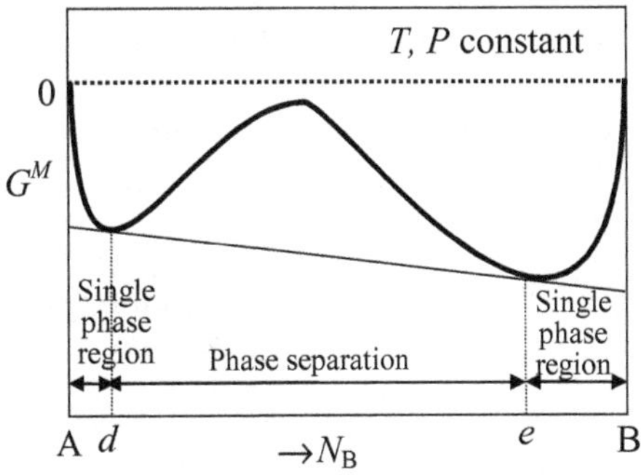


When the composition of A-B mixture lies in the "phase separation" region (between *d* and *e* in the figure), the mixture does not form a single solution, but two separate solutions; one with the composition *d* and the other with the composition *e*. For liquid, they will form two liquid solutions which are ***immiscible***. For solid, they will form a so-called ***miscibility gap***.

We now discuss more about the miscibility gap.

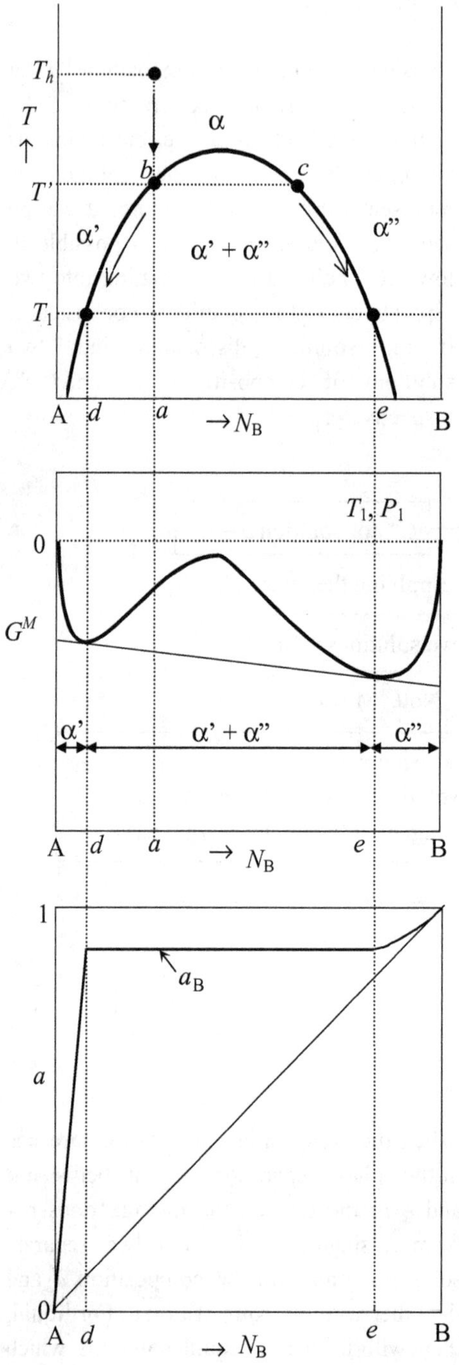


Suppose we have a solution of composition "*a*" at a sufficiently high temperature.

When the solution is slowly cooled, it maintains homogeneous α phase until the temperature reaches *T'* at which it begins to dissociate into phase α'(*b*) and α"(*c*).

As the temperature decreases further, the compositions of α' and α" are changed along the gap boundary curve.

At the temperature $T_1$

- Composition of α' : *d*
- Composition of α" : *e*
- Fraction of α' $= \dfrac{ae}{de}$
- Fraction of α" $= \dfrac{da}{de}$

When the average composition is changed, the proportions of α' and α" are changed accordingly, but the compositions of α' and α" stay at *d* and *e*, respectively, as long as the average composition lies between *d* and *e*.

Now, we examine the change in the activity of B, $a_B$, with the change in the overall composition. Recall that

$$G_B^M = RT \ln a_B$$

Since $G_B^M$ is constant between *d* and *e* (common tangent), $a_B$ is also constant in this composition range. This is obvious, because, although the average composition changes, the compositions of the individual phases (α' and α") remain unchanged.

More discussions on the miscibility gap are given in the chapter to follow.

Next, we discuss the plot of the Gibbs energy of solution versus composition with the following equations:

$$G^M = RT\left(N_A \ln a_A + N_B \ln a_B\right)$$

$$a_i = \gamma_i N_i$$

$$G^M = RT\left(N_A \ln N_A + N_B \ln N_B\right) + RT\left(N_A \ln \gamma_A + N_B \ln \gamma_B\right)$$

Term for ideal behavior — Excess term due to non-ideal behavior

If the solution behaves ideally, *i.e.*, $\gamma_i = 1$, the second term (excess term) vanishes as $\ln 1 = 0$.

If the solution exhibits a positive deviation from the ideal behavior, *i.e.*, $\gamma_i > 1$, the excess term in the above equation becomes positive and the Gibbs energy of solution is larger than that of the ideal solution.

On the other hand, if the solution exhibits a negative deviation, *i.e.*, $\gamma_i < 1$, the excess term becomes negative and thus the Gibbs energy of solution is smaller than that of the ideal solution.

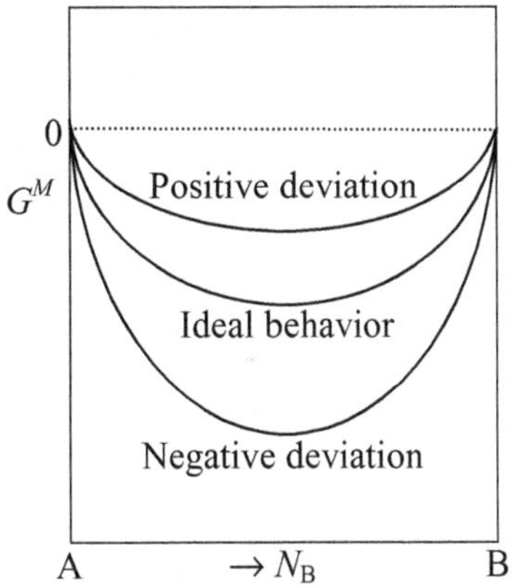


If a solution follows the regular solution model, we may modify the above equation accordingly.

$$G^M = RT\left(N_A \ln N_A + N_B \ln N_B\right) + RT\left(N_A \ln \gamma_A + N_B \ln \gamma_B\right)$$

For regular solution

$$RT \ln \gamma_A = \Omega N_B^2$$

$$RT \ln \gamma_B = \Omega N_A^2$$

$$G^M = \Omega N_A N_B + RT\left(N_A \ln N_A + N_B \ln N_B\right)$$

Knowing that,

$$G^M = H^M - TS^M$$

$$H^M = \Omega N_A N_B \text{ and, } S^M = -R\left(N_A \ln N_A + N_B \ln N_B\right)$$

As can be seen above that the regular solution exhibits non-zero $H^M$ and ideal $S^M$.

- If the formation of solution is exothermic, $\Omega < 0$.
- If the formation of solution is endothermic, $\Omega > 0$.

Graphical presentations of the above relationships for a regular solution are given below:

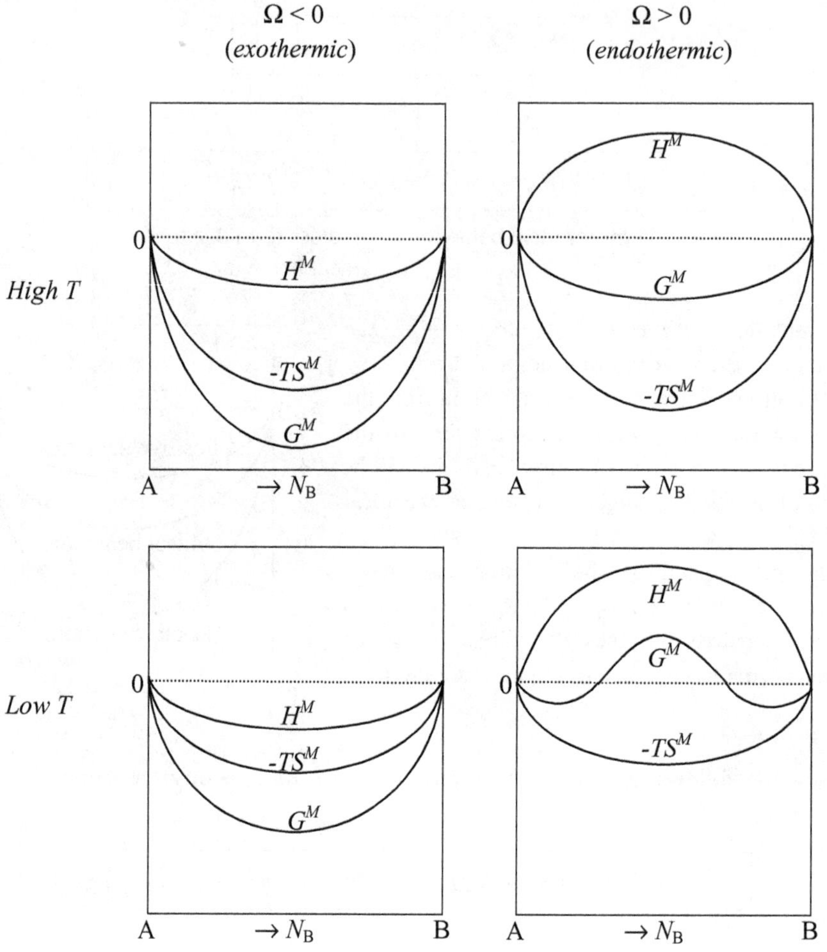


From the above figures,

- For exothermic solutions ($H^M < 0$, *i.e.*, $\Omega < 0$), the formation of solution results in a decrease in Gibbs energy at all temperatures: positive curvature ($\partial^2 G^M / \partial N_B^2 > 0$).
- For endothermic solutions ($H^M > 0$, *i.e.*, $\Omega > 0$), the Gibbs energy change for the formation of solution is rather complicated.
  - At high temperatures, $-TS^M$ term dominates, and thus Gibbs energy decreases by forming a solution: positive curvature ($\partial^2 G^M / \partial N_B^2 > 0$)
  - At low temperatures, $-TS^M$ term becomes smaller, and the Gibbs energy curve shows "convex upward" in the middle region of composition: negative curvature ($\partial^2 G^M / \partial N_B^2 < 0$). This leads to miscibility gap.

We now consider an A-B binary solution at temperature $T_2$ which is above the melting point of A ($T_{m,\text{A}}$), but below the melting point of B ($T_{m,\text{B}}$).

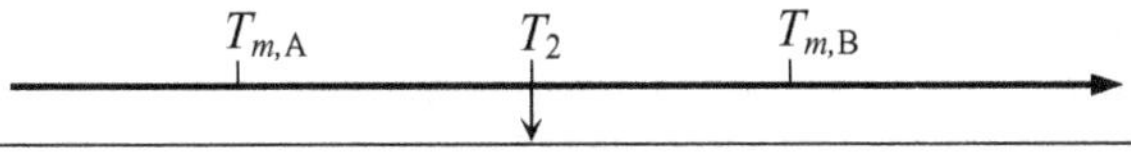


A: <u>Liquid is stable</u> → Pure *liquid* standard state : $G^M_{\text{A}(l,pure)} = 0$

Solid is unstable → Higher energy state than liquid: $G^M_{\text{A}(s,pure)} > 0$

B: <u>Solid is stable</u> → Pure *solid* standard state : $G^M_{\text{B}(s,pure)} = 0$

Liquid is unstable → Higher energy state than solid: $G^M_{\text{A}(l,pure)} > 0$

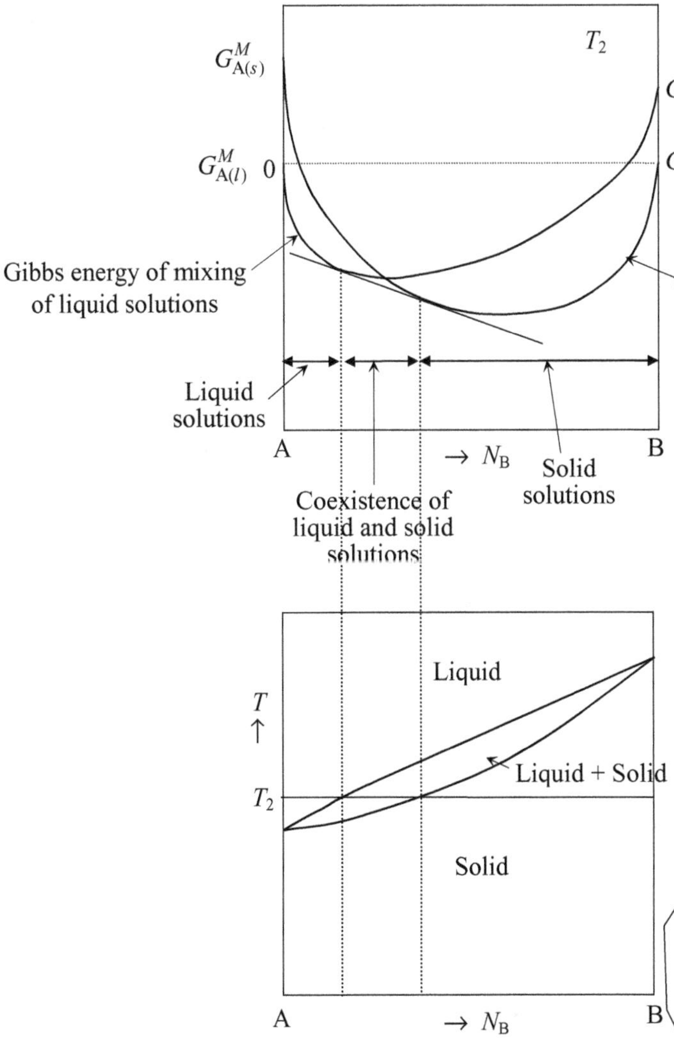


This is an example of phase diagrams when A and B have the same or similar crystal structures.

Complete solid solubility requires components to have the same crystal structure, and similar atomic size, electronegativity and valency. If any of these conditions are not met, a miscibility gap will occur in the solid state.

Consider a system which consists of components A and B that have differing crystal structures and has the phase diagram as shown below:

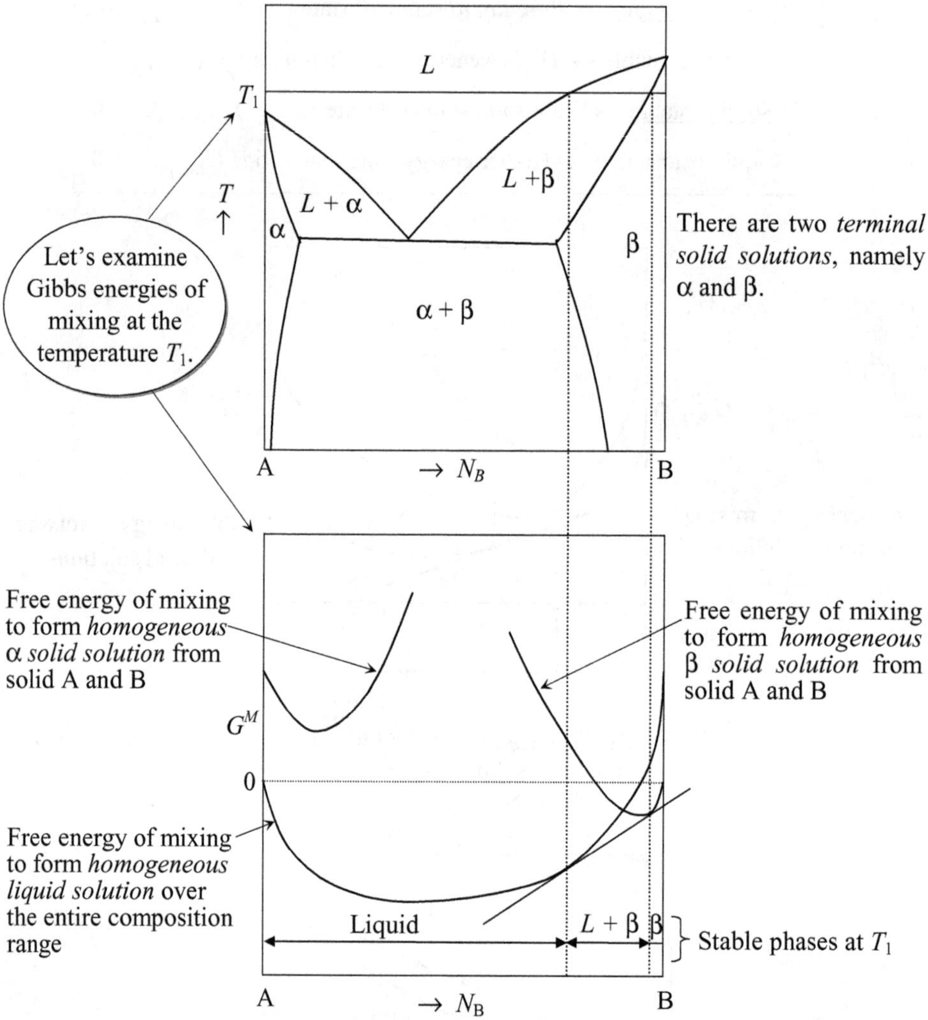
L
$T_1$
$L + \beta$
$T$ ↑
$L + \alpha$
α
β
There are two *terminal solid solutions*, namely α and β.
Let's examine Gibbs energies of mixing at the temperature $T_1$.
α + β
A
→ $N_B$
B
Free energy of mixing to form *homogeneous* α *solid solution* from solid A and B
Free energy of mixing to form *homogeneous* β *solid solution* from solid A and B
$G^M$
0
Free energy of mixing to form *homogeneous liquid solution* over the entire composition range
Liquid
$L + \beta$
β
Stable phases at $T_1$
A
→ $N_B$
B

> *Example 11.8*
>
> Can the Gibbs energy of solution be positive over the entire range of composition of a solution?

We will first check if addition of an infinitesimal amount of a solute to a pure substance increases or decreases the Gibbs energy of solution.

$$G^M = N_A G_A^M + N_B G_B^M$$

$G_i^M = RT\, ln\, a_i$, $a_i = \gamma_i N_i$

$$G^M = RT(N_A\, ln\, N_A + N_B\, ln\, N_B) + RT(N_A\, ln\, \gamma_A + N_B\, ln\, \gamma_B)$$

Knowing $N_A + N_B = 1$ and differentiation yields,

$$\left(\frac{\partial G^M}{\partial N_B}\right)_{P,T} = RT\, ln\left(\frac{N_B}{1-N_B}\right) - RT\, ln\, \gamma_A + RT\, ln\, \gamma_B + RTN_A \frac{\partial\, ln\, \gamma_A}{\partial N_B} + RTN_B \frac{\partial\, ln\, \gamma_B}{\partial N_B}$$

If B is infinitely dilute, *i.e.*, $N_B \to 0$,

then $\dfrac{N_B}{1-N_B} \to 0$, and hence

$ln\left(\dfrac{N_B}{1-N_B}\right) \to -\infty$

Thus, if B is infinitely dilute,

$$\left(\frac{\partial G^M}{\partial N_B}\right)_{P,T} \to -\infty$$

This equation clearly indicates that, at the very beginning of addition of the solute B, the Gibbs energy of solution $G^M$ sharply decreases with the slope of negative infinity, irrespective of the values of $\gamma_A$ and $\gamma_B$. Therefore, the Gibbs energy of solution should become negative, at least when the solution is sufficiently dilute. Even if the formation of a solution is extremely endothermic ( $H^M >> 0$), and thus the Gibbs energy of solution tends to be strong positive, it must be negative when the solution is extremely dilute. The figure here schematically shows the change of $G^M$ with composition when $H^M$ is large positive. The negative infinity of the slope at

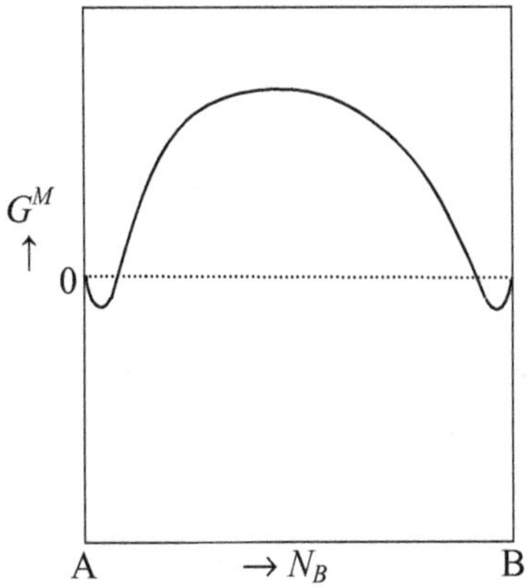

both ends implies strong tendency to form a solution. This adversely tells us the difficulty of purification of a material to the extreme.

*Example 11.9*

If the temperature is not much lower than the true melting point ($T_m$), the Gibbs energy of fusion can be estimated by the following equation:

$$\Delta G_f = \Delta H_f^{\circ}\left(1-\frac{T}{T_m}\right)$$ where $\Delta H_f^{\circ}$: enthalpy of fusion at $T_m$

Prove the above equation.

At the temperature T,

$$\Delta G_f = \Delta H_f - T\Delta S_f$$

At the true melting point.

$$\Delta G_f^{\circ} = \Delta H_f^{\circ} - T_m \Delta S_f^{\circ}$$

$$\Delta G_f^{\circ} = 0$$

$$\Delta S_f^{\circ} = \frac{\Delta H_f^{\circ}}{T_m}$$

Change in $\Delta H$ and $\Delta S$ with a small change in $T$ is negligibly small.
Thus,

$$\Delta H_f \cong \Delta H_f^{\circ}$$

$$\Delta S_f \cong \Delta S_f^{\circ}$$

$$\Delta G_f = \Delta H_f^{\circ}\left(1-\frac{T}{T_m}\right)$$

*Example 11.10*

The critical temperature, $T_C$, associated with the miscibility gap in the phase diagram shown below, is the temperature below which phase separation occurs.
If A-B binary solution follows the regular solution model, $T_C$ is given by

$$T_C = \frac{\Omega}{2R}$$

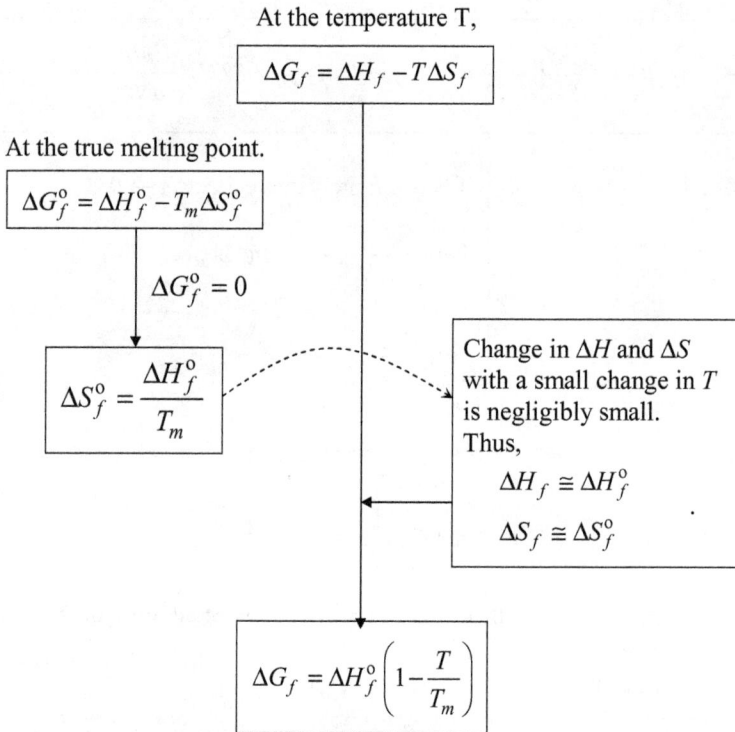

Using the figure shown below, prove the above equation.

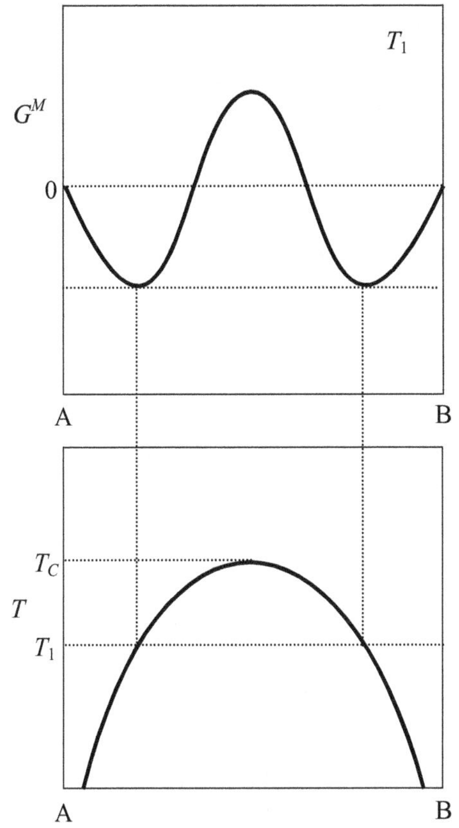


$$G^M = H^M - TS^M$$

For regular solution,

$$H^M = N_A N_B \Omega$$

$$S^M = -R(N_A \ln N_A + N_B \ln N_B)$$

$$G^M = \Omega N_A N_B + RT\left(N_A \ln N_A + N_B \ln N_B\right)$$

This is the equation which represents the Gibbs energy curve of the figure above. This figure shows some characteristic features that,

- Below $T_C$, the equation gives two minima and hence two inflection points.

***Inflection point***

A point on a curve at which the curvature (second derivative) changes sign. The curve changes from being concave upwards (positive curvature) to concave downwards (negative curvature), or vice versa

- Above $T_C$, the equation gives a curve convex downwards (or concave upwards).
- At $T_C$, the two minima and the two inflection points coincide.

Mathematically,

$$\frac{\partial G^M}{\partial N_B} = \frac{\partial^2 G^M}{\partial N_B^2}$$

$$G^M = \Omega N_A N_B + RT\left(N_A \ln N_A + N_B \ln N_B\right)$$

$$T = \frac{2N_B(1-N_B)\Omega}{R}$$

The maximum $T$ $\left(\frac{\partial T}{\partial N_B} = 0\right)$ occurs at $N_B = 0.5$.

$$T_C = \frac{\Omega}{2R}$$

*Example 11.11*

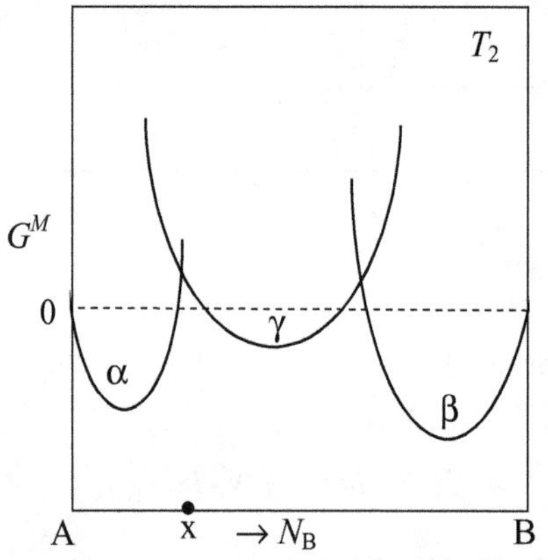


Shown is the Gibbs energy of mixing versus composition diagram of A-B binary system at the temperature $T$ and the pressure $P$.
The diagram shows two ***terminal phases***, α and β, and one ***intermediate phase*** γ.
If the overall composition of the system is given by the point x shown in the diagram, find the stable equilibrium phase(*s*) at $T$ and $P$.

The minimum value of the Gibbs energy that the system with the composition of x can have can be found by drawing the common tangent to the Gibbs energy curves of α and β phases. For the system at the overall composition of x, the equilibrium structure is the mixture of phase α and phase β with the following proportions (refer to the figure below):

$$\text{Fraction of } \alpha = \frac{xb}{ab} \qquad \text{Fraction of } \beta = \frac{ax}{ab}$$

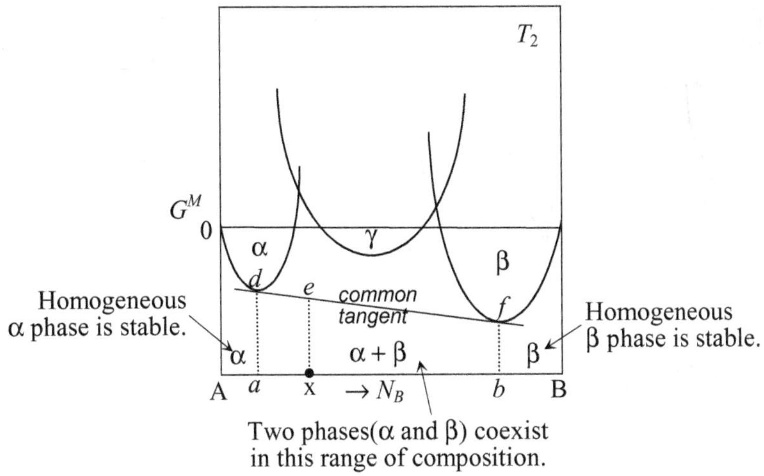


Note that the ***intermediate phase*** of γ does not appear as a stable phase at the present thermodynamic conditions. In other words, if γ phase is seen in the structure, then the system is not in equilibrium, but in non-equilibrium, ***meta-stable state***. However, if the thermodynamic state of the system is changed (*e.g.*, by changing temperature or pressure), γ phase may appear as a stable intermediate phase in a certain composition range. In the figure on the left shows a region where γ phase is stable. As the mean composition varies, the stable phase(s) will change: α, (α+γ), γ, (γ+β) and β.

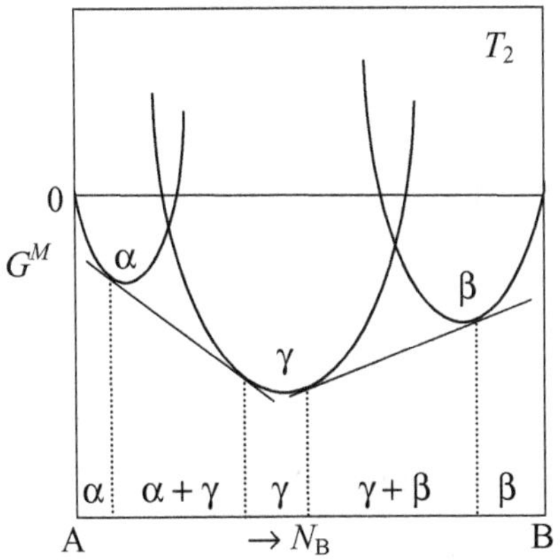


When a small deviation of compositions leads to a sharp change in $G^M$ as can be seen in the figure on the left, the phase tends to form a stoichiometrically fixed composition rather than a solution. If the Gibbs energy curve of γ phase is so narrow, and becomes practically a single vertical line, then a ***stoichiometric compound***, $A_mB_n$, where m and n are integers, is formed. For metallic systems it is usually termed ***intermetallic compound***.

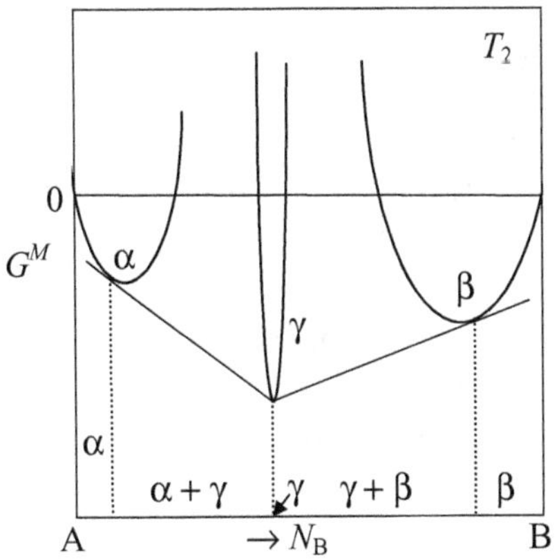


One thing to be noted is that the composition at which the Gibbs energy of a phase is at minimum does not necessarily appear as the stable phase. For instance, in the figure on

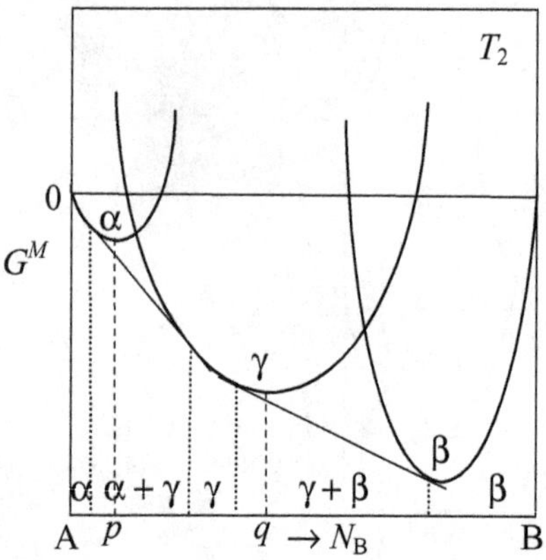


the left, the point $p$ represents the minimum point of the Gibbs energy curve of α phase, but it does not lie in the α phase stable region.

The same is true for the point $q$ which represents the minimum point of the Gibbs free energy curve of γ phase.

*Exercises*

11.10 The A-B binary system conforms to regular solution behavior and the interaction parameter Ω is 17,400 J mol$^{-1}$. Find the critical temperature $T_C$ below which the phase separation occurs. Calculate the compositions of α' and α" in equilibrium at 900 K.

11.11 At 1273 K, a copper-zinc alloy containing 16 mol% Zn lies on the solidus, and that on the liquidus contains 20.6 mol% Zn. The activity coefficient of Zn in liquid Cu-Zn alloys, relative to the *pure liquid zinc standard state*, is represented by

$$RT\, ln\, \gamma_{Zn} = -19{,}246 N_{Cu}^2 \qquad \text{where } R = 8.314\,\text{J mol}^{-1}\text{K}^{-1}$$

Find activity of copper in the alloy of the solidus composition, relative to the *pure solid copper standard state*. The standard free energy of fusion of copper is given by

$$\Delta G^{\circ}_{f,\text{Cu}} = 6{,}883 - 8.745\, ln\, T + 3.138\times10^{-3} T^2 + 53.739T, \qquad \text{J mol}^{-1}$$

11.12 Shown below is the diagram of Gibbs energy of mixing of liquid and solid solutions of the A-B binary system at temperature $T_1$.

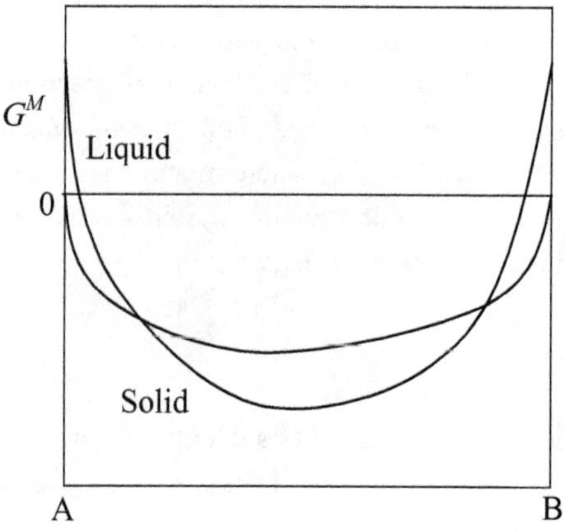


1) Is it true that $T_1 < T_{m(\text{A})}$ and $T_1 < T_{m(\text{B})}$ ? $T_{m(i)}$ is the melting point of $i$.

2) The diagram shown occurs when the heat of mixing, $H^M$, is positive. Prove that

$$H^M_{solid} > H^M_{liquid}$$

## 11.4 Influence of Interfaces on Equilibrium

A phase (a solid or liquid phase) is never unlimited in its physical size. It must have boundaries, which may be a free surface facing with its vapor phase and/or an interface facing with other phase. The energy state at the surface or interface differs from the energy state in the bulk of the phase, because the atom arrangement is different between the two. Up until now the effect of surface (or interface) has been neglected mainly because its effect is small and negligible in comparison with the variation of bulk properties.

> Various kinds of interfaces
>
> Liquids: liquid-vapor (usually called *surface*), immiscible liquid-liquid
> Solids: solid-gas (usually called *surface*), solid-liquid, solid-solid (two solids with different chemistry, two solids with different crystal structure, two solids with different crystal orientation)

The atoms at the *surface* are at higher energy state than those in the bulk as they have less number of neighboring atoms and thus are less bound to others than those in the bulk. The difference in the energy state between the *interface* and the bulk is mainly due to irregular arrangement of atoms at the interface.

As the relative proportion of the surface (or interface) to the bulk volume increases, the importance of the effect of the surface (or interface) increases accordingly. In particular the effect of the interface becomes extremely important in the early stage of phase transformation (for instance, solidification or solid-solid transformation) when the forming phase is still very fine in size.

The energy associated with the surface (excess energy with respect to that in the bulk; simply ***surface energy***) may be defined from the Gibbs energy expression of a system:

G = $f(T, P, A)$ for a closed system

↓ Total differentiation

$$dG = \left(\frac{\partial G}{\partial T}\right)_{P,A} dT + \left(\frac{\partial G}{\partial P}\right)_{T,A} dP + \left(\frac{\partial G}{\partial A}\right)_{T,P} dA$$

$$\left(\frac{\partial G}{\partial T}\right)_{P,A} = -S$$

$$\left(\frac{\partial G}{\partial P}\right)_{T,A} = V$$

$$\left(\frac{\partial G}{\partial A}\right)_{T,P} = \sigma$$ (***surface tension***: Gibbs energy change with change in the surface area)

↓

$$dG = -SdT + VdP + \sigma dA$$

↓

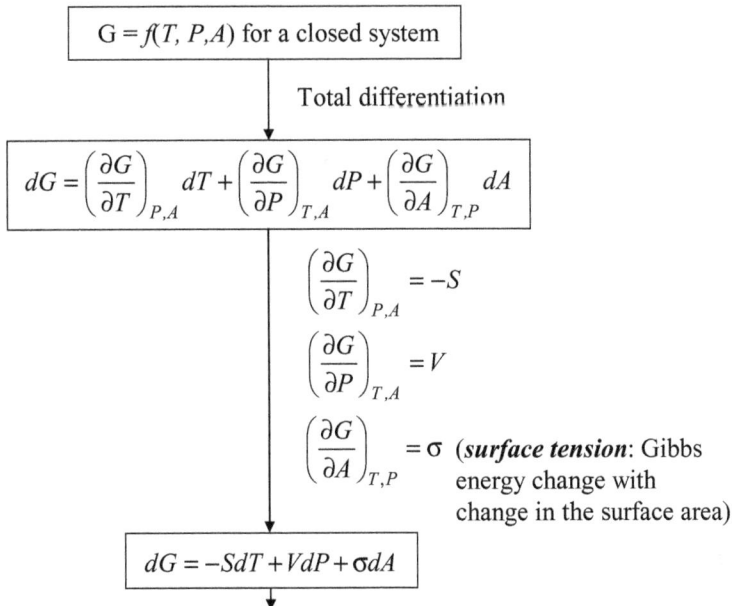

↓ At constant T and P

$$\sigma = \left(\frac{\partial G}{\partial A}\right)_{T,P}$$

The unit of surface tension is energy/area (J m$^{-2}$), but frequently given by newtons per meter (N m$^{-1}$: Recall that J = N m). The above equation is the thermodynamic definition of surface tension (or sometimes equivalently called surface energy). It can be seen from the above equation that, in the case that the surface tension is not different throughout the surface which is usually the case for liquid, the Gibbs energy decreases if the surface area decreases, and thus it is the natural tendency for the surface to decrease its area, *i.e.*, to contract.

Suppose that we have a spherical particle (or drop, or bubble) which is in equilibrium with its surroundings. Then there must be a force balance established between the particle and the surroundings:

| Internal pressure pushing the surface outwards. | = | Outside pressure pushing the surface inwards. | + | Squeezing force inwards due to surface tension |
|---|---|---|---|---|
| ↓ | | ↓ | | ↓ |
| $F_{in} = 4\pi r^2 P_{in}$ | | $F_{out} = 4\pi r^2 P_{out}$ | | $F_\sigma$ |

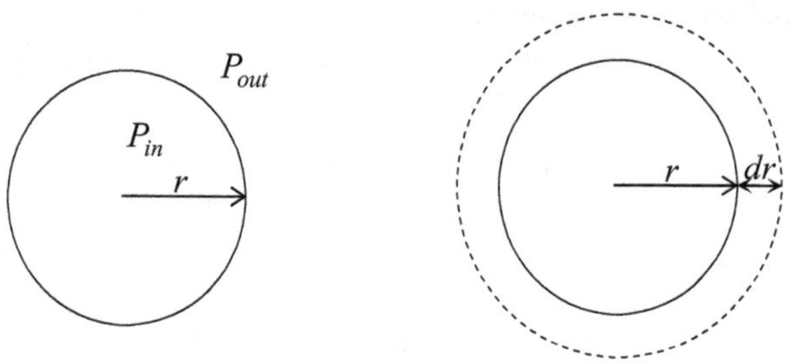


The force due to surface tension ($F_\sigma$) may be evaluated as follows:

The change in surface area ($A$) by changing the radius $r$ to $r+dr$

$$dA = 4\pi(r+dr)^2 - 4\pi r^2$$

↓ Ignoring $(dr)^2$

$$dA = 8\pi r dr$$

↓

↓

The amount of work to be done to stretch the surface by $dA$

$$dw = \sigma dA = 8\pi r\sigma dr$$

↓ (work = force × distance)

$$dw = F_\sigma dr$$

$$F_\sigma = 8\pi r\sigma$$

↓

$$F_{in} = F_{out} + F_\sigma$$

$$4\pi r^2 P_{in} = 4\pi r^2 P_{out} + 8\pi r\sigma$$

↓

$$P_{in} = P_{out} + \frac{2\sigma}{r}$$

↓

$$\Delta P = P_{in} - P_{out}$$

$$\Delta P = \frac{2\sigma}{r}$$

- If the particle (or drop, or bubble) is not spherical, but has a flat surface, then $r \to \infty$, and hence $\Delta P \to 0$. That is, there would be no effect of the surface and thus there would be no difference in pressure between the phase of interest and the surrounding phase.
- If the particle (or drop, or bubble) is spherical, on the other hand, the pressure inside is higher than the pressure outside by $\Delta P$ given by the above equation.

It is known that the Gibbs energy of the phase forming a particle (or drop, or bubble) is higher than the Gibbs energy of the surrounding phase by $V\Delta P$ (where $V$ is the molar volume):

$$\Delta G = V\Delta P$$

Combining the above two equations we have,

$$\Delta G = \frac{2\sigma V}{r}$$

The increase in Gibbs energy due to the surface tension (interfacial energy) is known as the ***capillary effect*** or the ***Gibbs Thomson effect***.

The Gibbs Thomson effect alters a number of physic-chemical properties of the phase:

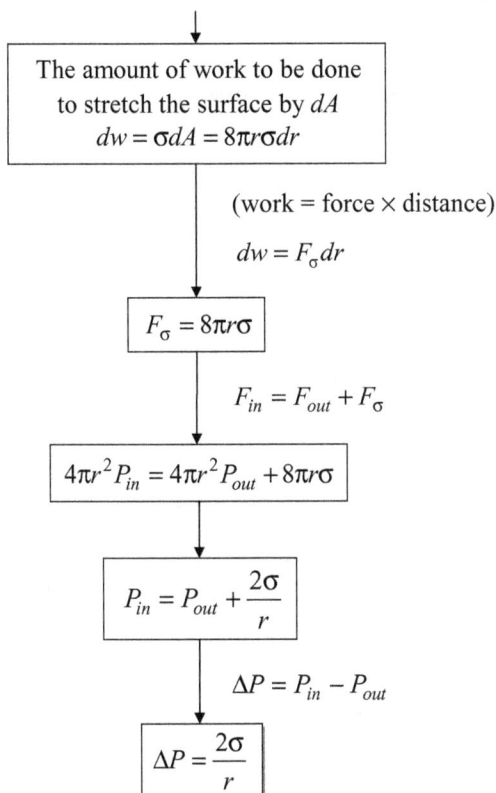

- It causes increase in the vapor pressure and melting point.
- It decreases the freezing point.
- It alters equilibrium compositions of phases involved.
- It causes small precipitates to dissolve and larger ones to grow (known as ***Ostwald ripening***).

***Capillary effect on vapor pressure***

If two phases α and β are in equilibrium,

$$\boxed{dG^{\alpha} = dG^{\beta}}$$

$dG = -SdT + VdP$

$dG = VdP$ at constant $T$

$$\boxed{V^{\alpha} dP^{\alpha} = V^{\beta} dP^{\beta}}$$

$P^{\alpha} - P^{\beta} = \frac{2\sigma}{r}$

Assuming $\gamma$ is constant

$dP^{\alpha} - dP^{\beta} = -\frac{2\sigma}{r^2} dr$

$$\boxed{\frac{(V^{\beta} - V^{\alpha})}{V^{\alpha}} dP^{\beta} = -\frac{2\sigma}{r^2} dr}$$

This equation is valid for equilibrium of any two phases. When we apply this equation to a condensed phase (α, liquid drop or solid particle) in equilibrium with its vapor (β):

$$V^{\beta} >> V^{\alpha} \text{ and } V^{\beta} = \frac{RT}{P^{\beta}} \text{ (assuming the vapor behaves ideally)}$$

Combining the above equations, we can get

$$\boxed{\frac{RT}{V^{\alpha}} \frac{dP^{\beta}}{P^{\beta}} = -\frac{2\sigma}{r^2} dr}$$

Assuming that the dependence of the molar volume on pressure is negligible, integration from $r = \infty$ to $r = r$ yields

$$\boxed{\frac{RT}{V^{\alpha}} \int_{r=\infty}^{r=r} \frac{dP^{\beta}}{P^{\beta}} = \int_{r=\infty}^{r=r} -\frac{2\sigma}{r^2} dr}$$

$$ln\left(\frac{P_r^{\beta}}{P_{\infty}^{\beta}}\right)=\frac{V^{\alpha}}{RT}\left(\frac{2\sigma}{r}\right)$$

This equation shows that,

- The vapor pressure of a phase is inversely proportional to its radius, and thus the smaller in size, the higher in the vapor pressure.
- When we have particles (or drops) in different sizes, smaller ones will shrink and larger ones will grow in size, because the vapor pressure at the smaller particle is higher than the vapor pressure at the larger particle, and thus gas molecules diffuse from the smaller particle side to the larger particle side.

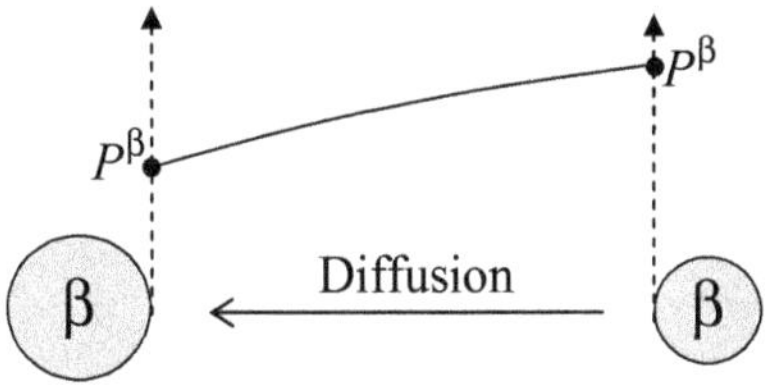


*Example 11.12*

At 900 K, liquid zinc droplets are suspended above the liquid zinc bath. Determine the vapor pressure in equilibrium with zinc droplets of the size of 1 μm, 0.1 μm, and 0.01 μm, respectively. The following data are given:
Surface tension of liquid zinc: 0.38 N $m^{-1}$
Molar volume of zinc: $9.5\times10^{-6}$ $m^3$ $mol^{-1}$
Equilibrium zinc vapor pressure above the zinc bath:

$$P_{Zn}^{o}=-\frac{6620}{T}-1.255\,log\,T+12.34\ \ \text{(mmHg)}$$

$$ln\left(\frac{P_r^{Zn}}{P_{\infty}^{Zn}}\right)=\frac{V^{Zn}}{RT}\left(\frac{2\sigma}{r}\right)$$

R = 8.314 J $mol^{-1}K^{-1}$
$J$ = kg $m^2sec^{-2}$
$r = 1\times10^{-6}$ m, $0.1\times10^{-6}$ m, $0.01\times10^{-6}$ m
$P_{Zn}^{o}$ = 1.277 mmHg = 0.0017 atm.

$$ln\left(\frac{P_r^{Zn}}{0.0017}\right)=\frac{9.5\times10^{-6}}{8.314\times900}\left(\frac{2\times0.38}{r}\right)$$

$$P_{\mathrm{Zn}} = 1.702 \times 10^{-3} \text{ atm. for } r = 1\ \mu\text{m}$$
$$P_{\mathrm{Zn}} = 1.717 \times 10^{-3} \text{ atm. for } r = 0.1\ \mu\text{m}$$
$$P_{\mathrm{Zn}} = 1.817 \times 10^{-3} \text{ atm. for } r = 0.01\ \mu\text{m}$$

***Capillary effect on melting temperature ($T_m$)***

If two phases of α and β are in equilibrium,

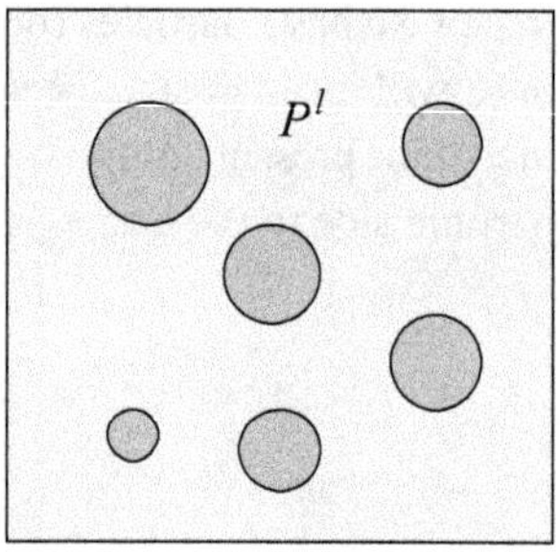


$$dG^{\alpha} = dG^{\beta}$$

$$dG = -SdT + VdP$$

$$-S^{\alpha}dT + V^{\alpha}dP^{\alpha} = -S^{\beta}dT + V^{\beta}dP^{\beta}$$

We apply this equation to solid($s$)-liquid($l$) equilibrium of a pure material:

$$-S^{s}dT + V^{s}dP^{s} = -S^{l}dT + V^{l}dP^{l}$$

If the solid is a particle with radius of $r$,

$$P^{s} = P^{l} + \frac{2\sigma_{s/l}}{r}$$

$$-S^{s}dT + V^{s}d\left(P^{l} + \frac{2\sigma_{s/l}}{r}\right) = -S^{l}dT + V^{l}dP^{l}$$

$$-S^{s}dT + V^{s}dP^{l} - \frac{2\sigma_{s/l}V^{s}}{r^2}dr = -S^{l}dT + V^{l}dP^{l}$$

Assuming that the pressure in the liquid phase ($P^l$) is not altered by the shape of the solid, and knowing that,

$S^l - S^s = \Delta S_f^{\circ}$ (Entropy change of fusion)

$$\Delta S_f^{\circ}dT = \frac{2\sigma_{s/l}V^{s}}{r^2}dr$$

Integration from $r = \infty$ to $r = r$,

$$\int_{T_m^o}^{T_{m,r}} \Delta S_f^o dT = \int_{r=\infty}^{r=r} \frac{2\sigma_{s/l} V^s}{r^2} dr$$

↓

$$T_{m,r} = T_m^o - \frac{2V^s \sigma_{s/l}}{\Delta S_f^o r}$$

↓

Knowing that at the melting temperature of the bulk material ($T_m^o$),

$$\Delta G_f^o = 0 = \Delta H_f^o - T_m^o \Delta S_f^o$$

↓

$$T_{m,r} = T_m^o \left(1 - \frac{2V^s \sigma_{s/l}}{\Delta H_f^o} \frac{1}{r}\right)$$

From the above equation,

- The melting temperature is inversely proportional to the radius of the solid particle.
- This equation is applicable not just to isolated spherical particles, but also to the general solid surface which is curved with radius $r$.

*Example 11.13*

A portion of the solid iron surface protrudes with the radius of 0.1 μm.

(1) Calculate the melting temperature at the tip of the protrusion.

(2) If this solid surface is in contact with liquid iron which is undercooled 5°C below the melting temperature of iron with a flat surface, determine whether the tip will melt or grow further into the liquid iron.

The following data are given:

Melting temperature of bulk iron ($T_m^o$): 1,536°C

Heat of fusion ($\Delta H_f^o$): 371,000 J mol$^{-1}$

Molar volume of iron ($V^{Fe}$): 8×10$^{-6}$ m$^3$ mol$^{-1}$

Interfacial tension ($\sigma_{s/l}$): 0.204 N m$^{-1}$

$$T_{m,r} = T_m^o \left(1 - \frac{2V^s \sigma_{s/l}}{\Delta H_f^o} \frac{1}{r}\right)$$

$$T_{m,r} = 1809 \times \left(1 - \frac{2 \times (8 \times 10^{-6}) \times 0.204}{371000} \times \frac{1}{0.1 \times 10^{-6}}\right)$$

$$T_{m,r} = 1808 \text{ K}$$

Thus the temperature at the tip is higher than the undercooled liquid ($T_l$ = 1809 – 5 = 1804*K*), and hence heat flows from the tip to the liquid, resulting in growth of the tip into the liquid. This is the basis of so-called ***dendrite growth*** in solidification of metals.

### *Capillary effect on solubility*

Consider that we have a two-phase A-B binary system consisting of β phase particles embedded in the α phase. We now examine two β particles; one with a spherical shape (radius = *r*) and the other with a flat shape (radius = ∞).

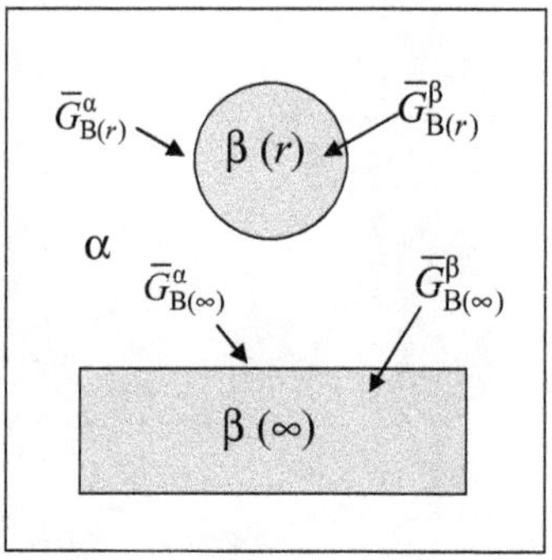


The figure shown here represents what is described above. Strictly speaking, this system cannot be in true total equilibrium.

- The partial molar Gibbs energy (or chemical potential) of a component in the spherical β phase is not the same as that in the flat β phase ($\bar{G}^{\beta}_{B(r)} \neq \bar{G}^{\beta}_{B(\infty)}$). This difference is known as the capillary effect or Gibbs-Thomson effect (we've seen this earlier.)
- However, the β phase particle (flat or spherical) is in ***local equilibrium*** with the α phase *locally* in the vicinity of the particle ($\bar{G}^{\alpha}_{B(r)} = \bar{G}^{\beta}_{B(r)}$ and $\bar{G}^{\alpha}_{B(\infty)} = \bar{G}^{\beta}_{B(\infty)}$).

Now we discuss the above in a more systematic way:

$$\Delta G^{\beta}_{B} = \bar{G}^{\beta}_{B(r)} - \bar{G}^{\beta}_{B(\infty)} = \frac{2V\sigma_{\alpha/\beta}}{r}$$

But $\bar{G}^{\alpha}_{B(r)} = \bar{G}^{\beta}_{B(r)}$ and $\bar{G}^{\alpha}_{B(\infty)} = \bar{G}^{\beta}_{B(\infty)}$

$$\Delta G^{\alpha}_{B} = \bar{G}^{\alpha}_{B(r)} - \bar{G}^{\alpha}_{B(\infty)} = \frac{2V\sigma_{\alpha/\beta}}{r}$$

Taking pure B with the flat surface as the standard state of B,

$$\bar{G}^{\alpha}_{B(\infty)} = G^{o}_{B} + RT\,ln\,a^{\alpha}_{B(\infty)}$$

$$\bar{G}^{\alpha}_{B(r)} = G^{o}_{B} + RT\,ln\,a^{\alpha}_{B(r)}$$

$$ln\left(\frac{a^{\alpha}_{B(r)}}{a^{\alpha}_{B(\infty)}}\right) = \frac{V}{RT}\frac{2\sigma_{\alpha/\beta}}{r}$$

Knowing that

$a_i = \gamma_i N_i$, where $\gamma_i$ is activity coefficient of component *i*

$$\ln\left(\frac{\gamma^{\alpha}_{B(r)}}{\gamma^{\alpha}_{B(\infty)}}\right)+\ln\left(\frac{N^{\alpha}_{B(r)}}{N^{\alpha}_{B(\infty)}}\right)=\frac{V}{RT}\frac{2\sigma_{\alpha/\beta}}{r}$$

If B in the α phase obeys the Henry's law,

$\gamma^{\alpha}_{B(r)}=\gamma^{\alpha}_{B(\infty)}=\gamma^{o}_{B}$ : constant

$$\ln\left(\frac{N^{\alpha}_{B(r)}}{N^{\alpha}_{B(\infty)}}\right)=\frac{V}{RT}\frac{2\sigma_{\alpha/\beta}}{r}$$

From the above results, we can draw some conclusions that,

- Terms at the right-hand side of the equation are all positive, and thus $N^{\alpha}_{B(r)} > N^{\alpha}_{B(\infty)}$, that is, the concentration (or solubility) of B of the α phase in the vicinity of the spherical β particle ( $N^{\alpha}_{B(r)}$ ) is higher than that in the vicinity of the flat β particle ( $N^{\alpha}_{B(\infty)}$ ).
- The smaller the radius of the particle, the higher solubility of B in the β phase.

As shown in the following figures, the above discussion can be visualized by use of the Gibbs energy of solution *vs.* composition curve together with the common tangent method.

In the case of β particle with the *flat* surface, the α phase of composition "a" is in *local equilibrium* with the β phase of composition "b". When the β phase is *spherical* with the radius of *r*, on the other hand, the α phase of composition "c" is in *local equilibrium* with the β phase of composition "d".

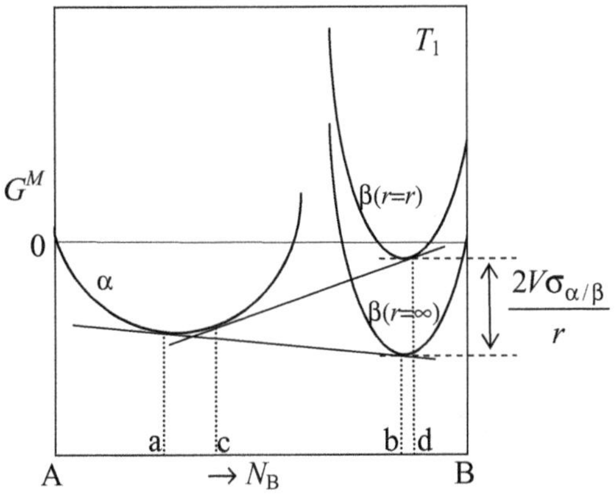


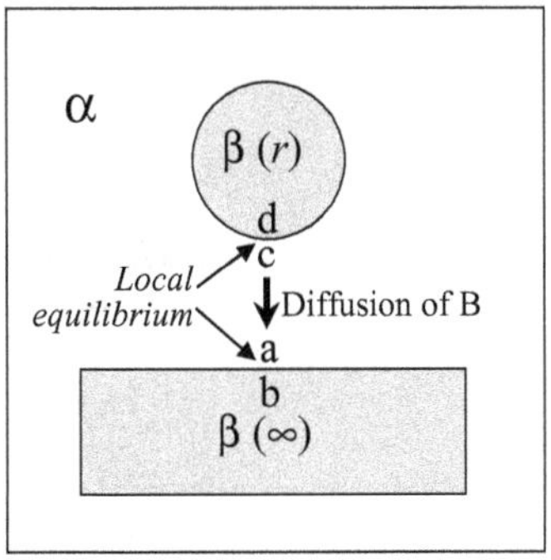


- Note that the composition "d" for the spherical β particle is richer in B than the composition of "b" for the flat particle.

- Note also that, in the $\alpha$ phase, the composition "c" is richer in B than the composition "a". Thus, if the mobility of atoms is high enough, which is usually the case at high temperatures, B atoms in the $\alpha$ phase will diffuse from the vicinity of the spherical $\beta$ particle to the flat $\beta$ particle.

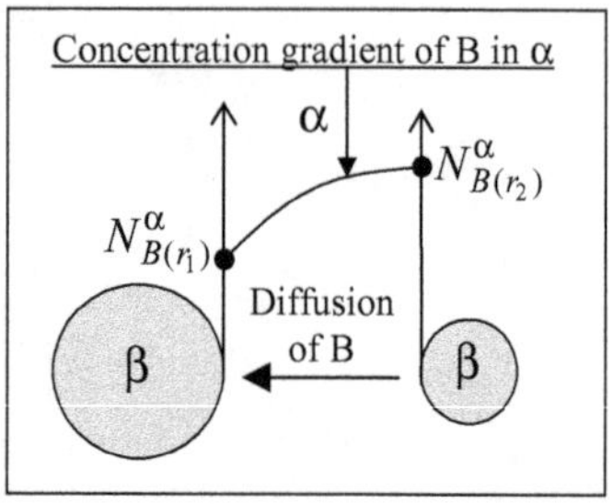


- This result can be extended to the case of two spherical particles of different radii: The concentration of B in the $\alpha$ phase is higher in the vicinity of smaller $\beta$ particle than in the vicinity of larger particle. This is schematically shown in the figure. If mobility of atoms is high enough, B atoms in the $\alpha$ phase will diffuse from the smaller particle side to the side of larger particle. This will lead to smaller particles dissolving and larger particles growing. This process is known as ***Ostwald ripening*** or ***coarsening***.

*Example 11.14*

Consider A-B binary system in that B is soluble in A ($\alpha$ phase) to a small extent, but A is virtually insoluble in B ($\beta$ phase). The solid solution of $\alpha$ phase follows the regular solution model. Derive an equation which shows the relationship between the concentration of B in the $\alpha$ phase in the vicinity of a large flat surface $\beta$ particle and that in the vicinity of a small spherical $\beta$ particle with the radius of $r$.

For the case of the large flat surface particle ($r = \infty$),

$$\bar{G}^{\alpha}_{B(\infty)} = G^{o}_{B} + RT\,ln\,a^{\alpha}_{B(\infty)}$$

$$\bar{G}^{\alpha}_{B(\infty)} = G^{o}_{B} + RT\,ln\,\gamma^{\alpha}_{B(\infty)} + RT\,ln\,N^{\alpha}_{B(\infty)}$$

For regular solution
$RT\,ln\,\gamma^{\alpha}_{B(\infty)} = \Omega(1 - N^{\alpha}_{B(\infty)})^2$

$$\bar{G}^{\alpha}_{B(\infty)} = G^{o}_{B} + \Omega(1 - N^{\alpha}_{B(\infty)})^2 + RT\,ln\,N^{\alpha}_{B(\infty)}$$

Similarly, for the case of the small spherical particle,

$$\bar{G}^{\alpha}_{B(r)} = G^{o}_{B} + \Omega(1 - N^{\alpha}_{B(r)})^2 + RT\,ln\,N^{\alpha}_{B(r)}$$

The difference of the partial molar Gibbs energy or chemical potential is then given by

$$\bar{G}^{\alpha}_{\mathrm{B}(r)} - \bar{G}^{\alpha}_{\mathrm{B}(\infty)} = \Omega\left[(1-N^{\alpha}_{\mathrm{B}(r)})^2 - (1-N^{\alpha}_{\mathrm{B}(\infty)})^2\right] + RT\,ln\left(\frac{N^{\alpha}_{\mathrm{B}(r)}}{N^{\alpha}_{\mathrm{B}(\infty)}}\right)$$

Knowing that

$$\bar{G}^{\alpha}_{\mathrm{B}(r)} - \bar{G}^{\alpha}_{\mathrm{B}(\infty)} = \frac{2V\sigma_{\alpha/\beta}}{r}$$

$$\Omega\left[(1-N^{\alpha}_{\mathrm{B}(r)})^2 - (1-N^{\alpha}_{\mathrm{B}(\infty)})^2\right] + RT\,ln\left(\frac{N^{\alpha}_{\mathrm{B}(r)}}{N^{\alpha}_{\mathrm{B}(\infty)}}\right) = \frac{2V\sigma_{\alpha/\beta}}{r}$$

The solubility of B in $\alpha$ is low, and thus $N^{\alpha}_{\mathrm{B}(\infty)} << 1$ and by ignoring the second order terms: ( $N^{\alpha}_{\mathrm{B}(\infty)}$ )$^2$ and ( $N^{\alpha}_{\mathrm{B}(r)}$ )$^2$,

$$2\Omega(N^{\alpha}_{\mathrm{B}(\infty)} - N^{\alpha}_{\mathrm{B}(r)}) + RT\,ln\left(\frac{N^{\alpha}_{\mathrm{B}(r)}}{N^{\alpha}_{\mathrm{B}(\infty)}}\right) = \frac{2V\sigma_{\alpha/\beta}}{r}$$

Rearrangement yields

$$RT\,ln\left(\frac{N^{\alpha}_{\mathrm{B}(r)}}{N^{\alpha}_{\mathrm{B}(\infty)}}\right) = \frac{2V\sigma_{\alpha/\beta}}{r} + 2\Omega(N^{\alpha}_{\mathrm{B}(r)} - N^{\alpha}_{\mathrm{B}(\infty)})$$

This equation relates the concentration of B of the $\alpha$ phase in the vicinity of a large flat surface $\beta$ particle ( $N^{\alpha}_{\mathrm{D}(\omega)}$ ) to that in the vicinity of a small spherical $\beta$ particle with the radius of $r$ ( $N^{\alpha}_{\mathrm{B}(r)}$ ), if the $\alpha$ phase follows the regular solution model. When compared with the equation derived earlier with the Henry's law assumption, this equation has an additional term of $2\Omega(N^{\alpha}_{\mathrm{B}(r)} - N^{\alpha}_{\mathrm{B}(\infty)})$ on the right-hand side. This term determines an additional capillary effect if the solution follows the regular solution behavior. If the value of $\Omega$ is positive, the capillary effect is deepened. If the value of $\Omega$ is negligibly small, the above equation reduces to the equation for the solution following the Henry's law.

*Example 11.15*

Discuss how the capillary effect alters the positions of phase boundaries of A-B binary phase diagram which has two phases ($\alpha$ and $\beta$) at the given $T$ and $P$.

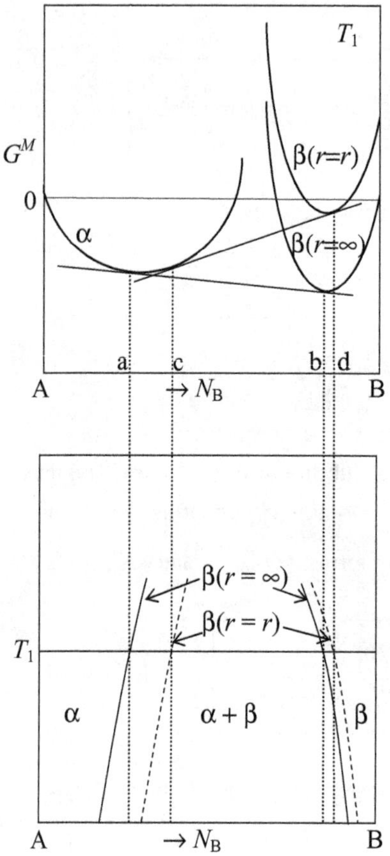


The phase boundaries shift towards the phase which shows the curvature of $r$ ($\beta$ phase in the present case).

As the radius *r decreases*,

- Solubility limit of B in the $\alpha$ phase increases.
- The $\beta$ phase becomes richer in B.
- This means that both $\alpha$ and $\beta$ phases become richer in B. (Violation of mass balance?)
- This is possible because the amount of the $\beta$ phase decreases and the amount of $\alpha$ phase increases (according to the lever rule).

**Three interfaces in equilibrium**

When two phases (or the same phase, but different crystal orientations) are in contact, they make a plane of interface. When three phases meet, they form a line; *that is*, three planes of interface meet at a line. If the line is sectioned at the right angle, it may look like the figure on the right. The lines here are the sections of the planes of phase interfaces and the point where three lines meet is actually the section of the three-phase contact line, which is normal to the plane of the paper. Now a question arises as to what determines the contact angles of the phases at the triple point.

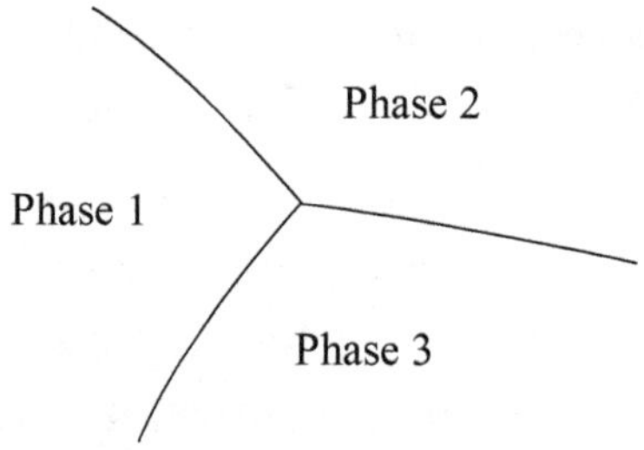


The angles forming at the contact point is in fact determined by the balance of the interfacial energies (or interfacial tensions) at the point. Each interface has its own interfacial tension. The figure here shows interfacial tensions acting

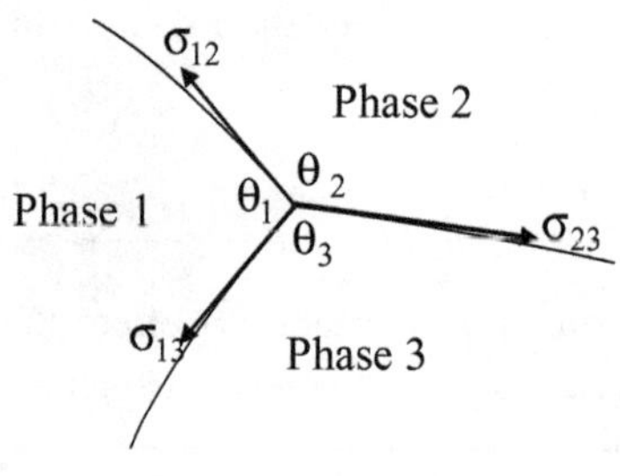

at the triple point. The vector of $\gamma_{12}$ indicates the interfacial tension between phases 1 and 2. The other two have similar definitions. Each vector is normal to the triple contact line. These vectors which represent the interfacial tension must be balanced each other in order to be in equilibrium, otherwise the triple point would move. There will be several ways of expressing the force balance at the triple point. One of them is illustrated in the figure below. The component of $\sigma_{12}$ normal to $\sigma_{23}$ ($\sigma_{12} \sin\theta_2$) must be balanced with the component of the $\sigma_{13}$ normal to the same $\sigma_{23}$ ($\sigma_{13} \sin\theta_3$).

$$\sigma_{12} \sin\theta_2 = \sigma_{13} \sin\theta_3$$

↓

$$\frac{\sigma_{12}}{\sin\theta_3} = \frac{\sigma_{13}}{\sin\theta_2}$$

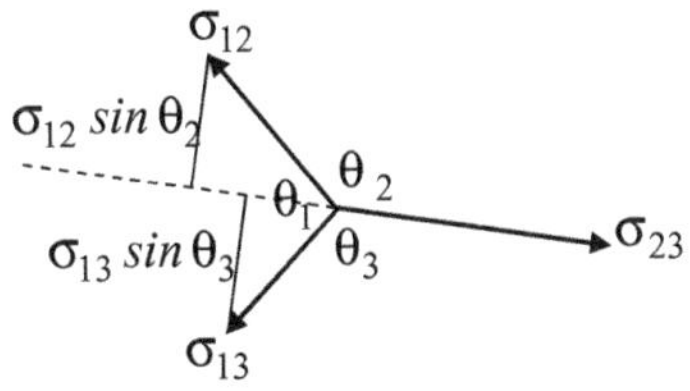


By similar analysis of force balance in other directions, we can get the following relationship:

$$\frac{\sigma_{12}}{\sin\theta_3} = \frac{\sigma_{13}}{\sin\theta_2} = \frac{\sigma_{23}}{\sin\theta_1}$$

When there are no forces other than the interfacial tensions acting at the interfaces, the above equation provides the condition on the balance of the interfacial tensions for three phases in contact to be in equilibrium (Strictly speaking, it is a *local* equilibrium unless the total area of interfaces of the entire system has been reduced to the minimum).

The interfacial energy between a liquid and a solid (which is inert to the liquid) can be estimated by locating the liquid drop on the solid substrate and equilibrating the liquid with its vapor. This method is known as the ***sessile drop technique***. The capillary effect together with gravitational effect results in the force balance as shown in the figure at the right. As the solid substrate is inert, it remains flat. The horizontal components of all the forces involved must be balanced at equilibrium.

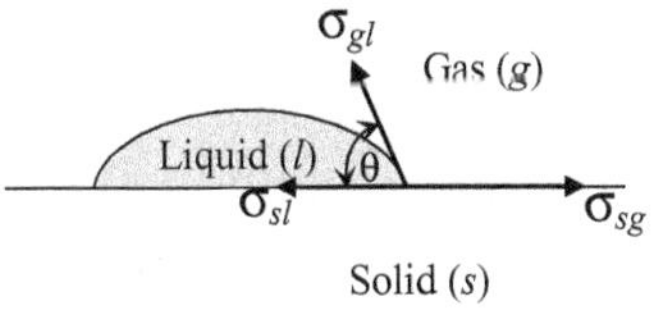


$$\sigma_{sg} = \sigma_{sl} + \sigma_{gl} \cos\theta$$

↓

$$\cos\theta = \frac{\sigma_{sg} - \sigma_{sl}}{\sigma_{gl}}$$

The angle θ in the above is called the ***contact angle***. If the contact angle is smaller than 90°, it is called that the liquid *wets* the solid. For values of θ greater than 90° it is called *non-wetting*.

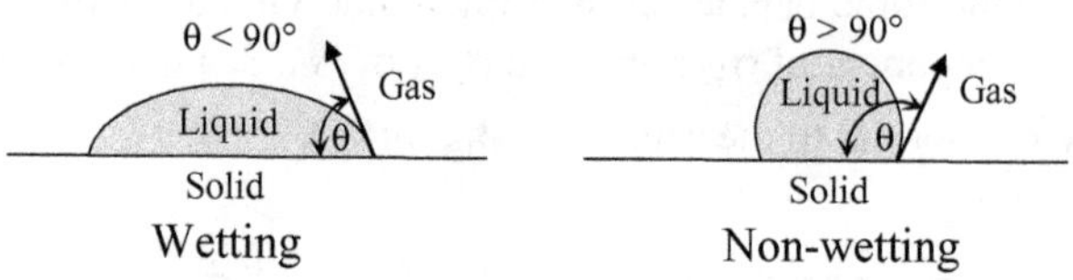


> *Example 11.16*
>
> A metal specimen is heated to 1,200°C and held at that temperature until equilibrium is reached with its vapor. The dihedral angle of a grain boundary which intersects the free surface at the right angle is found to be 158° as shown in the figure at the right. Determine the interfacial energy (or interfacial tension) of this grain boundary. The surface energy of the metal-vapor interface is 2.208 J m$^{-2}$ or 2.208 N m$^{-1}$. Assume that the surface energy of the metal is isotropic at the prevailing temperature.
>
> 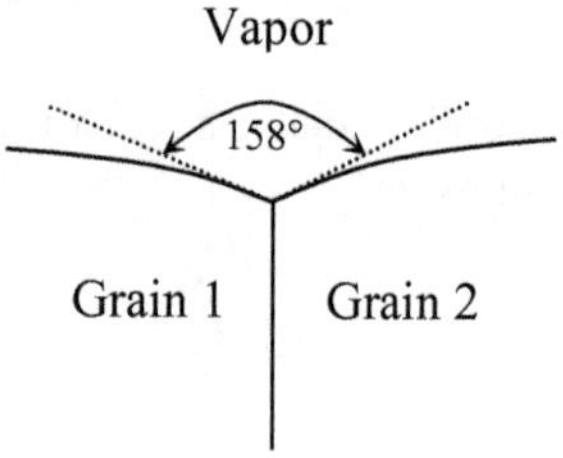
> 

From the force balance,

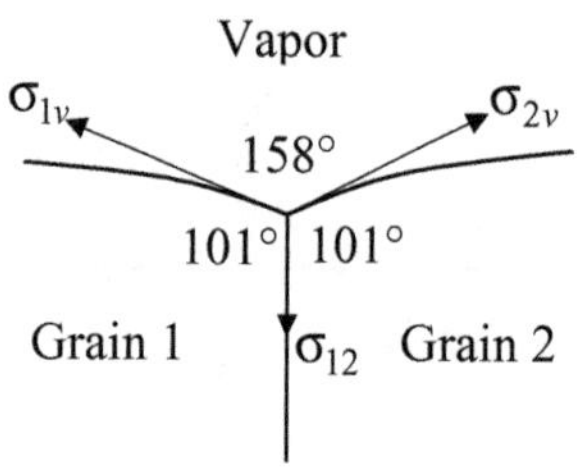


$$\frac{\sigma_{12}}{\sin 158^\circ} = \frac{\sigma_{1v}}{\sin 101^\circ} = \frac{\sigma_{2v}}{\sin 101^\circ}$$

$$\sigma_{1v} = \sigma_{2v} = 2.208 \text{ J m}^{-2}$$

$$\sigma_{12} = 0.843 \text{ J m}^{-2}$$

> *Example 11.17*
>
> A β phase particle may precipitate at the grain boundary of two α grains as shown in Figure (A) below. Sometimes it may happen that the β phase spread out the entire grain boundary as shown in Figure (B) below. Discuss in terms of the energy balance under what conditions the spreading occurs.

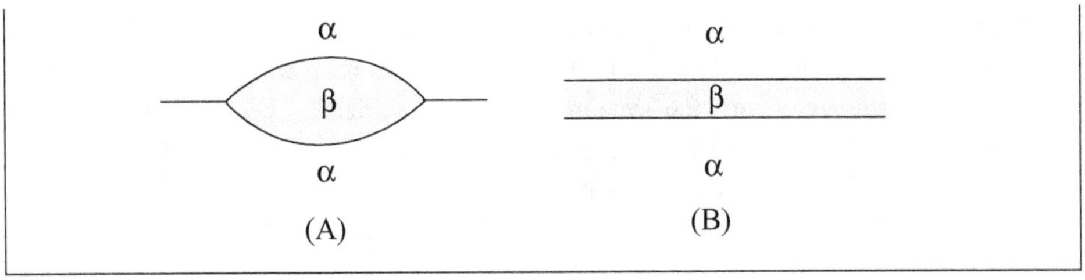


The balance in the horizontal direction yields,

$$\gamma_{\alpha\alpha} = 2\sigma_{\alpha\beta} \cos\frac{\theta}{2}$$

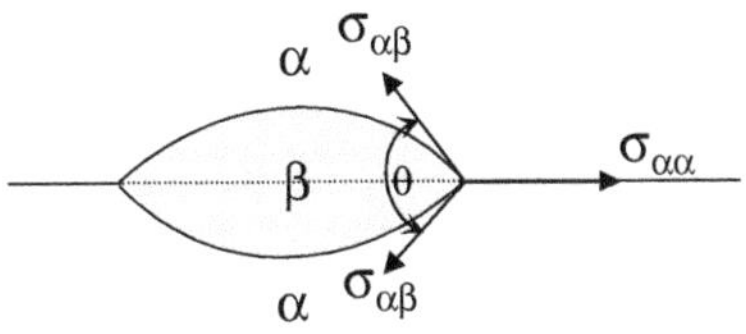


In principle, if the contact angle θ becomes zero, *that is*, $cos(\theta/2) = 1$, the β phase will spread out. In other words, if the grain boundary energy between the two α phases ($\sigma_{\alpha\alpha}$) is greater than two times of the grain boundary energy between α and β phases ($2\sigma_{\alpha\beta}$), the β phase will spread to form a thin film which separates the two α grains. It is because this spreading results in lowering the total surface energy.

*Exercises*

11.13 Prove that, if all grain boundaries in a polycrystalline have the same grain boundary energy independent of boundary orientation, the contact angles at the three-grain junctions are all 120° as shown in Figure (A) below. Prove that, even if the angle is all 120° at the junctions, grains having less than 6 boundaries will shrink as shown in Figure (B), but grains having more than 6 boundaries will grow as shown in Figure (C).

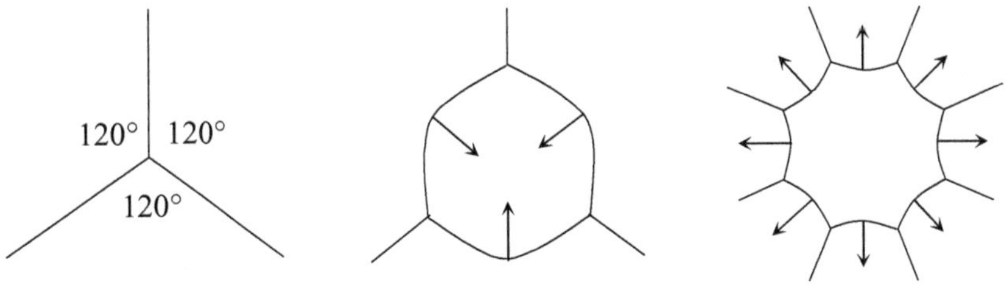


(A) (B) (C)

11.14 Solid spherical particles of copper (Cu) are to grow in a liquid copper. Determine the temperature of the liquid copper below its melting point in order for the copper particles to grow if the size of the particles is (A) 2 μm, (B) 20 nm. The following data are given: $T_{m,\mathrm{Cu}}$ = 1,085°C, atomic weight Cu = 63.5, density of Cu = 8,900 kg m$^{-3}$, Interfacial energy ($\sigma_{l/s\ \mathrm{Cu}}$) = 0.144 J m$^{-2}$, enthalpy of fusion ($\Delta H^{o}_{f,\mathrm{Cu}}$) = 13,300 J mol$^{-1}$.

11.15 In order to measure the interfacial energy (interfacial tension) between liquid iron and solid alumina ($Al_2O_3$) at 1,600°C, a liquid iron drop was kept on a flat solid alumina substrate until the system was fully stabilized. Then the contact angle between the iron droplet and the alumina substrate was measured to be 145°. Determine the interfacial energy between the liquid iron and the solid alumina at 1,600°C. Data given: the surface energy of liquid iron over its own vapor ($\sigma_{Fe}$) = 1.7 J $m^{-2}$, the surface energy of $Al_2O_3$ ($\sigma_{Al_2O_3}$) = 0.9 J $m^{-2}$.

# Chapter 12

# Phase Diagrams

A Phase diagram is a graphical representation to show thermodynamically distinct phases which occur at equilibrium. It shows regions of the phases which are stable at a given system, when the equilibrium is established in the system. Thermodynamic variables that can be used as axes for the graphical representation may include composition ($N_i$), temperature ($T$), pressure ($P$), and chemical potential ($\mu_i$). The choice of variables is a matter of convenience, but for metals and materials temperature and composition are commonly chosen.

## 12.1 One Component (Unary) Systems

### Temperature-Pressure Diagram

Phase changes are effected by three externally controllable variables: pressure, temperature and composition. In a one-component system, or unary system, the composition is not a variable, and thus there are only two variables which can vary: pressure and temperature. Every possible combination of temperature and pressure can be readily represented by points on a two-dimensional diagram.

Any vertical line on this *P*-*T* field represents the same temperature, and is called "***isotherm***".

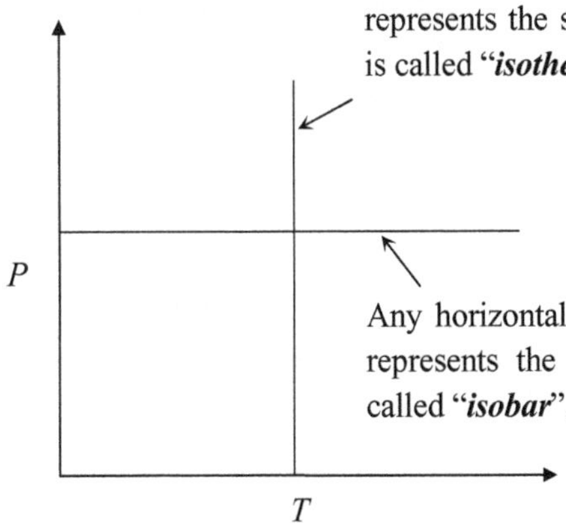


Any horizontal line on this *P*-*T* field represents the same pressure and is called "***isobar***".

The three states or phases, namely solid, liquid and gas, can be represented by the corresponding areas in the $P$-$T$ field:

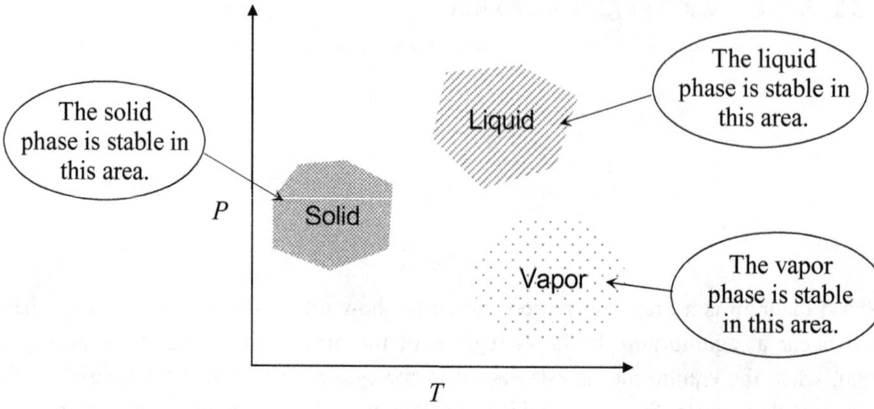


The areas of these phases will meet to form boundaries separating the phases, and the boundaries are called ***phase boundaries***.

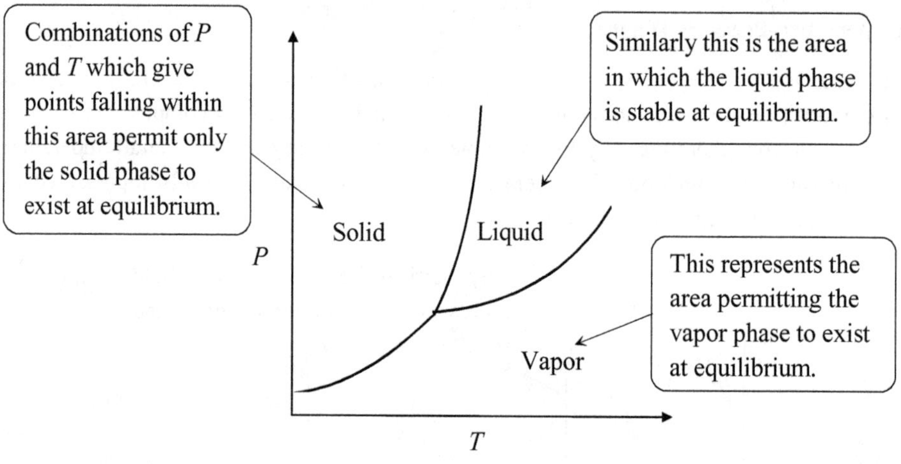


This is the phase diagram of one component system, and it enables to identify stable phase(s) thermodynamically at a given condition. Further interpretation of this diagram is given below (Refer to the diagram which follows):

Points falling directly on the phase boundaries represent conditions at which two neighboring phases coexist at equilibrium: *e.g.*, Point $a$: coexistence of the solid and gas phases, Point *b*: coexistence of the solid and liquid phases, Point *c*: coexistence of the liquid and gas phases

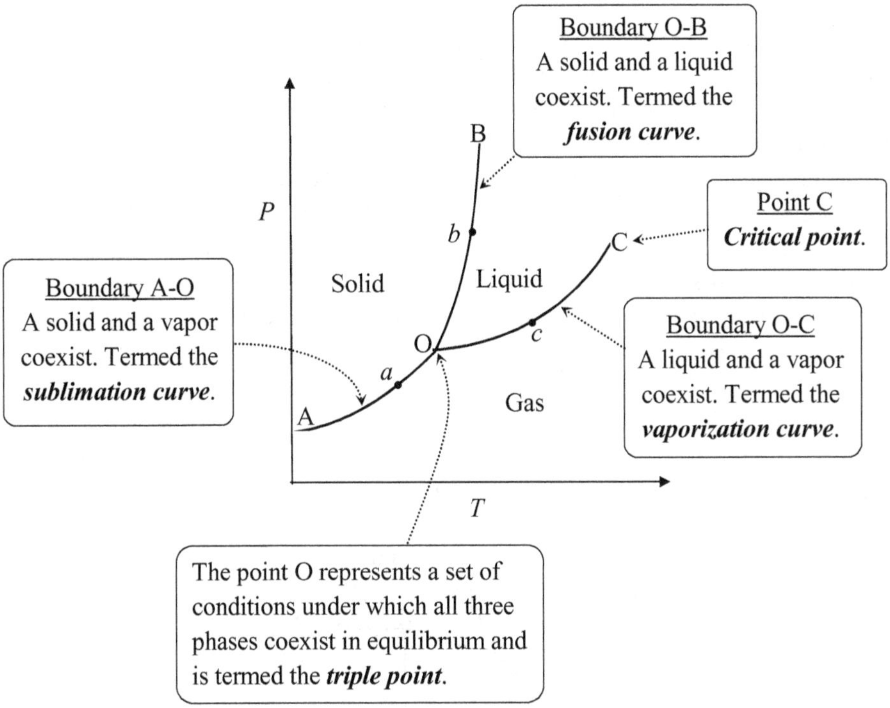


The vaporization curve (O-C) does not extend indefinitely, but ends abruptly at point C which is called the ***critical point***. Let's examine the physical significance of the critical point. Suppose that a liquid contained in a sealed vessel is in equilibrium with its vapor at the temperature $T_1$ (Fig. A). When the liquid is heated, the vapor pressure increases and hence the density of the vapor phase increases, but the quantity of the liquid decreases due to vaporization and the density of the liquid decreases due to thermal expansion (Fig. B). Eventually a stage at which the density of the vapor becomes equal to that of the remaining liquid reaches, and the interface between the two phases disappears and thus the two phases are indistinguishable (Fig. C).

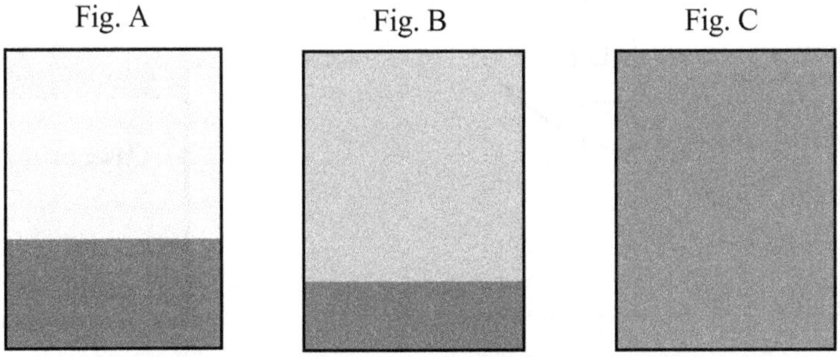

The temperature at which the interface disappears is called the ***critical temperature*** ($T_C$), and the corresponding vapour pressure is called the ***critical pressure*** ($P_C$). At and above $T_C$, therefore, the liquid phase does not exist.

Now we discuss dynamic changes which take place when a system under equilibrium is disturbed by change in temperature or pressure. Consider a system represented by point $k$, in which the liquid and vapor phases are in equilibrium at $T_1$ and $P_1$. If the temperature of the system is suddenly increased to $T_2$ while *keeping the total volume constant*, the initial equilibrium conditions are disturbed and the position of the system is now shifted to point $m$. This appears to put the system into a single phase (vapor phase) region. The position of $m$ is however not the equilibrium one for the system, because the volume of the system is kept constant. What should happen in reality is that the thermal energy supplied to raise temperature will cause the liquid to vaporise and hence increase the pressure of the system to $P_2$ so that a new equilibrium is established at point $n$.

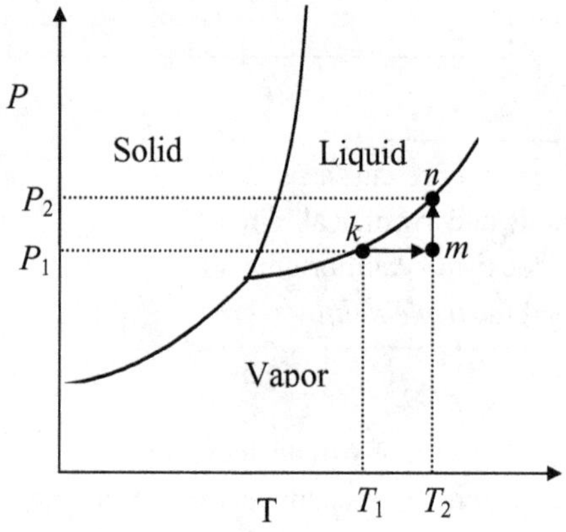


*Example 12.1*

The $P$-$T$ diagram for the substance A is given below:

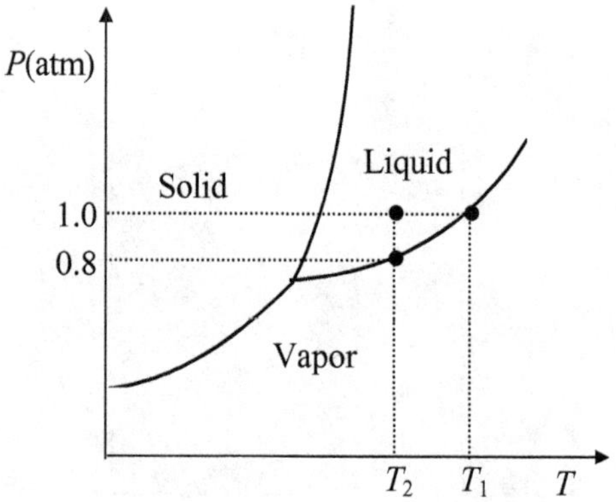


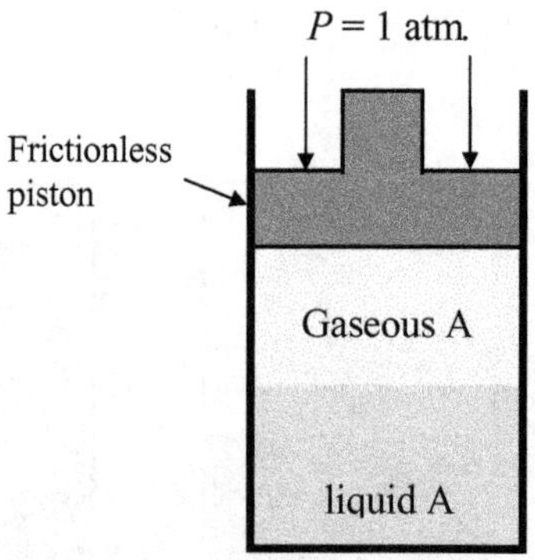

(1) If a cylinder (system) containing pure liquid A and pure gaseous A in equilibrium with the external pressure of 1 atm. at the temperature $T_1$ is brought to a new condition where the temperature is $T_2$ while the pressure is kept the same as 1 atm., what change would occur to the system?

(2) If the gaseous phase in the cylinder is initially not the pure A, but a mixture of A and an inert gas, what would be the pressure of A in the gas phase at the new condition?

(1) Since the external pressure of the cylinder is 1 atm., the vapor pressure of the gaseous A should be 1 atm. If conditions do not allow for the gaseous A to maintain the pressure at 1 atm., the gaseous A cannot exist under the conditions. From the *P-T* diagram given above, the gaseous A can exist up to 0.8 atm. of its pressure at $T_1$, and hence the gaseous A cannot exist in equilibrium with the liquid A. Therefore all the gaseous A will liquefy at 1atm. and $T_1$.

(2) This is not a true unary system, but a binary system of A and the inert gas. In this case, the vapor pressure of A will be adjusted so that the gaseous A at 0.8 atm. exists in equilibrium with the liquid A at $T_1$. The partial pressure of the inert gas will be 0.2 atm. to maintain the total pressure of 1 atm.

*Example 12.2*

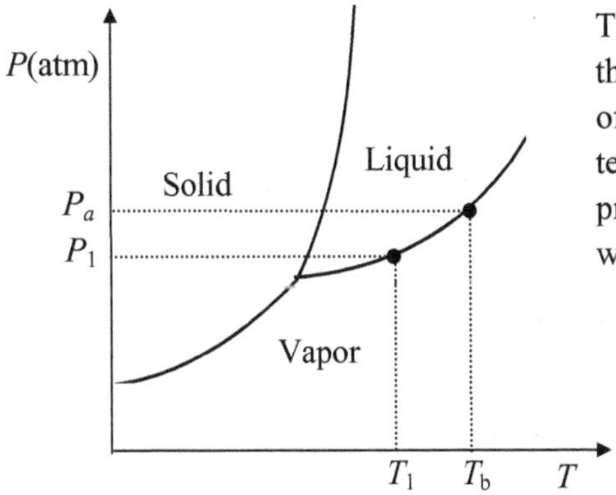


The figure shown is the phase diagram of the substance A. $P_1$ is the vapor pressure of A in equilibrium with the liquid A at the temperature $T_1$. $P_a$ indicates the ambient pressure. If the temperature is raised to $T_b$, what would happen to the liquid?

The conditions set in the question implicitly implies that the gas phase consists of the vapor of A at the pressure of $P_1$ and other gas (perhaps an inert gas) at the pressure of $(P_a - P_1)$ to make the total pressure of $P_a$. When the temperature is raised to $T_b$, the phase diagram shows that the vapor pressure of A becomes equal to the ambient pressure $P_a$. Therefore vaporization occurs throughout the bulk of the liquid A. The condition of free vaporization throughout the liquid is called ***boiling***. Note that, in the presence of an inert gas, a liquid does not suddenly start to form a vapor at its boiling temperature, but even at lower temperatures there is a vapor which is in equilibrium with the liquid (see *Example 12.1*).

*Example 12.3*

A liquid sample is allowed to cool at a constant pressure, and its temperature is monitored. The results are shown in Figure A. The phase diagram of this substance is given in Figure B. Which point in the phase diagram represents the process b → c in the cooling curve?

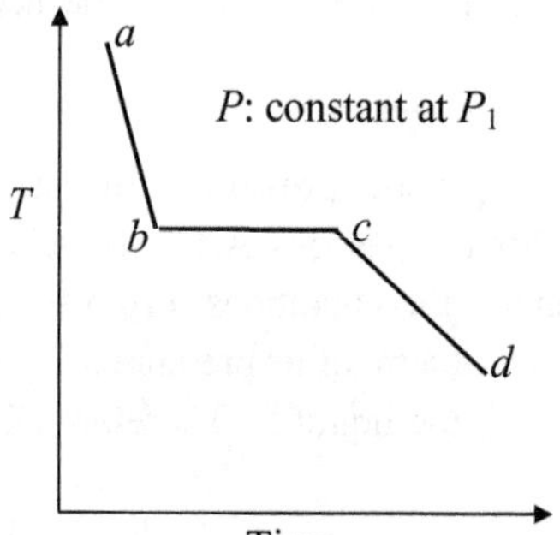

Fig A

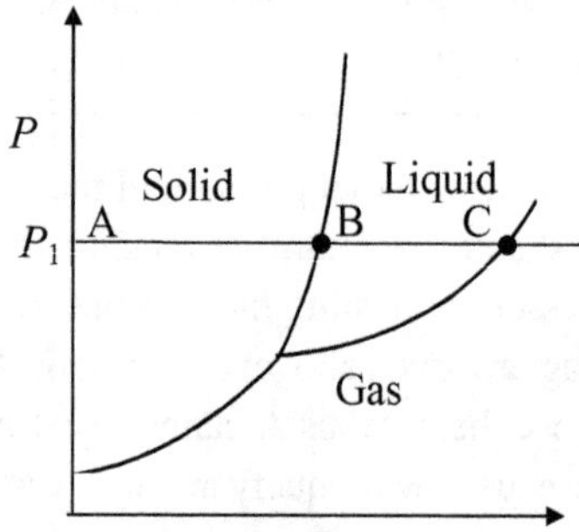

Fig B

① When a liquid sample is allowed to cool at a constant pressure $P_1$,

② the sample temperature will eventually hit the phase boundary at point B.

Fig A

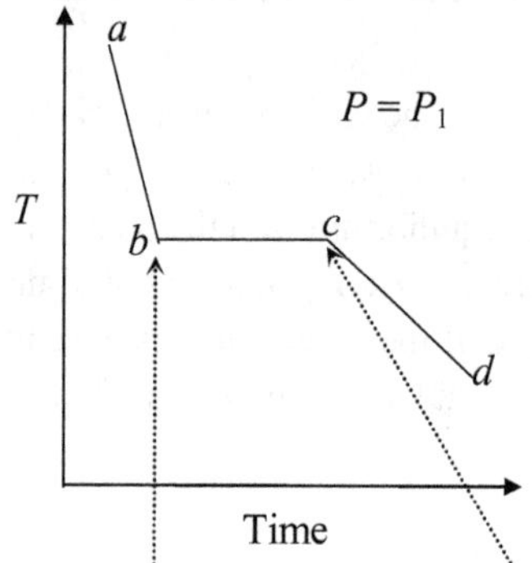


Fig B

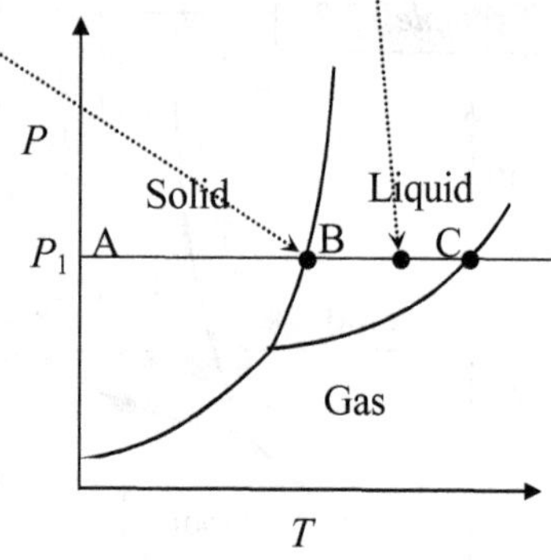


③ and then the sample begins to solidify. This is represented by the point *b* in the cooling curve.

④ During the phase transition from liquid to solid at the point B, heat is evolved and the cooling stops until the transition is complete at the point *c* in the cooling curve.

⑤ Once the transition has been complete at the point *c*, the sample temperature decreases again toward A from B in the phase diagram, or along the *c-d* in the cooling curve.

*Example 12.4*

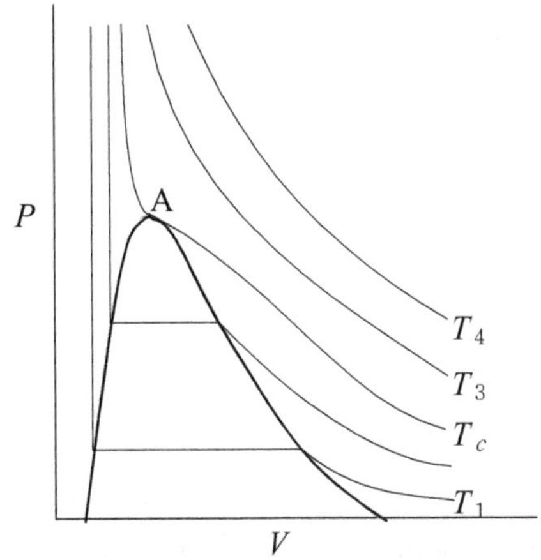


The relationship between pressure and volume is examined with a substance A. As the pressure is increased keeping the temperature at $T_1$, the volume of the gas decreases. When the pressure reaches a certain value, the volume suddenly reduces to the low value as the gas liquefies. A further increase in pressure causes little further reduction in volume as liquid is generally not compressible to a large extent. When the *P-V* relationship is examined at a number of different temperatures, one may obtain results shown in the diagram. Discuss the physical significance of the point A in the diagram.

At the point A the liquid and vapor are indistinguishable and the densities are identical. The point A is thus the critical point and the corresponding temperature $T_C$ is the critical temperature and $P_C$ is the critical pressure. All gases exhibit this type of behavior, but the values of critical pressure and critical temperature vary considerably from one substance to another.

*Example 12.5*

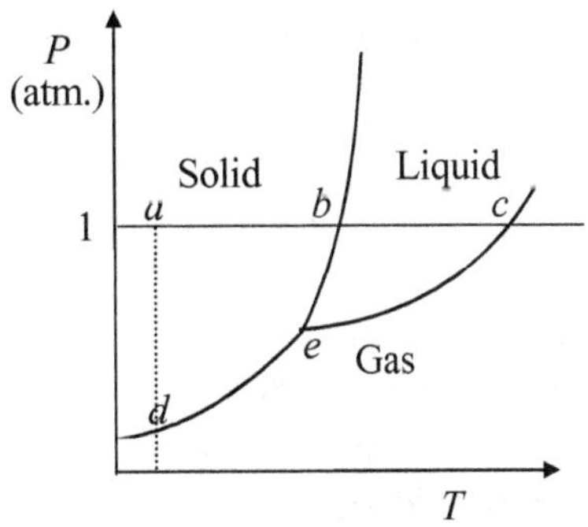


For the majority of metals the triple point lies far below atmospheric pressure and the critical point well above atmospheric pressure. The diagram given here is the *P-T* phase diagram of the metal M.

Discuss the change of the vapor pressure when the metal is gradually heated from the solid state at the point *a* to the temperature above the boiling point.

The behavior of the metal during heating under a constant pressure of 1 atm. can be predicted from the diagram:

- At point *a*, the solid metal is in equilibrium with the gas phase consisting of the vapor of M at the partial pressure of *d* and air (or inert gas) at the partial pressure of *a-d*, to make the total pressure of 1 atm.

- As the temperature is increased, the partial pressure (i.e., the vapor pressure) of the metal vapor increases along the curves *de* and then *ec*. Note that the metal M melts at *b* during heating.
- At point *c*, the vapor pressure of the metal M becomes identical to the ambient pressure (1 atm.), and hence the liquid metal boils.
- Further heating will result in the metal being completely in the vapor phase.

*Exercises*

12.1 Calculate the number of degrees of freedom at the triple point.

12.2 Calculate the number of degrees of freedom at the critical point.

12.3 A metal is sealed in a completely inert and pressure-tight container. Initially the metal fills half of the container, the balance of the space being filled with an inert gas at a pressure of 1 *atm*. What changes will occur within the container as the system is heated. The phase diagram of the metal is similar to that given in *Example 12.5*.

## Allotropy

Let's look at the fundamental difference between liquids and solids:

| | | Liquids | Solids |
|---|---|---|---|
| General atomic array | | Random arrays of atoms and molecules | Arrays of atoms or molecules in regular patterns |
| One component system | Atomic array | Only one type of random array is possible. | Many types of ordered arrays are possible. |
| | Number of phases | Only one liquid phase can exist. | A number of different solid phases may exist. |

***Allotropy*** is the property possessed by certain elements to exist in two or more distinct forms that are chemically identical but have different physical properties. The various crystalline forms in which a solid may exist are called ***allotropes*** or ***polymorphs.*** The transformation from one allotrope to another is called ***allotropic*** or ***polymorphic transformation.*** This transformation may occur either with pressure change or with temperature change. For instance, iron has one form of the crystal structure at room temperature and another at high temperature. When heated above 910°C the atomic structure changes from body centered cubic (*bcc*, termed ferrite) to face centered cubic (*fcc*, termed austenite).

For general discussion, suppose there are two possible solid phases, say $\alpha$ and $\beta$, in a given substance, and the Gibbs energies of $\alpha$ and $\beta$ vary as shown in the following figure:

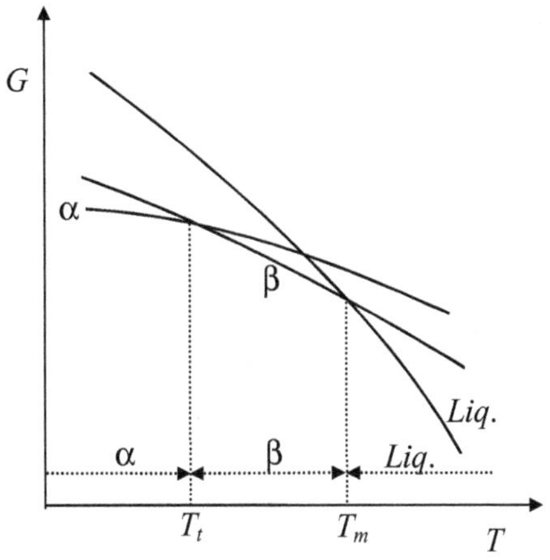


Note that

- $G_\alpha < G_\beta$ at $T < T_t$ and hence α phase is more stable,
- $G_\beta < G_\alpha$ at $T > T_t$ and hence β phase is more stable, and
- $G_{liq} < G_\beta$ at $T > T_m$ and hence the liquid phase is more stable.

The Gibbs energy curves which represent the stable phases over the range of the temperature will be

- The portion of the curve for α in the α phase stable region,
- The portion of the curve for β in the β phase stable region
- The portion of the curve for the liquid in the liquid phase stable region.

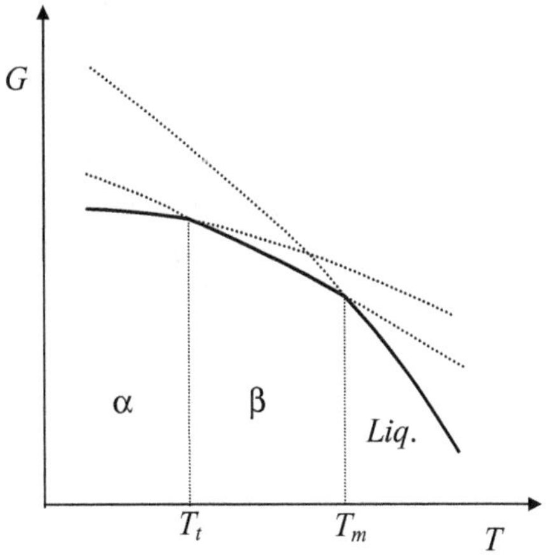


The above discussion may be linked to the *P-T* phase diagram of the substance as shown below:

$T_t$ is the temperature of allotropic transformation from α to β phase and $T_m$ is the melting point. By combining with the effect of pressure we can construct a *P-T* diagram for a substance of interest. At a given pressure, say, the atmospheric pressure, $P_a$, as shown in the diagram, when the α phase solid is heated, it will transform to the β phase at $T_t$, and the β phase, upon further heating, will melt at $T_m$ (melting point), and the liquid boils at $T_v$ (boiling point).

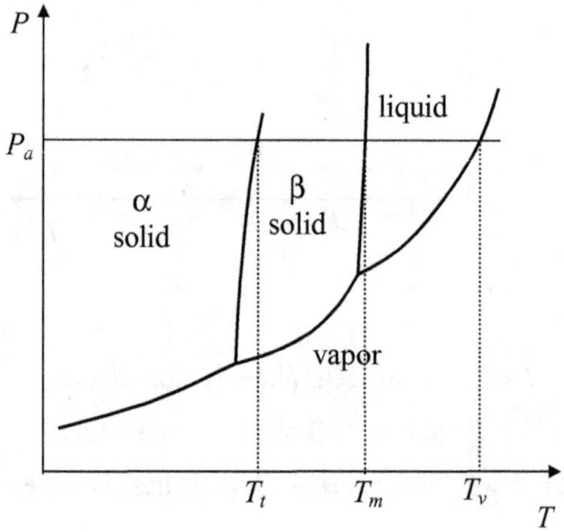


*Example 12.6*

α and β are two solid phases which are possible to appear in the substance A. At low temperatures α phase is more stable. Prove that one of the thermodynamic requirements for the appearance of allotropic transformation from α to β at constant pressure is

$$\text{Entropy of } \beta\ (S_\beta) > \text{Entropy of } \alpha\ (S_\alpha)$$

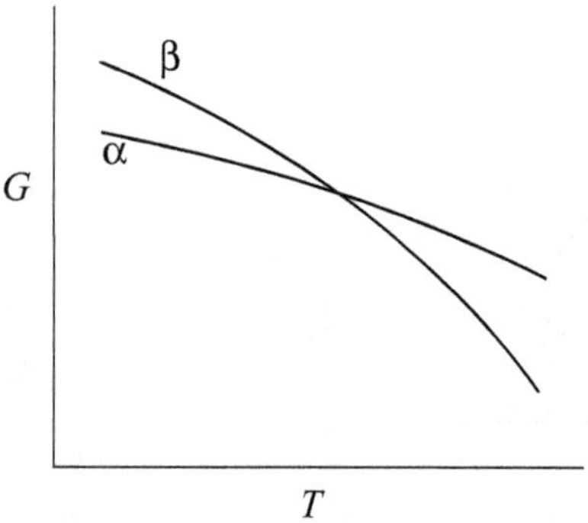


Different phases, say, α and β, in a given substance, in general have different Gibbs energies and entropies at a specific temperature and pressure. First, let's look at the α phase. Recall that the slope of the G-T curve is the negative entropy of the the phase.

$$\left(\frac{\partial G_i}{\partial T}\right)_P = -S_i$$

In order for β phase to appear at high temperatures, the Gibbs energy of β should fall more rapidly than that of α as the temperature is raised. In other words the Gibbs energy curve for β is should be steeper than that for α. This means that the entropy of the β phase should be larger than that of the α phase.

*Exercises*

12.4 The figure shown is the phase diagram of sulfur. The stable form of sufur at the room temperature is rhombic crystal structure. Discuss phase changes of sulfur upon heating.

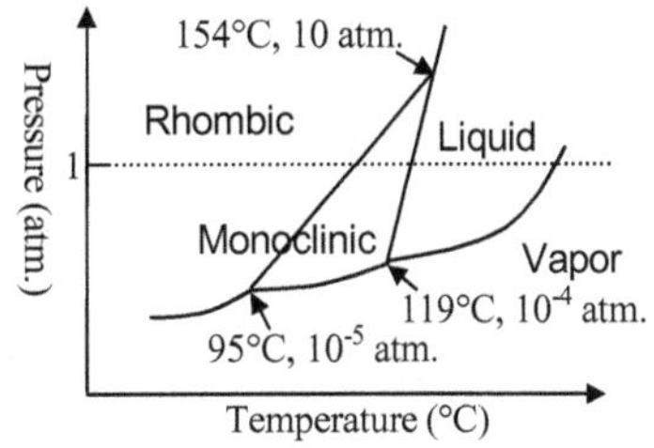


## 12.2 Binary Systems

### Binary Liquid Systems

When two liquids are brought together, they may

- totally dissolve in one another in all proportions (*total miscibility*),
- partially dissolve in one another (*partial miscibility*), or
- be completely immiscible (*immiscibility*).

<u>Total miscibility</u>

Consider the A-B binary liquid system in equilibrium with the vapor phase at a constant temperature. Is the composition of the vapor the same as that of the liquid? Not necessarily. Let's apply the Gibbs-Duhem equation to the liquid phase.

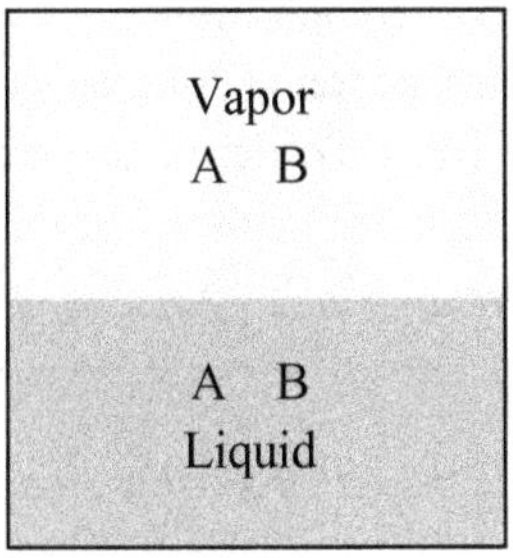


$$N_A d\, ln\, a_A + N_B d\, ln\, a_B = 0$$

By definition

$a_i = \dfrac{P_i}{P_i^*}$ : $P_i^*$ = sat. vapor pressure of $i$

$$N_A d\, ln\left(\frac{P_A}{P_A^*}\right) + N_B d\, ln\left(\frac{P_B}{P_B^*}\right) = 0$$

$P_i^*$ : constant at a given temperature

$$N_A d\ln P_A + N_B d\ln P_B = 0$$

$$\frac{N_A}{P_A}dP_A + \frac{N_B}{P_B}dP_B = 0$$

$P = P_A + P_B$ : $P$ = the total pressure
Differentiation with respect to $N_A$

$$\frac{dP}{dN_A} = \frac{dP_A}{dN_A} + \frac{dP_B}{dN_A}$$

$$\frac{dP}{dN_A} = \frac{dP_B}{dN_A}\left(1 - \frac{N_B P_A}{N_A P_B}\right)$$

First we apply this equation to an ideal solution. Recall that an ideal solution is the one in which all components obey Raoult's law, which states that the partial pressure of a component in solution is directly proportional to its molar concentration.
Let's assume that A is more volatile than B: *i.e.*, $P_A^* > P_B^*$ . Consider an arbitrary composition $N_A$ (Refer to the following figure). $P$ is the total vapor pressure which is in equilibrium with the liquid solution of composition $N_A$.
Now a question arises as to how the fraction of A in the vapor phase ($P_A/P$) is related to the fraction of A in the liquid phase ($N_A$). To derive a quantitative relationship, we need to make use of the last equation:

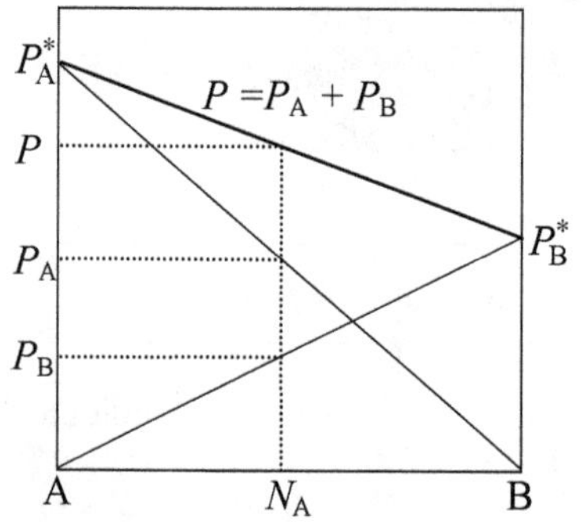


$$\frac{dP}{dN_A} = \frac{dP_B}{dN_A}\left(1 - \frac{N_B P_A}{N_A P_B}\right)$$

$\frac{dP}{dN_A} = (P_A^* - P_B^*)$ : Slope of the total $P$

$$\frac{dP_B}{dN_A} = -\frac{dP_B}{dN_B} = -P_B^* : \text{- (slope of } P_B \text{ curve)}$$

↓

$$P_A^* - P_B^* = -P_B^*\left(1 - \frac{N_B P_A}{N_A P_B}\right)$$

↓

$$\frac{P_B}{P_A} = \left(\frac{P_B^*}{P_A^*}\right)\frac{N_B}{N_A}$$

↓ Put $\alpha = \frac{P_B^*}{P_A^*}$ and add 1 to both sides

$$\frac{P_A + P_B}{P_A} = \frac{N_A + \alpha N_B}{N_A}$$

↓

$$\frac{P_A}{P} = \frac{N_A}{N_A + \alpha N_B}$$

↓ $\alpha = \frac{P_B^*}{P_A^*} < 1$, and

$N_A + N_B = 1$

Thus, $N_A + \alpha N_B < 1$

$$\frac{P_A}{P} > N_A$$

This last equation tells us that even for the ideal solution the mole fraction of a component in the vapor phase is not the same as that in the liquid phase. In the A-B binary solution of the above example in which A is more volatile than B, the vapor phase is richer in A than the liquid phase is

Now we establish a phase diagram which relates the compositions of the liquid and vapor phases at a given temperature. The following diagram indicates the step-by-step procedure of the establishment of the phase diagram:

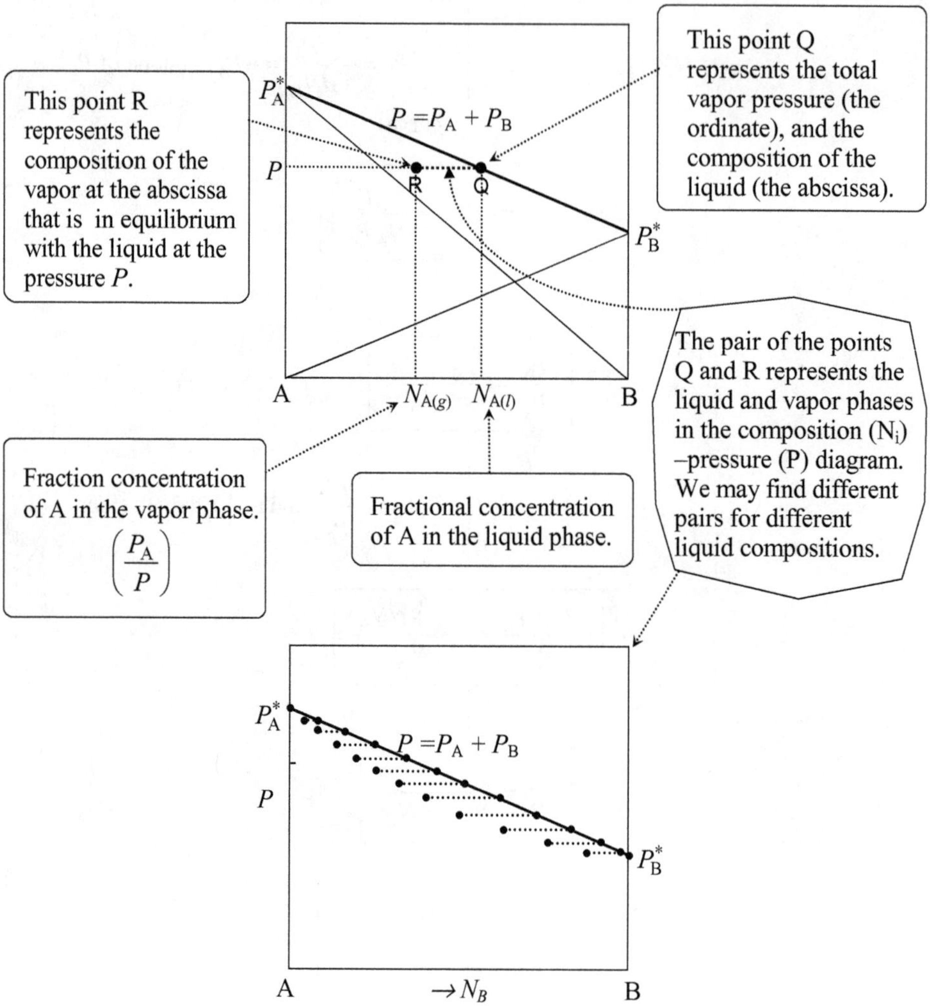
This point R represents the composition of the vapor at the abscissa that is in equilibrium with the liquid at the pressure P.
$P_A^*$
$P = P_A + P_B$
P
R
Q
This point Q represents the total vapor pressure (the ordinate), and the composition of the liquid (the abscissa).
$P_B^*$
A
$N_{A(g)}$
$N_{A(l)}$
B
The pair of the points Q and R represents the liquid and vapor phases in the composition ($N_i$) –pressure (P) diagram. We may find different pairs for different liquid compositions.
Fraction concentration of A in the vapor phase. $\left(\frac{P_A}{P}\right)$
Fractional concentration of A in the liquid phase.
$P_A^*$
$P = P_A + P_B$
P
$P_B^*$
A
$\rightarrow N_B$
B

When we connect all the pair points,

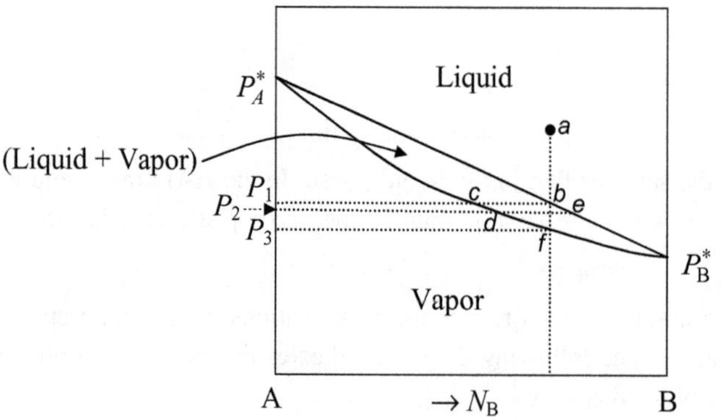
$P_A^*$
Liquid
a
(Liquid + Vapor)
$P_1$
$P_2$
$P_3$
c
b
e
d
f
$P_B^*$
Vapor
A
$\rightarrow N_B$
B

Consider lowering the pressure on a liquid solution of "a" in the above diagram.

- Until the pressure is lowered to $P_1$, the sample maintains a single liquid phase.
- When the total pressure reaches the point *b*, the vapor phase begins to appear, and thus the liquid coexists with its vapor of the composition *c*. According to the lever rule, the amount of the vapor phase is found to be negligibly small.
- Further decrease in the pressure results in the change in both the composition and the relative amount of each phase. At the pressure $P_2$, the compositions of the vapor and the liquid are given by the points *d* and *e*, respectively. Again, the relative amounts of the liquid and vapor are determined by the lever rule.
- If the pressure is further reduced to $P_3$, the composition of the vapor (*f*) becomes the same as the initial liquid composition. This means that at this pressure the liquid vaporizes completely and thus the amount of liquid is zero.
- Decrease in pressure below $P_3$ will bring the system to the region where only the vapor is present.

The above graphical expression of the pressure-composition relationship of equilibrium phases is a kind of phase diagram of the A-B system.

In a similar way a temperature-composition diagram can be developed as schematically shown below:

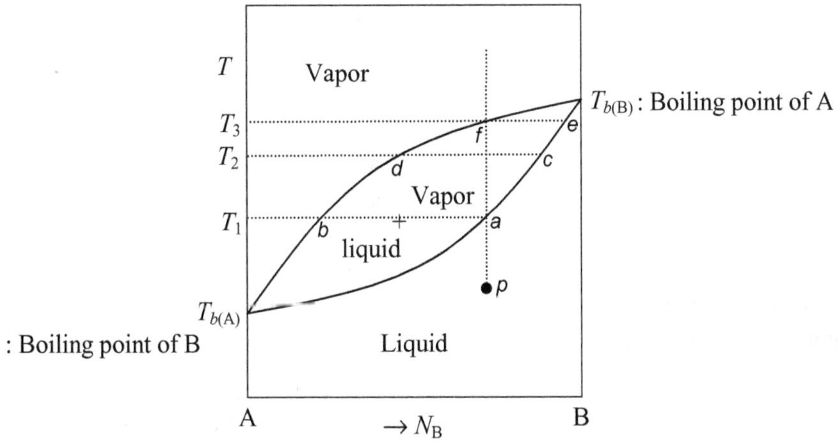


Consider a heating process of a liquid solution represented by the point *p* in the above diagram: Assuming the heating is slow enough that the system is virtually in equilibrium at every step of heating, the process will be,

- Until the temperature of the system reaches $T_1$, only the liquid phase exists.
- At $T_1$ the liquid begins to vaporizes. The compositions of the liquid and vapor phases are given by the points *a* and *b*, respectively. However, the amount of the vapor is negligibly small. Note that the vapor is richer in A than the liquid, because A is more volatile.

- On further heating, the compositions of both the liquid and the vapor are re-adjusted in such a way that the liquid composition moves from *a* toward *c* along the curve (***liquidus***), and the vapour composition from *b* toward *d* along the curve (***vaporus***).
- At $T_2$ the liquid of *c* coexists in equilibrium with the vapor of *d*. The proportion of each phase is determined by the lever rule.
- At $T_3$ the composition of the vapor (*f*) becomes equal to that of the original liquid. This means that the liquid vaporizes at this temperature virtually completely and thus the liquid of the composition *e* is present only as a trace.
- At temperatures higher than $T_3$ only vapor phase can exist.

Although temperature-composition phase diagrams of many liquids are similar to the one for an ideal solution shown above, there are a number of important solutions which exhibit a marked deviation. Recall the following equation which we have developed earlier:

$$\frac{dP}{dN_A} = \frac{dP_B}{dN_A}\left(1 - \frac{N_B P_A}{N_A P_B}\right)$$

When the total vapor pressure curve for a liquid solution shows a minimum or a maximum, *i.e.*,

$$\frac{dP}{dN_A} = 0 \text{ at } 0 < N_A < 1$$

then it is obvious that the boiling point curve in temperature-composition diagrams will show a maximum or minimum.

As $\frac{dP}{dN_A} \neq 0$, thus $\left(1 - \frac{N_B P_A}{N_A P_B}\right) = 0$

$$\frac{P_A}{P_B} = \frac{N_A}{N_B}$$

This last equation indicates that the composition of the vapor is the same as that of the liquid when the total vapour pressure shows a minimum or maximum. The following is the phase diagrams of the two cases:

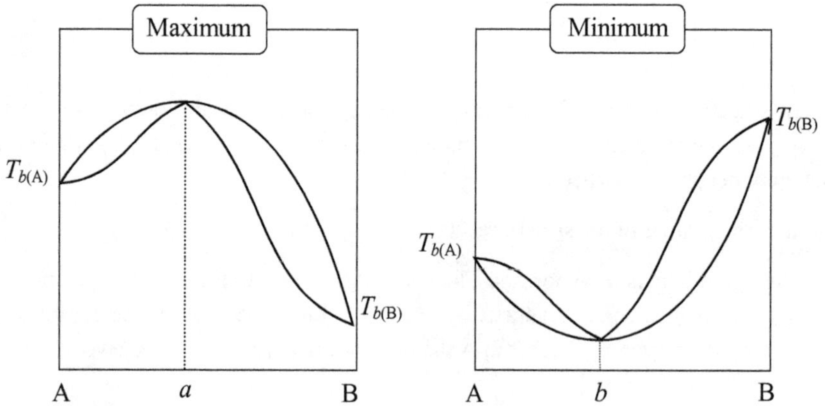

Note that at the maximum or minimum point the liquidus and vaporus curves are coincident and thus the composition of the vapor is the same as that of the liquid. In other words, evaporation of those liquid solutions denoted by *a* and *b* in the above diagrams occurs without change of composition. This type of mixture is said to form a ***azeotrope***.

Im miscibility

Certain liquids are completely immiscible in each other:

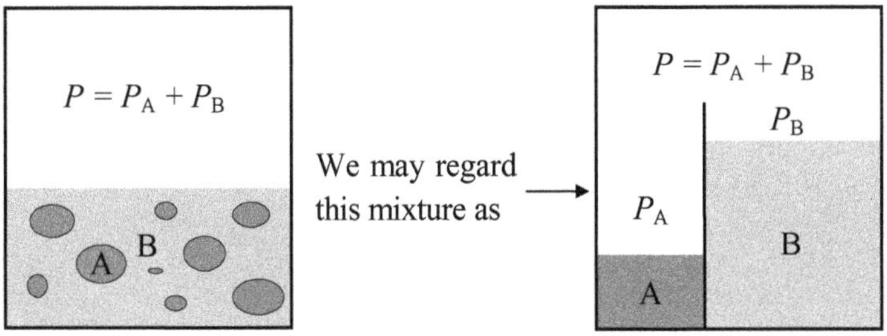


The liquid A is in equilibrium with its vapor at the vapor pressure of $P_A$, and the liquid B is also in equilibrium with its vapor at the vapor pressure of $P_B$. The total pressure ($P$) above the liquid mixture is thus the sum of the two vapor pressures:

$$P = P_A + P_B$$

When the mixture is heated in an open container, it boils at $P = 1$ *atm.*, not $P_A = 1$ *atm.* or $P_B = 1$ *atm.* This means that any mixture of immiscible liquids will boil at a temperature below the boiling point of either component. This is the basis of ***steam distillation***.

Partial miscibility

We have so far discussed two cases, namely complete solubility of liquids and complete immiscibility of liquids. But there are a number of liquid systems which show *partial miscibility*; *i.e.*, liquids that do not mix in all proportions at all temperatures.

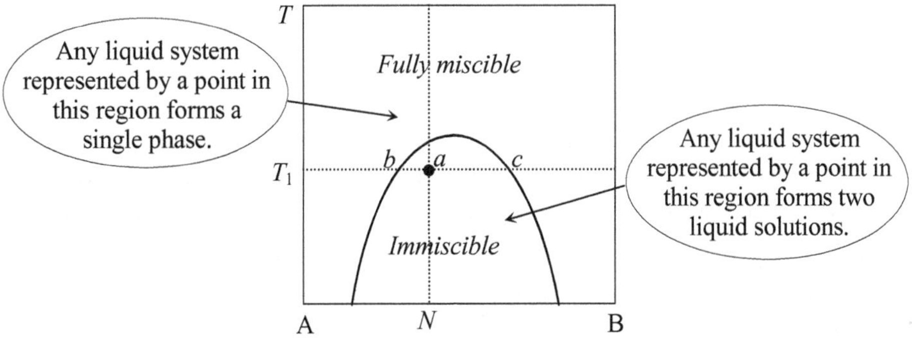

When a liquid mixture of the mean composition $N$ is prepared at the temperature $T_1$, two immiscible liquid solutions will form; one of the composition $b$ and the other of the composition $c$. This point $a$ in the diagram indicates merely the mean composition of the two immiscible solutions.

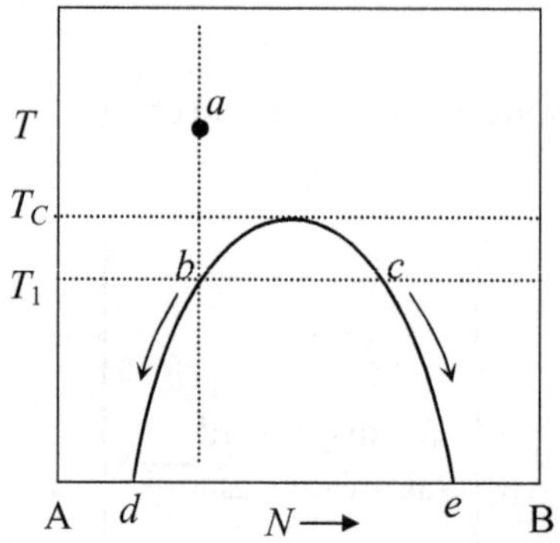


When a homogeneous liquid solution represented by the point $a$ in the above diagram is cooled, the phase separation into two immiscible solutions ($b$ and $c$) occurs at the temperature $T_1$. On further cooling compositions of both liquids change along the phase boundaries ($bd$ and $ce$). $T_C$ is the highest temperature at which phase separation can occur and is called the ***critical temperature***.

Given below are diagrams which are representative of partial miscibility or phase separation:

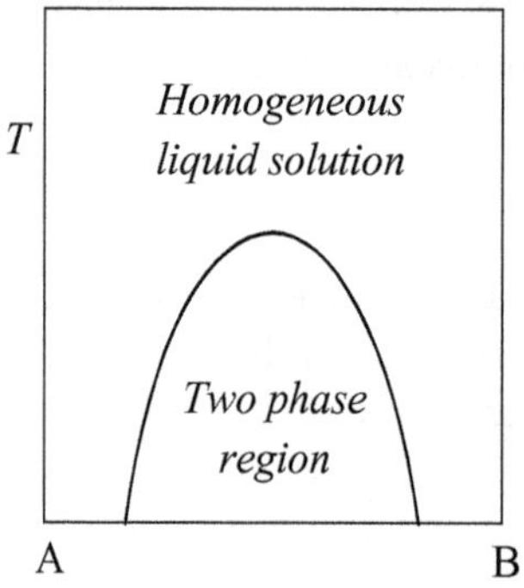


Fig A

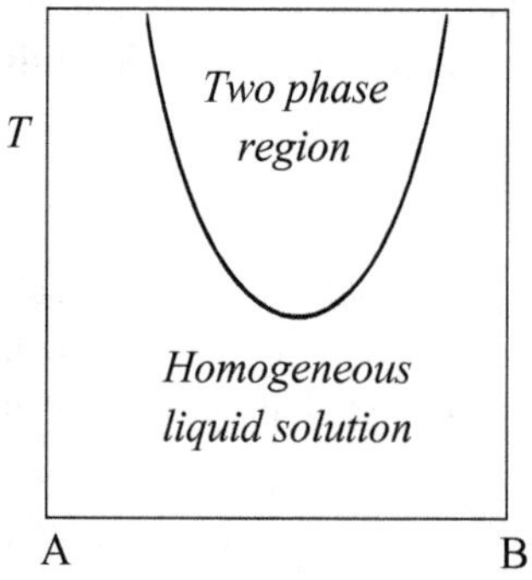


Fig B

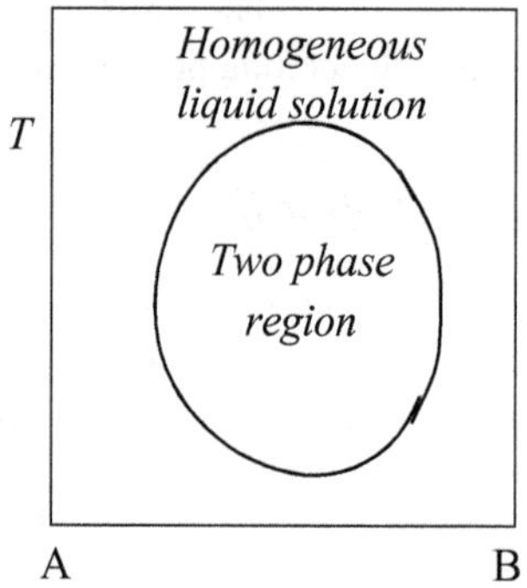


Fig C

As can be seen in Fig B, some systems show a ***lower critical temperature*** below which they form a homogeneous liquid solution in all proportions and above which they separate into two phases. Some systems have both ***upper*** and lower critical temperatures (Fig C).

We now consider systems which include a vapor phase together with liquid mixtures which exhibits a partial miscibility. The first example is a system which shows partial miscibility and a minimum boiling point (azeotrope). This example shows that the liquids become fully miscible before they boil.

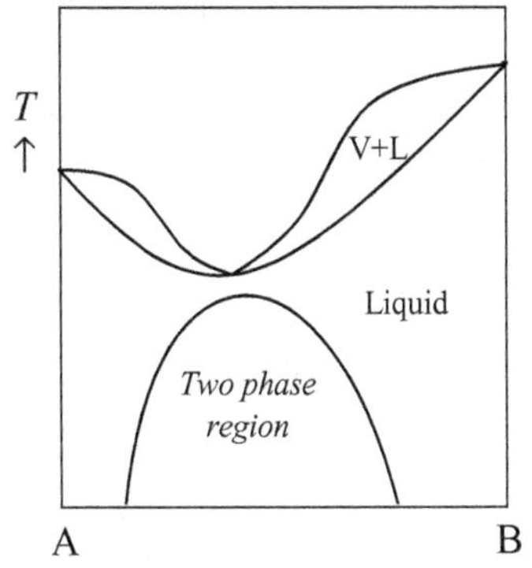


In some systems an upper critical temperature does not occur and the liquids boil before mixing is complete. In this case the phase diagram may look like

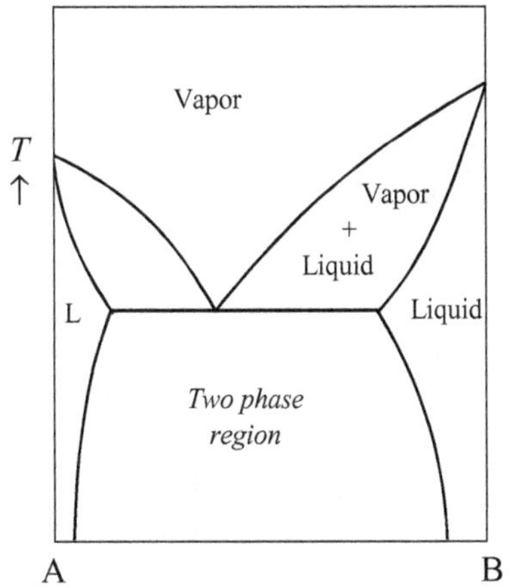


*Example 12.7*

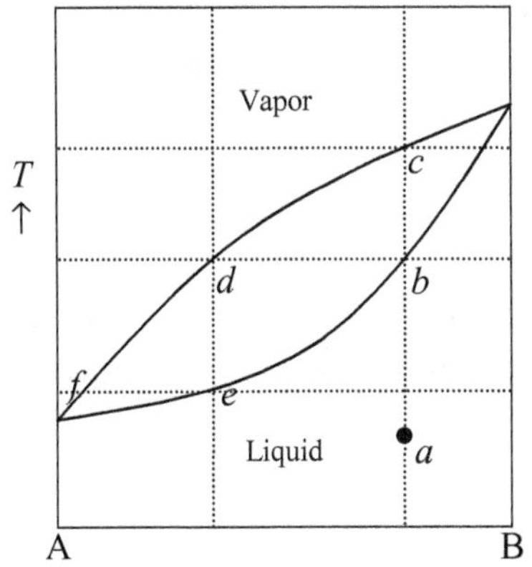


A liquid of composition $a$ is heated

1. Find the point which represents the temperature and composition of the liquid when it begins to boil.
2. Find the point which represents the temperature and composition of the vapor in equilibrium with the boiling liquid.
3. Now the vapor of the above question is withdrawn and completely condensed and then reheated. Find the point which represents the temperature and composition of the liquid and vapor at the new boiling point.

1. *b*, 2. *d*, 3. Liquid composition: *e*, vapor composition: *f*

If the boiling and condensation cycle is repeated successively, some interesting consequences will be resulted in. Notice in the following figure that the condensed liquid becomes richer in the more volatile component A as the cycle is repeated. This process is called ***fractional distillation***. Almost pure A may be obtained by repeating the cycle.

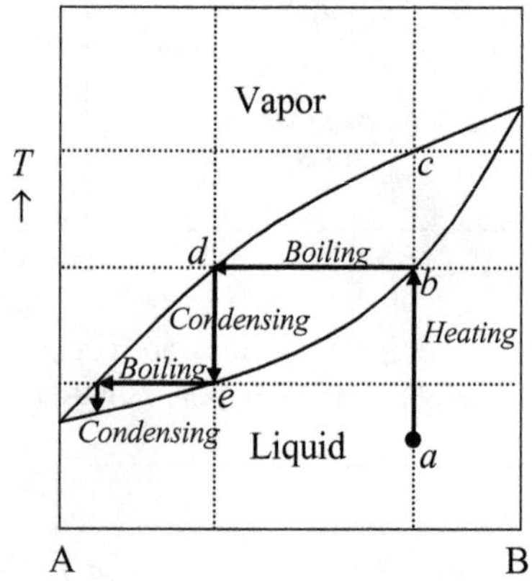


The efficiency of the fractional distillation depends very much on the shape of the phase diagram.

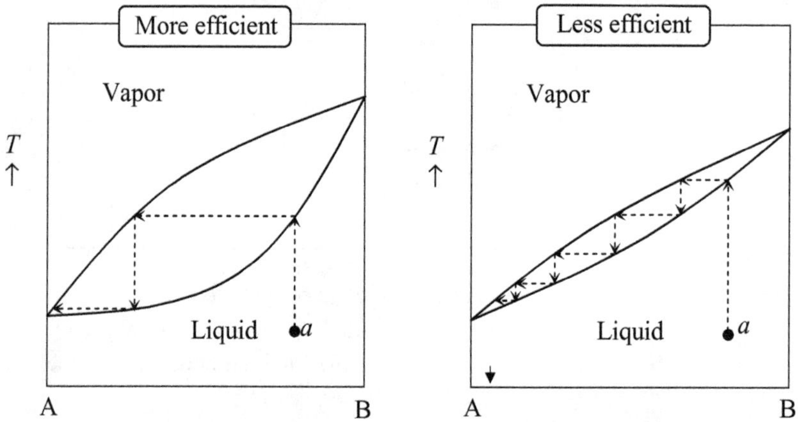


The efficiency is sometimes quantified by the number of the ***theoretical plates***, the number of effective vaporization and condensation steps that are required to achieve a condensate of given composition from a given distillate.

*Example 12.8*

Liquids A and B dissolve completely in one another in all proportions. If the change in vapor pressure $P$ (= $P_A + P_B$) with the fraction of A, $N_A$, is positive, *i.e.*,

$$\frac{dP}{dN_A} > 0$$

prove that the vapor phase is richer in component A than the liquid phase.

Recall the following equation:

$$\frac{dP}{dN_A} = \frac{dP_B}{dN_A}\left(1 - \frac{N_B P_A}{N_A P_B}\right)$$

Given $\frac{dP}{dN_A} > 0$

$\frac{dP_B}{dN_A} = -\frac{dP_B}{dN_B}$, but $\frac{dP_B}{dN_B} > 0$ always.

Thus, $\frac{dP_B}{dN_A} < 0$

$$1 - \frac{N_B P_A}{N_A P_B} < 0 \rightarrow \frac{P_A}{P_B} > \frac{N_A}{N_B}$$

*Exercises*

12.5 Liquids A and B exhibit a miscibility gap as shown in the following phase diagram. A mixture of 60 mol% of A and 40 mol% of B was prepared at 600°C. Calculate the mole fraction of the liquid rich in A.

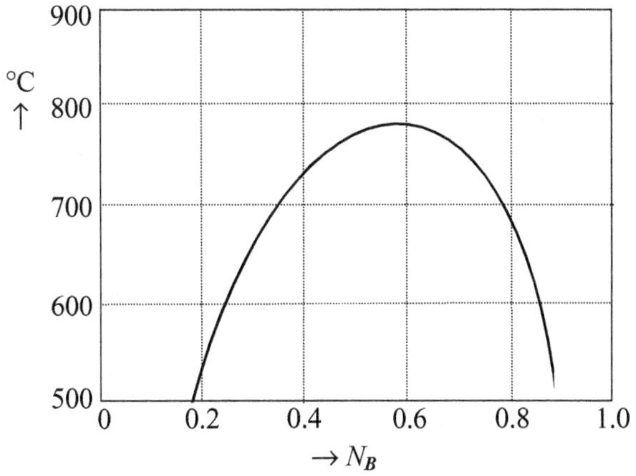

## Binary Systems without Solid Solution

### Eutectic Systems

Consider a system of two components, A and B, which are completely soluble in one another in the liquid state, but completely insoluble in one another in the solid state.
The melting point of a liquid (more correctly, the liquidus – to be explained later) is normally depressed if the liquid contains some other substance in solution.

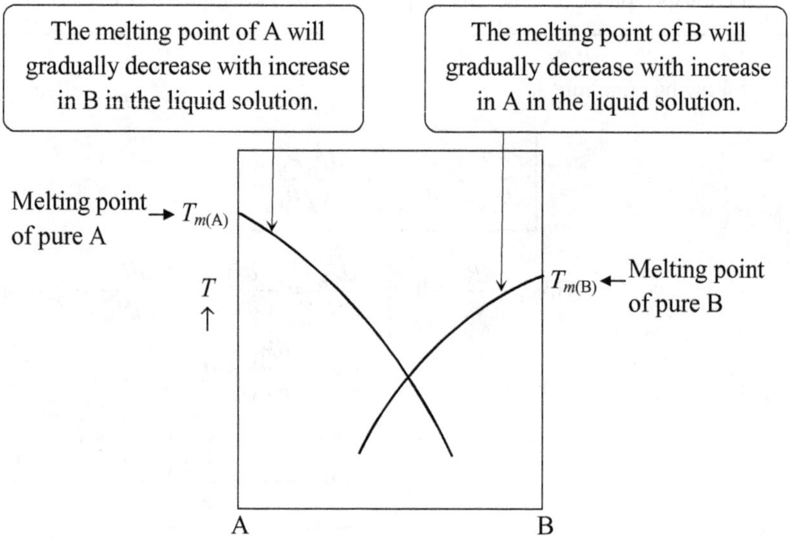
The melting point of A will gradually decrease with increase in B in the liquid solution.
The melting point of B will gradually decrease with increase in A in the liquid solution.
Melting point of pure A
$T_{m(A)}$
$T$ ↑
$T_{m(B)}$
Melting point of pure B
A
B

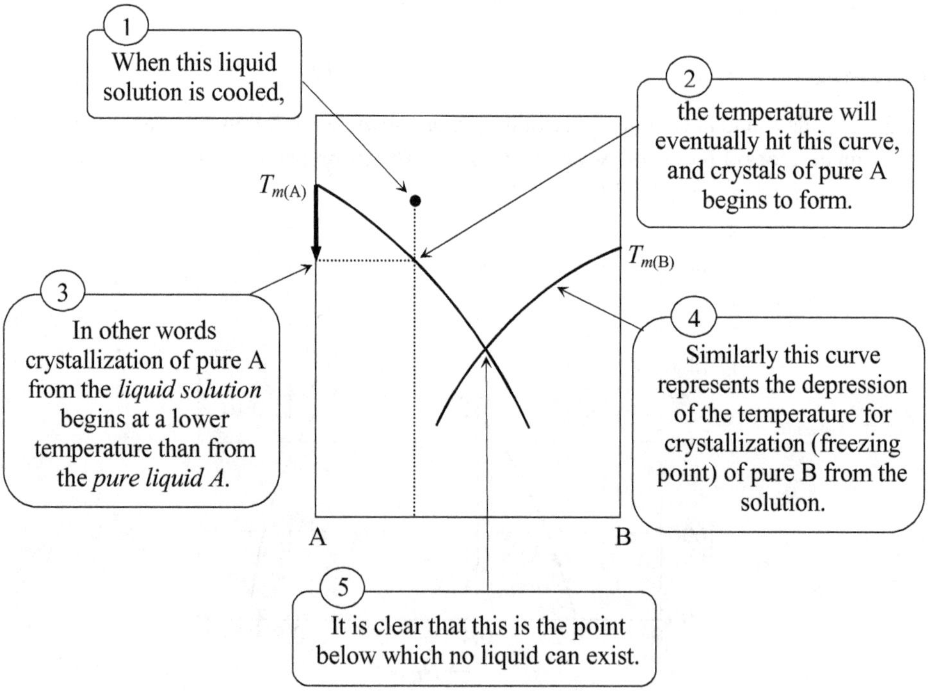
1
When this liquid solution is cooled,
2
the temperature will eventually hit this curve, and crystals of pure A begins to form.
$T_{m(A)}$
$T_{m(B)}$
3
In other words crystallization of pure A from the *liquid solution* begins at a lower temperature than from the *pure liquid A.*
4
Similarly this curve represents the depression of the temperature for crystallization (freezing point) of pure B from the solution.
A
B
5
It is clear that this is the point below which no liquid can exist.

Considering all the facts discussed above, we are able to draw a diagram which shows the temperature-stable phase relationship of a binary system in which no solid solution forms. The completed *equilibrium diagram* or *phase diagram* may look like the following figure:

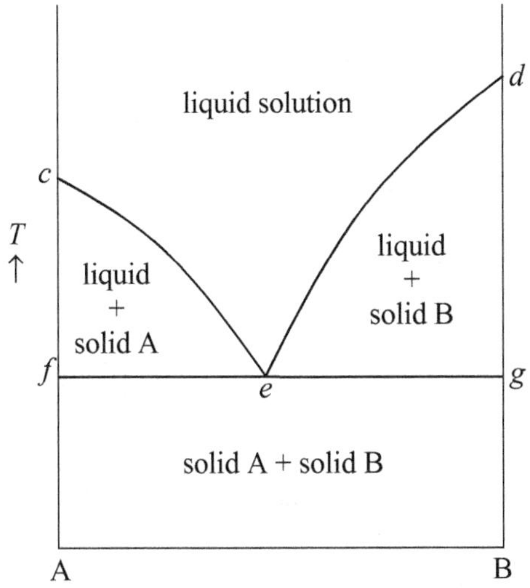


- The region above the curved line *ced* represents liquid solution. Any possible combination of temperature and composition which lies in this region will be completely molten. The lines *ce* and *ed* are called ***liquidus***.
- The region below the horizontal line *feg* represents the mixture of solid A and B. Any possible combination of temperature and composition which lies in this region will be completely solid. The line *feg*, or more precisely, *cfegd* is called ***solidus***.
- In the regions surrounded by the liquidus and solidus (*cfe* and *deg*), a solid phase coexists with a liquid phase. Consider the region *cfe*. Within this region, pure solid A can coexist with a number of different liquid solutions. Since solid A is the only solid present in this region, it is called the ***primary field*** of A. Similarly the region of *deg* is the *primary field* of B.
- The point *e* represents the lowest temperature at which a liquid solution can exist. At this point all remaining liquid solution solidifies. This point is called the ***eutectic point***.

We may extract a considerable amount of information from this type of phase diagram. Let's consider the case represented by the point *p* in the following figure. We know that this point is in the primary field of A in which pure solid A coexists with liquid. When an isothermal line passing the point of interest (*p*) is constructed, the point *r* represents the composition of the liquid phase which exists in equilibrium with pure solid A represented by the point *q*. The isothermal line *qr* which connects the compositions of the two phases that coexist in equilibrium is called the ***tie line***.

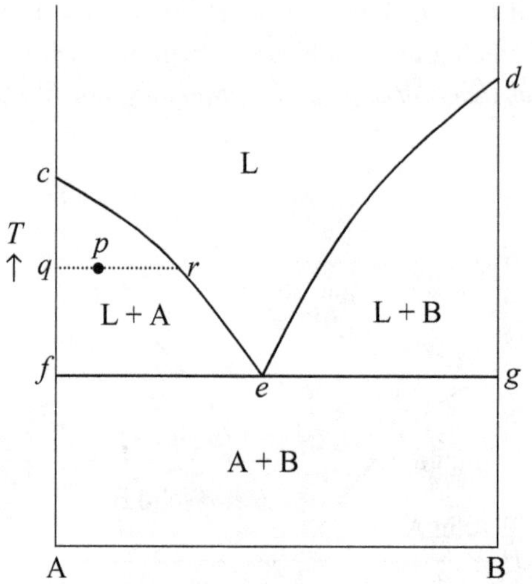


The relative amount of each phase is determined according to the lever rule:

$$\text{Fraction of solid A} = \frac{pr}{qr} \qquad \text{Fraction of liquid} = \frac{qp}{qr}$$

The microstructure at point *p* therefore would be the mixture of solid A and the liquid which may look like the figure given below:

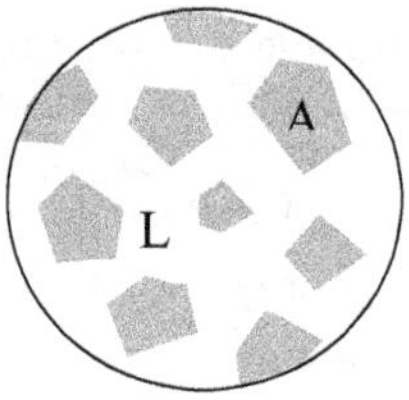


We now consider the eutectic point *e*. If the temperature is just above the eutectic temperature, it is a single liquid phase that is present. If the temperature is just below the eutectic temperature, however, two solid phases, solid A and solid B, are present. At the eutectic point, therefore, all three phases, *i.e.*, solid A, solid B and liquid, coexist in equilibrium. We may thus write a reaction which occurs at the eutectic point:

$$L(e) = A(s) + B(s)$$

This is called the *phase reaction* for the ***eutectic reaction***. On cooling at the eutectic point the reaction will proceed to the right. On heating the reaction will proceed to the left. The solid produced by the eutectic reaction is in general a fine grain mixture of A and B.

We now examine the cooling of an A-B binary solution from the liquid state to the complete solid state.

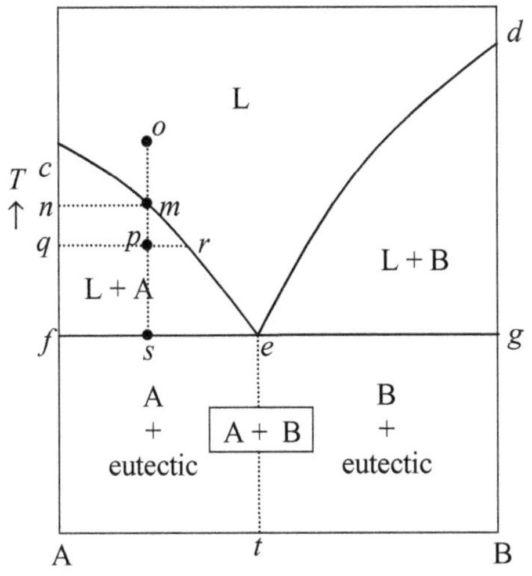


- When the liquid solution *o* is cooled, it remains in liquid until the temperature reaches the point *m*.
- At *m*, pure solid A(*n*) begins to precipitate.
- On further cooling the system enters the two phase region, and at *p* the system consists of pure solid A(*q*) and the liquid solution of *r*. Note that the liquid composition has been changed from *m* to *r* along the liquidus *mr*.
- At just above the eutectic temperature, the composition of the liquid in equilibrium with solid A is *e*. The fraction of the liquid is *fs* / *fe*.
- On further cooling to just below the eutectic temperature, the remaining liquid which has the eutectic composition will freeze immediately according to the eutectic reaction. The solid structure will thus be the mixture of the primary phase of A which has formed during cooling from *m* to *s* and the eutectic structure which is the fine mixture of A and B.
- The solid phase region *f*AB*g* thus can be divided into two sub-regions: *f*A*te* for solid A + eutectic and *et*B*g* for solid B + eutectic.

In a binary system two components may undergo a chemical reaction to form a compound. The compound formed may possess a definite or ***congruent*** melting point. If this is the case, the compound forms a separate phase and possesses a different crystal structure from those of the constituents. From many points of view a compound with a congruent melting point can be regarded as a pure substance. This type of compound will coexist in equilibrium with a liquid of identical composition. In other words the compound does not decompose below its melting point. The phase diagram of a binary system which forms a congruently-melting compound is somewhat different from the one shown above.

The phase diagram of a binary system shown below is effectively two simple eutectic diagrams linked together. Therefore there are two eutectic points in this kind of the system. A maximum point appears on the liquidus, which is the melting point of the congruently-melting compound.

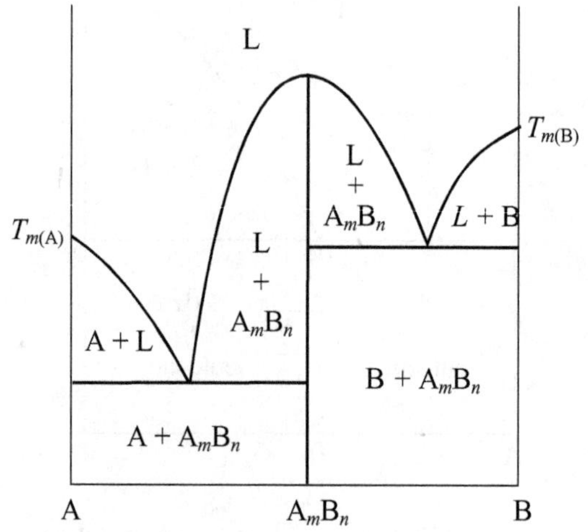


## Peritectic Systems

Some compounds do not have a sharp melting temperature, but rather decompose on heating into a liquid and another solid below the liquidus temperature. When the compound $A_mB_n$ is heated, it decomposes into solid A and liquid of $p$ at the temperature $T_P$. The compound $A_mB_n$ is stable only up to the temperature $T_P$. This type of compounds is called an ***incongruently-melting*** compound.

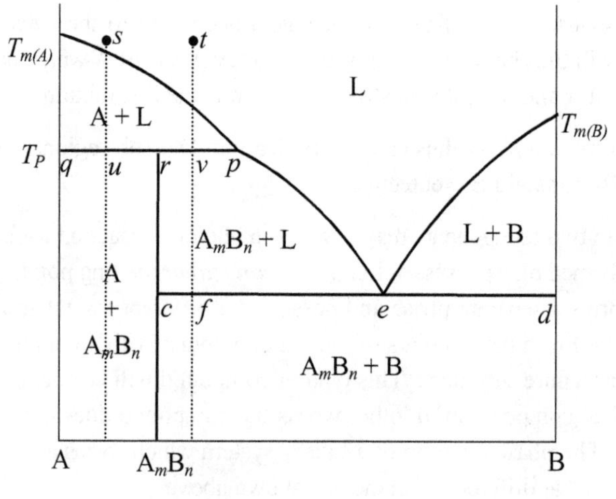

Consider the cooling of the liquid solution *s*.

(1) Until the temperature reaches the liquidus, the solution is fully molten.
(2) When the solution crosses the liquidus, primary crystals of pure A form.
(3) On further cooling the crystals of A grow and the liquid becomes richer in B.
(4) At a temperature just above $T_P$, the phases in equilibrium are pure solid A and the liquid of the composition of $p$. The relative amount of each phase is determined by the lever rule:

$$\text{Fraction of solid A} = \frac{up}{qp} \qquad \text{Fraction of liquid} = \frac{qu}{qp}$$

(5) At the temperature $T_P$ the crystallized A reacts with liquid and forms a compound $A_mB_n$.

$$\text{L(at } p) + \text{A(at } q) = A_mB_n\text{(at } r)$$

(6) This kind of reaction is called a ***peritectic reaction*** and the point $p$ is called the ***peritectic point***. The temperature $T_P$ is called the ***peritectic temperature***.
(7) The peritectic reaction proceeds until all the remaining liquid is exhausted.
(8) Just below the peritectic temperature, the coexistence of two solid phases, A at $q$ and $A_mB_n$ at $r$, is resulted in. Thus the microstructure of the system will consist of large primary crystals of A and small crystallites of $A_mB_n$.

Next, consider the cooling of the liquid solution *t*:

(1) Until the temperature reaches the liquidus, the solution is fully molten.
(2) When the solution crosses the liquidus, primary crystals of pure A form.
(3) On further cooling the crystals of A grow and the liquid becomes richer in B.
(4) At a temperature just above $T_P$, the phases in equilibrium are pure solid A and the liquid of the composition of $p$. The relative amount of each phase is determined by the lever rule:

$$\text{Fraction of solid A} = \frac{vp}{qp} \qquad \text{Fraction of liquid} = \frac{qv}{qp}$$

(5) At the temperature $T_P$ the crystallized A reacts with the liquid and forms compound $A_mB_n$.

$$\text{L(at } p) + \text{A(at } q) = A_mB_n\text{(at } r)$$

The peritectic reaction proceeds until all the primary solid A is completely exhausted. Since the quantity of B in the original composition is excessive for formation of $A_mB_n$ alone, the crystallization process does not end after the peritectic reaction, but continues to proceed by further cooling.

(6) Just below the peritectic temperature, the coexistence of the liquid phase at $p$ and $A_mB_n$ at $r$, is resulted in.
(7) On further cooling the crystals of $A_mB_n$ grow and the liquid becomes richer in B along the liquidus *pe*.
(8) At the temperature $T_e$, the remaining liquid (fraction $=cf/ce$) undergoes the *eutectic reaction.*

(9) The final solid consists of large crystals of $A_mB_n$ and small crystals of both $A_mB_n$ and B in a eutectic matrix.

## Monotectic Systems

The phase diagram given below shows a two-phase region consisting of two different liquids which are not miscible.

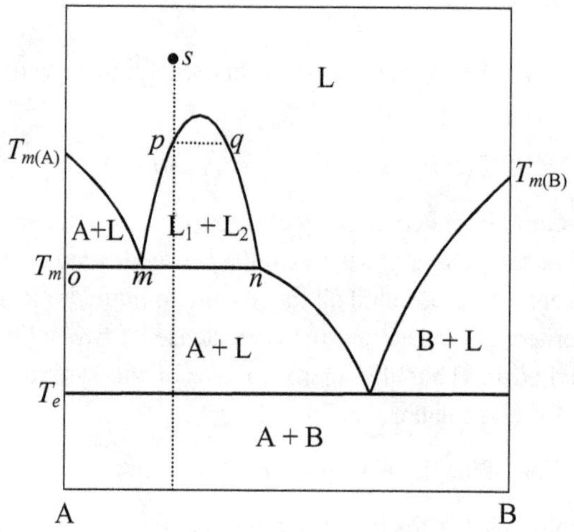


Consider the cooling of the liquid solution of *s*.

(1) At the liquidus (*p*), the melt begins to separate into two liquids, $L_1$ (at *p*) and $L_2$ (at *q*).
(2) As the temperature decreases, the compositions of the liquids alter along the liquidus *pm* and *qn*, respectively.
(3) At the temperature $T_m$, the liquid phase $L_1$ (at *m*) decomposes into pure solid A (at *o*) and liquid $L_2$ (at *n*). This reaction may be written as

$$L_1 \text{ (at } m) = \text{A (at } o) + L_2 \text{ (at } n)$$

This kind of reaction is called a ***monotectic reaction*** and the point *m* is referred to as the ***monotectic point***.

*Example 12.9*

It is possible, for any conceivable combination of temperature and total composition, to determine by inspection of the phase diagram exactly what phases will be present at equilibrium. It is also possible to determine the exact amount of each particular phase present under any given set of conditions.
Consider a simple eutectic system shown in the following figure. When the melt *s* is cooled, find
(1) the fraction of the primary solid phase of A just above the eutectic temperature $T_e$,
(2) the fraction of the eutectic structure just below $T_e$,

(3) the fraction of A in the eutectic structure below $T_e$,
(4) the fraction of A in total, i.e., the sum of A in the primary phase and the eutectic.

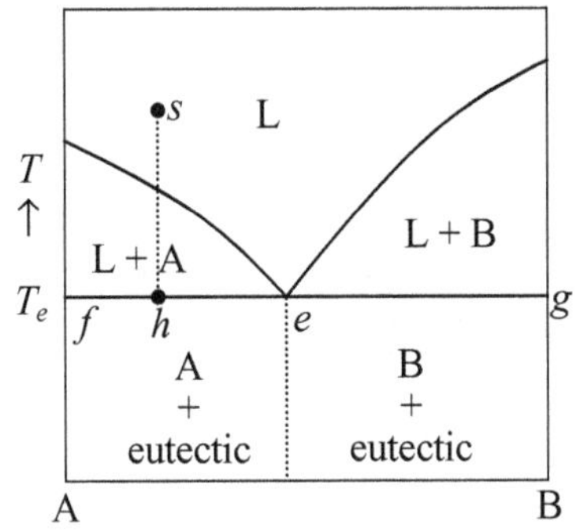


(1) According to the lever rule, the fraction of A $= he / fe$.
(2) The liquid fraction just above the eutectic temperature is converted into the eutectic structure when the temperature is lowered below the eutectic.

$$\text{Fraction of the eutectic structure} = \frac{fh}{fe}$$

(3) The liquid composition at the eutectic reaction is represented by $e$. This liquid is transformed into solid A and solid B by the eutectic reaction.

$$\text{Fraction of A in the eutectic structure} = \frac{eg}{fg}$$

(4) Fraction of the total A $= hg / fg$.

*Example 12.10*

Is the peritectic point in a binary system *invariant*?

Consider the Gibbs phase rule,

$$f = c - p + 2$$

$c - p$ : chemical contribution
2 : temperature and pressure
However, the pressure is fixed for the phase diagram. Thus 2 → 1.

↓

$$f = c - p + 1$$

$c = 2$ (A and B)
$p = 3$ (A, liquid and $A_mB_n$)

↓

$f = 0$ Invariant.

*Example 12.11*

If one of components or compounds in a binary system undergoes polymorphic transformations, there will be horizontal lines in the phase diagram separating the stable region of each polymorph. The figures given below represent two possible forms which a system can take when such transformations take place. Prove that conditions lying on these horizontal lines are univariant.

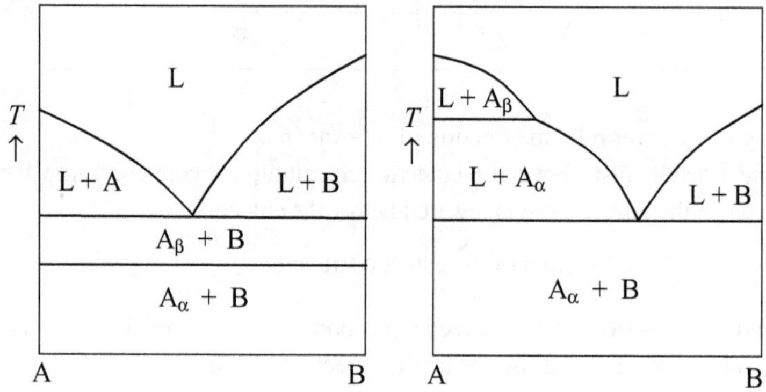


Fig 1 Fig 2

From the Gibbs phase rule,

$$f = c - p + 1$$

$c = 3$ ($A_\alpha$, $A_\beta$, B)
$p = 3$ ($A_\alpha$, $A_\beta$, B in Fig 1, $A_\alpha$, $A_\beta$, L in Fig 2)

$f = 1$ univariant

*Exercises*

12.6 In the eutectic alloy system AB, the compositions of the three conjugate (coexisting) phases of the eutectic are pure A, pure B and liquid of 80% B. Assuming equilibrium solidification of an alloy composed of 40% A and 60% B at a temperature just below the eutectic temperature, calculate the percentage of the primary A. Calculate the percentage of the total A.

## Binary Systems with Solid Solution

It is possible for solids to form a solution, *i.e.*, *solid solution*. The concept of either liquid solution or gaseous solution is familiar and easy to conceive of. By the term solid solution, it simply means that the solute component enters and becomes a part of the crystalline solvent, without altering its basic structure. This is not limited to solids involving elements, but applies equally to solids involving compounds.

There are two kinds of solid solutions, namely, ***substitutional*** solid solutions and ***interstitial*** solid solutions. In substitutional solid solutions, the solute element occupies a position of one of the solvent elements in the solvent crystal. In interstitial solid solutions, on the other hand, the solute element occupies one of the vacant spaces between solvent elements in the solvent crystal lattice without displacing a solvent element.

### *Total Solid Solubility*

Solid solutions with complete solid solubility, *i.e.*, solid solubility over the entire range of the composition, are possible to form, but always of a kind of the substitutional solid solution. For a metallic binary solution to exhibit a complete solid solubility, for instance, both metals must have the same type of crystal structure, because it must be possible to replace, progressively, all the atoms of the initial solvent with solute atoms without causing a change in crystal structure.

For a binary system in which two components are mutually soluble in all proportions in both the liquid and solid states, the possible phase diagram shapes are as shown below:

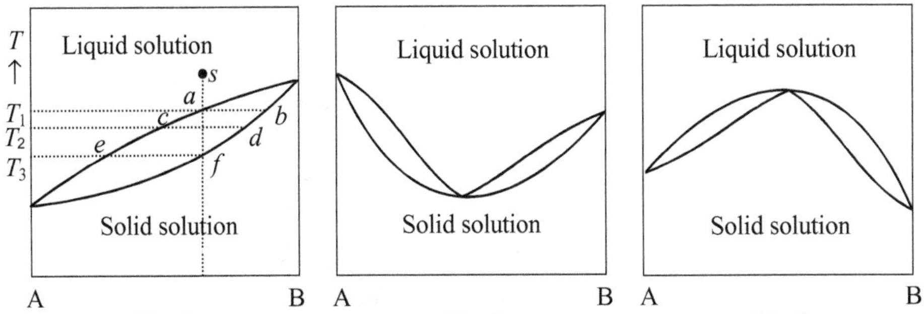


Fig. 1 Fig. 2 Fig. 3

First, consider the cooling of the liquid with the original composition of *s* in Fig.1.

(1) Freezing of the liquid solution will commence at $T_1$. Crystals which begin to form at this temperature are a solid solution of the composition *b*, but the amount of the solid solution forming at $T_1$ is infinitesimally small. Nevertheless, the liquid solution of *a* is in equilibrium with the solid solution *b* at $T_1$.

(2) As the temperature falls, the composition of the solid solution changes along the solidus and the composition of the liquid solution changes along the liquidus. At the temperature $T_2$, the liquid solution of *c* and the solid solution of *d* coexist in equilibrium. The relative amount of each phase is determined by the lever rule.

(3) At $T_3$, solidification is complete, and further cooling will bring the system to the solid phase.

Fig. 2 and 3 show the minimum and maximum melting points, respectively. The solid solution which corresponds to the minimum or maximum melting point behaves much like a pure component. It melts and freezes undergoing no changes in composition. In other words it melts *congruently*.

*Partial Solid Solubility*

In many cases, the atom size, crystal structure or other factors restrict solute atoms to dissolve in the solvent in the solid state. Thus it is much more common to find that solids are partly soluble in one another rather than be either completely soluble or completely insoluble. The following is an example of a phase diagram for a binary system which shows ***partial solid solubility***:

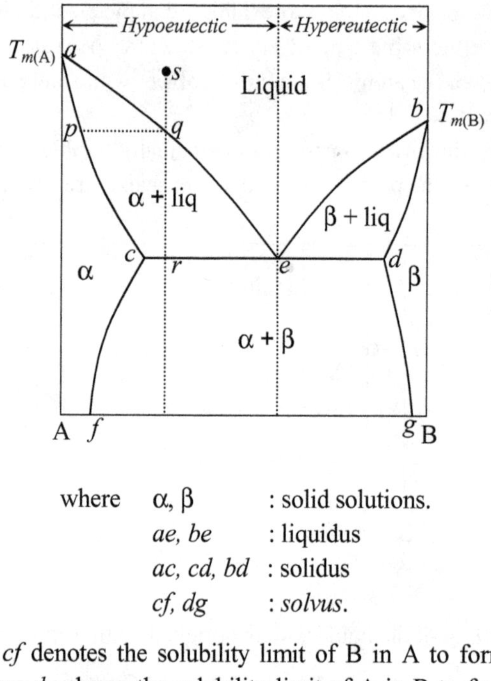


where α, β : solid solutions.
*ae*, *be* : liquidus
*ac*, *cd*, *bd* : solidus
*cf*, *dg* : *solvus*.

Note that the solvus *cf* denotes the solubility limit of B in A to form the α phase solid solution, and the solvus *dg* shows the solubility limit of A in B to form the β phase solid solution.

Consider the cooling of the solution *s*.

(1) At *q*, solidification begins. On further cooling the solid composition changes along the solidus *pc*, and the liquid composition along the liquidus *qe*.

(2) When the system arrives at the eutectic temperature, the liquid left (the fraction of liquid: *cr*/*ce*) undergoes the eutectic transformation:

$$\text{L (at } e) = \alpha \text{ (at } c) + \beta \text{ (at } d)$$

(3) On completion of the eutectic reaction, the resulting structure will be the mixture of the primary α phase and the *eutectic structure* which is the mixture of α and β phase.

In a binary phase diagram it is customary that the more common component is put on the left. Those structures which occur on the left side of the eutectic composition are called ***hypoeutectic***, and those on the right side are called ***hypereutectic***.

If a substance is allotropic this will affect the shape of phase diagrams for systems involving the substance. Consider a system which involves two allotropic substances, A and B. The following figure shows one of the possible diagrams which involve allotropic substances. The point *e* in the diagram is called the ***eutectoid point***, and the ***eutectoid reaction*** is

$$\gamma \text{ (at } e) = \alpha \text{ (at } c) + \beta \text{ (at } d)$$

Interpretation of the eutectoid phase diagram is generally the same as that of the eutectic phase diagram.

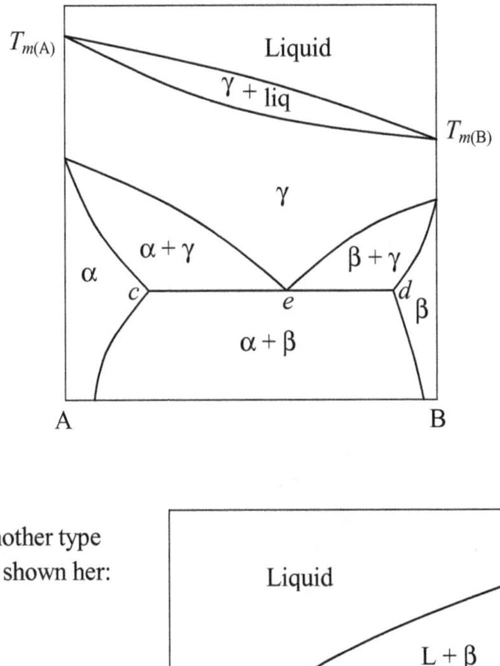


Now we consider another type of phase diagram as shown her:

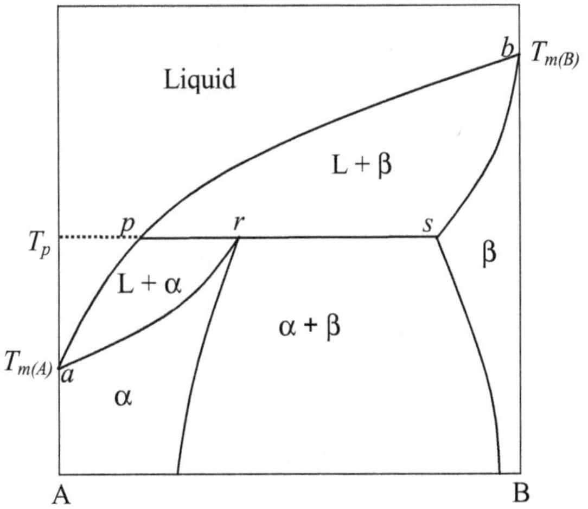

- As B in the α phase increases, the temperature at which liquid begins to form on heating rises along the solidus *ar*.
- The α phase at *r* decomposes upon heating into a liquid phase of *p* and the solid β phase at *s*. This reaction can be represented by

$$\alpha\,(\text{at } r) = \text{L}\,(\text{at } p) + \beta\,(\text{at } s)$$

- This reaction is called the peritectic reaction, and the point *p* is known as the peritectic point and $T_p$ the peritectic temperature.

Next we examine the cooling behaviour of several different total compositions with the figure given below.

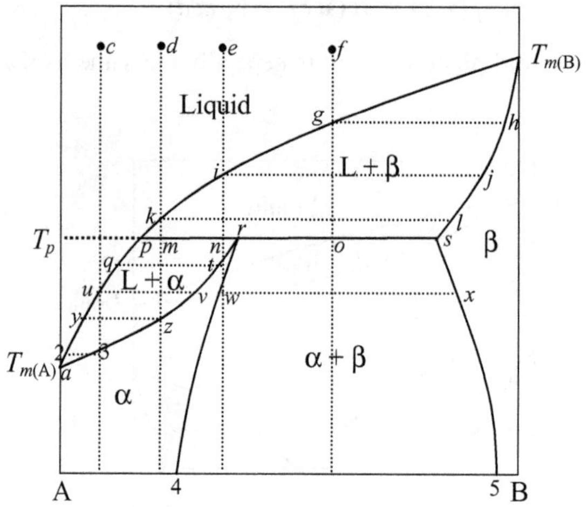


- Cooling of the liquid *c*
  - (1) $c \rightarrow u$ : Homogeneous liquid solution
  - (2) At *u* : Precipitation of solid α of the composition *v*
  - (3) $u \rightarrow 3$ : Increase in solid α phase. The compositions of the liquid and the α phase change along the *u*2 and *v*3, respectively.
  - (4) At 3 : Completion of solidification. The composition of α phase is given by 3 which is the same as the original liquid composition *c*.
  - (5) Below 3: Homogeneous α phase
- Cooling of the liquid *d*
  - (1) $d \rightarrow k$ : Homogeneous liquid solution
  - (2) At *k* : Precipitation of solid β phase of the composition *l*
  - (3) $k \rightarrow m$ : Increase in the solid β phase. The compositions of the liquid and the β phase change along *kp* and *ls*, respectively. The relative amount of each phase is determined by the lever rule.
  - (4) At *m* : Peritectic reaction: Portion of liquid *p* reacts with the solid β to form solid α at *r*.

$$L\,(\text{at } p) + \beta\,(\text{at } s) = \alpha\,(\text{at } r)$$

On completion of the peritectic reaction, the system consists of the liquid *p* and the solid α at *r*.

(5) $m \rightarrow z$ : Increase in the α phase. The compositions of the liquid and the α phase change along *py* and *rz*, respectively.

(6) At *z* : Completion of solidification.

(7) Below *z* : Homogeneous α phase

• Cooling of the liquid *e*

(1) $e \rightarrow i$ : Homogeneous liquid solution

(2) At *i* : Precipitation of solid β phase of the composition *j*

(3) $i \rightarrow n$ : Increase in the solid β phase. The compositions of the liquid and the β phase change along *ip* and *js*, respectively. The relative amount of each phase is determined by the lever rule.

(4) At *n* : Peritectic reaction : Portion of liquid *p* reacts with the solid β to form solid α at *r*.

$$L\,(\text{at } p) + \beta\,(\text{at } s) = \alpha\,(\text{at } r)$$

On completion of the peritectic reaction, the system consists of the liquid *p* and the solid α at *r*.

(5) $n \rightarrow t$ : Increase in the α phase. The compositions of the liquid and the α phase change along *pq* and *rt*, respectively.

(6) At *t* : Completion of solidification.

(7) $t \rightarrow w$ : Homogeneous α phase

(8) At *w* : Precipitation of β phase of the composition *x*

(9) Below *w*: Mixture of the α and β phases. The compositions of the α and β phases change along the solvus *w*4 and *x*5, respectively. The relative amount of each phase is determined by the lever rule.

• Cooling of the liquid *f*

(1) $f \rightarrow g$ : Homogeneous liquid solution

(2) At *g* : Precipitation of solid β phase of the composition *h*

(3) $g \rightarrow o$ : Increase in the solid β phase. The compositions of the liquid and the β phase change along *gp* and *hs*, respectively. The relative amount of each phase is determined by the lever rule.

(4) At *o* : Peritectic reaction : All liquid *p* reacts with a portion of the solid β to form solid α at *r*.

$$L\,(\text{at } p) + \beta\,(\text{at } s) = \alpha\,(\text{at } r)$$

On completion of the peritectic reaction, the system consists of the solid β at *s* and the solid α at *r*.

(5) Below *o*: Mixture of the α and β phases. The compositions of the α and β phases change along the solvus *r*4 and *s*5, respectively. The relative amount of each phase is determined by the lever rule.

Given below are some other types of binary phase diagrams

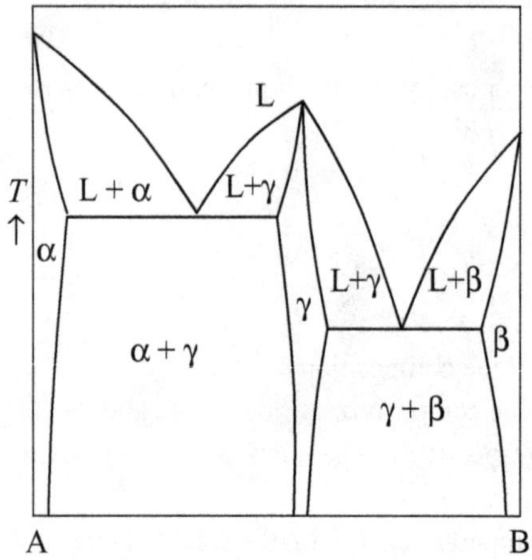


Fig. 1

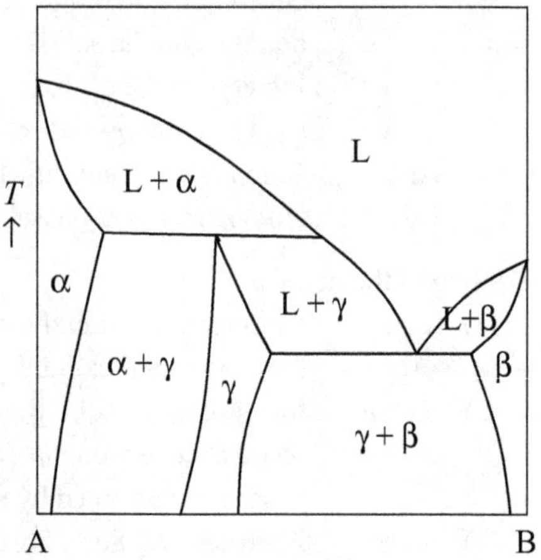


Fig. 2

where α and β phases : ***terminal solid solutions***
γ phase : ***intermediate solid solution***
γ in Fig 1 : congruently melting
γ in Fig 2 : incongruently melting

Interpretation of these diagrams is much the same as those discussed previously.

*Example 11.12*

The Fig. 1 in the following shows part of the phase diagram of the A-B binary system. When a liquid sample of the composition $x$ was cooled to the room temperature, it was found that each crystal or grain was *richer in A toward the center* (Fig. 2). Discuss the solidification process which would enable the formation of this non-uniform concentration grain structure and determine whether the structure shown in Fig. 2 is the equilibrium one.

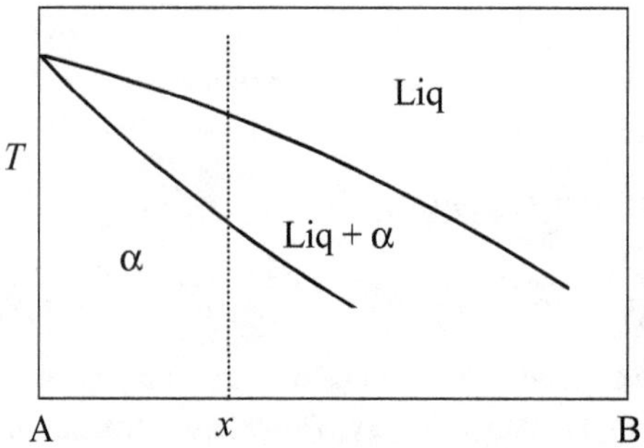


Fig. 1

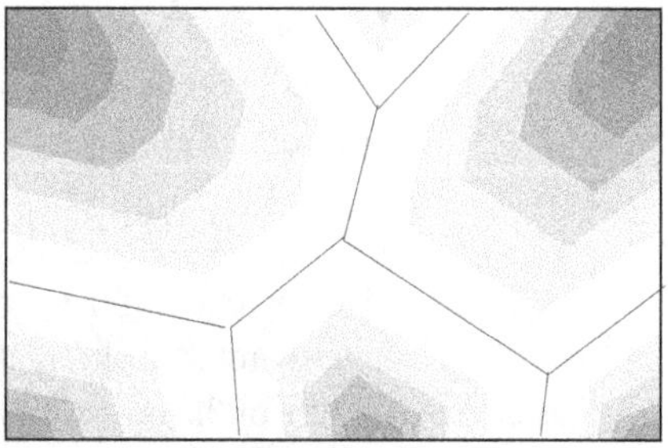

Fig. 2

True equilibrium solidification is almost not attainable in practice, because it requires both phases, the liquid and the solid, to be homogeneous throughout at all solidification times. This is possible only if a sufficiently long time is given at each infinitesimal step of the decrement of temperature. Therefore equilibrium solidification requires infinitesimally slow rates of heat extraction. With ordinary cooling rates, certain departures from equilibrium are unavoidable.

Consider the cooling of the liquid solution of *s* in the following diagram:

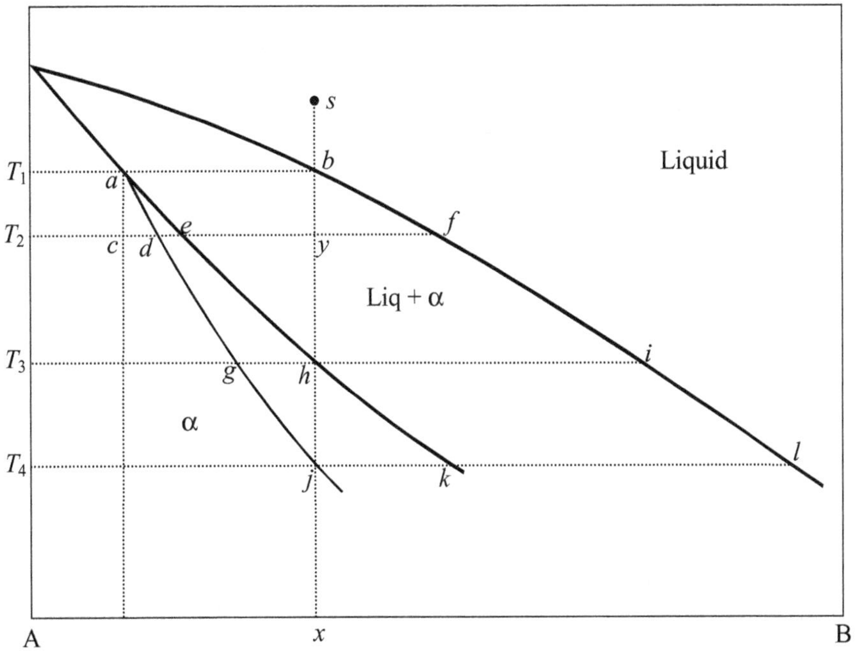


- The liquid remains homogeneous upon cooling from *s* to *b*.
- Solidification begins at $T_1$ with the deposition of crystals of the composition *a*.
- As cooling proceeds, the liquid composition changes along the liquidus.
- Solid forming at $T_2$ will have the composition of *e* and the liquid has the composition of *f*.
- During cooling from $T_1$ to $T_2$, however, a number of new nuclei form, each with the composition given along the solidus *ae* at its formation temperature. These nuclei will grow at the expense of the liquid.
- At the temperature $T_2$, there may be a number of nuclei having precipitated at different compositions lying somewhere between *ce*. None other than crystals having the composition *e* is in equilibrium with the liquid. However, not enough time is available for the compositions within the solid to change fully to the equilibrium values. Therefore the solid will have compositions ranging from *c* to *e* with an average somewhere between the two, say, *d*.
- On further cooling the liquid composition changes along the liquidus *fi*, the equilibrium solid composition along the solidus *eh*, and the average composition of the solid in a real practice along the line *dg*.

- Solidification would be completed at $T_3$, should equilibrium be maintained during the cooling. In reality, however, there is still liquid remaining.

$$\text{Fraction of liquid remained} = \frac{gh}{gi}$$

- On further cooling, the liquid composition changes along the liquidus *il* and the average solid composition along the line *gj*.
- At $T_4$, solidification is completed. It is thus seen that nonequilibrium freezing is characterised by
    1) increased temperature range over which liquid and solid are present
    2) $T_1 \leftrightarrow T_3$ for equilibrium solidification
    3) $T_1 \leftrightarrow T_4$ for nonequilibrium solidification
    4) a composition range remaining in the solids ( at least $j \leftrightarrow k$).
- In summary, the solidification begins with the formation of numerous solid nuclei and the growth of these nuclei follows. Each nucleus has a gradient of composition from its centre to the periphery. This nonequilibrium effect is referred to as ***coring***.

*Example 12.13*

Consider the A-B binary system. If A and B form a random solid solution with, say, 10 atom percent B, the probability of finding a B atom on any specific lattice site is just 0.1. Under certain conditions, however, B atoms may favor certain specific sites than the rest. B atoms will then preferentially position themselves on these specific sites. The probability of finding B atoms in these sites will greatly increase. This type of arrangement is referred to as an ***ordered structure***. The process in which a random disordered solid solution is rearranged into an ordered solid solution is called an ***order-disorder transition***.
Discuss the order-disorder transition using the phase diagram given :

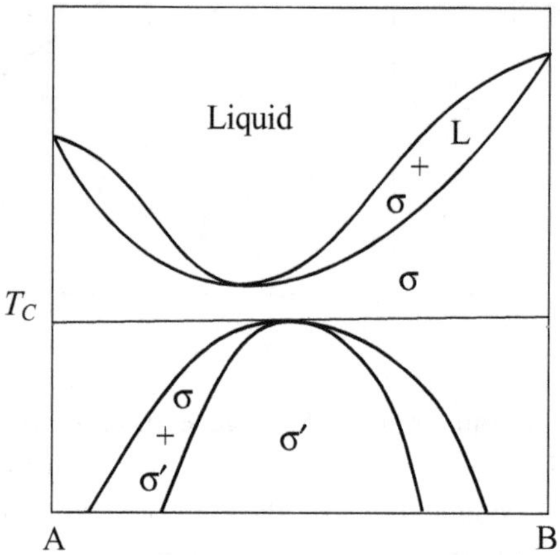


where
σ : disordered phase
σ' : ordered phase

- Ordered structure in the σ' field
- Disordered structure in the σ field
- Coexistence of ordered and disordered structures in the (σ + σ') field
- Disordered structure only at temperatures above $T_C$

*Exercises*

12.7 Consider the A-B binary peritectic system as shown in the following diagram:

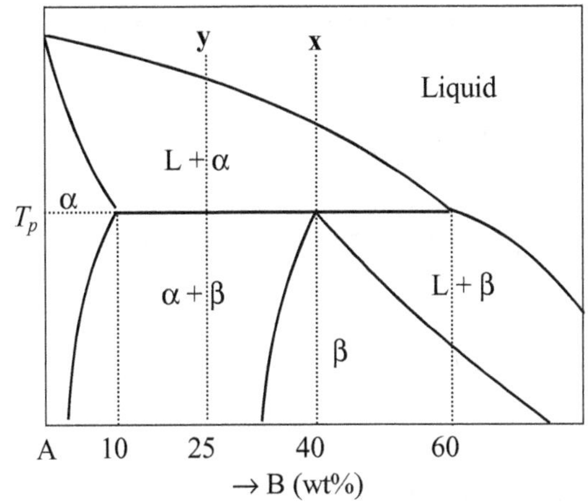


When liquid **x** is cooled maintaining equilibrium conditions, calculate wt% α that exists just above the peritectic temeperature $T_p$. When liquid **y** is cooled maintaining equilibrium conditions, calculate wt% β that exists just below the peritectic temperature $T_p$.

12.8 Phase diagram is one way of expressing thermodynamic equilibrium of a system. Every part of a phase diagram therefore has to conform to thermodynamic principles. Find errors in the phase diagrams shown below. Justify your answer.

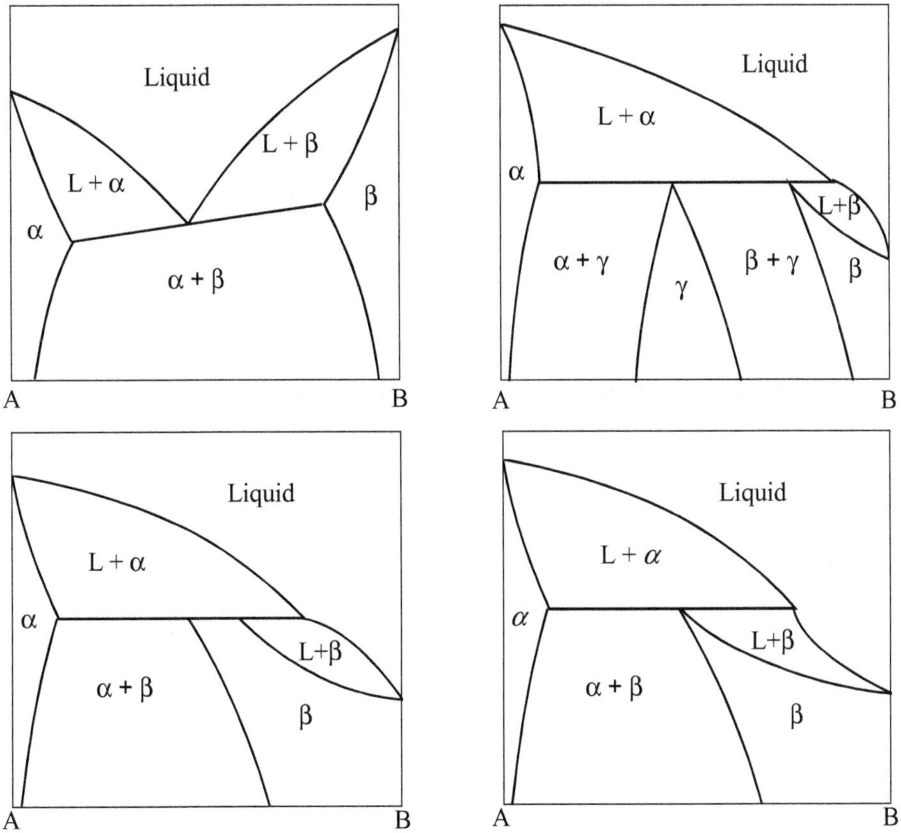

## 12.3 Thermodynamic Models

**Ideal Solutions**

Recall that the partial molar free energy, or chemical potential, of the component $i$ in a solution is given by

$$\overline{G}_i = G_i^{\mathrm{o}} + RT\, ln\, a_i$$

If the solution behaves ideally over the entire range of composition,

$$a_i = N_i$$

$$\overline{G}_i = G_i^{\mathrm{o}} + RT\, ln\, N_i$$

Consider the A-B binary system consisting of liquid solution and α solid solution phases.

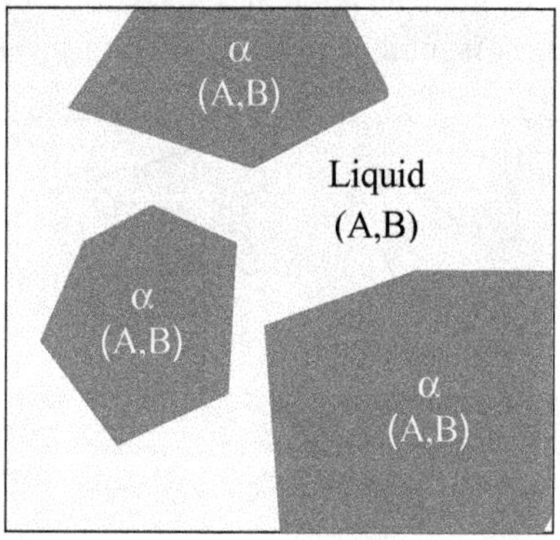


Chemical potentials of components of the liquid and α phase are given as follows:

Liquid

$$\overline{G}_{\mathrm{A}(l)} = G_{\mathrm{A}(l)}^{\mathrm{o}} + RT\, ln\, N_{\mathrm{A}(l)}$$
$$\overline{G}_{\mathrm{B}(l)} = G_{\mathrm{B}(l)}^{\mathrm{o}} + RT\, ln\, N_{\mathrm{B}(l)}$$

α phase

$$\overline{G}_{\mathrm{A}(\alpha)} = G_{\mathrm{A}(s)}^{\mathrm{o}} + RT\, ln\, N_{\mathrm{A}(\alpha)}$$
$$\overline{G}_{\mathrm{B}(\alpha)} = G_{\mathrm{B}(s)}^{\mathrm{o}} + RT\, ln\, N_{\mathrm{B}(\alpha)}$$

Recall that at equilibrium the chemical potential of a component must be the same in all phases throughout the system. Therefore

**Component A**

$$\bar{G}_{A(l)} = \bar{G}_{A(\alpha)}$$

$$G^{o}_{A(l)} + RT\, ln\, N_{A(l)} = G^{o}_{A(s)} + RT\, ln\, N_{A(\alpha)}$$

$$\frac{N_{A(\alpha)}}{N_{A(l)}} = exp\left(\frac{G^{o}_{A(l)} - G^{o}_{A(s)}}{RT}\right)$$

**Component B**

$$\bar{G}_{B(l)} = \bar{G}_{B(\alpha)}$$

$$G^{o}_{B(l)} + RT\, ln\, N_{B(l)} = G^{o}_{B(s)} + RT\, ln\, N_{B(\alpha)}$$

$$\frac{N_{B(\alpha)}}{N_{B(l)}} = exp\left(\frac{G^{o}_{B(l)} - G^{o}_{B(s)}}{RT}\right)$$

$$\frac{1 - N_{A(\alpha)}}{1 - N_{A(l)}} = exp\left(\frac{G^{o}_{B(l)} - G^{o}_{B(s)}}{RT}\right)$$

*Note that*

$G^{o}_{A(l)} - G^{o}_{A(s)} = \Delta G^{o}_{f,A(T)}$ : Standard Gibbs energy of fusion of A at $T$,

$G^{o}_{B(l)} - G^{o}_{B(s)} = \Delta G^{o}_{f,B(T)}$ : Standard Gibbs energy of fusion of B at $T$.

$$\frac{N_{A(\alpha)}}{N_{A(l)}} = exp\left(\frac{\Delta G^{o}_{f,A(T)}}{RT}\right)$$

$$\frac{1 - N_{A(\alpha)}}{1 - N_{A(l)}} = exp\left(\frac{\Delta G^{o}_{f,B(T)}}{RT}\right)$$

Recall that, neglecting differences in heat capacities of the pure solid and liquid phases for both A and B,

$$\Delta G^{o}_{f,A(T)} = \Delta H^{o}_{f,A}\left(1 - \frac{T}{T_{m(A)}}\right)$$

$$\Delta G^{o}_{f,B(T)} = \Delta H^{o}_{f,B}\left(1 - \frac{T}{T_{m(B)}}\right)$$

$\Delta H^{o}_{f,A}$ and $\Delta H^{o}_{f,B}$ : enthalpies of fusion of A and B, respectively, at their true melting points.

$$\frac{N_{A(\alpha)}}{N_{A(l)}} = exp\left(\frac{\Delta H^{o}_{f,A}}{R}\left[\frac{1}{T} - \frac{1}{T_{m(A)}}\right]\right)$$

$$\frac{1 - N_{A(\alpha)}}{1 - N_{A(l)}} = exp\left(\frac{\Delta H^{o}_{f,B}}{R}\left[\frac{1}{T} - \frac{1}{T_{m(B)}}\right]\right)$$

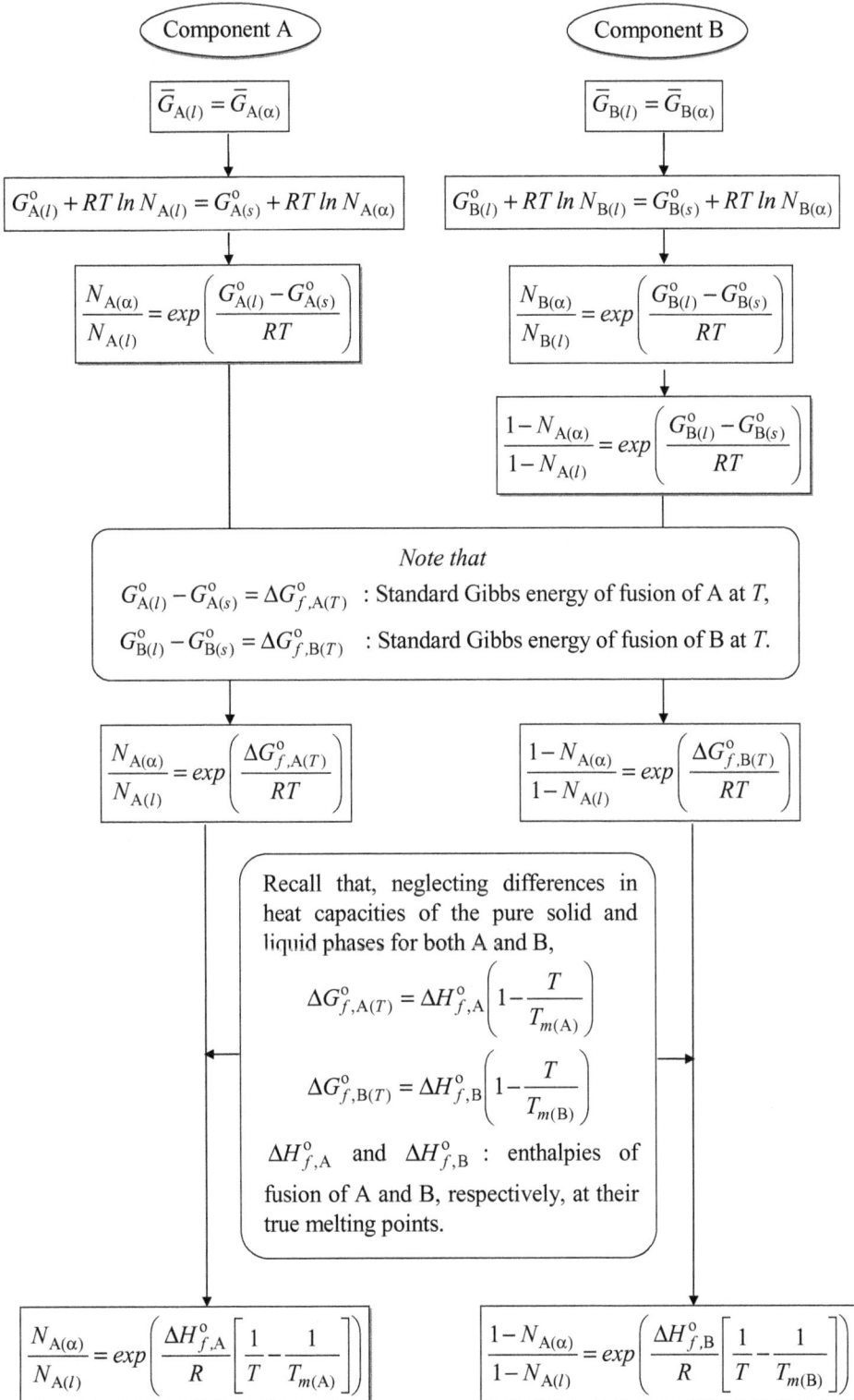

Thus in the ideal solution model of a two-phase field, a knowledge of the enthalpies of fusion of the pure components at their respective melting points allows simultaneous solution of these two equations for the two unknowns, $N_{A(L)}$ and $N_{A(\alpha)}$ at the temperature of interest.

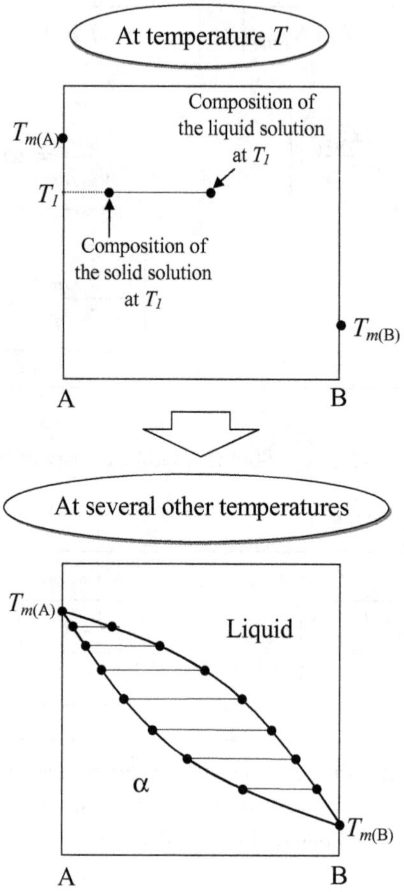


Given below is a typical form of phase diagram for an ideal binary solution:

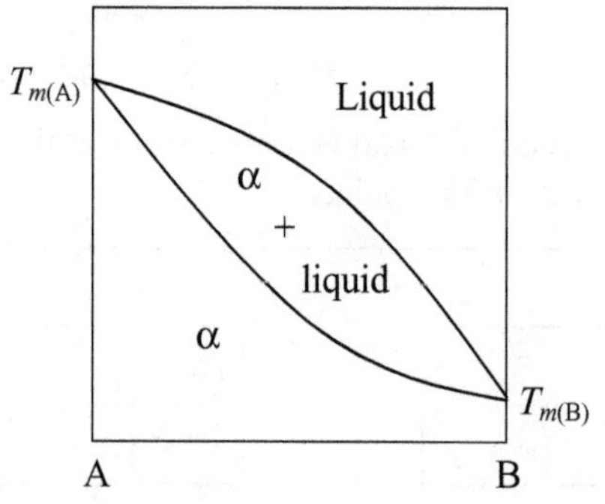

Pattern of a phase diagram depends on values of the enthalpies of fusion of the components in the solution, i.e., $\Delta H^{\circ}_{f,\mathrm{A}}$ and $\Delta H^{\circ}_{f,\mathrm{B}}$, and some examples are shown below:

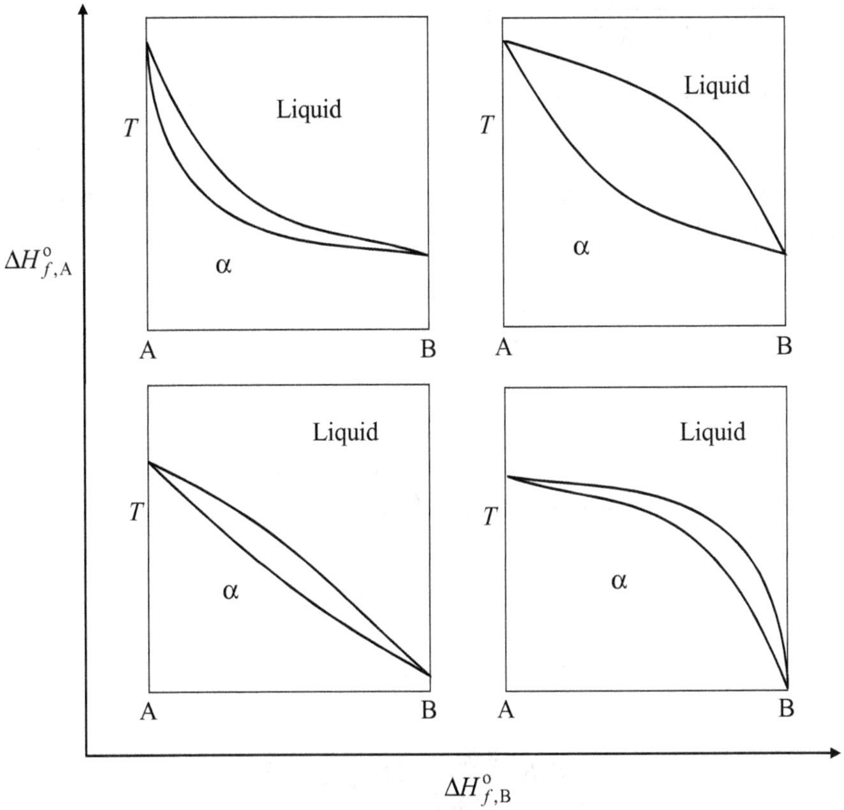


**Non-ideal Solutions**

Recall that the partial molar free energy, or chemical potential, of the component $i$ in a solution is given by

$$\bar{G}_i = G_i^{\circ} + RT\, ln\, a_i$$

Consider the A-B binary system consisting of α and β solid solutions.

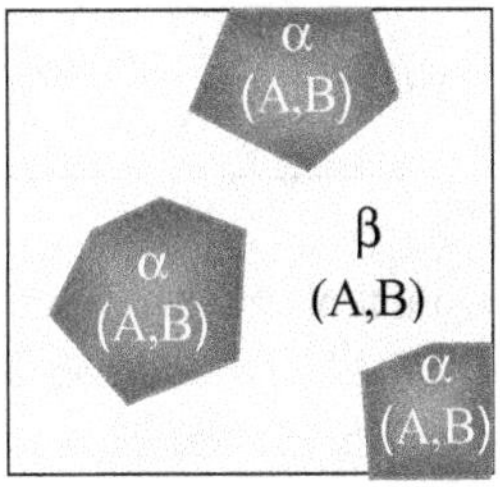

Chemical potentials of the components in $\alpha$ and $\beta$ phases are

| $\alpha$ phase | $\beta$ phase |
|---|---|
| $\bar{G}_{A(\alpha)} = G^{o}_{A(\alpha)} + RT\,ln\,a_{A(\alpha)}$ | $\bar{G}_{A(\beta)} = G^{o}_{A(\beta)} + RT\,ln\,a_{A(\beta)}$ |
| $\bar{G}_{B(\alpha)} = G^{o}_{B(\alpha)} + RT\,ln\,a_{B(\alpha)}$ | $\bar{G}_{B(\beta)} = G^{o}_{B(\beta)} + RT\,ln\,a_{B(\beta)}$ |

Recall that at equilibrium the chemical potential of a component must be the same in all phases throughout the system. Therefore,

Component A

$$\bar{G}_{A(\alpha)} = \bar{G}_{A(\beta)}$$

$$G^{o}_{A(\alpha)} + RT\,ln\,a_{A(\alpha)} = G^{o}_{A(\beta)} + RT\,ln\,a_{A(\beta)}$$

Standard Gibbs energy of transformation of A from $\alpha$ to $\beta$ phase

$$G^{o}_{A(\beta)} - G^{o}_{A(\alpha)} = \Delta G^{o}_{t,A(\alpha\to\beta)}$$

$$RT\,ln\left(\frac{a_{A(\alpha)}}{a_{A(\beta)}}\right) = \Delta G^{o}_{t,A(\alpha\to\beta)}$$

$$a_i = \gamma_i N_i$$

$$RT\,ln\left(\frac{\gamma_{A(\alpha)}}{\gamma_{A(\beta)}}\right) + RT\,ln\left(\frac{N_{A(\alpha)}}{N_{A(\beta)}}\right) = \Delta G^{o}_{t,A(\alpha\to\beta)}$$

In a similar way for the component B,

$$RT\,ln\left(\frac{\gamma_{B(\alpha)}}{\gamma_{B(\beta)}}\right) + RT\,ln\left(\frac{1-N_{A(\alpha)}}{1-N_{A(\beta)}}\right) = \Delta G^{o}_{t,B(\alpha\to\beta)}$$

We now have two equations for two unknowns, $N_{A(\alpha)}$ and $N_{A(\beta)}$, provided that the standard Gibbs energies of transformation are known and the activity coefficients are given or expressed in terms of compositions. The values of $N_{A(\alpha)}$ and $N_{A(\beta)}$ thus found from the equations are the phase boundary compositions.

As real solutions may depart from ideality in a number of different ways, the activity coefficients in the above equations may take various expressions. We here discuss one simple type of departure from ideality, namely, the *regular solution model*, for the purpose of illustration.

For the A-B binary regular solution, we have seen in the section 2.3.5 that

$$RT\,ln\,\gamma_A = \Omega\,N_B^2 \qquad RT\,ln\,\gamma_B = \Omega\,N_A^2$$

Combination yields,

$$\Omega_{(\alpha)}(1-N_{A(\alpha)})^2 - \Omega_{(\beta)}(1-N_{A(\beta)})^2 + RT\,ln\left(\frac{N_{A(\alpha)}}{N_{A(\beta)}}\right) = \Delta G^{o}_{t,A(\alpha\to\beta)}$$

$$\Omega_{(\alpha)}N_{A(\alpha)}^2 - \Omega_{(\beta)}N_{A(\beta)}^2 + RT\,ln\left(\frac{1-N_{A(\alpha)}}{1-N_{A(\beta)}}\right) = \Delta G^{o}_{t,B(\alpha\to\beta)}$$

Parameters which are the input components for the calculation of phase boundary compositions, $N_{A(\alpha)}$ and $N_{A(\beta)}$, in the above equations are,

| | |
|---|---|
| $\Omega_{(\alpha)}$ | : Interaction parameter for the $\alpha$ phase standard state |
| $\Omega_{(\beta)}$ | : Interaction parameter for the $\beta$ phase standard state |
| $\Delta G^{o}_{t,A(\alpha\to\beta)}$ | : Standard Gibbs energy of phase transformation of A from $\alpha$ to $\beta$ |
| $\Delta G^{o}_{t,B(\alpha\to\beta)}$ | : Standard Gibbs energy of phase transformation of B from $\alpha$ to $\beta$ |

The first two parameters ( $\Omega_{(\alpha)}$ and $\Omega_{(\beta)}$ ) are independent of temperature, and the last two are constant at a given temperature. Thus all of these parameters are constant at any given temperature.

The regular solution model is much more flexible than the ideal solution model and a wide variety of phase diagrams can be produced with this model.

The procedure of the phase boundary calculation can thus be represented by the algorithm given the diagram below.

Topographical changes in the phase diagram for a system A-B with regular solid and liquid phases brought about by systematic changes in the interaction parameters, $\Omega_{(s)}$ and $\Omega_{(l)}$, are illustrated in the following diagram.*

Melting points of pure A and B are assumed 800 K and 1200 K, respectively, and standard entropies of fusion of A and B are assumed to be 10 J mol$^{-1}$K$^{-1}$.

<Calculation Algorithm of the regular solution model>

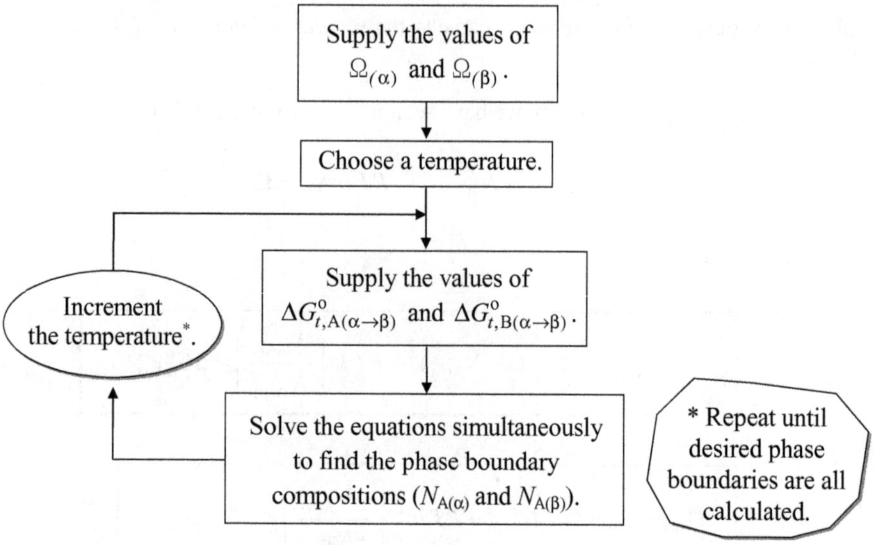


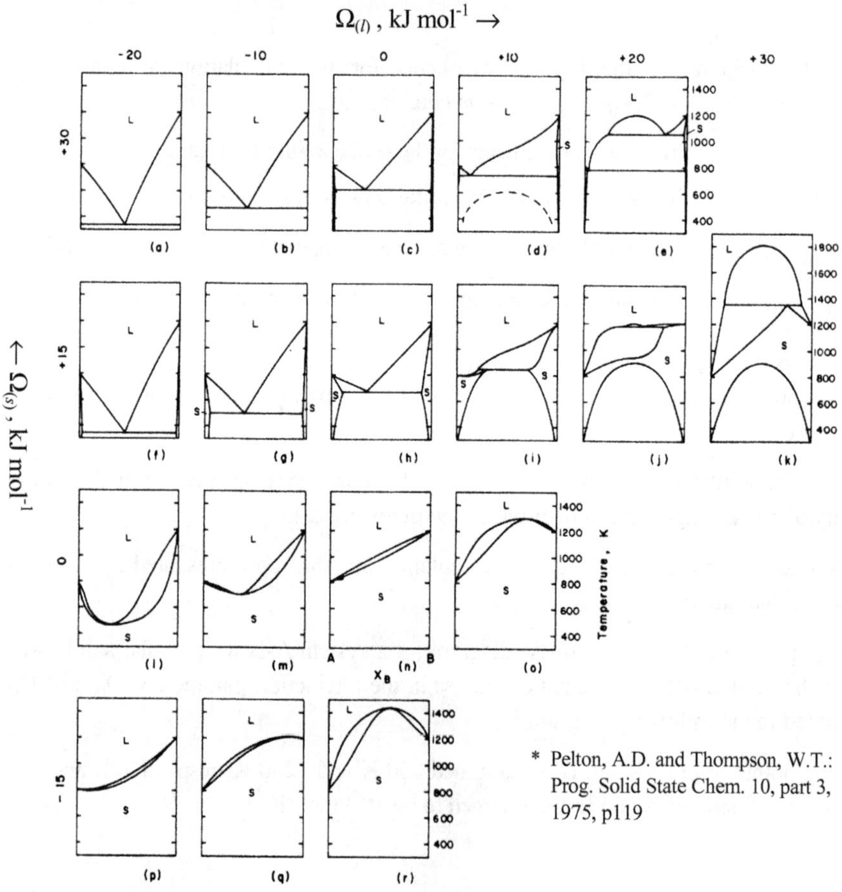


* Pelton, A.D. and Thompson, W.T.: Prog. Solid State Chem. 10, part 3, 1975, p119

*Example 12.14*

Prove that, if the heat capacity of pure liquid phase is the same as that of pure solid phase for both A and B in the A-B binary system, the equations of the liquidus and solidus are

$$ln\left(\frac{\gamma_{A(\alpha)}}{\gamma_{A(l)}}\right)+ln\left(\frac{N_{A(\alpha)}}{N_{A(l)}}\right)=\frac{\Delta H^{o}_{f,A(T_m)}}{R}\left(\frac{1}{T}-\frac{1}{T_{m(A)}}\right)$$

$$ln\left(\frac{\gamma_{B(\alpha)}}{\gamma_{B(l)}}\right)+ln\left(\frac{N_{B(\alpha)}}{N_{B(l)}}\right)=\frac{\Delta H^{o}_{f,B(T_m)}}{R}\left(\frac{1}{T}-\frac{1}{T_{m(B)}}\right)$$

$$RT\,ln\left(\frac{\gamma_{A(\alpha)}}{\gamma_{A(l)}}\right)+RT\,ln\left(\frac{N_{A(\alpha)}}{N_{A(l)}}\right)=\Delta G^{o}_{f,A(T)}$$

*At T*

$$\Delta G^{o}_{f,A(T)}=\Delta H^{o}_{f,A(T)}-T\Delta S^{o}_{f,A(T)}$$

*At the melting point,* $T_{m(A)}$

$$\Delta G^{o}_{f,A(T_m)}=0=\Delta H^{o}_{f,A(T_m)}-T_{m(A)}\Delta S^{o}_{f,A(T_m)}$$

$$\Delta S^{o}_{f,A(T_m)}=\frac{\Delta H^{o}_{f,A(T_m)}}{T_{m(A)}}$$

If $C_{P,A(L)} = C_{P,A(S)}$

$$\Delta H^{o}_{f,A(T)}=\Delta H^{o}_{f,A(T_m)}$$

$$\Delta S^{o}_{f,A(T)}=\Delta S^{o}_{f,A(T_m)}$$

$$\Delta G^{o}_{f,A(T)}=\Delta H^{o}_{f,A(T_m)}\left(1-\frac{T}{T_{m(A)}}\right)$$

$$ln\left(\frac{\gamma_{A(\alpha)}}{\gamma_{A(l)}}\right)+ln\left(\frac{N_{A(\alpha)}}{N_{A(l)}}\right)=\frac{\Delta H^{o}_{f,A(T_m)}}{R}\left(\frac{1}{T}-\frac{1}{T_{m(A)}}\right)$$

For B, in a similar way,

$$\ln\left(\frac{\gamma_{B(\alpha)}}{\gamma_{B(L)}}\right)+\ln\left(\frac{N_{B(\alpha)}}{N_{B(L)}}\right)=\frac{\Delta H^{o}_{f,B(T_m)}}{R}\left(\frac{1}{T}-\frac{1}{T_{m(B)}}\right)$$

If the assumption that the heat capacities of the pure liquid and solid A are equal is unwarranted, the expression for the standard Gibbs energy of fusion of A at $T$ must be corrected:

$$\Delta G^{o}_{f,\mathrm{A}(T)} = \Delta H^{o}_{f,\mathrm{A}(T)} - T\Delta G^{o}_{f,\mathrm{A}(T)}$$

$$\Delta H^{o}_{f,\mathrm{A}(T)} = H^{o}_{\mathrm{A}(L),T} - H^{o}_{\mathrm{A}(S),T}$$

$$\Delta S^{o}_{f,\mathrm{A}(T)} = S^{o}_{\mathrm{A}(L),T} - S^{o}_{\mathrm{A}(S),T}$$

$$\Delta H^{o}_{f,\mathrm{A}(T)} = \Delta H^{o}_{f,\mathrm{A}(T_m)} + \int_{T_{m(\mathrm{A})}}^{T} \Delta C_{P,\mathrm{A}}\,dT$$

$$\Delta S^{o}_{f,\mathrm{A}(T)} = \Delta S^{o}_{f,\mathrm{A}(T_m)} + \int_{T_{m(\mathrm{A})}}^{T} \frac{\Delta C_{P,\mathrm{A}}}{T}\,dT$$

where

$$\Delta C_{P,\mathrm{A}} = C_{P,\mathrm{A}(L)} - C_{P,\mathrm{A}(S)}$$

$$\Delta G^{o}_{f,\mathrm{A}(T)} = \left(\Delta H^{o}_{f,\mathrm{A}(T_m)} + \int_{T_{m(\mathrm{A})}}^{T} \Delta C_{P,\mathrm{A}}\,dT\right) - T\left(\Delta S^{o}_{f,\mathrm{A}(T_m)} + \int_{T_{m(\mathrm{A})}}^{T} \frac{\Delta C_{P,\mathrm{A}}}{T}\,dT\right)$$

*Exercises*

12.9 Metals A and B behave ideally in both liquid and solid solutions. Calculate compositions of the solid and liquid solutions in equilibrium at 1250*K*. The following information is known:

| | | |
|---|---|---|
| A | : $T_m$ = 1350 K | $\Delta H^{o}_{f,\mathrm{A}} = 1{,}500\ \mathrm{J\,mol^{-1}}$ |
| B | : $T_m$ = 700 K | $\Delta H^{o}_{f,\mathrm{B}} = 3{,}300\ \mathrm{J\,mol^{-1}}$ |

12.10 Components A and B behave regularly in both liquid and solid solutions. Find the compositions of the liquid and solid solutions in equilibrium at 1300*K*. The following data are available:

| | | |
|---|---|---|
| A | : $T_m$ = 1350 K | $\Delta H^{o}_{f,\mathrm{A}} = 9{,}330\ \mathrm{J\,mol^{-1}}$ |
| B | : $T_m$ = 1150 K | $\Delta H^{o}_{f,\mathrm{B}} = 9{,}760\ \mathrm{J\,mol^{-1}}$ |
| $\Omega_{(L)} = -5{,}600\ \mathrm{J\,mol^{-1}}$ | | $\Omega_{(S)} = -11{,}400\ \mathrm{J\,mol^{-1}}$ |

12.11 The A-B binary system behaves ideally in both its liquid and solid solutions. The element A melts at 1500 K with the heat of fusion of 14,700 J mol$^{-1}$, and the heat capacity difference, $\Delta C_{P,\mathrm{A}}$ (= $C_{P,\mathrm{A}(L)}$ - $C_{P,\mathrm{A}(S)}$), is approximately constant and equal to 4.6 J mol$^{-1}$K$^{-1}$. The element B melts at 2300 K, but the heat of fusion is unknown. The heat capacities of the pure liquid and solid B are equal. It was found in an experiment with a liquid solution of $N_\mathrm{B}$ = 0.22 that in cooling the first solid crystals appeared at 1700 K. Calculate the heat of fusion of B.

12.12 A and B have negligible mutual solid solubility in the solid state and their phase diagram shows a eutectic transformation. The liquid phase at 1 *atm.* is represented by the equation

$$G^E = 2100 N_\mathrm{A} N_\mathrm{B}\left(1 - \frac{T}{3000}\right) \quad \text{J mol}^{-1}$$

The following information is known:

| | | | |
|---|---|---|---|
| A: | $T_m$ = 1500 K | $\Delta H^\circ_{f,A} = 15{,}100$ J mol$^{-1}$ | $\Delta C_{P(S\to L)} = 0$ |
| B: | $T_m$ = 1200 K | $\Delta H^\circ_{f,\mathrm{B}} = 11{,}100$ J mol$^{-1}$ | $\Delta C_{P(S\to L)} = 0$ |

Develop equations of the liquidus of the above system and explain how to find the eutectic temperature and composition using the equations developed.

## 12.4 Ternary Systems

### Composition Triangles

Ternary systems are those possessing three components. Therefore there are four independent variables in the A-B-C ternary system: temperature, pressure and two composition variables (The third one is not independent since the sum of the mole (or mass) fractions is unity: $N_\mathrm{A} + N_\mathrm{B} + N_\mathrm{C} = 1$).

Construction of a complete diagram which represents all these variables would require a four dimensional space. However, if the pressure is assumed constant (customarily at 1 *atm.*), the system can be represented by a three-dimensional diagram with three independent variables, *i.e.*, temperature and two composition variables. In plotting three dimensional diagrams, it is customary that the compositions are represented by triangular coordinates in a horizontal plane and the temperature in a vertical axis.

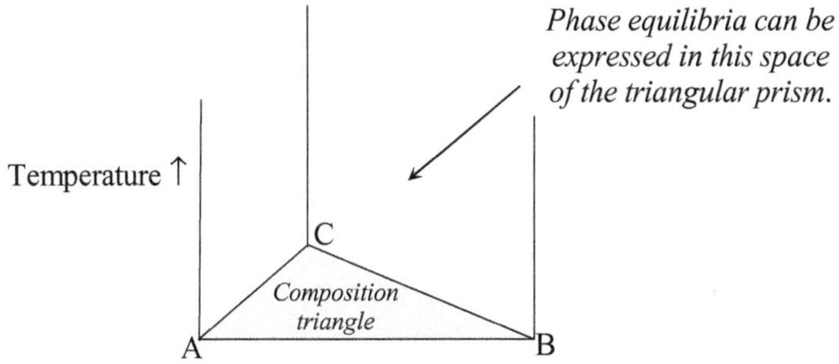

Phase equilibria can be expressed in this space of the triangular prism.
Temperature ↑
C
Composition triangle
A
B

For plotting ternary compositions, it is common to employ an ***equilateral composition triangle*** with coordinates in terms of either mole fraction or weight percent of the three components.

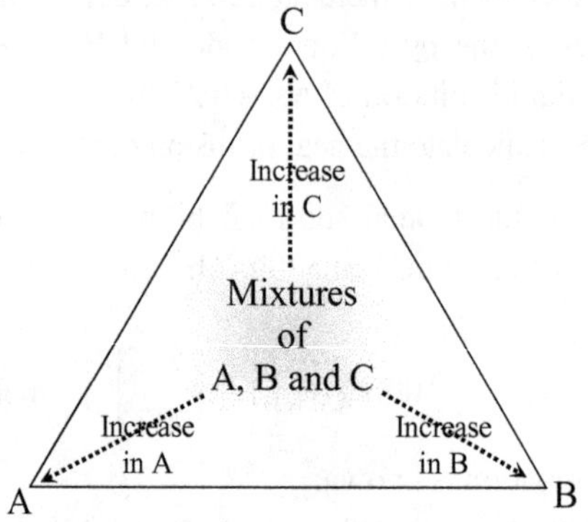


- Three pure components are represented by the apices, A, B and C.
- Binary compositions are represented along the edges: *e.g.*, any point on the line B-C is composed entirely of components B and C without A.
- Points inside the triangle represent mixtures of all three components.

We now discuss several different methods of determining the proportions of three components represented by a point in the triangle.

To determine the composition of the mixture represented by the point $P$ in the following figures,

*Method 1*

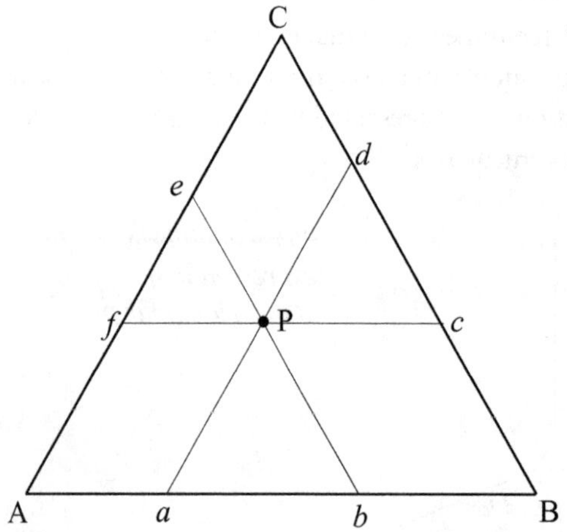


- Draw lines through P parallel to each of the sides of the triangle.

$$\text{Proportion of A} = \frac{\text{B}b}{\text{AB}} = \frac{e\text{C}}{\text{AC}}$$

$$\text{Proportion of B} = \frac{\text{A}a}{\text{AB}} = \frac{d\text{C}}{\text{BC}}$$

$$\text{Proportion of C} = \frac{\text{A}f}{\text{AC}} = \frac{c\text{B}}{\text{BC}}$$

*Method 2*

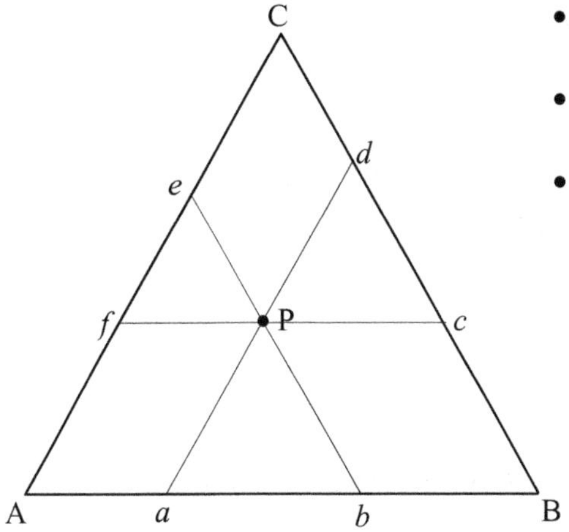


- Draw lines through P parallel to each of the sides of the triangle.
- Notice that each side is now divided into three parts.
- If the side A-B is chosen,

$$\text{Proportion of A} = \frac{b\text{B}}{\text{AB}}$$

$$\text{Proportion of B} = \frac{\text{A}a}{\text{AB}}$$

$$\text{Proportion of C} = \frac{ab}{\text{AB}}$$

The composition can also be found from the sides B-C and A-C. The two end parts of each line represent the proportions of the components at the opposite ends and the middle part represents the proportion of the third component."

*Method 3*

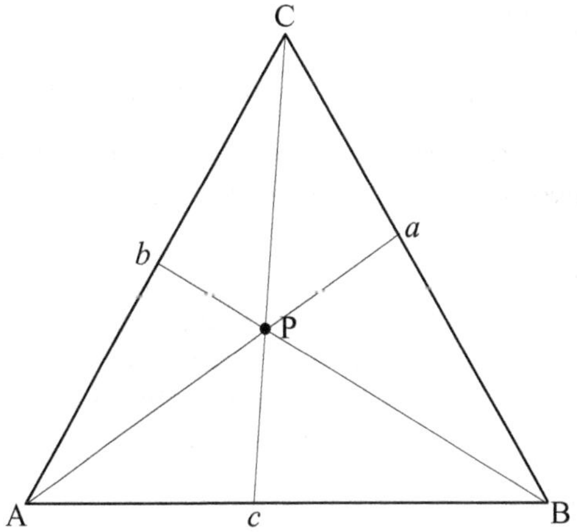


- Draw lines from apices through P to the opposite sides of the triangle.

$$\frac{\text{Proportion of A}}{\text{Proportion of B}} = \frac{c\text{B}}{\text{A}c}$$

$$\frac{\text{Proportion of B}}{\text{Proportion of C}} = \frac{\text{C}a}{a\text{B}}$$

$$\frac{\text{Proportion of C}}{\text{Proportion of A}} = \frac{\text{A}b}{b\text{C}}$$

All the methods presented above are based on the same principle: *i.e.*, the material balance using the *lever rule*. Therefore these are not limited to equilateral triangles, but equally valid for scalene triangles which frequently appear when dealing with subsystems.

Consider the subsystem XYZ within the system ABC.

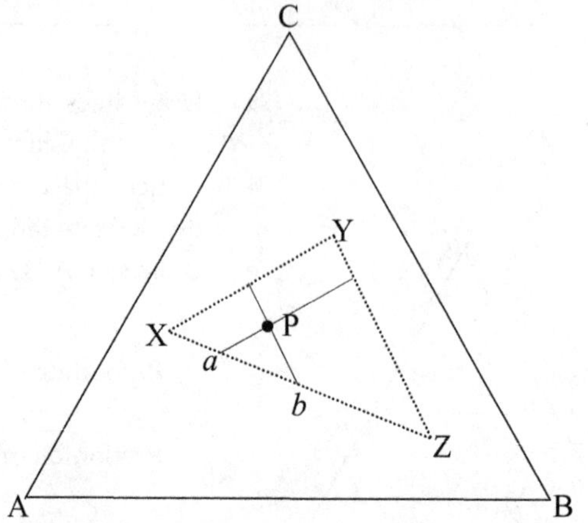
C
Y
X
P
a
b
Z
A
B

- Points X, Y and Z are mixtures of A, B and C and the composition of each point can be determined by one of the methods described above.
- Since the point P is inside the subsystem XYZ, it may be considered as a mixture of X, Y and Z:

$$\text{Proportion of X} = \frac{b\text{Z}}{\text{XZ}}$$

$$\text{Proportion of Y} = \frac{ab}{\text{XZ}}$$

$$\text{Proportion of Z} = \frac{\text{X}a}{\text{XZ}}$$

Another important relationship which can be drawn from composition triangles is that,

"*If any two mixtures (or solutions) or components are mixed together, the composition of the resultant mixture lies on the straight line which joins the original two compositions.*"

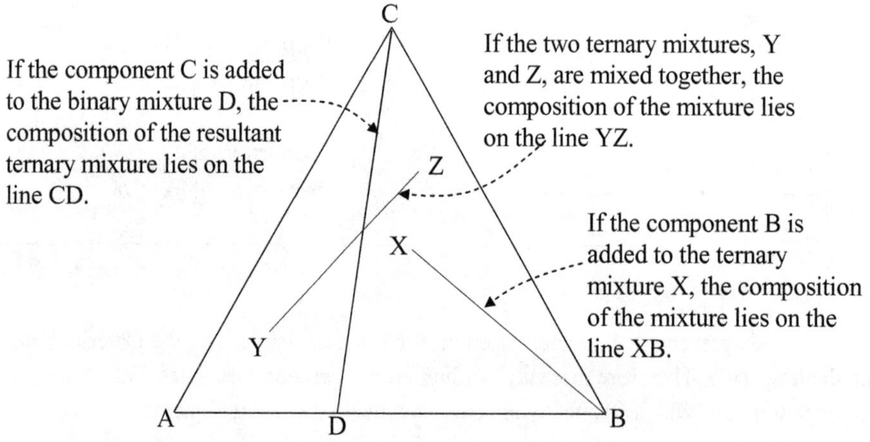
C
If the component C is added to the binary mixture D, the composition of the resultant ternary mixture lies on the line CD.
If the two ternary mixtures, Y and Z, are mixed together, the composition of the mixture lies on the line YZ.
Z
X
If the component B is added to the ternary mixture X, the composition of the mixture lies on the line XB.
Y
A
D
B

In all cases, the position of the resultant mixture on the line is determined by the lever rule.

As explained earlier, the temperature is represented by the axis perpendicular to the plane of the composition triangle. The point S in the diagram below left represents a ternary mixture of the composition P at the temperature $T_1$. (Recall that the pressure is assumed constant in this type of composition-temperature coordinates.). The diagram below right is an example of the three dimensional ternary phase diagram drawn using the composition triangle - temperature coordinates.

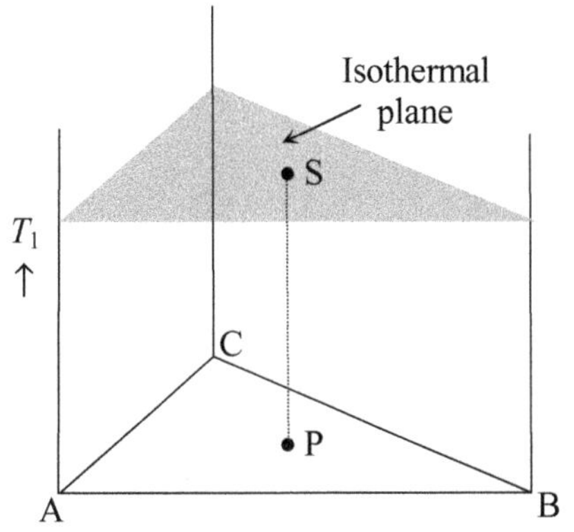


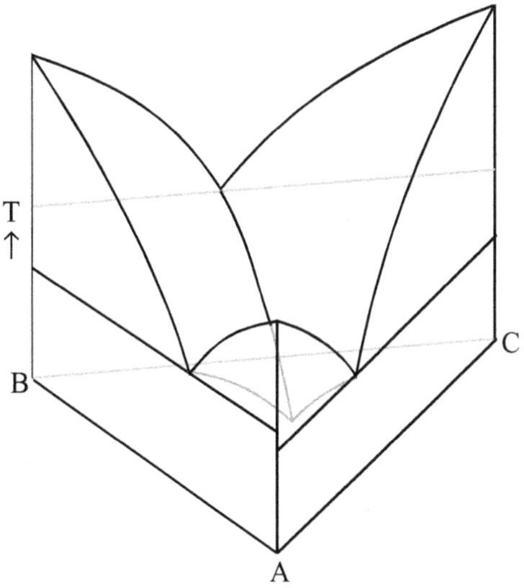


*Example 12.15*

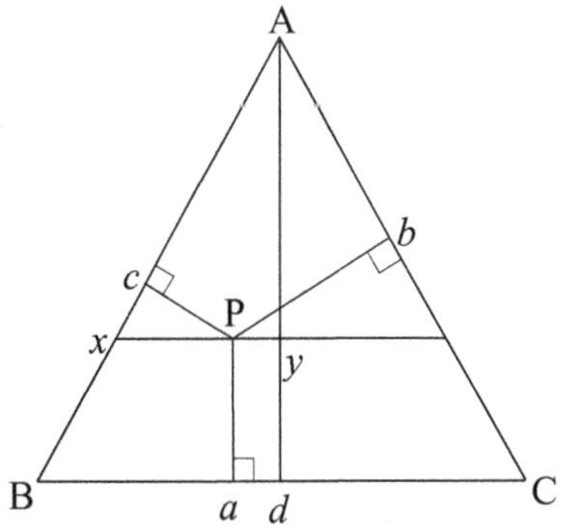


The point P in the equilateral composition diagram represents a ternary mixture of components A, B and C. Prove that

$$\text{Proportion of A} = \frac{\text{P}a}{\text{A}d}$$

$$\text{Proportion of B} = \frac{\text{P}b}{\text{A}d}$$

$$\text{Proportion of C} = \frac{\text{P}c}{\text{A}d}$$

Proportion of A $= \dfrac{Bx}{AB} = \dfrac{yd}{Ad} = \dfrac{Pa}{Ad}$. The proportions of B and C can be obtained in a similar way.

*Example 12.16*

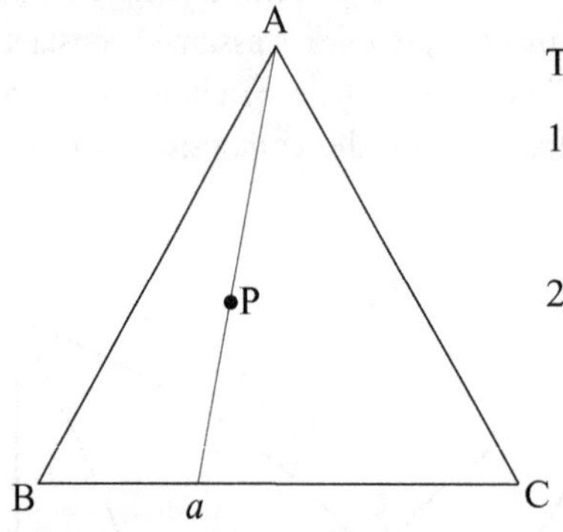


The point P represents a ternary alloy.

1) When pure A is added to the alloy, in which direction does the overall composition change?
2) When pure A precipitates out, in which direction does the composition of the remaining alloy change?

1) From P to A  2) From P to a

*Example 12.17*

The Gibbs phase rule is of use in phase equilibrium studies of multicomponent systems.

1) Determine the maximum number of phases which can coexist in equilibrium in a ternary system.
2) For condensed systems the effect of pressure is negligible in many cases. Therefore, when the pressure is fixed at 1 *atm.*, the number of variables of the system is reduced by one unit. When equilibrium conditions of a ternary system at a constant pressure are represented in the space of the composition triangle - temperature prism, prove the following:
   a) Four phase equilibria are represented by points.
   b) Three phase equilibria are represented by lines.
   c) Two phase equilibria are represented by surfaces.
   d) Single phase equilibria are represented by spaces.

1) From the Gibbs phase rule,

$$f = c - p + 2$$

$c = 3$ (ternary system)
$f = 0$ (the maximum number of phases occurs at zero degree of freedom.)

$$p = 5$$

A maximum of five phases can coexist in equilibrium.

2) If the pressure is fixed at a constant value,

$$f = c - p + 1$$

| | $p$ | $c$ | $f$ | Explanation |
|---|---|---|---|---|
| (a) | 4 | 3 | 0 | No degree of freedom: Neither composition nor temperature can be chosen freely. In other words four phases can exist together in equilibrium only at a fixed composition and temperature : *invariant* |
| (b) | 3 | 3 | 1 | One degree of freedom: One variable (either the concentration of one of the components or temperature) can be freely varied, and then all others are fixed: *univariant.* |
| (c) | 2 | 3 | 2 | Two degrees of freedom: Two variables are at our discretion, and the rest are then fixed : *bivariant.* |
| (d) | 1 | 3 | 3 | Three degrees of freedom: Three variables can be varied freely: *e.g.*, after choosing a composition of the ternary system, temperature can still be varied while maintaining single phase state : *trivariant.* |

*Exercises*

12.13 Fig. A below shows the composition triangle of the ABC ternary system. Determine the composition of the mixture represented by the point P.

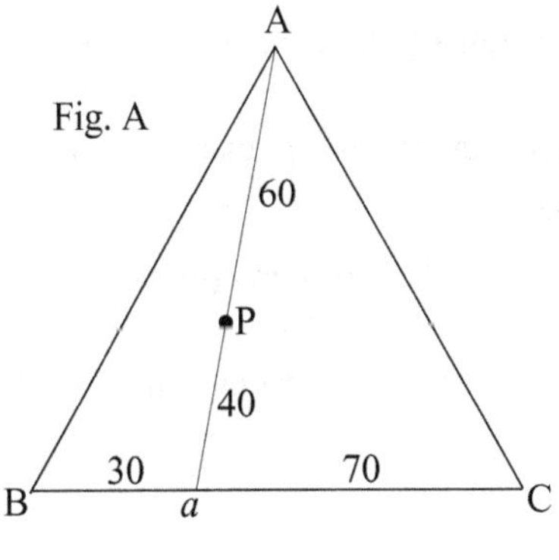


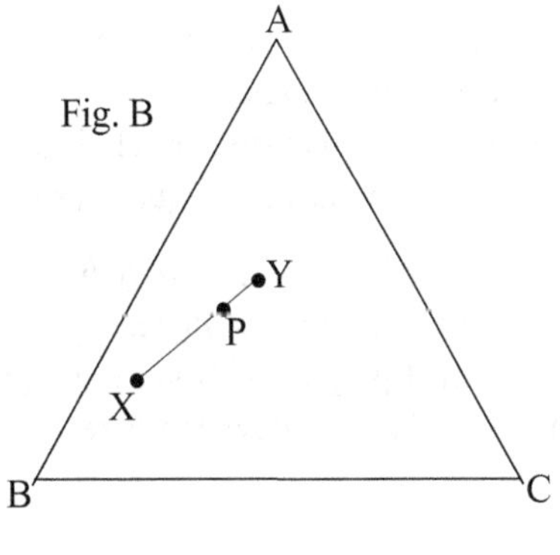


12.14 A mixture represented by the point P in Fig. B above is to be prepared by mixing the mixtures X and Y. Determine the ratio of X to Y to obtain the right composition. The composition of each point is given in the following table:

| | A | B | C |
|---|---|---|---|
| P | 35% | 40% | 25% |
| X | 20% | 70% | 10% |
| Y | 40% | 30% | 30% |

## Polythermal Projections

This figure represents a simple ternary phase diagram. However, it has the disadvantage that lines in the figure are not seen in true length, and hence it is difficult to obtain quantitative information.

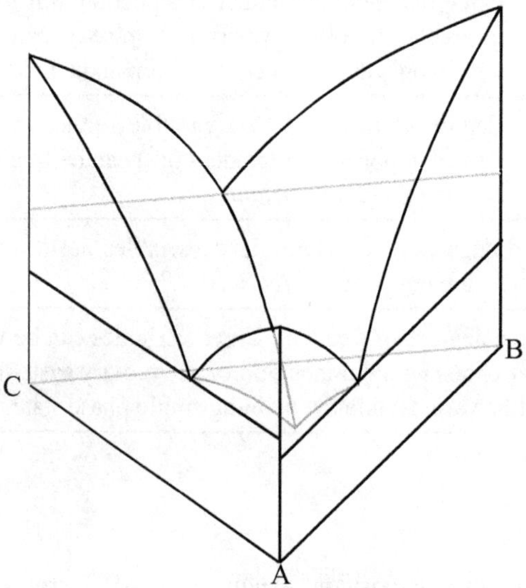


There are two ways to solve this problem:

(1) A two dimensional representation of the ternary liquidus surface on the base composition triangle.
(2) Two dimensional isothermal diagrams which represent isothermal plane intersections with various surfaces (liquidus, solidus, *etc*).

The first method consists of a ***polythermal projection*** of all features (liquidus, *etc*) down onto the base composition triangle.
The polythermal projection is generally given with constant temperature lines (isothermal lines) as shown in the following figure:

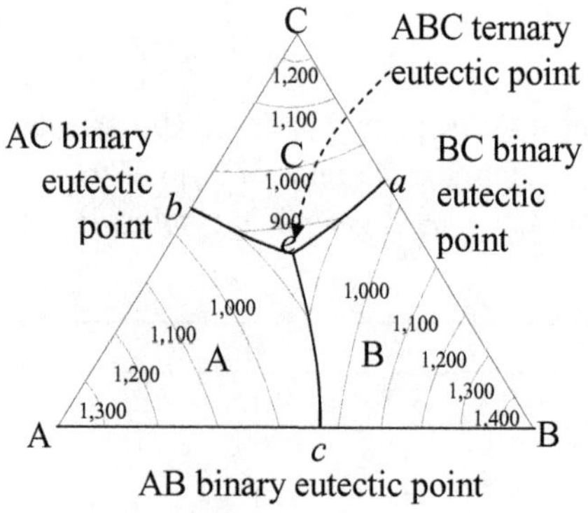

These lines are called ***liquidus isotherms***. The intersections of adjoining liquidus surfaces like *ae*, *be* and *ce* are called the ***boundary curves***. When a liquid whose composition lies in the region surrounded by A*ceb* is cooled, the first crystalline phase that appears is A, and hence A is called the ***primary phase*** and the region A*ceb* is the ***primary field*** of A. In this field, solid A is the last solid to disappear when any composition within this field is heated. Similarly, B and C are primary phases in their respective primary fields, B*aec* and C*aeb*.

Now, let's examine crystallization paths of the simple ternary system. Suppose that a liquid of composition *p* is allowed to cool.

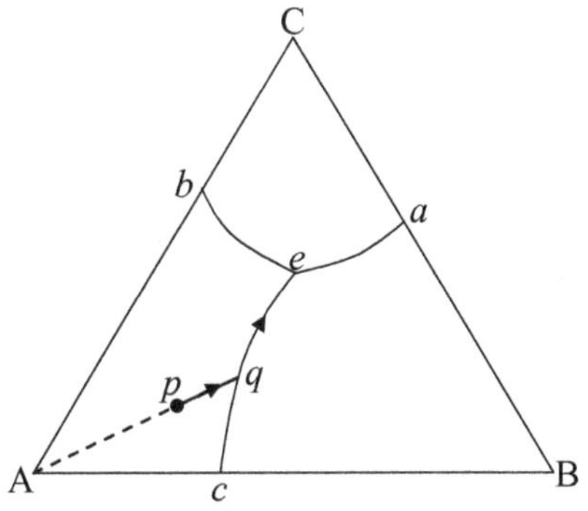


- The liquid solution remains as liquid until the system temperature reaches the liquidus surface (the curved surface A*ceb*).
- At the liquidus temperature (*i.e.*, when the temperature hits the surface A*ceb*), pure solid A begins to crystallize.
- As the temperature decreases further, solid A continues to precipitate out of the liquid, and thus the liquid continues to be depleted in A and the liquid composition changes along the line *pq* (moving from *p* toward *q*).
- At *q*, the second phase B begins to precipitates as well, and the liquid composition now moves along the curve *qe*. Until it reaches the ternary eutectic point *e*, both A and B crystallize.
- At *e* (the ternary eutectic point), solid phases A, B and C crystallize and the temperature remains unchanged until all liquid has completely exhausted.
- The final product will consist of large crystals of A and B which have crystallized before reaching the point *e*, and small crystals of eutectic structure of A, B and C which have crystallized at the point *e*.

In multicomponent systems, stoichiometric compounds are frequently formed between components. The phase diagrams shown below are for ABC ternary system forming a binary compound AB which melts *congruently*, which, in other words, is stable until the temperature reaches its melting point:

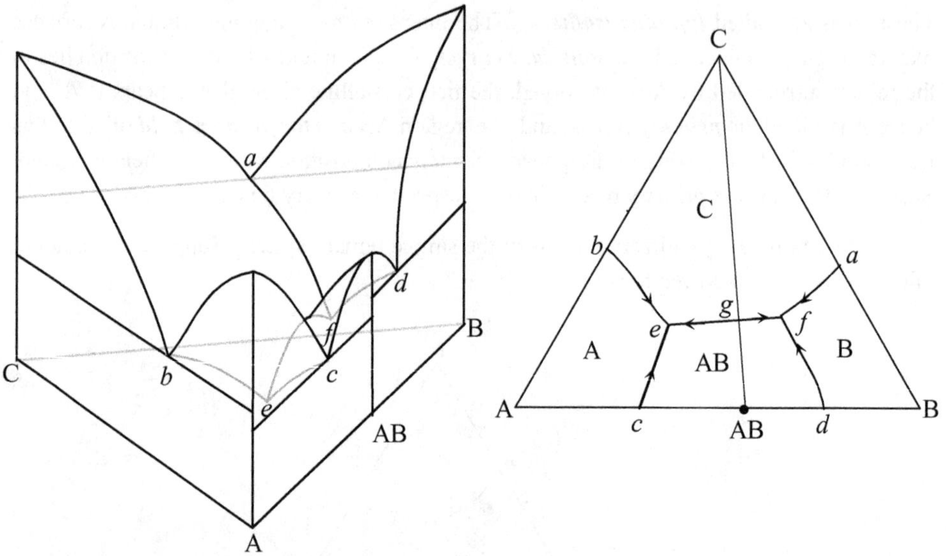


- The straight line C-AB is called an ***Alkemade line***.
- An Alkemade line divides a ternary composition triangle into two sub-composition triangles.
- The final phases produced by equilibrium crystallization of any composition within one of these sub-triangles are those indicated by the apices of the triangle. For instance, any composition within the composition triangle A-C-AB results in producing phases A, C and AB at equilibrium.
- The crossing point (*g*) on the boundary curve *ef* by the Alkemade line is the maximum in temperature on the curve *ef*.
- The points *e* and *f* are ternary eutectic.
- Each sub-composition triangle can be treated as a true ternary system.
- The Alkemade line in this case represents a true binary system of C and AB.
- The arrows in the diagram indicate directions of decreasing temperature.

The following phase diagrams are for ABC ternary system forming a binary compound AB which melts *incongruently*, which, in other words, is thermally unstable and decomposes into other phases during heating:

In this case, the Alkemade line C-AB does not cross the boundary curve (*ef*) between these primary phases. Now, we state the ***Alkemade theorem*** in a more general form:

- The direction of falling temperature on the boundary curve of two intersecting primary phase areas is always away from the Alkemade line.

- If the Alkemade line intersects the boundary curve, the point of intersection represents a temperature maximum on the boundary curve.
- If the Alkemade line does not intersect the boundary curve, then temperature maximum on the curve is represented by the end which, if prolonged, would intersect the Alkemade line.

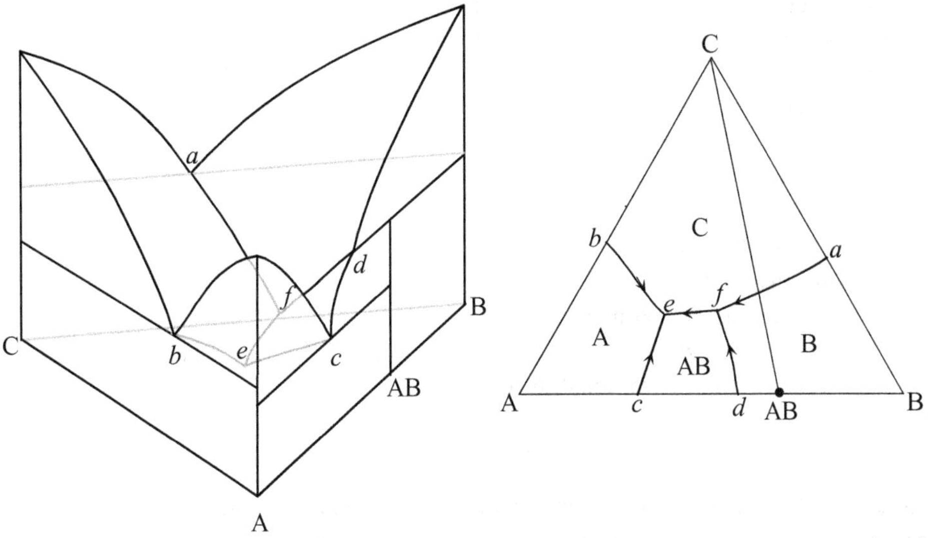


Now, it is obvious from the Alkemade theorem that for the case of the incongruently-melting compound as shown above only the point *e* is eutectic. It is also apparent by examining the above phase diagrams for the congruently-melting compound and the incongruently-melting compound that,

- The composition of the compound lies *within* the primary field of the compound if it has a congruent melting point, and
- The composition of the compound lies *outside* the primary field of the compound if it has an incongruent melting point.

The ternary invariant points (*e.g.*, *e* and *f* in the above diagrams) that appear in a system without solid solution are either ***ternary eutectics*** or ***ternary peritectics***. Whether it is eutectic or peritectic is determined by the directions of falling temperatures along the boundary curves.

- If an invariant point is the minimum point in temperature along all three boundary curves, it is a ternary eutectic.
- If the point is not the minimum point, it is a ternary peritectic.

In the previous diagrams, points *e* and *f* for the congruently-melting compound are both ternary eutectic. On the other hand, for the incongruently-melting compound the point *e* is the ternary eutectic, whereas the point *f* is the ternary peritectic.

Alkemade lines are also called in many different ways including ***conjugation lines*** and ***joins***. Both Alkemade lines and ***Alkemade triangles*** (composition triangles produced by Alkemade lines) are of use in the understanding of ternary systems. They play an essential role in understanding crystallization or heating paths:

*Example 12.18*

This figure is a simple ternary phase diagram which shows a ternary eutectic at the point *e*.
Explain the crystallization path of each of liquids represented by points *p*, *q*, and *r*.
Explain also the change in

(1) liquid composition,
(2) mean solid composition,
(3) solid phase(s),
(4) instantaneous composition of solids crystallizing, and
(5) change in the ratio of liquid to solid phases.

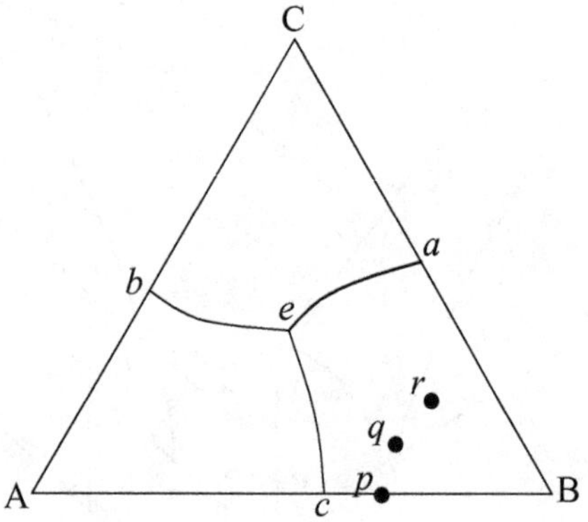


As the whole diagram is an Alkemade triangle and point *e* is the ternary eutectic, all crystallization paths of this ternary system should terminate at this ternary eutectic. However, the binary liquid composition such as point *p* terminates at its binary eutectic *c*.

(1) *Point p*

- The system remains liquid until the temperature reaches liquidus.
- At the liquidus, solid B begins to crystallize.
- As the temperature decreases, solid B continues to crystallize and the liquid composition changes toward point *c* along the line *pc*.
- The ratio of solid (B) to liquid at the moment the liquid composition arrives at point *c* is represented by the lever rule, *cp*/*p*B.
- At point *c*, both A and B co-crystallize forming eutectic structure until all liquid is consumed. The temperature remains unchanged until the eutectic reaction given below has been completed:

$$L \rightarrow A_{(s)} + B_{(s)}$$

(2) *Point q*

- At the liquidus solid B begins to crystallize.
- As the temperature decreases, solid B continues to crystallize and the liquid composition moves straight away from B along the line *qu*.

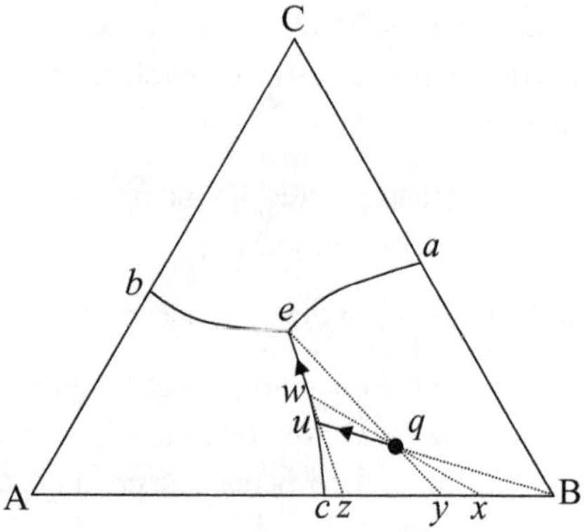

- At point $u$, the second phase A begins to appear.
- With further cooling, the crystallization path follows the boundary curve $ue$ with crystallization of both A and B.
- At point $w$ which is an arbitrary point on the boundary curve $ue$, the mean composition of the solid is represented by the point $x$. Proportions of solid A and solid B are determined by the lever rule: (solid A)/(solid B) = $x$B/A$x$. The ratio of liquid to solid is also given by the lever rule: $qx/wq$.
- The instantaneous composition of the solid phases crystallizing at $w$ can be determined by drawing the tangent to the curve $ue$ at $w$ and finding the intersection of the tangent with line AB. Therefore, point $z$ represents the instantaneous composition of the solid phases.
- At point $e$, the eutectic crystallization occurs :

$$L \rightarrow A_{(s)} + B_{(s)} + C_{(s)}$$

- During the eutectic reaction, the mean composition of the solid phases changes along the line $yq$ (from $y$ to $q$). The solid composition reaches point $q$ when the liquid is completely consumed by the eutectic crystallization.
- In summary, the liquid composition changes along the path $quwe$, whereas the mean composition of solid phases follows the path B$xyq$.
- The final structure of the system after complete solidification consists of large crystals of A and B which have been crystallized during the path $quwe$ and a mixture of small crystals of A, B and C (eutectic structure).

(3) *Point r*

- The crystallization path can be determined in the same manner as the above case. The difference is that the second phase crystallizing in this case is C instead of B.

*Example 12.19*

This figure is a phase diagram of ABC ternary system which forms two binary compounds, AC and BC.

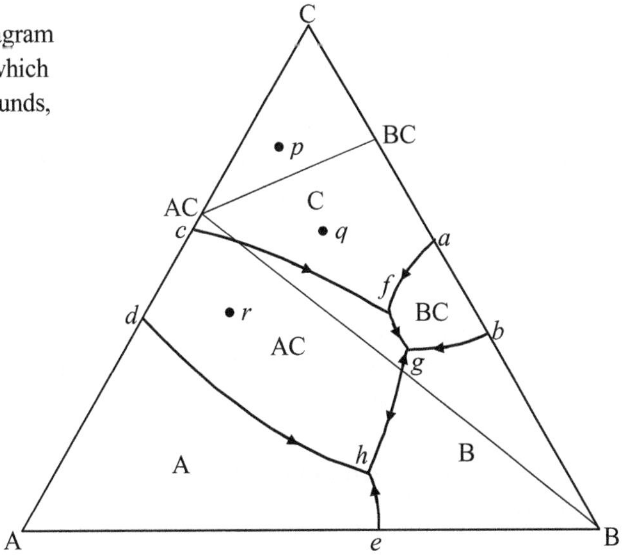

(1) Why is an Alkemade line connecting A and BC not drawn?
(2) Describe crystallization paths of liquid compositions *p*,*q* and *r*.
(3) Describe heating paths of mean solid compositions of *p*,*q* and *r*.
(4) Would the compounds AC and BC melt congruently or incongruently?

(1) Because the primary fields of A and BC are not in contact with each other, and hence these two fields do not form a boundary curve.

(2) *Point p*

As point *p* lies in the Alkemade triangle C-AC-BC, the final solid phases in equilibrium should be C, AC and BC. From the figure given below,

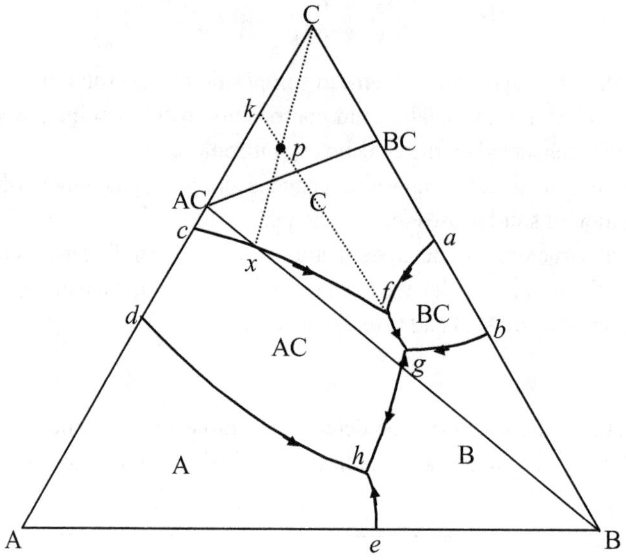


- On touching the liquidus at *p*, solid C begins to crystallize.
- From *p* to *x*, solid C continues to crystallize.
- At *x*, solid compound AC appears.
- From *x* to *f*, both AC and C crystallize.
- As the mean composition of the solid crystallized out along the cooling path *p*-*x*-*f* is given by point *k*, there is some liquid left when crystallization path arrives at *f* (the liquid/solid ratio is *kp*/*pf*).
- At *f*, the final solidification takes place through a peritectic reaction: AC and BC crystallize out together at the expense of liquid and portion of C. This can easily be seen from the Alkemade triangle C-AC-BC. During the peritectic reaction, the mean solid composition moves from *k* to *p*.
- At completion of the solidification at point *f*, the final solid phases in equilibrium are C, AC and BC, since the mean composition of the system lies within the Alkemade triangle C-AC-BC as mentioned earlier.

*Point q*

As point $q$ lies in the Alkemade triangle AC-BC-B, the final solid phases in equilibrium should be AC, BC and B. Refer to the figure given below:

- On touching the liquidus at $q$, solid C begins to crystallize.
- From $q$ to $x$, solid C continues to crystallize.
- At $x$, solid AC appears.
- From $x$ to $f$, AC and C crystallize out together.
- As the mean composition of the solid crystallized out along the cooling path $q$-$x$-$f$ is given by point $k$, there is some liquid left when crystallization path arrives at $f$ (the liquid/solid ratio is $kq/qf$).
- At $f$, a peritectic reaction takes place: AC and BC crystallize out together at the expense of C and portion of liquid. During the peritectic reaction, the mean solid composition moves from $k$ to $m$.

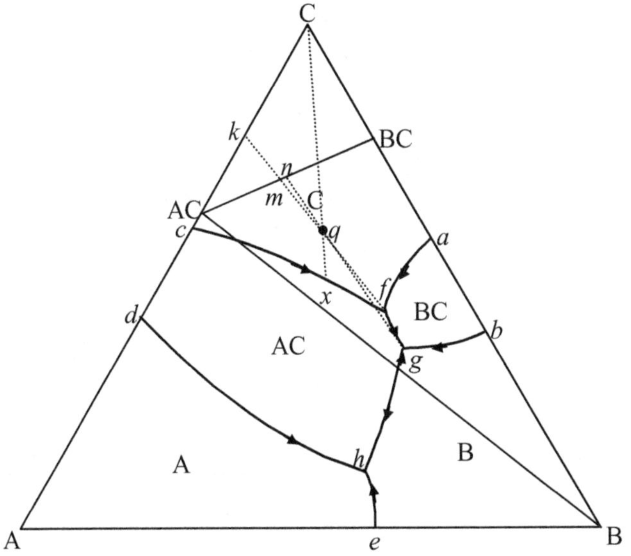


- From $f$ to $g$, AC and BC crystallize out together, and the mean solid composition moves from $m$ to $n$.
- At $g$, a eutectic reaction takes place until the last portion of liquid is completely consumed: three solid phases AC, BC and B crystallize out together to form a eutectic structure.
- During the eutectic reaction at $g$, the mean solid composition changes from $n$ to $q$.
- At completion of the solidification at point $g$, the final solid phases in equilibrium are AC, BC and B since the mean composition of the system lies within the Alkemade triangle AC-BC-B as mentioned earlier.

*Point r*

As point $r$ lies in the Alkemade triangle AC-A-B, the final solid phases in equilibrium should be AC, A and B. Referring to the figure given below:

- On touching the liquidus at *r*, solid AC begins to crystallize.
- From *r* to *z*, solid AC continues to crystallize.
- At *z*, solid A appears.
- From *z* to *h*, both AC and A crystallize.
- As the mean composition of the solid crystallized out between *r* and *h* is given by point *k*, there is some liquid left when crystallization path arrives at *h*.
- At *h*, the final solidification takes place through a eutectic reaction: AC, A and B crystallisz out together to form a eutectic structure. During the eutectic reaction, the mean solid composition moves from *k* to *r*.

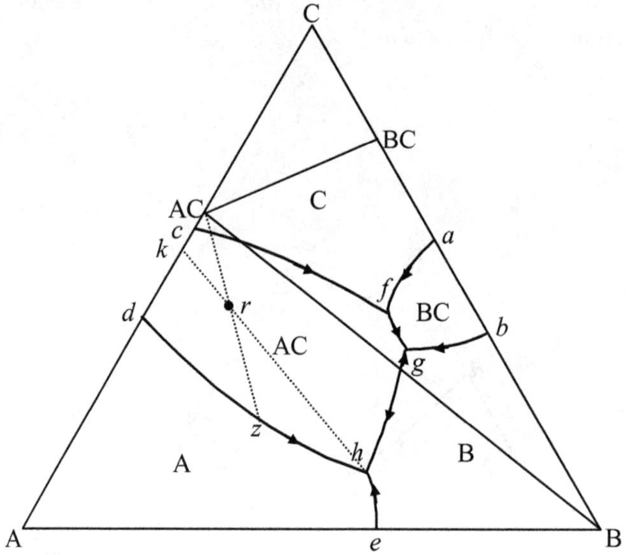


(3) Heating from below the liquidus can be considered as the opposite to cooling from above the liquidus. The heating path of composition *p* is illustrated below. Others may be determined in the similar manner.

- At the peritectic point *f*, the peritectic reaction of AC and BC reacting form both C and liquid phase proceeds until BC is consumed completely. This occurs isothermally. The composition of the liquid phase formed by this peritectic reaction is that of point *f*.
- Above the peritectic temperature, solids C and AC continue to dissolve into the liquid phase, and the liquid composition changes from *f* to *x*.
- When the liquid reaches point *x*, solid AC has completely dissolved and C is the only solid phase left.
- As the temperature increases, solid C continues to dissolve and the liquid composition moves from *x* to *p*.

- When the liquid reaches point $p$, solid C has completely dissolved and the system becomes a liquid phase of the composition of $p$.

(4) Both AC and BC melt incongruently, as the stoichiometric compositions of these compounds lie outside their respective primary fields.

*Example 12.20*

Referring to the figure giver

(1) Determine whether bin compound δ and ternaı ε and φ melt congruenı incongruently.

(2) Draw all Alkemade lin

(3) Determine the crystalli a liquid of compositior

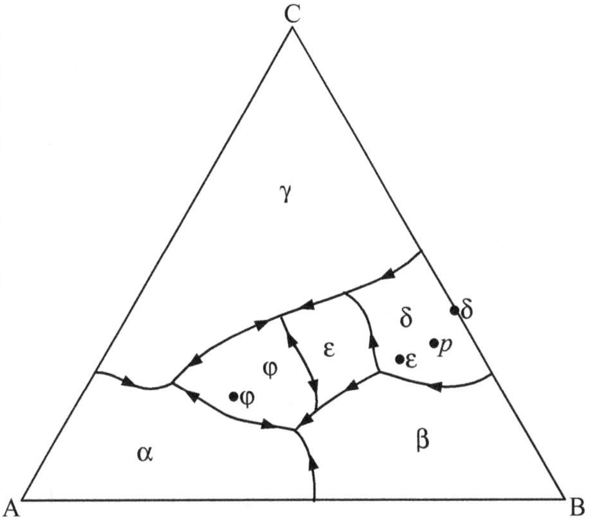


(1) The binary compound δ and the ternary compound $\varphi$ melt congruently, since they are within their respective primary fields, but the ternary compound ε melts incongruently as it is outside its primary field.

(2) All the Alkemade lines are given in the following diagram.

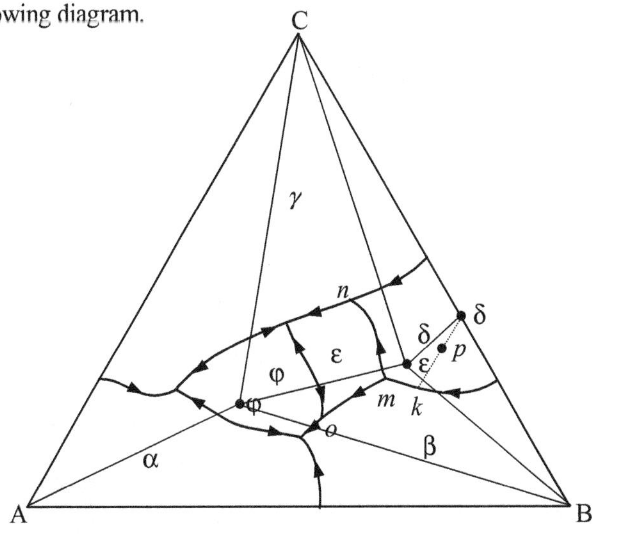

(3) Crystallization path (Refer to the figure given above):

- When the temperature reaches the liquidus, δ phase begin to crystallize.
- Phase δ continues to crystallize out until the liquid composition contacts the boundary curve between δ and β at point *k*.
- At *k*, phase β appears, and thereafter δ and β crystallize together with the liquid composition changing along the path *km*.
- At point *m*, a ternary peritectic reaction occurs isothermally.

$$L + \delta + \beta \rightarrow \varepsilon$$

- One might think that there would be three possible results after completion of the above peritectic reaction:
  - The liquid is exhausted before either δ or β, and hence solidification is completed at point *m*.
  - δ phase is exhausted first, and crystallization continues along the path *mo*.
  - β phase is exhausted first, and crystallization continues along the path *mn*. Which is the case depends on in which Alkemade triangle the total composition lies. Since point *p* lies within the triangle δ-ε-B(β), solidification should terminate at point *m*.

*Exercises*

12.15 Discuss the crystallisation paths of the overall liquid compositions *p* and *q* in the following ternary phase diagram:

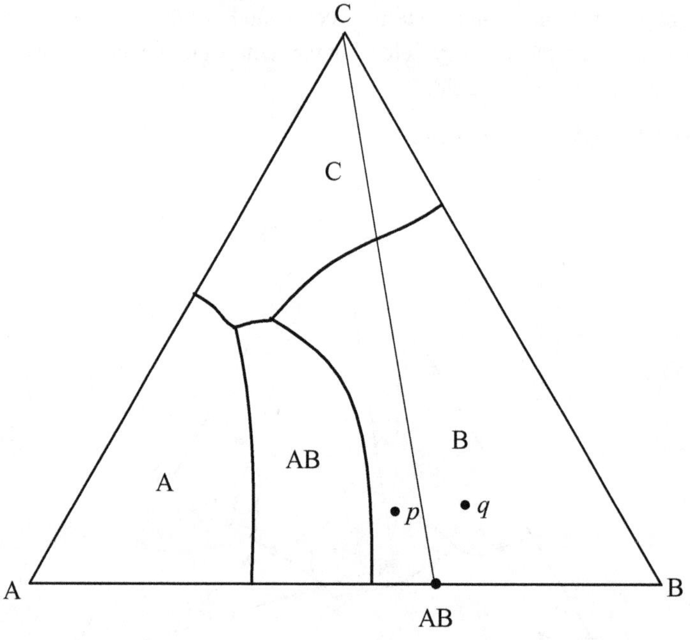

12.16 Shown below is the phase diagram of the $SiO_2$-$CaO$-$Al_2O_3$ system.* Discuss solidification paths for the compositions *p*, *q* and *r* indicated on the diagram.

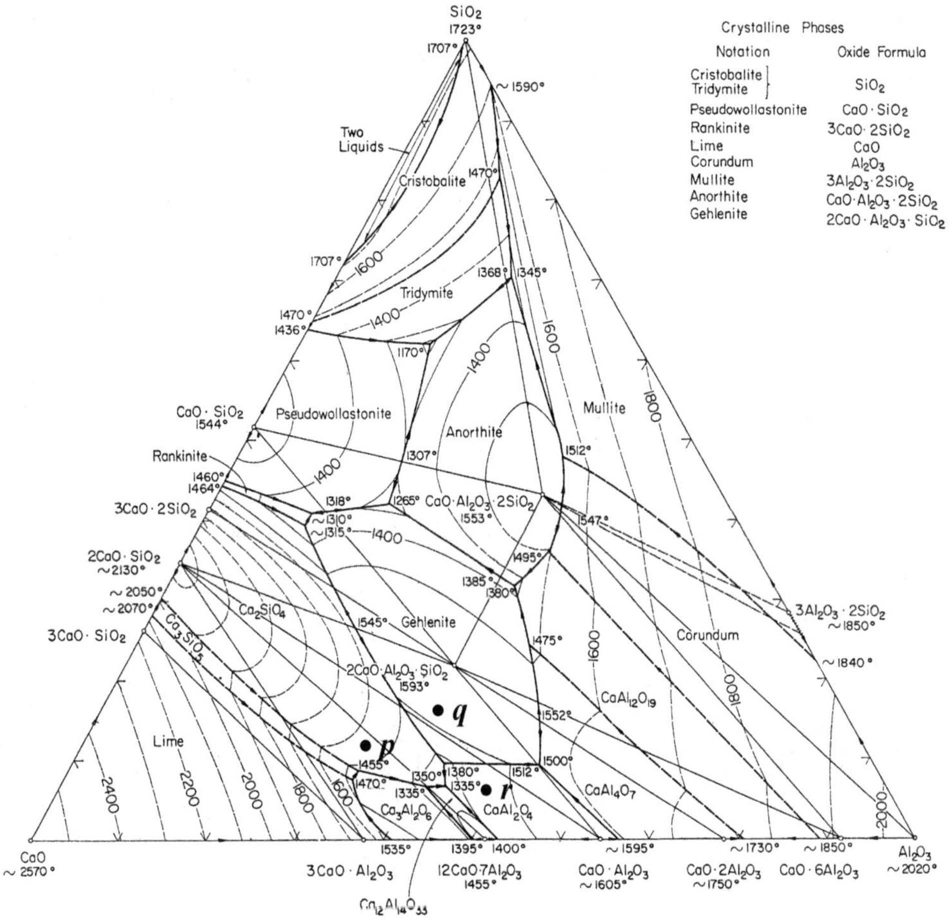


* "*Phase diagrams for ceramists*", E.M. Levin, C.R. Robbins and H.F. McMurdie, The American Ceramic Society, Inc. (1964), p219

## Isothermal Sections

The following figure represents a simple ternary phase diagram. However, it has the disadvantage that lines in the figure are not seen in true length, and hence it is difficult to obtain quantitative information. In order to solve this problem, the polythermal projection method was discussed until now. From now on, the other method, namely, two dimensional isothermal diagrams are discussed.

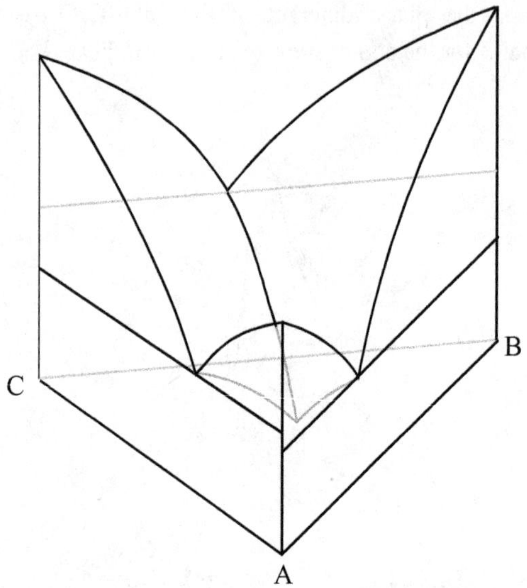


First, we discuss a simple ternary eutectic system without solid solution as shown above. The following figures show isothermal sections at a number of different temperatures:

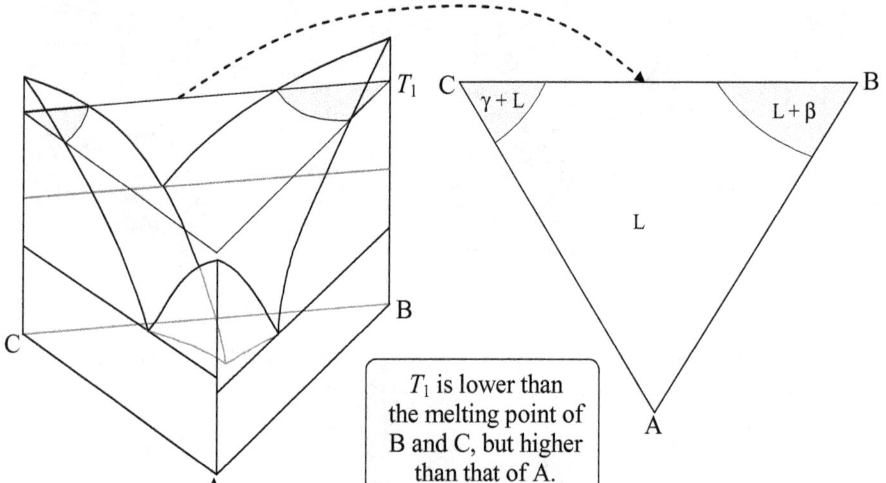


In the above isothermal section, β and γ represent the crystal structures of B and C, respectively. As the prevailing temperature $T_1$ is lower that the melting point of both B and C, the solid phases of β and γ precipitate at their respective corners. Further cooling makes these two-phase areas enlarged.

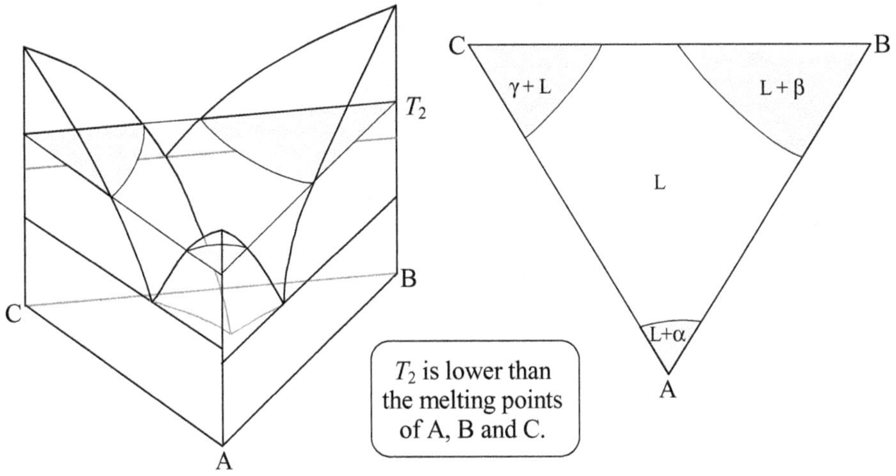

C
B
A
$T_2$
C
B
A
γ + L
L + β
L
L+α
$T_2$ is lower than the melting points of A, B and C.


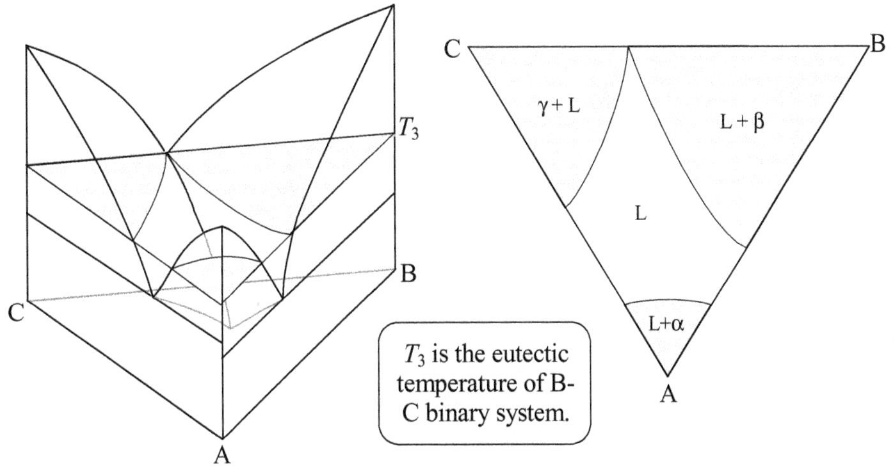

C
B
A
$T_3$
C
B
A
γ + L
L + β
L
L+α
$T_3$ is the eutectic temperature of B-C binary system.


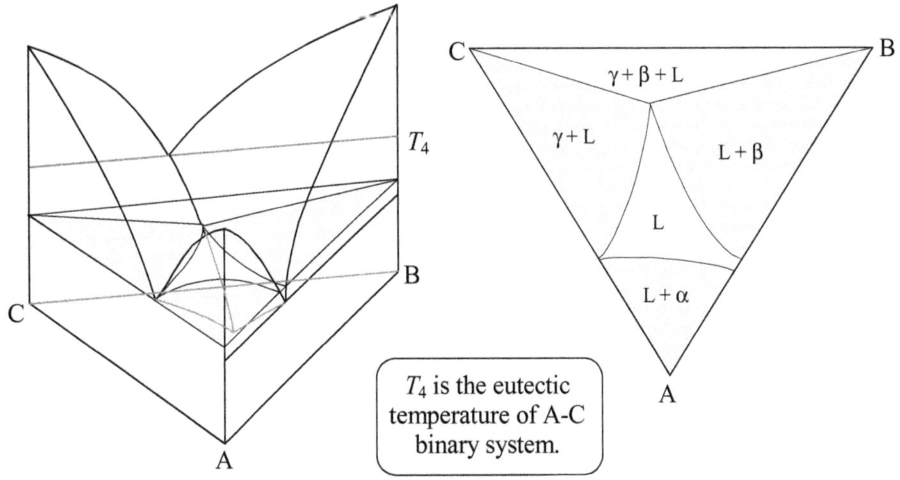

C
B
A
$T_4$
C
B
A
γ + β + L
γ + L
L + β
L
L + α
$T_4$ is the eutectic temperature of A-C binary system.

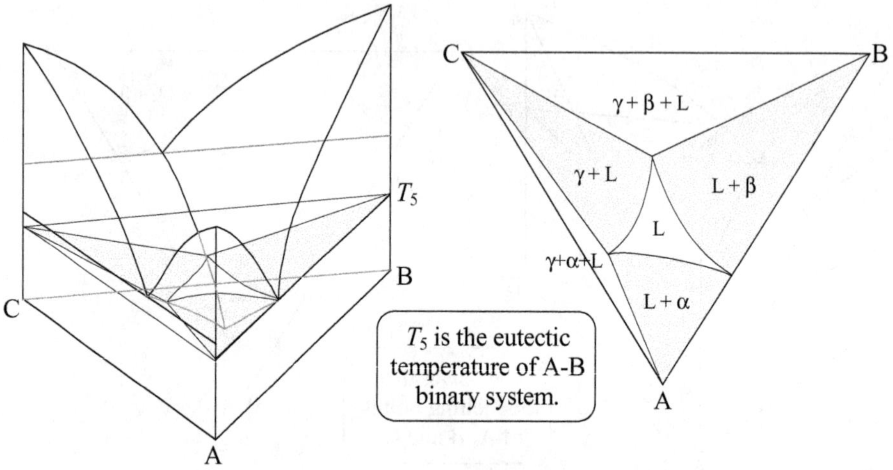
$T_5$
B
C
A
$T_5$ is the eutectic temperature of A-B binary system.
C
B
γ + β + L
γ + L
L + β
L
γ+α+L
L + α
A

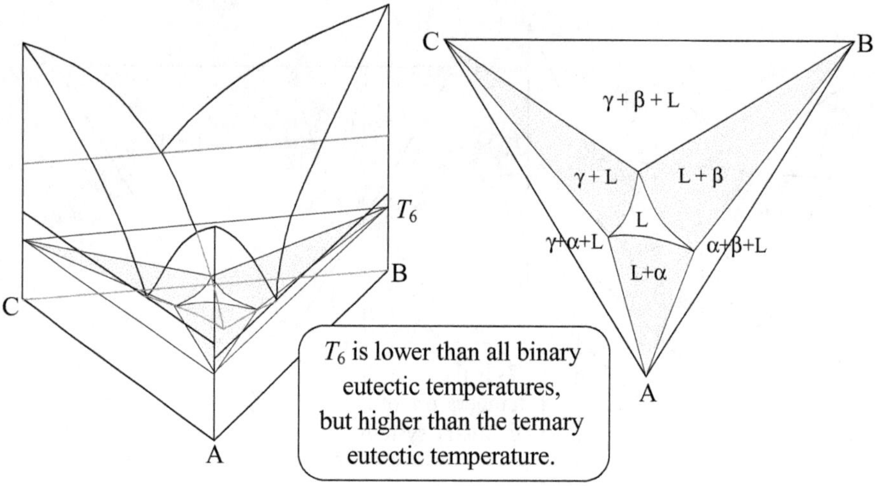
$T_6$
B
C
A
$T_6$ is lower than all binary eutectic temperatures, but higher than the ternary eutectic temperature.
C
B
γ + β + L
γ + L
L + β
L
γ+α+L
α+β+L
L+α
A

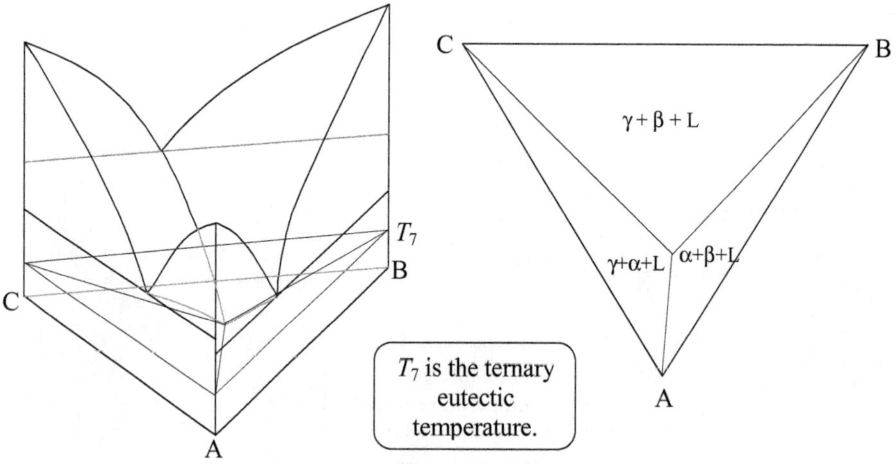
$T_7$
B
C
A
$T_7$ is the ternary eutectic temperature.
C
B
γ + β + L
γ+α+L
α+β+L
A

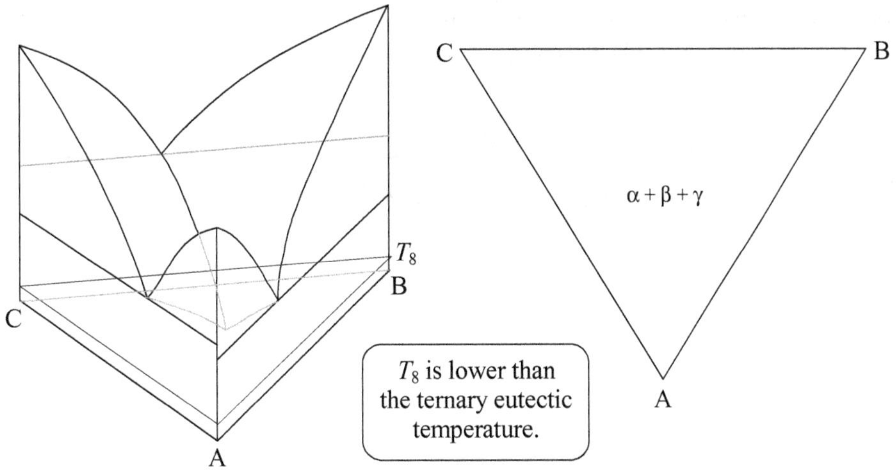


Polythermal projections of the liquidus discussed previously do not provide information on the compositions of solid phases if solid solutions or non-stoichiometric compounds are formed at equilibrium. In this case, the method of isothermal section is particularly useful. The following figure represents a simple ternary eutectic system with *terminal solid solutions* formed.

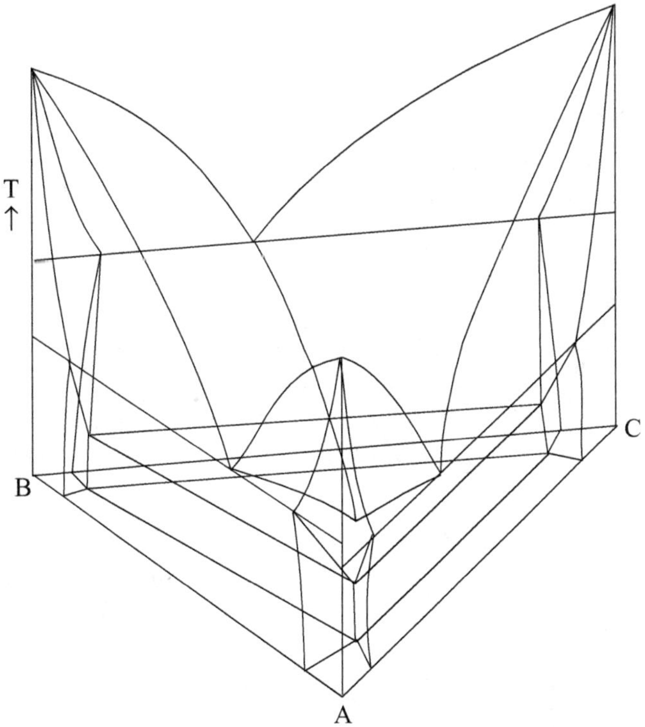

Shown in the following are the isothermal sections at a number of different temperatures of an hypothetical system ABC like the one given above:

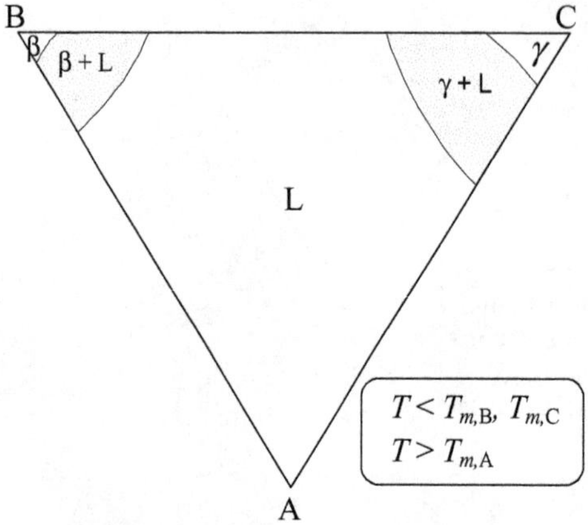


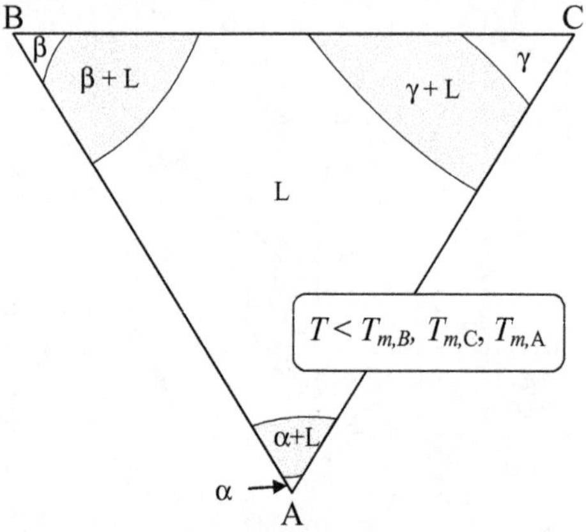


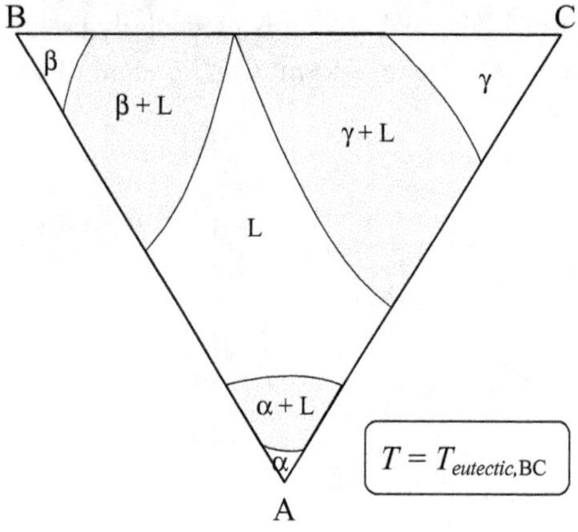


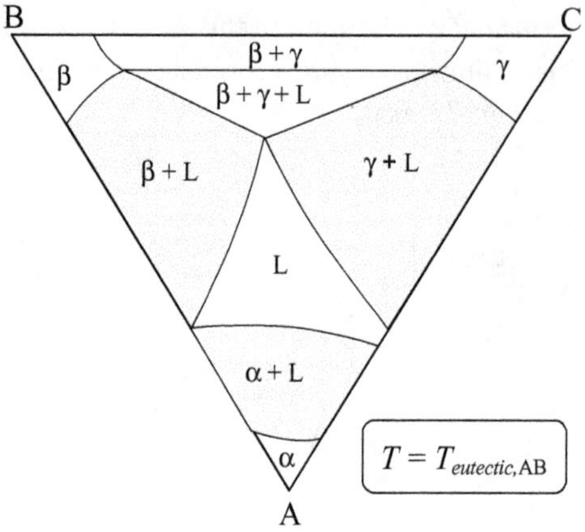


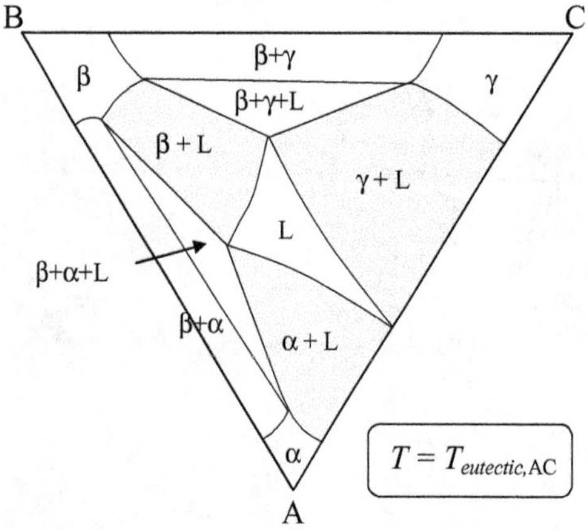


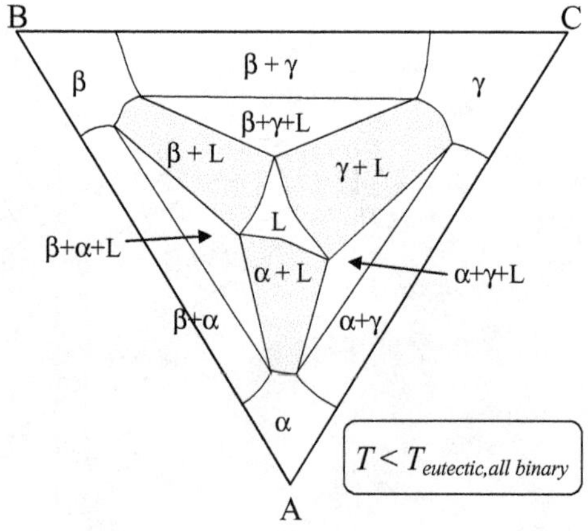

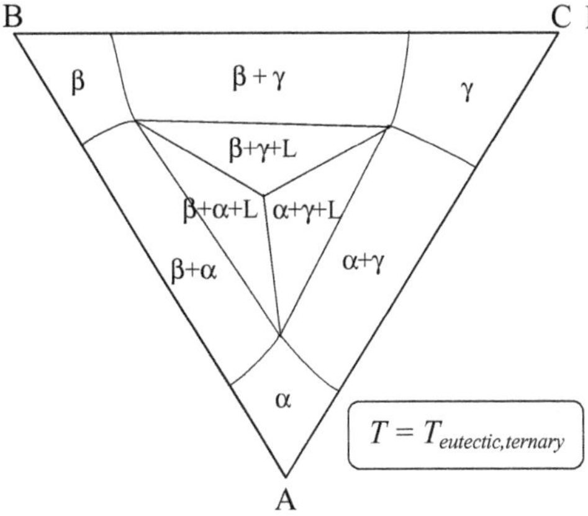


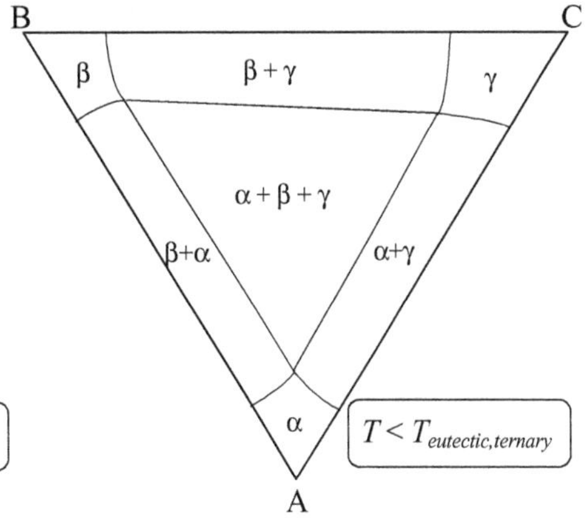


- If the overall composition of an alloy lies within the liquid phase region at a particular temperature, the alloy will exist as liquid.
- If the overall composition lies within one of the solid solution regions, the alloy will exist as the corresponding solid solution.
- If the composition lies within one of two-phase regions, say, β and liquid phases, both β and liquid phases will coexist.

Now a question arises as to how to determine the compositions of β phase and the liquid phase which are in equilibrium with each other. The usual practice is to include *tie lines* in the isothermal sections, which join the composition points of ***conjugate phases*** which coexist in equilibrium at a given temperature and pressure.

Tie lines are in fact common tangents to the Gibbs energy surfaces of the phases that coexist in equilibrium. Let's assume that α and β phases, both of which form solid solutions, coexist. The following figure shows schematic of the Gibbs energy surfaces for α and β:

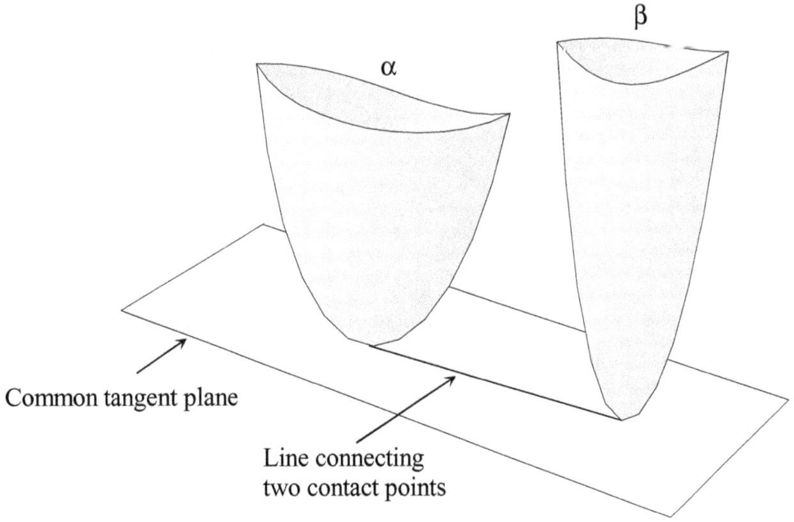

A common tangent plane shown above produces a pair of contact points. By rolling the plane on the surfaces, an infinite number of pairs of contact points are generated. Mapping on a reference plane of lines which connect selected conjugate pairs will show how contact points on one surface correspond to those on the other surface.

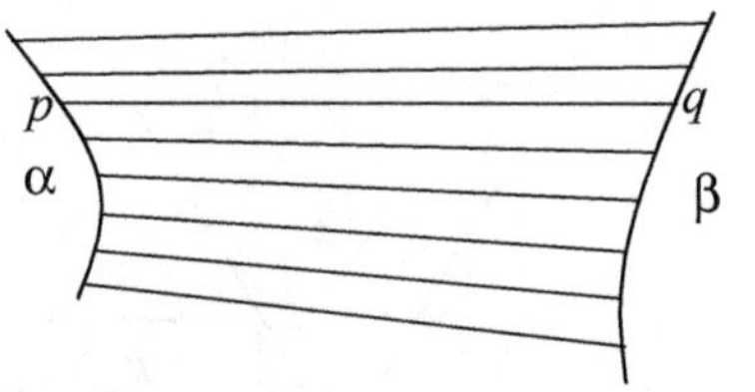


For instance, the line $pq$ is one of the tie lines, which is the common tangent of the Gibbs energy surfaces for α and β. This indicates that the α phase at the composition $p$ is in equilibrium with the β phase at the composition $q$.

At a given temperature, the Gibbs free energy of each phase in a ternary system may be represented in a graphical form with the composition triangle as base and the Gibbs energy as vertical axis. Then it would look like this:

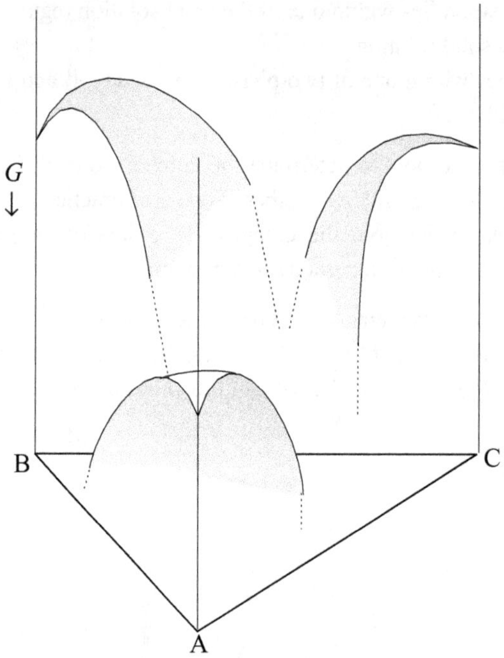


The phases that would exist at equilibrium and the compositions of the phases are determined by the contact points of a common tangent plane to their Gibbs energy surfaces. Connection of the contact points forms a ***tie line***. In a ternary system, a common tangent plane can contact two Gibbs energy surfaces at an infinite number of points, and hence an infinite number of tie lines are generated. However, a common tangent plane to three Gibbs energy surfaces generates only one set of contact points. Connection of these points forms a ***tie-triangle***.

The following is an example of an isothermal section showing a number of tie lines:

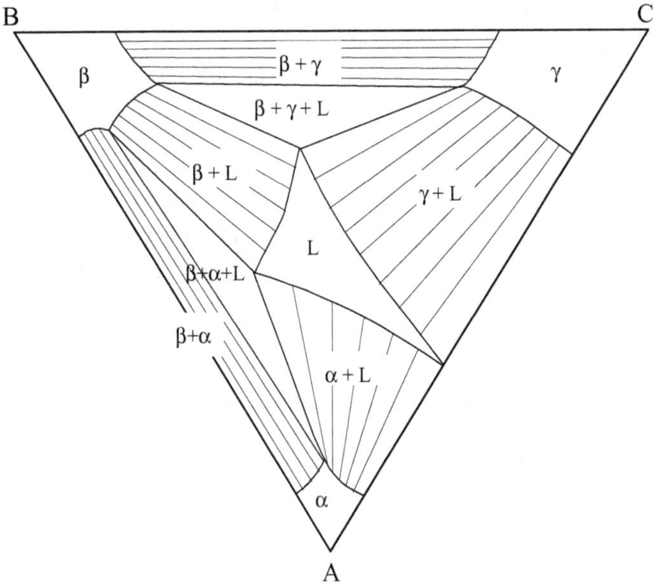


*Example 12.21*

In construction of isothermal sections or isothermal ternary phase diagrams a tie triangle together with contacting single phase and two-phase areas play an important role. In fact it may be considered as a building block of isothermal phase diagrams. Isothermal ternary phase diagrams are composed of a number of these building blocks. Shown is an example of a tie-triangle with adjacent single and two phase areas:

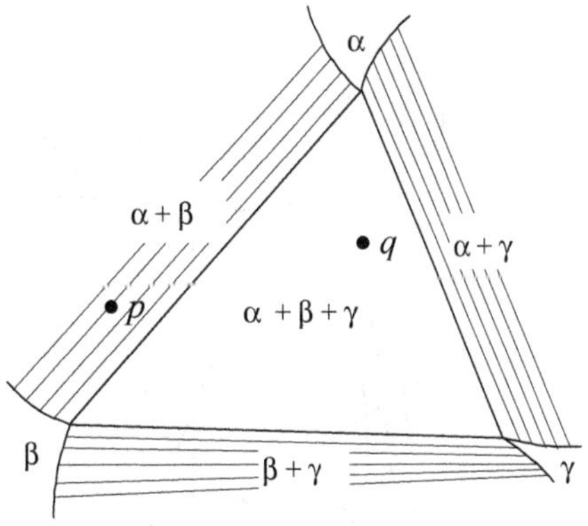


1. Determine relative proportion of each phase that exists at equilibrium for a system of overall composition of $p$ in the diagram.
2. Determine the fraction of α phase that exists at equilibrium for the overall composition of $q$.
3. Are the boundary curves at the α phase corner correct?

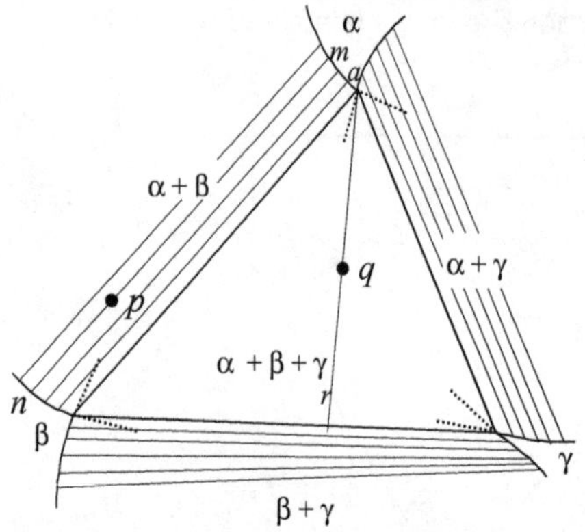


1) Fraction of α phase = $\frac{pn}{mn}$
2) Fraction of α phase = $\frac{qr}{ar}$
3) Both extensions of the boundary curves of the single phase areas must project either into the triangle (γ corner), or outside triangle (β corner), but not in mix. The α corner is wrong.

*Exercises*

12.17 Prove that tie lines must not cross each other within any two phase region.

12.18 ABC ternary system forms three binary eutectics and a ternary eutectic as shown below. Discuss equilibrium cooling paths for the overall compositions *p, q* and *r* indicated in the diagram. Discuss also the change in microstructure that should occur during cooling.

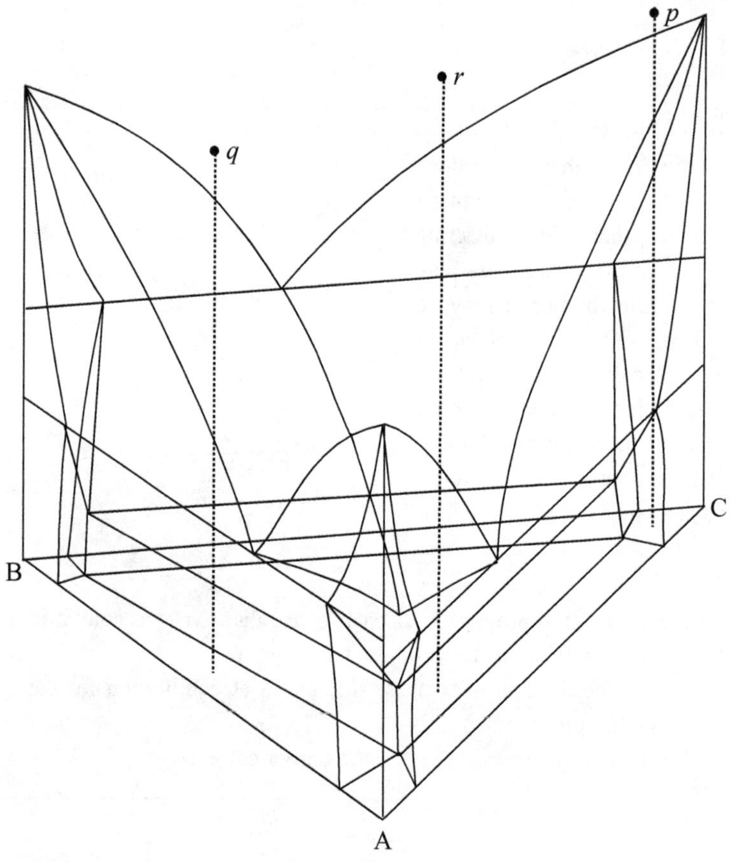

## 12.5 Predominance Diagrams

Phase diagrams are usually represented in terms of the composition and temperature of the system. There may be many other ways to graphically represent stable phases under different thermodynamic conditions. We may choose the chemical potential or partial pressure of the component(s) in the system. These alternative diagrams may find many useful applications, particularly, in high temperature corrosion and minerals processing areas. These diagrams are collectively termed ***predominance diagrams*** or ***stability diagrams***.

We now construct the predominance diagram for the Ni-S-O system. The possible phases in the system at 1000$K$ will be,

- Solid phases: NiS, NiO, $NiSO_4$, $Ni_3S_2$, Ni
- Gas Phase: $O_2$, $S_2$, $SO_2$, $SO_3$

Four species in the gas phase are not all independent, but they are related by the following reaction equilibria:

$$1/2S_2 + O_2 = SO_2$$
$$1/2S_2 + 3/2O_2 = SO_3$$

Therefore, two gas species out of four may suffice to represent the gas phase. The selection of two species is arbitrary and a matter of convenience. We now choose $O_2$ and $SO_2$.

For the solid phases, we first assume for simplicity that there are only four species: NiS, NiO and $NiSO_4$.

If we choose $P_{O_2}$ and $P_{SO_2}$ for the abscissa and ordinate for the diagram, the next step will be to check which solid species is more stable over other solids for a given set of $P_{O_2}$ and $P_{SO_2}$.

(1) NiO *vs.* NiS:

$$NiS + 3/2O_2 = NiO + SO_2 \qquad \Delta G_1 = -485800 + 115.3T, \text{J}$$

(2) NiO *vs.* $NiSO_4$:

$$NiO + SO_2 + 1/2O_2 = NiSO_4 \qquad \Delta G_2 = -345200 + 293.2T, \text{J}$$

(3) NiS *vs.* $NiSO_4$:

$$NiS + 2O_2 = NiSO_4 \qquad \Delta G_3 = -831000 + 408.5T, \text{J}$$

From the equilibrium constant expression at 1000$K$ and assuming that all solid species are in their respective pure states

$$K_{(1)} = 2.26\times10^{19} = \frac{a_{NiO}P_{SO_2}}{a_{NiS}P_{O_2}^{3/2}} = \frac{P_{SO_2}}{P_{O_2}^{3/2}}, \text{ and thus} \qquad \text{(a) } \log P_{SO_2} = 19.35 + 1.5\log P_{O_2}$$

$$K_{(2)} = 5.25\times10^{2} = \frac{a_{NiSO_4}}{a_{NiO}P_{SO_2}P_{O_2}^{1/2}} = \frac{1}{P_{SO_2}P_{O_2}^{1/2}}, \text{ and thus (b) } \log P_{SO_2} = -2.72 - 0.5\log P_{O_2}$$

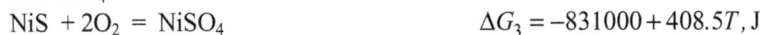

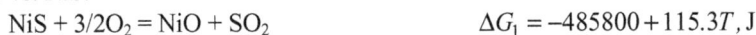

$$K_{(3)} = 1.17\times10^{22} = \frac{a_{NiSO_4}}{a_{NiS}P_{O_2}^2} = \frac{1}{P_{O_2}^2}\text{, and thus} \qquad \text{(c) } \log P_{O_2} = -11.04$$

When we plot the above three equations of (a), (b) and (c) with variables of $\log P_{O_2}$ as the abscissa and $\log P_{SO_2}$ as the ordinate, they will look like the ones given below:

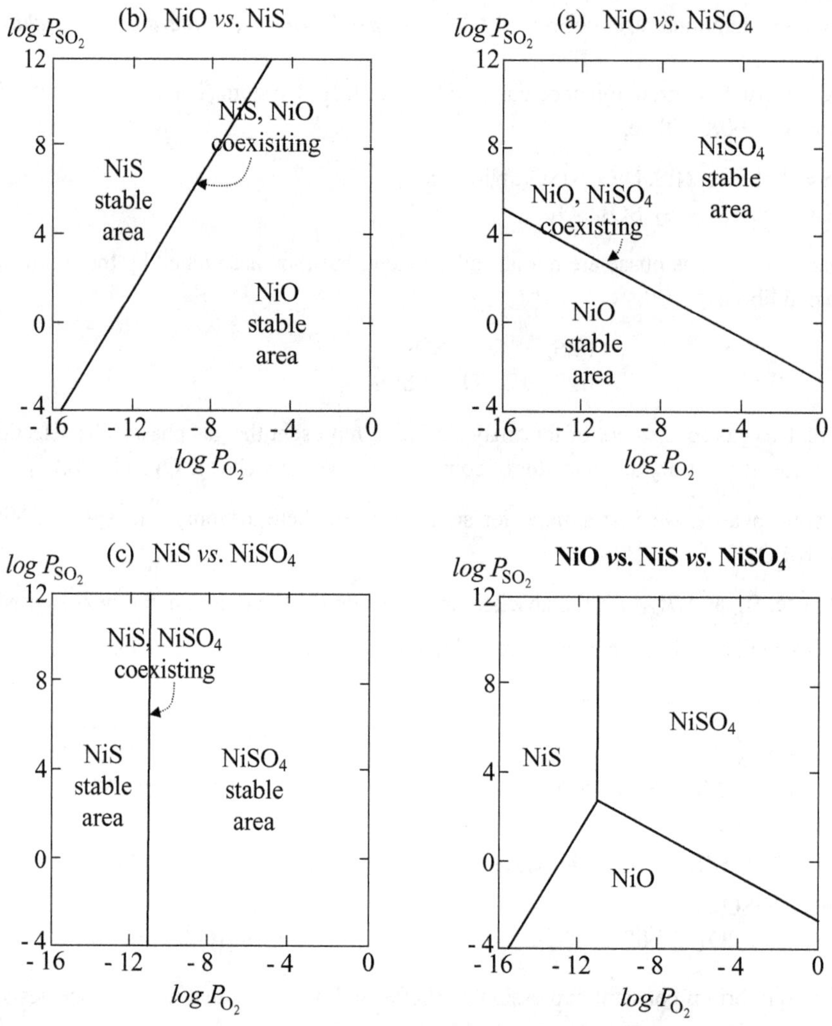


The last diagram for NiO-NiS-$NiSO_4$ is the combination of the other three diagrams which are for comparison of individual two solids. This last diagram is called the *predominance diagram* or *stability diagram*. The diagram shows the fields of stability for NiO, NiS and $NiSO_4$. The triple point indications the condition at which all three solid phases exist together in equilibrium.

As mentioned earlier, in the real Ni-S-O system, two more solids, $Ni_3S_2$ and metallic Ni can also be stable phases under certain gas compositions. These solids must be included to complete the diagram. The diagram which includes these two additional solids can be constructed by the same principle as applied in the above. The result is shown in the following diagram:

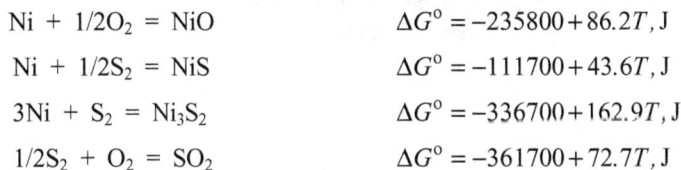

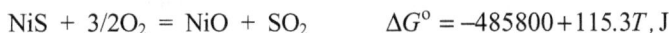

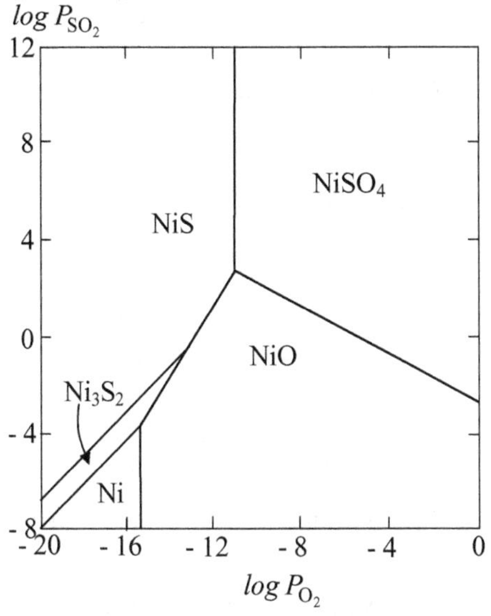


*Example 12.22*

The gas compositions in the reactor which contained initially NiS at 1000 K were measured to find $P_{O_2} = 10^{-12}$ atm. and $P_{SO_2} = 10^{-1}$ atm.

Assuming that the reactor has reached the equilibrium state, determine which condensed phase will prevail in the reactor. The following data are given:

| | |
|---|---|
| $Ni + 1/2O_2 = NiO$ | $\Delta G^o = -235800 + 86.2T$, J |
| $Ni + 1/2S_2 = NiS$ | $\Delta G^o = -111700 + 43.6T$, J |
| $3Ni + S_2 = Ni_3S_2$ | $\Delta G^o = -336700 + 162.9T$, J |
| $1/2S_2 + O_2 = SO_2$ | $\Delta G^o = -361700 + 72.7T$, J |

First, NiO and NiS are compared to find more stable one. Combining the first two and the last reactions,

$$NiS + 3/2O_2 = NiO + SO_2 \qquad \Delta G^o = -485800 + 115.3T, \text{J}$$

$$\Delta G = \Delta G^o + RT\, ln\left(\frac{P_{SO_2}}{P_{O_2}^{3/2}}\right)$$

T=1000 K

$P_{O_2} = 10^{-12}$ atm. and $P_{SO_2} = 10^{-1}$ atm.

$$\Delta G = -45000 \text{ J}$$

Since the Gibbs energy change of the reaction is negative, it is clear that NiO is more stable than NiS under the specified conditions.

Next, NiO and $Ni_3S_2$ are compared. Proper combination of the given reactions will produce the following relationship:

$$Ni_3S_2 + 7/2O_2 = 3NiO + 2SO_2 \qquad \Delta G^o = -1094100 + 241.4T, \text{J}$$

$$\Delta G = \Delta G^o + RT\,ln\left(\frac{P_{SO_2}^2}{P_{O_2}^{7/2}}\right)$$

T=1000 K

$P_{O_2} = 10^{-12}$ atm. and $P_{SO_2} = 10^{-1}$ atm.

$$\Delta G = -49000 \text{ J}$$

As the Gibbs energy change of the reaction is negative, NiO is more stable than $Ni_3S_2$ under the specified conditions.

Overall, NiO is more stable than NiS and $Ni_3S_2$ and thus the dominating species in the reactor will be NiO.

The same conclusion can be drawn from the predominance diagram we constructed earlier. The predominance diagram is of convenience for this type of task.

The Ellingham diagram which was discussed in the earlier chapter (10.8) may also be considered a kind of the predominance diagram. It gives areas where individual species are stable. For example, the Ellingham diagram for the following oxidation reaction can be modified in the form of the predominance diagram:

$$M + O_2 = MO_2 \qquad \Delta\tilde{G}_f^o = RT\,ln\,P_{O_2}$$

Ellingham diagram

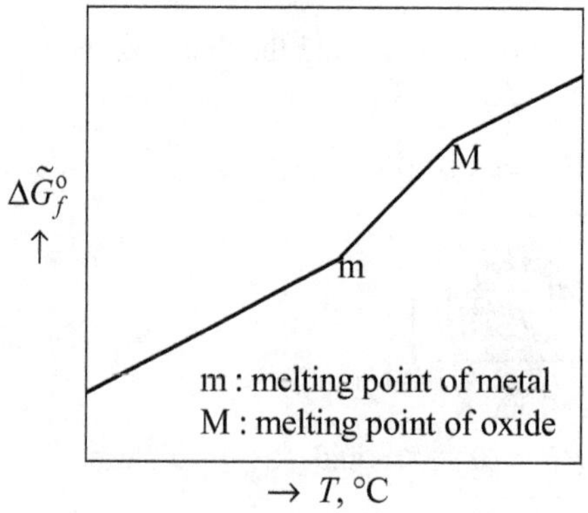


Predominance diagram

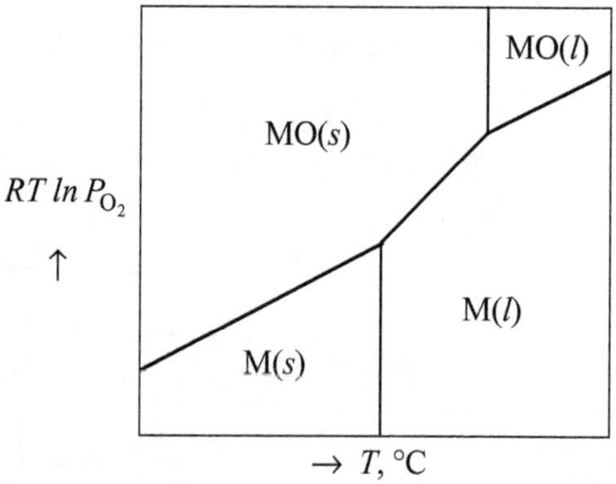

# Chapter 13

# Electrochemistry

## 13.1 Basic Electrochemical Concepts

Virtually all chemical processes combine change with exchange of electrons, which leads to change in the valence state of some or all of the participating elements. Reactions which involve electron transfer are of the general category of oxidation-reduction or **redox reactions**.

| *Oxidation* | *Reduction* |
|---|---|
| The element involved undergoes a *loss of electrons*. (e.g., $Cu \rightarrow Cu^{2+}$, $2O^{2-} \rightarrow O_2$) | The element involved undergoes a *gain of electrons*. (e.g., $Cu^{2+} \rightarrow Cu$, $O_2 \rightarrow 2O^{2-}$) |

Since electrons are transferred, oxidation and reduction must occur simultaneously.

The electrons lost by one species are taken up by the other.

$$FeO \quad + \quad C \quad = \quad Fe \quad + \quad CO$$

Reduction
$$Fe^{2+} + 2e \rightarrow Fe$$

Oxidation
$$C \rightarrow C^{2+} + 2e$$

$$O^{2-} \rightarrow O^{2-}$$

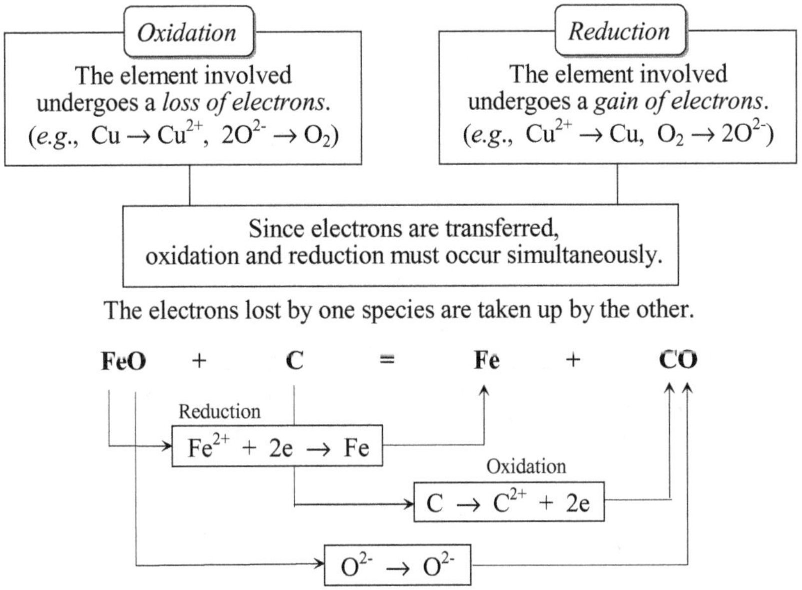

The stoichiometry of a reaction which involves electron transfer is related to the electrical quantities determined by *Faraday's law* which states;

- One equivalent of product is produced by passage of 96,487 coulombs of charge, or

- One equivalent of chemical change produces 96,487 coulombs of electricity.

One equivalent of a substance undergoing oxidation produces *Avogadro's number* ($N$), or one mole of electrons:

$$M \quad \rightarrow \quad M^+ \quad + \quad e$$

| | | |
|---|---|---|
| $N$ atoms of M | $N$ ions of $M^+$ | $N$ numbers of electrons |

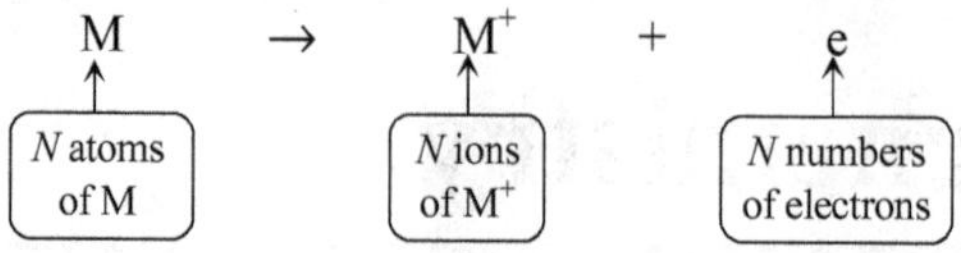

According to Faraday's law, therefore, the passage of 96,487 coulombs of electricity corresponds to the passage of Avogadro's number, $N = 6.023 \times 10^{23}$, of electrons. Faraday's law is understandable because an Avogadro's number of electrons added to or removed from a reagent will produce an equivalent of product. The quantity of charge that corresponds to a chemical equivalent is known as the *Faraday, F*:

$$F = 96,487 \text{ coulombs}$$

Suppose that a redox reaction takes place in such a way that the reaction produces a detectable electric current. This will be possible only if the reaction can be separated physically so that electrons are lost in one part of the system and gained in the other part.

$$A + BX \rightarrow AX + B$$

$$A + B^+X^- \rightarrow A^+X^- + B$$

| $A \rightarrow A^+ + e$ | $X^- \rightarrow X^-$ | $B^+ + e \rightarrow B$ |
|---|---|---|
| Oxidation | | Reduction |

| $e \longrightarrow$ | |
|---|---|
| A, AX<br>$(A \rightarrow A^+ + e)$ | B, BX<br>$(B^+ + e \rightarrow B)$ |
| $\longleftarrow X^-$ | |

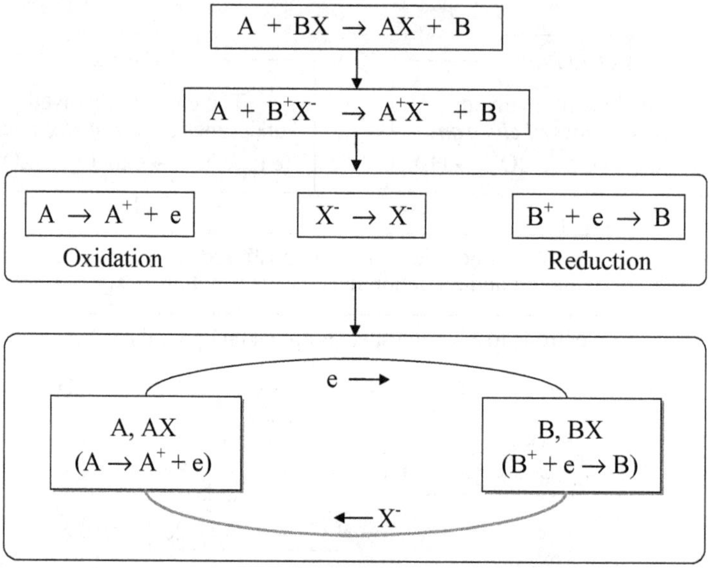

In the above device, a mixture of A and AX, and a mixture of B and BX are physically separated, but joined by two connections, *i.e.*,

- An electronic conductor which allows the passage of electrons only
- An ionic conductor which allows the passage of $X^-$ ions only.

With the arrangement shown above, the reaction proceeds spontaneously, in which electrons move from left to right and $X^-$ ions from right to left so that the electroneutrality is maintained. This type of reactions which take place in an electrochemical manner is called *electrochemical reaction*.

A device like the one shown above, which permits a spontaneous electrochemical reaction to produce a detectable electric current, is termed a ***galvanic cell***. In other words, the galvanic cell is an arrangement in which a chemical reaction produces electrical work, not through the intermediary of thermal energy transfer.

As shown in the above figure, oxidation occurs in one ***half-cell*** and reduction takes place in the other half-cell. The part at which oxidation occurs is referred to as the ***anode***, while the part at which reduction occurs is termed ***cathode***. The part at which either oxidation occurs (the anode) or reduction occurs (the cathode) is termed an ***electrode***

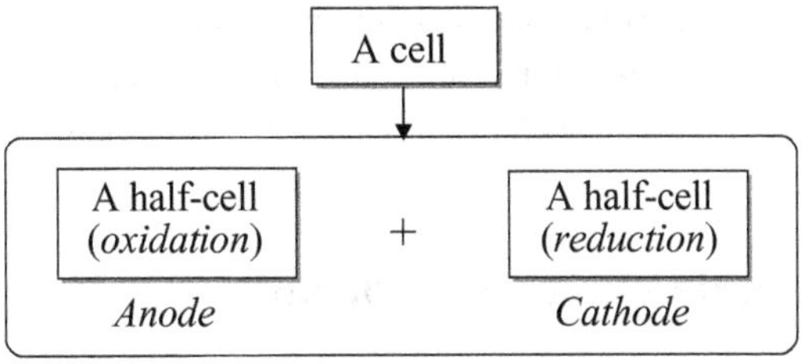


If the Gibbs energy change for the reaction

$$A + BX \rightarrow AX + B \qquad \Delta G_r$$

is negative, the reaction proceeds from left to right as written. As the electrochemical reaction is basically the same as the above chemical reaction (but in different arrangement in which electrons move from the anode to the cathode), the *driving forces* for the chemical reaction ($\Delta G_r$) and that for the electrochemical reaction (*i.e.*, driving force for the electron movement) must be related to each other.

Electrons flow if there is an electric potential difference or an electrical voltage. If a galvanic cell is set up in conjunction with an appropriate external circuit so that the electric potential difference which exists in the cell is opposed by an identical, but opposite in direction, voltage from an external source, the electrochemical reaction of the cell ceases to proceed. The electrochemical reaction is now brought into equilibrium. This type of equilibrium is referred to as ***electrochemical equilibrium***. That is, when the chemical driving force ($\Delta G_r$) is balanced with the opposing electrical driving force, the reaction is in equilibrium electrochemically even though $\Delta G_r \neq 0$.

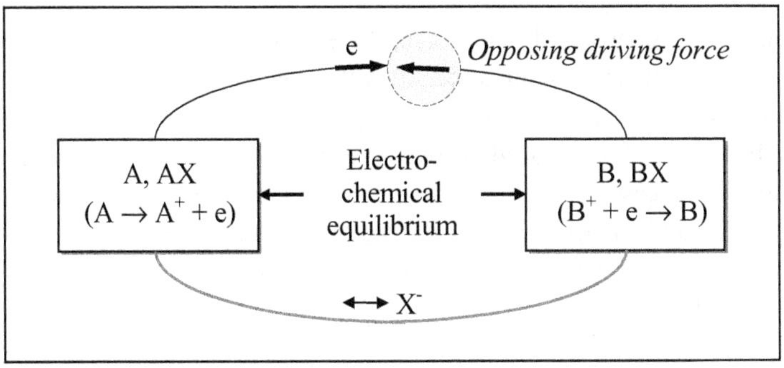

The application of thermodynamic principles to this type of electrochemical processes is discussed in the next section (5.1.2).

*Exercises*

13.1 Calculate the number of electrons released when one gram mole of calcium is oxidized. Calculate the coulombs of electricity generated by the oxidation.

13.2 The galvanic cell performs an electric work when the electrochemical reaction of the cell occurs. As work is not a state function, the amount of work which a system does depends on the path it takes for given initial and final states. Under what conditions does the amount of work the cell performs become maximum?

## 13.2 Electrochemical Cell Thermodynamics

Consider the reaction

$$A + BX \rightarrow AX + B \qquad \Delta G_r < 0$$

If all reactants and products are mixed in a same reactor, the above chemical reaction will proceed from left to right with a driving force equivalent to the Gibbs energy change of $\Delta G_r$. The reaction will continue until the Gibbs energy change becomes zero ($\Delta G_r = 0$).
If a galvanic cell is constructed as shown below, the same reaction will occur, but in an electrochemical manner.

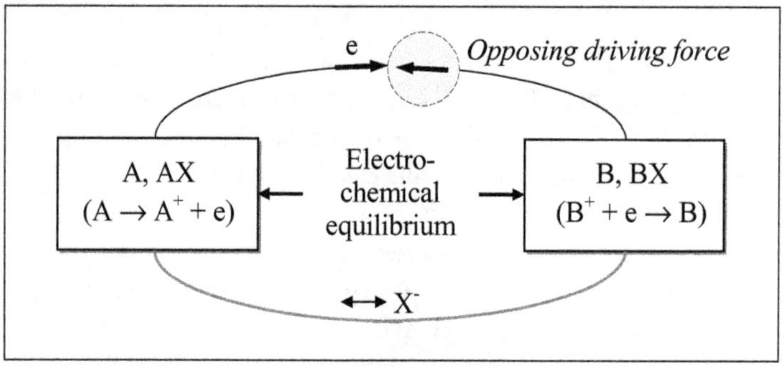


In the absence of the opposing driving force, the electron transfer from left to right, which enables the electrochemical reaction to occur, will enjoy the full driving force equivalent to $\Delta G_r$. If an external force opposing to this driving force is applied, the net driving force for the electron transfer will be reduced accordingly. The electrochemical reaction of the cell will then suffer a decrease in the driving force. When the opposing external force (voltage) is increased to exactly balance the driving force of the electrochemical reaction of the cell, the electrons will cease to flow and so will the cell reaction.

Now we examine the amount of work the cell performs under various conditions. To help understand the concept of work of a cell, the case of gas expansion in a cylinder is reviewed below:

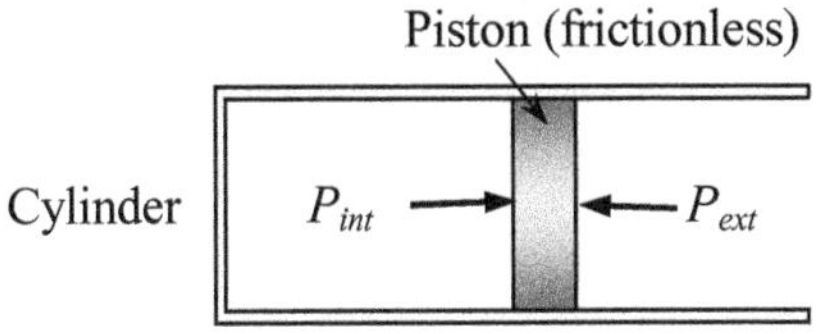


- If $P_{ext} = 0$, the piston will move to the right without doing work, because there is nothing to work against.
- If $P_{ext}$ is applied and increased, the piston will move to the right, but by doing work against $P_{ext}$.
- If $P_{ext}$ is only infinitesimally smaller than $P_{int}$, *i.e.*, $P_{ext} = P_{int} - dP$, the piston will move to the right by doing the maximum work, $w_{max}$, against $P_{ext}$.
- If $P_{ext}$ is infinitesimally larger than $P_{int}$, *i.e.*, $P_{ext} = P_{int} + dP$, the direction of the piston movement will be reversed.
- The last two cases depict the conditions for reversibility of gas expansion, and the concept of the maximum work.

*Work under reversible conditions* $(w_{rev})$ = *the maximum work* $(w_{max})$

$$w_{rev} = w_{max}$$

Work of electrochemical cells has analogy with work of gas expansion discussed above.

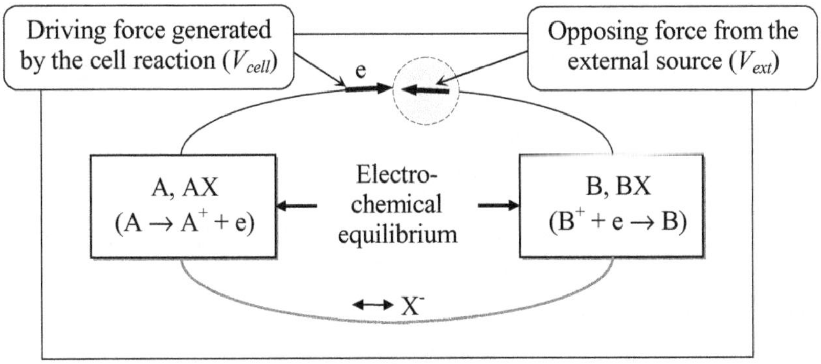


- If there is no opposing force from an external source, i.e., $V_{ext} = 0$, electrons will move without doing work, because there is nothing to work against.
- If the external opposing force ($V_{ext}$) is applied and increased, the electrons move with doing work against $V_{ext}$. The net driving force for the electron transfer ($\Delta V$) is

$$\Delta V = V_{cell} - V_{ext}$$

- If $V_{ext}$ is only infinitesimally smaller than $V_{cell}$, *i.e.*,

$$V_{ext} = V_{cell} - dV$$

the electrons move with doing maximum work, $w_{max}$, against $V_{ext}$. Since the net driving force for the electron transfer is infinitely small ($dV$), the cell reaction will proceed at an infinitely slow rate, but performing the maximum work.

- If $V_{ext}$ is only infinitesimally larger than $V_{cell}$, *i.e.*,

$$V_{ext} = V_{cell} + dV$$

the direction of the electron transfer is reversed. In other words, the direction of the cell reaction can be reversed by an infinitesimal change in the external opposing force.

- The last two cases depict the conditions of reversibility of the electrochemical reaction of the cell.

*Work under reversible conditions* ($w_{rev}$) = *the maximum work* ($w_{max}$)

$$\boxed{w_{rev} = w_{max}}$$

In general the ***electromotive force*** of a cell operating under reversible conditions is referred to as ***emf*** (ε), while the force observed when conditions are irreversible is termed a ***voltage***. In other words, *emf*, ε, is the maximum possible voltage that a galvanic cell can produce.

**Sign convention for *emf***

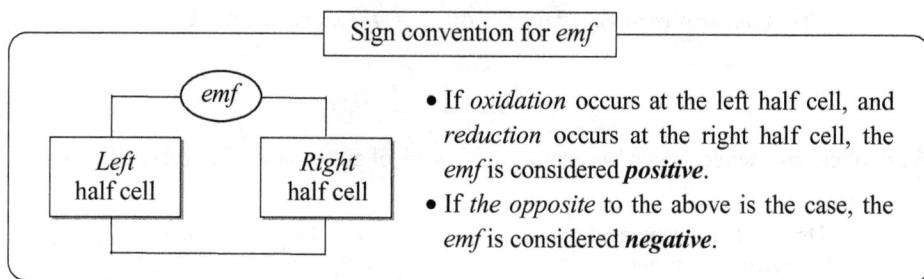

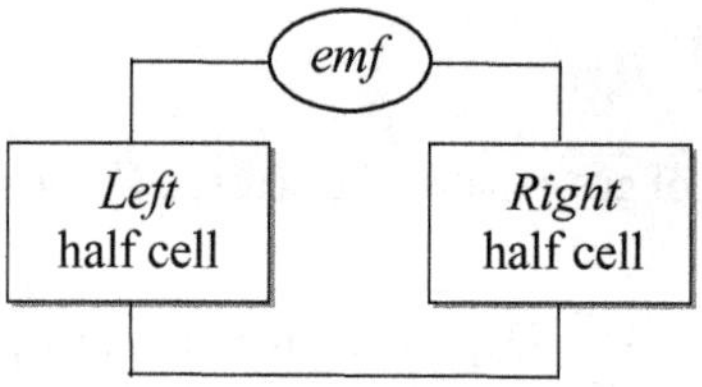


- If *oxidation* occurs at the left half cell, and *reduction* occurs at the right half cell, the *emf* is considered ***positive***.
- If *the opposite* to the above is the case, the *emf* is considered ***negative***.

Electrical work ($w_{elec}$) per mole of reactant of a galvanic cell is the product of the charge transported ($q$, coulombs/mol ) and the voltage ($V$, volts) .

$$\boxed{w_{elec} = -qV, \quad J\,mol^{-1}}$$

The negative sign is due to the fact that the work is done to the surroundings.

Electrical charge ($q$) is the product of the number of coulombs per equivalent (*Faraday*, $F$) and the valency ($n$)

$$q = nF$$

$$\boxed{w_{elec} = -nFV}$$

If the transportation of the charge is carried out reversibly, *that is*, the voltage is the electromotive force (*emf*, ε), the work has the maximum value.

$$w_{elec,\,max} = -nF\varepsilon$$

Recall that, when a system undergoes a reversible process at constant $T$ and $P$,

$$dG = dw_{add}$$
$$\text{or } \Delta G = w_{add}$$

$w_{add}$ : reversible (or maximum) work exclusive of $PV$ work

For the galvanic cell, electrical work is virtually only work other than $PV$ work.

$$w_{add} = w_{elec,\,max}$$

$$\Delta G = w_{elec,\,max}$$

$$\Delta G = -nF\varepsilon$$

This equation is known as the ***Nernst equation*** and significant in that,

- It enables us to measure $\Delta G$ for the reaction by applying an opposing electric potential of magnitude ($-\varepsilon$) externally, which results in no current to flow.
- It means that when the chemical driving force ($\Delta G$) is exactly balanced by the external opposing voltage ($-\varepsilon$), the whole system (the cell) is in equilibrium, *i.e.*, electrochemical equilibrium.
- The measured $\Delta G$ can be used as a criterion of whether the process (or reaction) will take place spontaneously when all the species involved in the cell reactions are mixed in a reactor under otherwise same conditions.

The electrochemical equilibrium may be manifested with the aid of a graphical representation as follows:

Recall that the Gibbs energy change of a process in equilibrium is given,

$$\Delta G = w_{add} \quad \text{or} \quad \Delta G - w_{add} = 0$$

where $w_{add}$ is the additional work other than $PV$ work.

We now consider two different cases: one with no additional work ( $\Delta G = 0$ ), and the other with additional work of $-nF\varepsilon$ ( $\Delta G + nF\varepsilon = 0$ ).

When $\Delta G = 0$

When $\Delta G + nF\mathcal{E} = 0$

$G$

System

Process path

$G$

System

$\Delta G$

$nF\mathcal{E}$

Process path

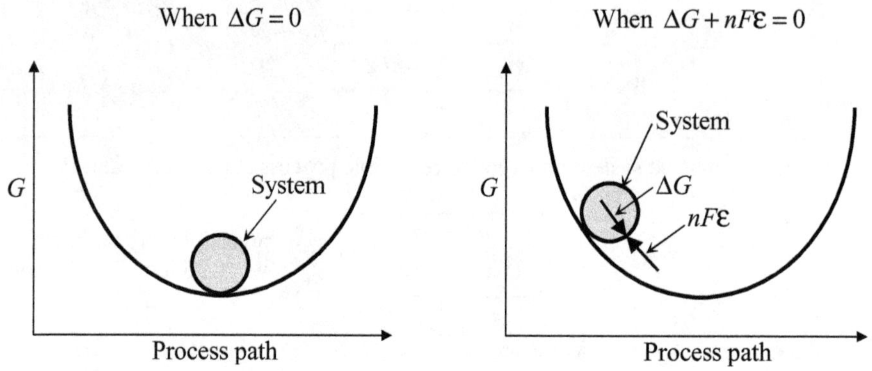

Now consider a general reaction,

$$a\text{A} + b\text{B} = m\text{M} + n\text{N} \qquad\qquad \Delta G$$

$$\Delta G = \Delta G^\circ + RT\,ln\left(\frac{a_\text{M}^m a_\text{N}^n}{a_\text{A}^a a_\text{B}^b}\right)$$

If this reaction takes place electrochemically,

$$\Delta G = -nF\mathcal{E}$$
$$\Delta G^o = -nF\mathcal{E}^o$$

where $\mathcal{E}^\circ$: *emf* when all species are present in their respective standard states.

$$\mathcal{E} = \mathcal{E}^o - \frac{RT}{nF}\,ln\left(\frac{a_\text{M}^m a_\text{N}^n}{a_\text{A}^a a_\text{B}^b}\right)$$

This equation is a variation of the Nernst equation previously derived and also known as the **Nernst equation**. The above equation plays a central role in electrochemistry.

- This equation enables us to determine how the *emf* of a cell should vary with compositions.
- The *emf* at the standard states ($\mathcal{E}^\circ$) can be determined by constructing a cell with all the reagents at unit activity.
- Activities or activity coefficients of reagents can be obtained.

We now derive several equations which relate some important thermodynamic functions to the electromotive force (*emf*):

$$\left(\frac{\partial \Delta G}{\partial T}\right)_P = -\Delta S$$

$\downarrow$ $\Delta G = -nF\varepsilon$

$$\Delta S = nF\left(\frac{\partial \varepsilon}{\partial T}\right)_P$$

$\downarrow$ $\Delta G = \Delta H - T\Delta S$

$$\Delta H = -nF\left[\varepsilon - T\left(\frac{\partial \varepsilon}{\partial T}\right)_P\right]$$

Where $\left(\frac{\partial \varepsilon}{\partial T}\right)_P$ is called the ***temperature coefficient***.

The value of $\Delta S$ is independent of temperature to a good approximation:

$$\Delta S = nF\left(\frac{\partial \varepsilon}{\partial T}\right)_P$$

$\downarrow$ Integration by keeping $\Delta S$ constant

$$\varepsilon_{T_2} = \varepsilon_{T_1} + \frac{\Delta S}{nF}(T_2 - T_1)$$

It can be seen in the above equation that the *emf* is a linear function of temperature. Thus, by measuring $\varepsilon$ and the temperature coefficient, we can obtain thermodynamic properties of the cell ($\Delta G$, $\Delta H$ and $\Delta S$).

*Example 13.1*

The figure shown below depicts the following redox reaction which occurs in an electrochemical manner:

$$Zn + CuSO_4 = Cu + ZnSO_4$$

(1) If $|\varepsilon_{cell}| > |V_{ext}|$, will Zn be oxidized or reduced?
(2) If $|\varepsilon_{cell}| < |V_{ext}|$, in which direction is work done, on the cell or by the cell?
(3) Find conditions for the cell to perform the maximum work.
(4) The standard Gibbs energy change of the above reaction is -213,000 *J* at 25°C. If all species involved in the reaction are at their respective standard states, calculate the maximum amount of work the cell can perform per mole of Zn.

(5) Calculate the standard *emf* ($\varepsilon^o$) at 25°C.

(6) If the cell consists of pure solid Zn and pure solid Cu,
and $CuSO_4$(aqueous solution)
and $ZnSO_4$(aqueous solution), both with the same activity,
calculate the *emf* of the cell at 25°C.

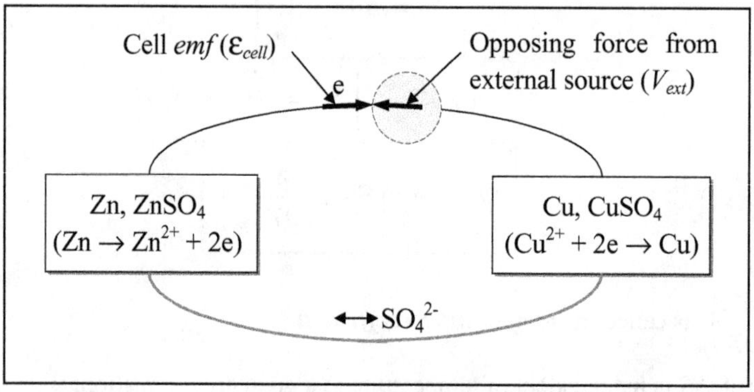


(1) The reaction will proceed from left to right with the net driving force of

$$\Delta V = |\varepsilon_{cell}| - |V_{ext}| > 0.\ \text{For Zn, thus, } Zn \rightarrow Zn^{2+}\text{: } \textit{oxidation}.$$

(2) As the external voltage is larger than the cell *emf*, work is done on the system.

$$\Delta V = |\varepsilon_{cell}| - |V_{ext}| < 0.\ \text{For Zn, thus, } Zn^{+2} \rightarrow Zn\text{: } \textit{reduction}.$$

This is an ***electrolysis*** process: *i.e.*, the current is consumed rather than produced, and metallic Zn is produced from the zinc solution.

(3) The cell performs the maximum work when $|\varepsilon_{cell}| = |V_{ext}|$

(4) $\Delta G = w_{max}$, and thus $w_{max} = -213{,}000\ \text{J mol}^{-1}$

(5) $\Delta G^o = -nF\varepsilon^o \qquad n = 2,\ F = 96{,}487$ coulombs, $\rightarrow \varepsilon^o = 1.104$ volts

(6) Using the Nernst equation,

$$\varepsilon = \varepsilon^o - \frac{RT}{nF} ln\left(\frac{a_{Cu}a_{ZnSO_4}}{a_{Zn}a_{CuSO_4}}\right)$$

↓

$$\varepsilon = 1.104 - \frac{(8.314)(298)}{(2)(96{,}487)} ln\left(\frac{(1)\times a_{ZnSO_4}}{(1)\times a_{CuSO4}}\right)$$

↓ $a_{ZnSO_4} = a_{CuSO_4}$

$$\varepsilon = 1.104 \text{ volts}$$

*Example 13.2*

Is the following statement true?
"If the physical size of an electrochemical cell is doubled, the cell *emf* will also be doubled."

We check the validity of the statement with the following example:

$$Zn + CuSO_4 = Cu + ZnSO_4 \qquad \Delta G^o = -213{,}000 \text{ J at } 298 \text{ K}$$

$$\Delta G = \Delta G^o + RT\,ln\left(\frac{a_{Cu}a_{ZnSO_4}}{a_{Zn}a_{CuSO_4}}\right)$$

Suppose that
$a_{Zn} = 1, a_{Cu} = 1,$
$a_{ZnSO_4} = 50a_{CuSO_4}$

$$\Delta G = -213{,}000 + (8.314)(298)\,ln\left(\frac{1\times 50}{1\times 1}\right) = -203{,}300J$$

$\Delta G = -nF\varepsilon$
$n = 2$

$$\varepsilon = 1.054 \text{ volts}$$

We now double the stoichiometry of the reaction,

$$2Zn + 2CuSO_4 = 2Cu + 2ZnSO_4 \qquad \Delta G^o = -426{,}000\,J \text{ at } 298K$$

$$\Delta G = \Delta G^o + RT\,ln\left(\frac{a_{Cu}^2 a_{ZnSO_4}^2}{a_{Zn}^2 a_{CuSO_4}^2}\right)$$

$a_{Zn} = 1, \quad a_{Cu} = 1, \quad a_{ZnSO_4} = 50a_{CuSO_4}$

$$\Delta G = -426{,}000 + (8.314)(29\,8)\,ln\left(\frac{1\times 50^2}{1\times 1}\right) = -406{,}600J$$

$\Delta G = -nF\varepsilon$
$n = 4$

$$\varepsilon = 1.054 \text{ volts}$$

As can be seen in the above, the electromotive force of a cell is an intensive property. Like temperature and pressure it is independent of the size of the system.

*Exercises*

13.3 The standard *emf* of the cell reaction

$$Zn + CuSO_4 = Cu + ZnSO_4$$

is 1.104 volts at 25°C. Calculate the equilibrium constant of the above reaction at 25°C.

13.4 The standard *emf* of the cell reaction

$$Cd(s) + Hg_2Cl_2(s) = 2Hg(l) + CdCl_2(aq)$$

is given as a function of temperature:

$$\varepsilon^o = 0.487 + (13.3 \times 10^{-4}T - 2.4 \times 10^{-6}T^2, \text{ (volts)}$$

Calculate the values of $\Delta G^o$, $\Delta H^o$, and $\Delta S^o$ of the reaction at 35°C.

13.5 Calculate the *emf* at 25°C of the cell represented by the following reaction:

$$Zn(s) + CdSO_4(aq) = Cd(s) + ZnSO_4(aq)$$

The activities of $CdSO_4(aq)$ and $ZnSO_4(aq)$ are $7.0 \times 10^{-3}$ and $3.9 \times 10^{-3}$, respectively, and cadmium is in the form of alloy with more noble metal ($a_{Cd} = 0.6$). The standard *emf* of the cell at 25°C is 0.36 volts.

## 13.3 Electrochemical Cells and Electrodes

The following figure gives a typical example of electrochemical cells (galvanic cells):

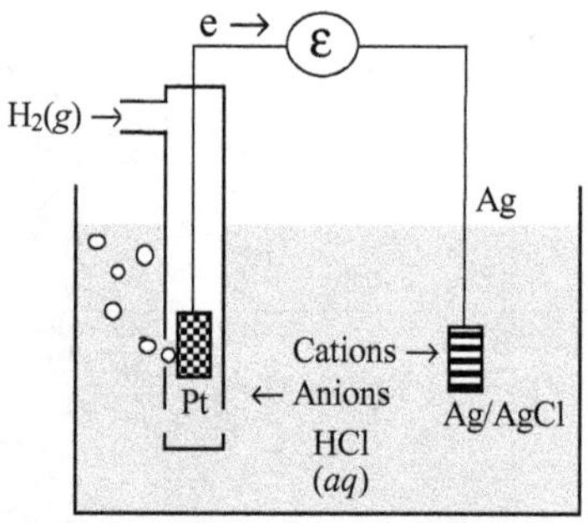


The cell consists of

- a hydrogen electrode,
- a silver-silver chloride electrode, and
- an aqueous solution of HCl.

The two electrodes are immersed in the HCl solution and connected to a potentiometer which measures the *emf* of the cell. The reactions which occur in the cell may be represented by a diagram given below:

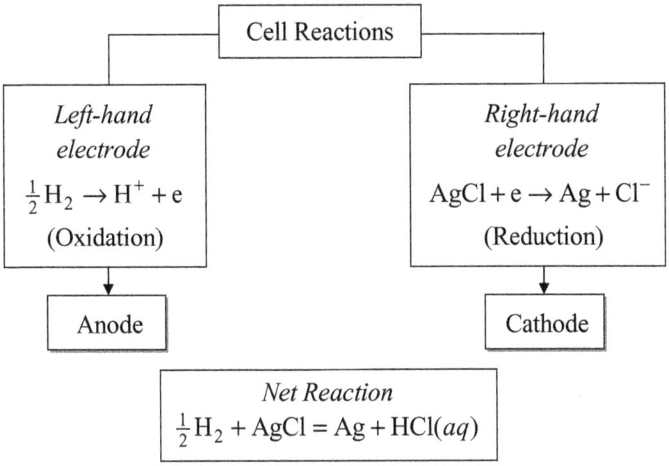


The convention is that,

- Since the left-hand electrode (anode) is in excess of electrons due to oxidation reaction, it is also terms a ***negative electrode***, and
- Since the right-hand electrode (cathode) is in deficit of electrons due to reduction reaction, it is also terms a ***positive electrode***.

The substance like the aqueous solution of HCl in the above cell is called the ***electrolyte***. In general, a compound $M_mX_X$ is an electrolyte if it dissociates, partially or completely, into its cations and anions when dissolved in water or other appropriate solvents. The compound may be an *acid* (in this case M is hydrogen, $H^{+1}$), a *base* (in this case X is the hydroxyl group, $OH^{-1}$), or a *salt* (in this case M is usually a metal and X a non-metal, for example, NaCl)

It is awkward to present a cell in such a manner as producing a figure like the one given above. The cell assembly is thus, by convention, represented by a simple diagram as shown below:

$$\text{Pt} \mid H_2(g, P = p \text{ atm.}) \mid \text{HCl}(aq, a = m) \mid \text{AgCl} \mid \text{Ag}$$

where *P*: pressure at *p* atm., *aq*: aqueous solution, *a*: activity of the value of *m*, the vertical lines: the phase boundaries.

The above expression may be further simplified as,

$$\text{Pt} \mid H_2 \mid \text{HCl}(aq) \mid \text{AgCl} \mid \text{Ag}$$

When $P_{H_2} = 1$ atm. and $a_{HCl} = 1$, the above cell drives electrons from the hydrogen electrode to the silver electrode with the *emf* of 0.222 volts at 25°C. This means that oxidation occurs at the left-hand electrode, and reduction at the right-hand electrode. Therefore the *emf* of the cell is *positive*: $\varepsilon = +0.222$ volts.

If the cell were described in the opposite way, *i.e.*,

$$\text{Ag} \mid \text{AgCl} \mid \text{HCl}(aq) \mid \text{H}_2 \mid \text{Pt}$$

the *emf* would be *negative* ($\varepsilon = -0.222$ V) because reduction would occur at the left-hand electrode.

In summary, the convention of representation of the electrochemical cell is

anode | solution | cathode

*An important feature of electrochemical cells:*
The reaction must be capable of being separated physically so that the direct chemical reaction is prevented from occurring.

*Example*

$\text{Pt} \mid \text{H}_2 \mid \text{HCl}(aq) \mid \text{AgCl} \mid \text{Ag}$

Anode Electrolyte Cathode

Reactants $H_2$ and AgCl are isolated at the separate electrodes, but maintain the electrical contact with each other through the aqueous HCl solution (electrolyte).
In some cases, two different solutions are used to prevent direct chemical reaction. A typical example is the ***Daniell cell*** illustrated below:

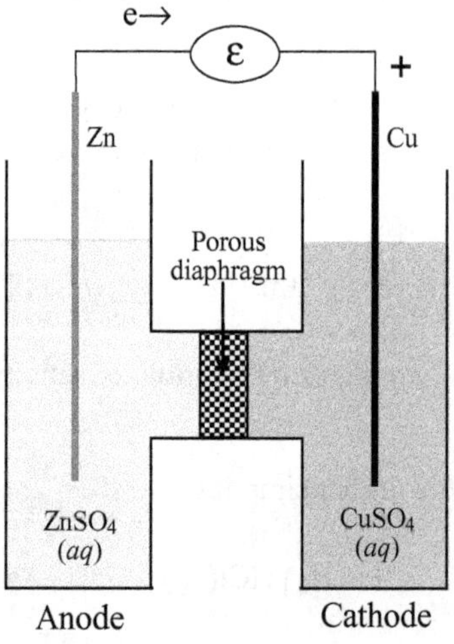

The cell comprises

- a zinc anode dipping into $ZnSO_4$ solution,
- a copper cathode dipping into $CuSO_4$ solution, and
- a porous diaphragm which prevents the two solutions from mixing, but allows electrical contact by the passage of $SO_4^{2-}$ ions through it.

The two solutions constitute the electrolytes of the cell, which are the media through which ionic current flows.

<u>Cell reactions</u>:

Anode : $Zn + SO_4^{2-}(aq) = ZnSO_4(aq) + 2e$, or $Zn = Zn^{2+}(aq) + 2e$

Cathode : $CuSO_4(aq) + 2e = Cu + SO_4^{2-}(aq)$, or $Cu^{2+}(aq) + 2e = Cu$

Net : $Zn + CuSO_4(aq) = ZnSO_4(aq) + Cu$, or $Zn + Cu^{2+}(aq) = Zn^{2+}(aq) + Cu$

Again, it is not convenient to draw the cell figure and write the cell reactions as shown above. The cell is thus conventionally represented by the following diagram:

$$Zn \mid ZnSO_4(aq) \vdots CuSO_4(aq) \mid Cu$$

or

$$Zn \mid Zn^{2+}(aq) \vdots Cu^{2+}(aq) \mid Cu$$

where the dashed vertical line represents the porous diaphragm separating the two aqueous solutions.

The *emf* of the cell depends on the activities of $ZnSO_4$ and $CuSO_4$ (recall the Nernst equation). Suppose all the reagents of the cell are in their standard states at 25°C, *i.e.*, Zn (pure), Cu(pure), $ZnSO_4$(*aq*, saturated with solid $ZnSO_4$) and $CuSO_4$(*aq*, saturated with solid $CuSO_4$).

$$Zn + CuSO_4(aq) = ZnSO_4(aq) + Cu, \quad \Delta G^o = -213{,}000 \text{ J at } 25°C$$

$$\boxed{\Delta G = -nF\varepsilon^o}$$

↓ $\Delta G^o = -213{,}000$ J at 25°C; $n = 2$

$$\boxed{\varepsilon^o = 1.104 \text{ volts}}$$

Thus,

$$Zn \mid Zn^{2+}(aq) \vdots Cu^{2+}(aq) \mid Cu, \quad \varepsilon^o = 1.104 \text{ volts}$$

If the reaction is written in the opposite direction, i.e.,

$$Cu \mid Cu^{2+}(aq) \vdots Zn^{2+}(aq) \mid Zn, \quad \varepsilon^o = -1.104 \text{ volts}$$

The *emf* of the cell at a non-standard state may be determined by the Nernst equation:

$$\varepsilon = \varepsilon^o - \frac{RT}{nF} ln\left(\frac{a_{Cu}a_{ZnSO_4}}{a_{Zn}a_{CuSO_4}}\right)$$

The standard *emf* can be evaluated for all possible cells by one of the following two ways:

- Direct measurements of $\varepsilon^o$ under standard conditions
- Measurements of $\varepsilon$ under non-standard conditions and calculations of $\varepsilon^o$ using the Nernst equation.

A complete list of the *emf's* of all possible cells, however, would be inordinately long and impractical. It would be much more convenient to develop some means of expressing the tendency of reduction (or *electron-accepting power*) of the individual electrodes. This can be done by using a ***reference electrode*** against which the electron-accepting power of other electrodes is compared. It is agreed that the hydrogen electrode comprising $H_2$ gas at 1 *atm.* pressure and an aqueous solution containing hydrogen ions at unit activity (*i.e.*, ***standard hydrogen electrode***) is chosen as the reference electrode.

When we construct the following cell with all the reagents at their standard states, the standard emf of the cell at 25°C is found to be + 0.337 volts:

$$H_2 \,|\, H^+ \,|\, Cu^{2+} \,|\, Cu \qquad \varepsilon^o = +0.337 \text{ volts}$$

Meaning of the *positive emf*

- Oxidation occurs at the left-hand electrode.

$$H_2(g) \rightarrow 2H^+ + 2e$$

- Reduction occurs at the right-hand electrode.

$$2Cu^{2+} + 2e \rightarrow Cu$$

- Thus, Cu is *stronger* than $H_2$ in terms of *electron-accepting* tendency.

When we construct the following cell with all the reagents at their standard states, the standard *emf* of the cell at 25°C is found to be - 0.763 volts.

$$H_2 \,|\, H^+ \,|\, Zn^{2+} \,|\, Zn, \qquad \varepsilon^o = -\,0.763 \text{ volts}$$

Meaning of the *negative emf*

- Reduction occurs at the left-hand electrode:

$$2H^+ + 2e \rightarrow H_2$$

- Oxidation occurs at the right-hand electrode:

$$Zn \rightarrow Zn^{2+} + 2e$$

- Thus, Zn is weaker than $H_2$ in terms of *electron-accepting* tendency

If the above results may be presented using a diagram as follows:

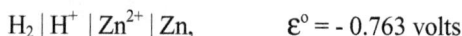

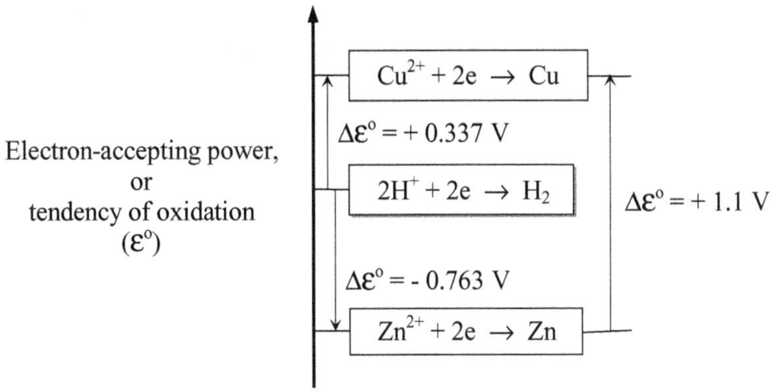


From the above figure one can easily see that Cu is stronger than Zn in terms of the electron-accepting power and hence, if we construct a cell with Zn and Cu as follows, the standard *emf* of the cell will be found as follows:

$$Zn \mid Zn^{2+}(aq) \vdots Cu^{2+}(aq) \mid Cu$$

The overall cell reaction:

$$Zn + Cu^{2+} = Zn^{2+} + Cu$$

This reaction can be considered to be the sum of the following two cell reactions:

$$\begin{array}{lll} & Zn + 2H^+ = Zn^{2+} + H_2 & \varepsilon^o = +0.763\ V \\ +) & Cu^{2+} + H_2 = Cu + 2H^+ & \varepsilon^o = +0.337\ V \\ \hline & Zn + Cu^{2+} = Zn^{2+} + Cu & \varepsilon^o = +1.1\ V \end{array}$$

Although it is impossible to measure the *emf* of a single half-cell or electrode like $H_2 \rightarrow 2H^+ + 2e$ and $Zn \rightarrow Zn^{2+} + 2e$, it is possible to assign an *emf* value for a cell, not an absolute value, but a value relative to the standard hydrogen electrode. This last statement is equivalent to saying that the ***standard hydrogen electrode*** (***SHE***) is assigned a standard *emf* of zero. The value of the *emf* of a single electrode, or half-cell, which is relative to the standard hydrogen electrode, is called the ***standard electrode potential*** or the ***standard half-cell potential***.

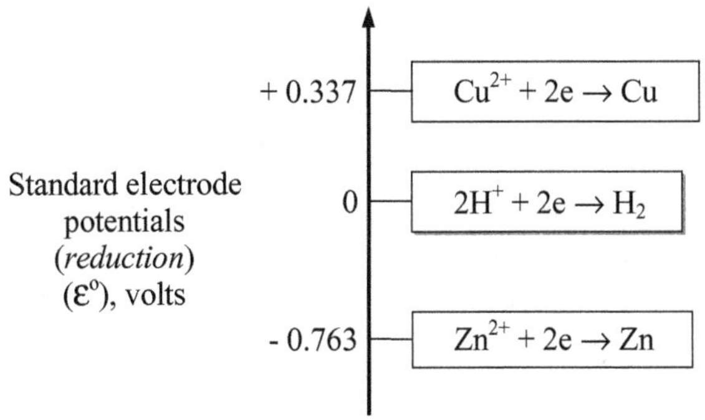

If we are concerned with the electron-donating power, *i.e.*, the tendency of reduction, the *emf* is the same in the numerical value as the one for the reduction, but opposite in sign.

The standard reduction potentials of a number of half-cells are given below, and a fuller list is given in Appendix III.

Some selected half-cell reactions and elective motive forces (*emf*)

| Half-cell reactions (reduction) | *emf* (volts) |
|---|---|
| $Li^+ + e = Li$ | -3.045 |
| $Ca^{2+} + 2e = Ca$ | -2.87 |
| $Na^+ + e = Na$ | -2.714 |
| $Al^{3+} + 3e = Al$ | -1.66 |
| $Zn^{2+} + 2e = Zn$ | -0.763 |
| $Fe^{2+} + 2e = Fe$ | -0.44 |
| $Ni^{2+} + 2e = Ni$ | -0.25 |
| $Pb^{2+} + 2e = Pb$ | -0.126 |
| $2H^+ + e = H_2$ | 0.0000 |
| $Cu^{2+} + 2e = Cu$ | 0.337 |
| $Ag^+ + e = Ag$ | 0.799 |
| $O_2 + 4H^+ + 4e = 2H_2O$ | 1.229 |
| $Cl_2 + 2e = 2Cl^-$ | 1.3595 |
| $PbO_2 + 4H^+ + 2e = Pb^{2+} + 2H_2O$ | 1.455 |
| $Co^{3+} + e = Co^{2+}$ | 1.82 |
| $F_2 + 2e = 2F^-$ | 2.89 |

*Example 13.3*

Consider two hypothetical half-cells

$$A^{a+} + ae^- \rightarrow A \qquad \varepsilon_A^o$$

$$B^{b+} + be^- \rightarrow B \qquad \varepsilon_B^o$$

When a cell is built from these two half-cells to give the following cell reaction

$$bA + aB^{b+} \rightarrow bA^{a+} + aB$$

prove the standard *emf* of the cell ($\varepsilon^o$) is given

$$\varepsilon^o = \varepsilon_B^o - \varepsilon_A^o$$

Recall that the *emf* is related to the free energy change by $\Delta G = -nF\varepsilon$.

$$(1)\ A^{a+} + ae^- \rightarrow A \qquad \Delta G_A^o = -aF\varepsilon_A^o$$

$$(2)\ B^{b+} + be^- \rightarrow B \qquad \Delta G_B^o = -bF\varepsilon_B^o$$

$\downarrow$ $a\times(2) - b\times(1)$

$$bA + aB^{b+} \rightarrow bA^{a+} + aB \qquad \Delta G^{o} = -(ab)F\varepsilon^{o}$$

Here $(ab)$ is because the number of electrons involved in this reaction is "*ab*".

When we add or subtract chemical equations, we also add and subtract changes in thermodynamic functions like $U$, $H$, $S$ and $G$.

$$\Delta G^{o} = a\Delta G_{B}^{o} - b\Delta G_{A}^{o}$$

$$\Delta G_{A}^{o} = -aF\varepsilon_{A}^{o}$$

$$\Delta G_{B}^{o} = -bF\varepsilon_{B}^{o}$$

$$\Delta G^{o} = -(ab)F\varepsilon^{o}$$

$$\varepsilon^{o} = \varepsilon_{B}^{o} - \varepsilon_{A}^{o}$$

The above example shows that half-cell potentials are directly combined *without* taking stoichiometric coefficients into consideration.
However, calculation of free energy change *must take* stoichiometric coefficients into account. *Why?*

There is a significant difference between $\Delta G$ and $\varepsilon$:

- $G$ is an *extensive property* of the system so that when the number of moles is changed, $G$ must be adjusted accordingly.
- On the other hand, $\varepsilon$ is an *intensive property* of the system so that it is independent of the size of the system.

$$\Delta G = -nF\varepsilon$$

$\Delta G$: Extensive property; $n$: Extensive quantity; $\varepsilon$: Intensive property

*Example 13.4*

Calculate $\varepsilon^{o}$ and $\Delta G^{o}$ for the cell at 25°C.

$$Li \mid Li^{+}(aq) \mid Cd^{2+}(aq) \mid Cd$$

From the table of the standard half-cell potentials

| | | |
|---|---|---|
| | $Li^{+} + e = Li$ | $\varepsilon_{Li}^{o} = -3.045$ V |
| | $Cd^{2+} + 2e = Cd$ | $\varepsilon_{Cd}^{o} = -0.403$ V |
| Cell reaction | $2Li + Cd^{2+} = 2Li^{+} + Cd$ | $\varepsilon^{o}$ |

1) Direct calculation from the standard half-cell *emf*'s

$$\varepsilon^{o} = \varepsilon^{o}_{Cd} - \varepsilon^{o}_{Li} = -0.403 - (-3.045) = 2.642\ \text{V}$$

2) Calculation from $\Delta G^{o}$

$$\Delta G^{o} = nF\varepsilon^{o}$$

$$\Delta G^{o}_{Li} = -(1)(96{,}487)(-3.045) = 293{,}800\ \text{J}$$
$$\Delta G^{o}_{Cd} = -(2)(96{,}487)(-0.403) = 77{,}770\ \text{J}$$
$$\Delta G^{o} = \Delta G^{o}_{Cd} - 2\Delta G^{o}_{Li} = 77{,}770 - (2)(293{,}800) = -509{,}830\ \text{J}$$

$n = 2$

$$\varepsilon^{o} = 2.642\ \text{V}$$

*Example 13.5*

Calculate the standard potential of the half-cell

$$Cr^{2+} + 2e \rightarrow Cr \qquad \varepsilon^{o}$$

Given: (1) $Cr^{3+} + 3e \rightarrow Cr$ $\quad \varepsilon^{o}_{Cr^{3+}/Cr} = -0.74\ \text{V}$

(2) $Cr^{3+} + e \rightarrow Cr^{2+}$ $\quad \varepsilon^{o}_{Cr^{3+}/Cr^{2+}} = -0.41\ \text{V}$

$$\Delta G^{o} = -nF\varepsilon^{o}$$

$$\Delta G^{o}_{i} = -nF\varepsilon^{o}_{i}$$
$$\Delta G^{o}_{1} = -(3)\times(96487)\times(-0.74) = 214200J$$
$$\Delta G^{o}_{2} = -(1)\times(96487)\times(-0.41) = 39560\ J$$
$$\Delta G^{o} = \Delta G^{o}_{1} - \Delta G^{o}_{2} = 214200 - 39560 = 174640J$$

$n = 2$

$$\varepsilon^{o} = -0.905\ \text{V}$$

Let's try to find $\varepsilon^{o}$ directly from the half-cell *emf*'s:

Direct calculation: $\varepsilon^{o} = \varepsilon^{o}_{Cr^{3+}/Cr} - \varepsilon^{o}_{Cr^{3+}/Cr^{2+}} = \underline{-0.33\ \text{V}}$ : *Incorrect*

*Important point*: When half-cell potentials are combined to produce a new half-cell, the potentials are not additive. When in doubt, find $\varepsilon$ by calculating it from $\Delta G$.

*Example 13.6*

Calculate the *emf* of a reversible propane-oxygen fuel cell which is operated under standard conditions at 25°C. The two half-cell reactions in the fuel cell are given in the following:

$$C_3H_8(g) + 6H_2O(l) \rightarrow 3CO_2(g) + 20H^+(aq) + 20e$$
$$5O_2(g) + 20H^+(aq) + 20e \rightarrow 10H_2O(l)$$

The overall cell reaction can be found from the above two half-cell reactions:

$$C_3H_8(g) + 5O_2(g) = 3CO_2(g) + 4H_2O(l)$$

If we can find $\Delta G_r^o$ of the above reaction, the standard *emf* of the above reaction, $\varepsilon^o$, can be calculated using the Nernst equation, $\Delta G_r^o = -nF\varepsilon^o$.

$$\Delta G_r^o = \left(3\Delta G_{f,CO_2}^o + 4\Delta G_{f,H_2O}^o\right) - \left(\Delta G_{f,C_3H_8}^o\right)$$

From data sources (Appendix II),

$\Delta G_{f,CO_2}^o = -394550 \text{ J mol}^{-1}$

$\Delta G_{f,H_2O}^o = -230770 \text{ J mol}^{-1}$

$\Delta G_{f,C_3H_8}^o = -23500 \text{ J mol}^{-1}$

$$\Delta G_r^o = -2083230 \text{ J}$$

$\Delta G_r^o = -nF\varepsilon^o$

$n = 20$

$F = 96487$ coulomb

$$\varepsilon^o = 1.08 \text{ volts}$$

*Example 13.7*

The reversible *emf* of the following cell at 25°C was measured to find 0.307 volts. The anode of the cell was the standard hydrogen electrode.

$$Pt|H_2|H^+||Cu^{2+}|Cu$$

Calculate the activity of $Cu^{2+}$ in the electrolyte. The standard *emf* (reduction) of $Cu^{2+}/Cu$ is 0.337 volts.

The cell reaction is,

$$H_2 + Cu^{2+} = 2H^+ + Cu$$

Applying the Nernst equation,

$$\varepsilon = \varepsilon^{\circ} - \frac{RT}{nF} ln\left(\frac{a^2_{H^+} a_{Cu}}{P_{H_2} a_{Cu^{2+}}}\right)$$

SHE $\rightarrow a_{H^+} = 1,\ P_{H_2} = 1$

$a_{Cu} = 1$

$n = 2$

$\varepsilon^{\circ} = 0.337$ volts

$$0.307 = 0.337 - \frac{8.314 \times 298}{2 \times 96487} ln\left(\frac{1}{a_{Cu^{2+}}}\right)$$

$$a_{Cu^{2+}} = 0.096$$

*Example 13.8*

Corrosion of a metal proceeds basically in an electrochemical manner. For instance, the likelihood of corrosion of iron (Fe) in various aqueous solutions may be found by comparing the half-cell potentials of the following half-cell reactions:

Iron (Fe) half-cell:

$Fe^{2+}(aq) + 2e \rightarrow Fe$ $\quad \varepsilon^{\circ} = -0.44$ volts

Half-cells of aqueous solutions (acidic):

$2H^+(aq) + 2e \rightarrow H_2(g)$ $\quad \varepsilon^{\circ} = 0.0$ volts

$4H^+(aq) + O_2(g) + 4e \rightarrow 2H_2O(l)$ $\quad \varepsilon^{\circ} = 1.229$ volts

Half-cell of aqueous solution (basic):

$2H_2O(l) + O_2(g) + 4e \rightarrow 4OH^-(aq)$ $\quad \varepsilon^{\circ} = 0.401$ volts

Discuss the importance of the acidity ($a_{H^+}$ or $a_{OH^-}$) of the aqueous solution.

Three different cells may be considered:

| | Anode (Oxidation) | Cathode (Reduction) |
|---|---|---|
| (1) | | $2H^+(aq) + 2e \rightarrow H_2(g)$ |
| (2) | $Fe \rightarrow Fe^{2+}(aq) + 2e$ | $4H^+(aq) + O_2(g) + 4e \rightarrow 2H_2O(l)$ |
| (3) | | $2H_2O(l) + O_2(g) + 4e \rightarrow 4OH^-(aq)$ |

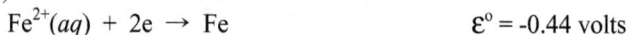

The cell reactions of the above three cases and their respective standard *emf*'s will be,

| | Cell reactions | Standard *emf* ($\varepsilon^o$), V |
|---|---|---|
| (1) | $Fe + 2H^+(aq) \rightarrow Fe^{2+}(aq) + H_2(g)$ | 0.440 |
| (2) | $2Fe + 4H^+(aq) + O_2(g) \rightarrow 2Fe^{2+}(aq) + 2H_2O(l)$ | 1.669 |
| (3) | $2Fe + 2H_2O(l) + O_2(g) \rightarrow 2Fe^{2+}(aq) + 4OH^-(aq)$ | 0.841 |

The standard *emf*'s, $\varepsilon^o$, of all three cases are all positive, meaning that iron will be corroded in all of the above aqueous environments, provided that all the species are in their respective standard states ($\Delta G^o = -nF\varepsilon^o < 0$).

What if some species in the reactions are not at the standard states? In these cases, we have to consult the Nernst equation:

(1) $$\varepsilon = 0.44 - \frac{RT}{2F} ln\left(\frac{a_{Fe^{2+}} P_{H_2}}{a_{Fe} a_{H^+}^2}\right)$$

(2) $$\varepsilon = 1.669 - \frac{RT}{4F} ln\left(\frac{a_{Fe^{2+}}^2 a_{H_2O}^2}{a_{Fe}^2 a_{H^+}^4 P_{O_2}}\right)$$

(3) $$\varepsilon = 0.841 - \frac{RT}{4F} ln\left(\frac{a_{Fe^{2+}}^2 a_{OH^-}^4}{a_{Fe}^2 a_{H_2O}^2 P_{O_2}}\right)$$

It can be seen that the value of *emf*, $\varepsilon$, varies with the terms in the bracket of the logarithm. As long as $\varepsilon$ is positive, the corrosion will proceed, as least, from the thermodynamic point of view. The most varying terms in the above equations are $a_{H^+}$ and $a_{OH^-}$ which represent the acidity of the aqueous solution.

*Exercises*

13.6 Write the electrode reactions and the cell reactions for the following galvanic cells, and calculate the standard *emf*'s of the cells at 25°C. Determine the positive electrodes:

(1) $Cu \mid CuCl_2(aq) \mid Cl_2(g)$, Pt

(2) Ag, AgCl | HCl(*aq*) ||Hbr(*aq*) | AgBr, Ag

13.7 Devise galvanic cells in which the cell reactions are the following:

(1) $Fe + CuSO_4(aq) = FeSO_4(aq) + Cu$

(2) $Pb(s) + PbO_2(s) + 2H_2SO_4(aq) = 2PbSO_4(s) + 2H_2O(l)$

Indicate electrode reactions in each case. Calculate the standard *emf* of each of the cells at 25°C.

13.8 Calculate the standard *emf* ($\varepsilon^{o}$) and the equilibrium constant ($K$) at 25°C for the reaction

$$Tl + Ag^{+} = Tl^{+} + Ag$$

## 13.4 Concentration Cells

When two half-cells are connected, an electrochemical cell is formed. The connection is made by bringing the solutions in the half-cells into contact so that ions can pass between them.

- If these two solutions are the same, there is no liquid junction, and we have *a cell without transference.*
- If these two solutions are different, transport of ions across the junction will cause irreversible changes in the two electrolytes, and we have *a cell with transference.*

The driving force of a cell may come from a chemical reaction or from a physical change. When the driving force of a cell is changes in composition (concentration) of species (or of gas pressures), the cell is called ***concentration cell***. The change in concentration can occur either in the electrolyte or in the electrodes.

The electrochemical cells can therefore be classified as follows:

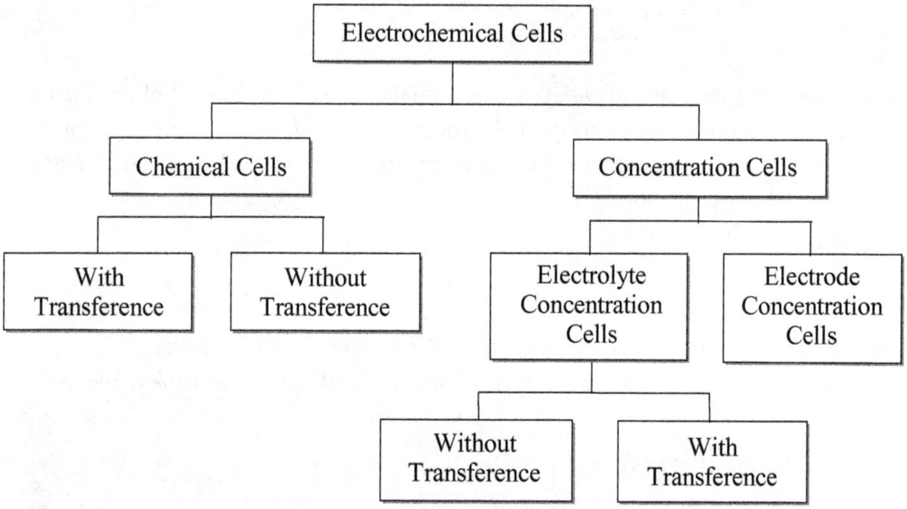


To form an ***electrode concentration cell*** the electrode material must have a variable concentration. Amalgam and gaseous electrodes fall into this classification. An example of electrode concentration cells is the one in which two amalgam electrodes of different concentrations dip into a solution containing the solute metal ions.

$$\text{Hg-Cd}(a_{Cd(1)})\,|\,CdSO_4(\text{solution})\,|\,\text{Hg-Cd}(a_{Cd(2)})$$

*Left-hand electrode*
(anode, oxidation)
$Cd(a_{Cd(1)}) \rightarrow Cd^{2+} + 2e$

*Right-hand electrode*
(cathode, reduction)
$Cd^{2+} + 2e \rightarrow Cd(a_{Cd(2)})$

*Net cell reaction*
$Cd(a_{Cd(1)}) \rightarrow Cd(a_{Cd(2)})$

Applying the Nernst equation,

$$\varepsilon = \varepsilon^{o} - \frac{RT}{nF} ln\left(\frac{a_{Cd(2)}}{a_{Cd(1)}}\right)$$

$n = 2$
$\varepsilon^{o} = 0$
(because at the standard state the cell reactions at both sides are the same)

$$\varepsilon = -\frac{RT}{2F} ln\left(\frac{a_{Cd(2)}}{a_{Cd(1)}}\right)$$

As seen above, no chemical change occurs, but what occurs is the process which is the transfer of cadmium from an amalgam to the other. Cadmium will tend to move spontaneously from the amalgam with high cadmium activity to that of low cadmium activity. From the last equation,

- If $a_{Cd(1)} > a_{Cd(2)}$, $\varepsilon > 0$ and hence the reaction proceeds as indicated in the cell reaction.

- If $a_{Cd(1)} < a_{Cd(2)}$, $\varepsilon < 0$ and hence the reaction proceeds in the opposite direction.

The above type of arrangement of a concentration cell is thermodynamically significant in that, if the activity of cadmium in one amalgam is known, the activity of cadmium in the other amalgam can be determined by measuring the *emf* of the cell.

An electrode concentration cell can be constructed using two electrodes which consist of a gas at different partial pressures.

The following is an example of the hydrogen electrode concentration cell:

$$Pt \mid H_2(P_1) \mid HCl(aq) \mid H_2(P_2) \mid Pt$$

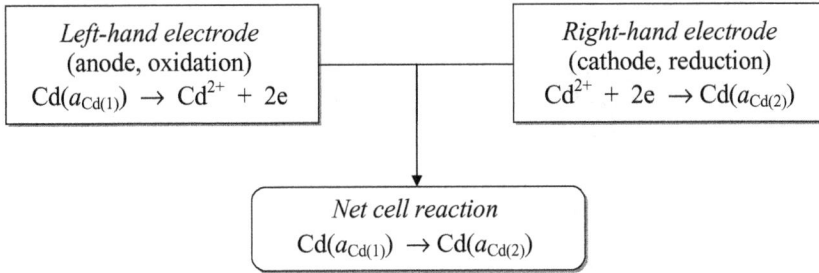

**Cell Process**

| *Left-hand electrode* (anode) | *Right-hand electrode* (cathode) |
|---|---|
| $H_2(P_1) \rightarrow 2H^+ + 2e$ | $2H^+ + 2e \rightarrow H_2(P_2)$ |

*Overall Change*

$$H_2(P_1) \rightarrow H_2(P_2)$$

$$\varepsilon = -\frac{RT}{2F} ln \frac{P_2}{P_1}$$

The *emf* of this cell is thus determined by the hydrogen pressures.

It has been found that $ZrO_2(s)$ stabilized with CaO or $YO_{1.5}$ is an ionic conductor of oxygen ions in certain ranges of the oxygen pressure and temperature. Using this material as the *solid electrolyte*, an oxygen concentration cell can be constructed:

$$\text{Pt, } O_2(g,\ P_{O_2(a)})\ |\ ZrO_2{+}CaO\ |\ O_2(g,\ P_{O_2(c)}),\text{ Pt}$$

**Cell Process**

| *Left-hand electrode* (anode) | *Right-hand electrode* (cathode) |
|---|---|
| $O^{2-} \rightarrow \frac{1}{2}O_2(P_{O_2(a)}) + 2e$ | $\frac{1}{2}O_2(P_{O_2(c)}) + 2e \rightarrow O^{2-}$ |

Overall cell reaction

$$\tfrac{1}{2}O_2(P_{O_2(c)}) \rightarrow \tfrac{1}{2}O_2(P_{O_2(a)})$$

$$\varepsilon = -\frac{RT}{4F} ln \frac{P_{O_2(a)}}{P_{O_2(c)}}$$

The oxygen concentration cell finds a number of applications in measurements of several thermodynamics properties at high temperatures. The following is the schematic diagram of the oxygen concentration cell:

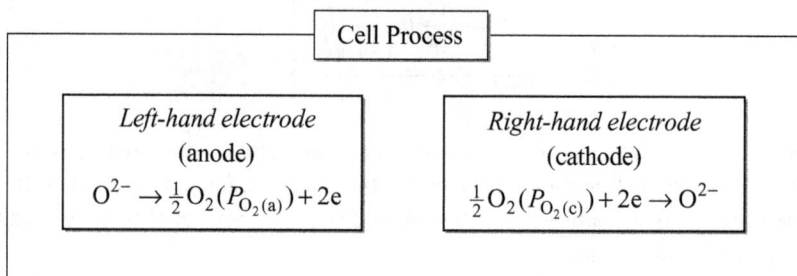

Schematic diagram of oxygen concentration cell
(*Oxygen sensor*)

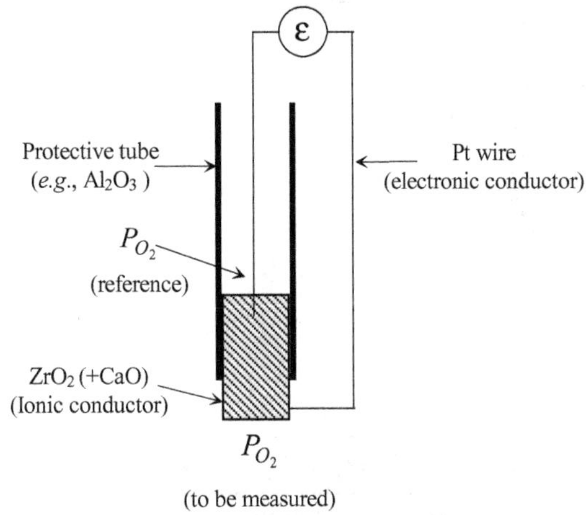


This type of oxygen concentration cell which is usually called the ***oxygen sensor*** is of considerable use in high temperature processing of metals and materials and in combustion processes. The oxygen sensor with a proper design enables *in situ* measurements of

- oxygen partial pressure in a gas mixture
- equilibrium oxygen potential in a metal-metal oxide mixture
- oxygen content in a liquid metal
- activity of a metal in an alloy
- activity of an oxide in an oxide solution.

The oxygen pressure of the reference electrode ($P_{O_2(ref)}$) can be fixed either by using a gas mixture containing oxygen of known pressure (*e.g.*, air), or by using a metal-metal oxide mixture.

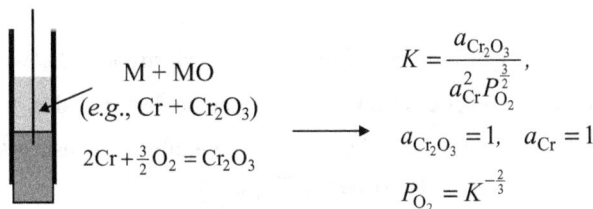

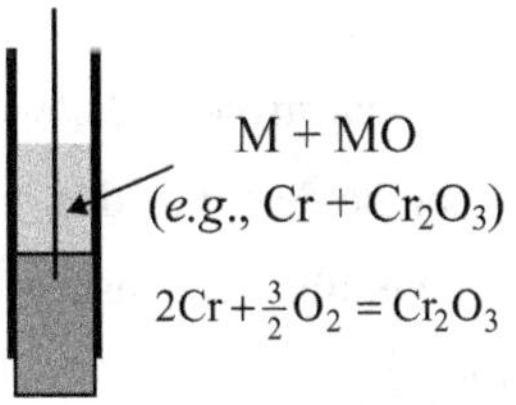


$$2Cr + \tfrac{3}{2}O_2 = Cr_2O_3 \longrightarrow K = \frac{a_{Cr_2O_3}}{a_{Cr}^2 P_{O_2}^{\frac{3}{2}}},$$

$$a_{Cr_2O_3} = 1, \quad a_{Cr} = 1$$

$$P_{O_2} = K^{-\frac{2}{3}}$$

Suppose that a cell is constructed by using a metal-metal oxide couple for both electrodes:

$$Cr,\ Cr_2O_3 \mid ZrO_2(+CaO) \mid Ni,\ NiO$$

It is now assumed that,

- Metals and oxides in both electrodes are pure and hence their activities are all unity
- The chemical equilibrium has been established in both half-cells.

Then the following thermodynamic analysis becomes possible:

$$Cr, Cr_2O_3 \mid ZrO_2(+CaO) \mid Ni, NiO$$

$$2Cr + \tfrac{3}{2}O_2 = Cr_2O_3$$

$$K_{Cr} = \frac{a_{Cr_2O_3}}{a_{Cr}^2 P_{O_2(Cr)}^{\frac{3}{2}}}$$

$$P_{O_2(Cr)} = K_{Cr}^{-\frac{2}{3}}$$

$$Ni + \tfrac{1}{2}O_2 = NiO$$

$$K_{Ni} = \frac{a_{NiO}}{a_{Ni} P_{O_2(Ni)}^{\frac{1}{2}}}$$

$$P_{O_2(Ni)} = K_{Ni}^{-2}$$

*Electrochemical Process*

$$Cr, Cr_2O_3 \mid ZrO_2(+CaO) \mid Ni, NiO$$

(anode) (electrolyte) (cathode)

$$O^{2-} \rightarrow \tfrac{1}{2}O_2 + 2e \quad \left(P_{O_2(Cr)}\right)$$

$$O^{2-} \rightarrow O^{2-}$$

$$\tfrac{1}{2}O_2 + 2e \rightarrow O^{2-} \quad \left(P_{O_2(Ni)}\right)$$

*Net Process*

$$\tfrac{1}{2}O_2(P_{O_2(Ni)}) \rightarrow \tfrac{1}{2}O_2(P_{O_2(Cr)})$$

*Emf of the cell*

$$\varepsilon = -\frac{RT}{4F} ln\left(\frac{P_{O_2(Cr)}}{P_{O_2(Ni)}}\right)$$

Provided that the oxygen pressure of the reference electrode is known (say, $P_{O_2(Cr)}$), the oxygen pressure of the other electrode ($P_{O_2(Ni)}$) can be determined by measuring the *emf* of the cell. This makes it possible to determine the equilibrium constant ($K_{Ni}$) and hence the Gibbs energy of formation of the oxide ($\Delta G^{o}_{NiO}$).

A cell in which the emf is derived only from the free energy change of dilution of the electrolyte is called the ***electrolyte concentration cell.***
Consider a simple cell;

$$Pt \mid H_2 \mid HCl(aq) \mid AgCl \mid Ag$$

The net cell reaction is

$$\tfrac{1}{2}H_2 + AgCl \rightarrow Ag + HCl(aq)$$

If two such cells are electrically connected in the opposed manner, the combination constitutes a cell that may be written

$$\text{Ag} \mid \text{AgCl} \mid \text{HCl}(aq, a_1) \mid \text{H}_2 \mid \text{HCl}(aq, a_2) \mid \text{AgCl} \mid \text{Ag}$$

The overall change in this cell is simply the difference between the changes in the two separate cells.

Cell Reactions

| Left-hand electrode | Right-hand electrode |
|---|---|
| $\text{HCl}(aq,a_1) + \text{Ag} \rightarrow \text{AgCl} + \frac{1}{2}\text{H}_2$ | $\text{AgCl} + \frac{1}{2}\text{H}_2 \rightarrow \text{HCl}(aq,a_2) + \text{Ag}$ |

Net Reactions

$$\text{HCl}(aq, a_1) \rightarrow \text{HCl}(aq, a_2)$$

Note that there is no direct transference of the electrolyte (HCl) from one side to the other. HCl is removed from the left-hand side by the left-hand electrode reaction and it is added to the right-hand side by the right-hand electrode reaction. This cell is an example of a ***electrolyte concentration cell without transference.***

If two electrolytes of different concentrations are directly in contact with each other, this gives rise to a ***junction potential***. An example of a concentration cell with a liquid junction is

$$\text{Pt} \mid \text{H}_2\,(1\,atm.) \mid \text{HCl}\,(aq, c_1) \vdots \text{HCl}\,(aq, c_2) \mid \text{H}_2\,(1\,atm.) \mid \text{Pt}$$

Cell Reaction

| Anode reaction | Cathode reaction |
|---|---|
| $\frac{1}{2}\text{H}_2(1\,atm.) \rightarrow \text{H}^+(c_1) + \text{e}$ | $\text{H}^+(c_2) + \text{e} \rightarrow \frac{1}{2}\text{H}_2(1\,atm.)$ |

Net reaction

$$\text{H}^+(c_2) \rightarrow \text{H}^+(c_1)$$

In the above cell, the two HCl solutions are in electrolytic contact, but are prevented from mechanical mixing. In general, this is done by means of a porous diaphragm or by stiffening one of the solutions at its point of contact with the other by agar-agar or gelatine.

Since the direct contact between solutions of different concentrations is not a balanced state, as required for reversible processes, the system is not directly susceptible to thermodynamic analysis. The liquid junction problem may be circumvented by connecting the different solutions by means of a bridge containing a saturated KCl solution.

*Example 13.9*

Concentration cells have been used extensively to determine thermodynamic properties of metallic solutions. The following cell is built to measure the thermodynamic properties of Mg in Mg-M alloys:

$$\text{Mg}(l,\ \text{pure})\ |\ \text{MgCl}_2,\ \text{CaCl}_2\ |\ \text{Mg-M}(l)$$

Choose the correct one in the following:

1) The alloying element M should be *more* noble than Mg.
2) The alloying element M should be *less* noble than Mg.

The element M has to be more noble than M. In other words, the chloride of M should be much less stable than either $MgCl_2$ or $CaCl_2$. Otherwise the element M would react with $MgCl_2$ or $CaCl_2$ in preference of Mg.
The cell process of the above is

$$\text{Mg}(l) \rightarrow \text{Mg(in alloy)}$$

$$\varepsilon = -\frac{RT}{2F} ln\left(\frac{a_{\text{Mg(alloy)}}}{a_{\text{Mg}}}\right)$$

$$a_{Mg} = 1$$

$$a_{\text{Mg(alloy)}} = exp\left(-\frac{2F\varepsilon}{RT}\right)$$

As can be seen here, the activity of Mg in the alloy can be found by measuring the *emf* of the cell. Once the activity of Mg has been found, thermodynamic properties of mixing can then be determined:

$$G_{\text{Mg}}^{M} = RT\, ln\, a_{\text{Mg(alloy)}}$$

$$a_{\text{Mg(alloy)}} = exp\left(-\frac{2F\varepsilon}{RT}\right)$$

$$G_{\text{Mg}}^{M} = -2F\varepsilon$$

$$S_{\text{Mg}}^{M} = -\left(\frac{\partial G_{\text{Mg}}^{M}}{\partial T}\right)_P \qquad H_{\text{Mg}}^{M} = G_{\text{Mg}}^{M} + TS_{\text{Mg}}^{M}$$

$$S_{\text{Mg}}^{M} = 2F\left(\frac{\partial \varepsilon}{\partial T}\right)_P \qquad H_{\text{Mg}}^{M} = -2F\varepsilon + 2FT\left(\frac{\partial \varepsilon}{\partial T}\right)_P$$

As we can see in the above, if we measure the *emf* in a range of temperatures, we can determine $S_{\text{Mg}}^{M}$ and $H_{\text{Mg}}^{M}$.

> *Example 13.10*
>
> The standard Gibbs energy of formation of an oxide can be determined by measuring the *emf* of a cell appropriately designed. Consider the following cell:
>
> $$\text{M,M}_\text{a}\text{O}_\text{b}|\text{ZrO}_2(+\text{CaO})|\text{O}_2(g,P_{\text{O}_2(\text{c})}) \qquad \varepsilon$$
>
> Prove that the free energy of formation of $\text{M}_a\text{O}_b$, $\Delta G^o_{f,\text{M}_\text{a}\text{O}_\text{b}}$, is given by the equation
>
> $$\Delta G^{\text{o}}_{f,\text{M}_\text{a}\text{O}_\text{b}} = -4\left(\frac{b}{2}\right)F\varepsilon + \left(\frac{b}{2}\right)RT\,ln\,P_{\text{O}_2(\text{c})}$$

$$\text{M},\text{M}_a\text{O}_b|\,\text{ZrO}_2(+\text{CaO})|\text{O}_2(g,P_{\text{O}_2(c)}) \qquad \varepsilon$$

*Anodic Reaction*

$$\text{M}_a\text{O}_b = a\text{M} + \tfrac{b}{2}\text{O}_2$$

or

$$b\text{O}^{2-} = \tfrac{b}{2}\text{O}_2 + (2b)\text{e}$$

*At the electrolyte*

$$\text{O}^{2-} \leftarrow \text{O}^{2-}$$

*Cathodic Reaction*

$$\tfrac{1}{2}\text{O}_2 + 2\text{e} = \text{O}^{2-}$$

or

$$\tfrac{b}{2}\text{O}_2 + (2b)\text{e} = b\text{O}^{2-}$$

*Net Cell Reaction*

$$\text{O}_2(P_{\text{O}_2(c)}) \rightarrow \text{O}_2(P_{\text{O}_2(a)})$$

$$\varepsilon = -\frac{RT}{4F}ln\left(\frac{P_{\text{O}_2(\text{a})}}{P_{\text{O}_2(\text{c})}}\right)$$

*From the anodic reaction,*

$$\Delta G^{\text{o}}_{f,\text{M}_\text{a}\text{O}_\text{b}} = -RT\,ln\,K$$

$$K = \frac{a_{\text{M}_\text{a}\text{O}_\text{b}}}{a^a_\text{M}P^{\frac{b}{2}}_{\text{O}_2(\text{a})}} = P^{-\frac{b}{2}}_{\text{O}_2(\text{a})}$$

$$\Delta G^{\text{o}}_{f,\text{M}_\text{a}\text{O}_\text{b}} = \tfrac{b}{2}RT\,ln\,P_{\text{O}_2(\text{a})}$$

$$\Delta G^{\text{o}}_{f,\text{M}_\text{a}\text{O}_\text{b}} = -4\left(\frac{b}{2}\right)F\varepsilon + \left(\frac{b}{2}\right)RT\,ln\,P_{\text{O}_2(\text{c})}$$

If a metal-metal oxide couple (*e.g.*, Me-MeO) is used for the reference electrode (the cathode in the present example), $P_{O_2(c)}$ in the above equation must be replaced by the equilibrium oxygen pressure of the couple:

$$\mathrm{Me}+\tfrac{1}{2}\mathrm{O}_2=\mathrm{MeO} \qquad K=\frac{a_{\mathrm{MeO}}}{a_{\mathrm{Me}}P_{\mathrm{O}_2(c)}^{\frac{1}{2}}} \quad \rightarrow \quad P_{\mathrm{O}_2(c)}=K^{-2}$$

**Example 13.11**

Galvanic cells with solid $ZrO_2$ (+CaO) electrolyte are used as the oxygen sensor for high temperature applications. The activity of a metal element in an alloy can be determined using the oxygen sensor. Consider the following cell:

Ni, NiO | $ZrO_2$ (+CaO) | <u>M</u> (in alloy) | MO

Derive an equation which relates the activity of M in the alloy to the cell *emf.*

First, we will derive an equation through a rigorous analysis of the system, and then find a simpler way to derive the same equation.

The above cell may be rewritten in the form given below:

$P_{O_2(a)}$ (in equilibrium with Ni/NiO) | $ZrO_2$ (+CaO) | $P_{O_2(c)}$ (in equilibrium with <u>M</u>/MO)

$$\varepsilon=-\frac{RT}{4F}ln\left(\frac{P_{\mathrm{O}_2(a)}}{P_{\mathrm{O}_2(c)}}\right)$$

The cell reaction in the anode,

$$\mathrm{Ni}+\tfrac{1}{2}\mathrm{O}_2=\mathrm{NiO}, \quad \Delta G^{o}_{\mathrm{Ni}}$$

$$K_{\mathrm{Ni}}=\frac{a_{\mathrm{NiO}}}{a_{\mathrm{Ni}}P_{\mathrm{O}_2(a)}^{\frac{1}{2}}}$$

$$P_{\mathrm{O}_2(a)}=K_{\mathrm{Ni}}^{-2}$$

The cell reaction in the cathode

$$\mathrm{M}+\tfrac{1}{2}\mathrm{O}_2=\mathrm{MO}, \quad \Delta G^{o}_{\mathrm{M}}$$

$$K_{\mathrm{M}}=\frac{a_{\mathrm{MO}}}{a_{\mathrm{M}}P_{\mathrm{O}_2(c)}^{\frac{1}{2}}}$$

$$P_{\mathrm{O}_2(c)}=K_{\mathrm{M}}^{-2}a_{\mathrm{M}}^{-2}$$

$$\varepsilon=-\frac{RT}{2F}ln\left(\frac{K_{\mathrm{M}}a_{\mathrm{M}}}{K_{\mathrm{Ni}}}\right)$$

$$\varepsilon=-\frac{RT}{2F}ln\left(\frac{K_{\mathrm{M}}}{K_{\mathrm{Ni}}}\right)-\frac{RT}{2F}ln\left(a_{\mathrm{M}}\right)$$

$$K_{\mathrm{M}}=exp\left(-\frac{\Delta G^{o}_{\mathrm{M}}}{RT}\right) \qquad K_{\mathrm{Ni}}=exp\left(-\frac{\Delta G^{o}_{\mathrm{Ni}}}{RT}\right)$$

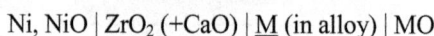

$$\varepsilon = -\frac{1}{2F}\left(\Delta G^{o}_{\text{Ni}} - \Delta G^{o}_{\text{M}}\right) - \frac{RT}{2F} ln\left(a_{\text{M}}\right)$$

$$\text{Ni} + \text{MO} = \text{NiO} + \text{M}, \quad \Delta G^{o}_{\text{Ni/M}}$$

$$\Delta G^{o}_{\text{Ni}} - \Delta G^{o}_{\text{M}} = \Delta G^{o}_{\text{Ni/M}}$$

$$\varepsilon = -\frac{1}{2F}\left(\Delta G^{o}_{\text{Ni/M}}\right) - \frac{RT}{2F} ln\left(a_{\text{M}}\right)$$

$$\Delta G^{o}_{\text{M/Ni}} = -2F\varepsilon^{o}$$

$$\varepsilon = \varepsilon^{o} - \frac{RT}{2F} ln\left(a_{\text{M}}\right)$$

Now we derive the same equation in a simpler way directly using the Nernst equation.

$$\text{Ni(s)} + \text{MO(s)} = \text{NiO(s)} + \underline{\text{M}} \quad \Delta G^{o}_{\text{Ni/M}}$$

$$\Delta G_{\text{Ni/M}} = \Delta G^{o}_{\text{Ni/M}} + RT\, ln\left(\frac{a_{\text{M}} a_{\text{NiO}}}{a_{\text{Ni}} a_{\text{MO}}}\right)$$

$$\Delta G_i = -nF\varepsilon_i$$

$$\varepsilon = \varepsilon^{o} - \frac{RT}{2F} ln\left(\frac{a_{\text{M}} a_{\text{NiO}}}{a_{\text{Ni}} a_{\text{MO}}}\right)$$

$$a_{NiO} = 1, \quad a_{MO} = 1, \quad a_{Ni} = 1$$

$$\varepsilon = \varepsilon^{o} - \frac{RT}{2F} ln\, a_{\text{M}}$$

*An important point*: For the analysis described above to be valid, all components other than M in the metal alloy must be considerably more noble than M so that these are practically inert under the oxygen potential prevailing in the electrode.

*Example 13.12*

The method described in the example 3 can also be used to determine the activity of an oxide in the oxide solution. Consider the following cell:

M(*l*) | MO(*l*) | $ZrO_2$ (+CaO) | Me(*l*) | MeO(*l*, in solution)

Derive an equation which relates the activity of MeO in the solution to the cell *emf*.

Cell Reaction

$$M + MeO = Me + MO, \quad \Delta G^o$$

$$\Delta G = \Delta G^o + RT\, ln\left(\frac{a_{Me}a_{MO}}{a_M a_{MeO}}\right)$$

$$\Delta G_i = nF\varepsilon_i$$

$$\varepsilon = \varepsilon^o - \frac{RT}{2F} ln\left(\frac{a_{Me}a_{MO}}{a_M a_{MeO}}\right)$$

$$a_M = 1, \quad a_{MO} = 1, \quad a_{Me} = 1$$

$$\varepsilon = \varepsilon^o + \frac{RT}{2F} ln\, a_{MeO}$$

*Example 13.13*

The oxygen sensor is capable of in situ measurement of oxygen potential in liquid metals. Consider the following cell:

M, MO | $ZrO_2$ (+CaO) | $\underline{O}$ (*dissolved in a metal*)

Derive an equation which relates the activity of oxygen dissolved in the metal ($\underline{O}$) to the cell *emf*.

The above cell may be rewritten in the form given below:

$P_{O_2(a)}$ (in equilibrium with M/MO) | $ZrO_2$ (+CaO) | $P_{O_2(c)}$ (in equilibrium with $\underline{O}$ in the metal)

$$\varepsilon = -\frac{RT}{4F} ln\left(\frac{P_{O_2(a)}}{P_{O_2(c)}}\right)$$

The cell reaction in the anode,

$$M + \tfrac{1}{2}O_2 = MO, \quad \Delta G_M^o$$

$$K_M = \frac{a_{MiO}}{a_M P_{O_2(a)}^{\frac{1}{2}}}$$

The cell reaction in the cathode

$$\tfrac{1}{2}O_2 = \underline{O}, \quad \Delta G_O^o$$

$$K_O = \frac{a_O}{P_{O_2(c)}^{\frac{1}{2}}}$$

$$P_{O_2(a)} = K_M^{-2} \qquad P_{O_2(c)} = K_O^{-2} a_O^2$$

$$\varepsilon = -\frac{RT}{2F} ln\left(\frac{K_O}{K_M a_O}\right)$$

$$\varepsilon = -\frac{RT}{2F} ln\left(\frac{K_O}{K_M}\right) + \frac{RT}{2F} ln(a_O)$$

$$K_O = exp\left(-\frac{\Delta G_O^o}{RT}\right) \qquad K_M = exp\left(-\frac{\Delta G_M^o}{RT}\right)$$

$$\varepsilon = -\frac{1}{2F}\left(\Delta G_M^o - \Delta G_O^o\right) + \frac{RT}{2F} ln(a_O)$$

$$M + \underline{O} = MO, \quad \Delta G_{M/O}^o$$

$$\Delta G_M^o - \Delta G_O^o = \Delta G_{M/O}^o$$

$$\varepsilon = -\frac{1}{2F}\left(\Delta G_{M/O}^o\right) + \frac{RT}{2F} ln(a_O)$$

$$\Delta G_{M/O}^o = -2F\varepsilon^o$$

$$\varepsilon = \varepsilon^o + \frac{RT}{2F} ln(a_O)$$

A simpler way of deriving the equation:

Cell Reaction

$$M + \underline{O} = MO$$

$$\varepsilon = \varepsilon^o - \frac{RT}{2F} ln\left(\frac{a_{MO}}{a_M a_O}\right)$$

$$a_M = 1, \qquad a_{MO} = 1$$

$$\varepsilon = \varepsilon^o + \frac{RT}{2F} ln\, a_O$$

*Exercises*

13.9 The *emf* of the following cell

$$Mg(l) \mid MgCl_2\text{-}CaCl_2(l) \mid Mg_2Si(s), Si(s)$$

was found to be

$$\varepsilon^{o} = 0.21767 - 8.607\times10^{-5}T, \quad (V)$$

Express the standard free energy of formation of $Mg_2Si(s)$ as a function of temperature.

13.10 The *emf* of the following cell was measured at a number of different temperatures:

$$Cu_2O(s), CuO(s) \mid ZrO_2(+CaO) \mid O_2(g, air)$$

The results were reported as follows:

| *T* (*K*) | 973 | 1023 | 1073 | 1123 | 1173 | 1223 | 1273 |
|---|---|---|---|---|---|---|---|
| ε (mV) | 170.0 | 143.9 | 117.7 | 91.6 | 65.5 | 39.3 | 13.2 |

Calculate the standard free energy of formation of $CuO(s)$ as a function of temperature. The Gibbs energy of formation of $Cu_2O$ is given:

$$\Delta G^{o}_{f,Cu_2O} = -168{,}400 + 71.25T, \qquad \text{J mol}^{-1}$$

13.11 The activity of chromium in liquid Ni-Cr alloys was measured using the following cell:

$$Pt \mid Cr(s), Cr_2O_3(s) \mid ZrO_2(+CaO) \mid Ni\text{-}Cr(l), Cr_2O_3 \mid Pt$$

The cell *emf* was measured to be 125 mV at 1,600°C for the Ni-Cr alloy of $N_{Cr}$ = 0.109. Calculate the activity of Cr in the alloy.

13.12 The oxygen content of molten iron can be measured using the following cell:

$$Pt \mid Cr(s), Cr_2O_3(s) \mid ZrO_2(+CaO) \mid \underline{O}\text{ (in liquid Fe)} \mid Mo$$

Derive an equation which relates the measured *emf* to the oxygen content (*wt*%) in the Fe melt at 1,600°C. The following data are given:

$$2Cr + \tfrac{3}{2}O_2 = Cr_2O_3 \qquad \Delta G^{o}_{Cr} = -1{,}110{,}100 + 247.32T, \qquad \text{J mol}^{-1}$$

$$\tfrac{1}{2}O_2(g) = \underline{O}(1\text{wt\%}) \qquad \Delta G^{o}_{O} = -117{,}150 - 2.887T, \qquad \text{J mol}^{-1}$$

1 wt% in Fe melt standard state.

## 13.5 Activities in Aqueous Solutions

Thermodynamic properties of an electrolytic solution are generally described by using the activities of different ionic species present in the solution. The problem of defining activities is however somewhat more complicated in electrolytic solutions than in solutions of non-electrolytes, because the requirement of overall electrical neutrality in the solution prevents any increase in charged ions by keeping oppositely charged ions unchanged. Consider the 1:1 electrolyte AB which dissociates into $A^+$ ions and $B^-$ ions in the aqueous solution.

$$AB = A^+ + B^-$$

The partial molar Gibbs energies of the two ions, $\bar{G}_{A^+}$ , $\bar{G}_{B^-}$ , are,

$$\bar{G}_{A^+} = G^o_{A^+} + RT\, ln\, a_{A^+}$$

and

$$\bar{G}_{B^-} = G^o_{B^-} + RT\, ln\, a_{B^-}$$

From the definition of the partial molar Gibbs energy,

$$\bar{G}_{A^+} = \left(\frac{\partial G}{\partial m_{A^+}}\right)_{m_{B^-},T,P}$$

$$\bar{G}_{B^-} = \left(\frac{\partial G}{\partial m_{B^-}}\right)_{m_{A^+},T,P}$$

where $m_{A^+}$ and $m_{B^-}$ are the molalities of $A^+$ and $B^-$ ions, respectively, in the solution. The ***molality*** is defined as the number of moles of solute in 1,000g of water ($H_2O$).

As the molality of an ion ( $m_{A^+}$ or $m_{B^-}$ ) cannot be altered independently, it is not possible to measure either $\bar{G}_{A^+}$ or $\bar{G}_{B^+}$ . In order to overcome this difficulty, we introduce ***mean thermodynamics properties*** of two ions.

Suppose that *n* moles of AB are dissolved in water:

$$nAB = nA^+ + nB^-$$

$$\boldsymbol{G} = n\bar{G}_{A^+} + n\bar{G}_{B^-}$$

↓

$$\bar{G}_{A^+} = G^o_{A^+} + RT\, ln\, a_{A^+}$$

$$\bar{G}_{B^-} = G^o_{B^-} + RT\, ln\, a_{B^-}$$

$$\boldsymbol{G} = nG^o_{A^+} + nG^o_{B^-} + nRT\, ln\,(a_{A^+} a_{B^-})$$

***G*** in the above equation is the total Gibbs energy of $2n$ moles of ions ($n$ moles of $A^+$ ion and $n$ moles of $B^-$ ions). If we define ***the mean partial molar Gibbs energy of ion*** ($\overline{G}_\pm$) as

$$\overline{G}_\pm = \frac{\boldsymbol{G}}{n_+ + n_-}$$

where $n_+$ is the number of moles of positive ions and $n_-$ is the number of moles of negative ions.

Combining the above two equations by knowing $n_+ + n_- = 2n$,

$$\overline{G}_\pm = \frac{\overline{G}_{A^+} + \overline{G}_{B^-}}{2} = \frac{G^o_{A^+} + G^o_{B^-}}{2} + RT\,ln\,(a_{A^+}a_{B^-})^{\frac{1}{2}}$$

*Mean partial molar Gibbs energy of ion* | *Mean standard molar Gibbs energy of ion* | ***Mean activity of ion*** ($a_\pm$)

$$a_\pm = (a_{A^+}a_{B^-})^{\frac{1}{2}}$$

$$a_{A^+} = \gamma_{A^+}m_{A^+}$$
$$a_{B^-} = \gamma_{B^-}m_{B^-}$$

$$a_\pm = (\gamma_{A^+}\gamma_{B^-})^{1/2}(m_{A^+}m_{B^-})^{1/2}$$

Activities in aqueous solution are generally based on the ***1 molality standard state.***

1 molality standard state

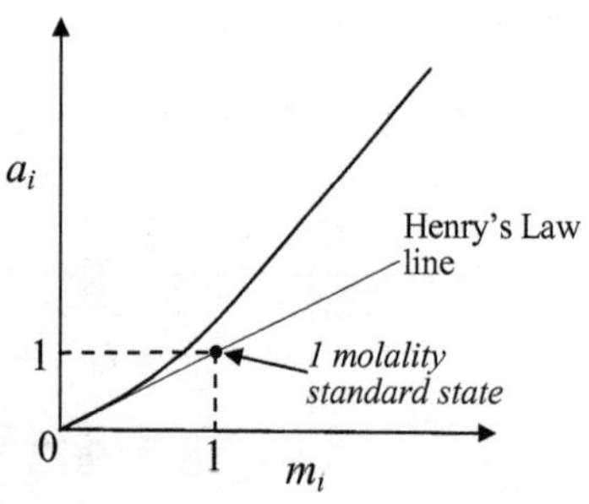


The ***mean ionic molality*** ($m_\pm$) and the ***mean ionic activity coefficient*** ($\gamma_\pm$) are respectively defined as

$$m_{\pm} = (m_{A^+} m_{B^-})^{1/2} \qquad \gamma_{\pm} = (\gamma_{A^+} \gamma_{B^-})^{1/2}$$

$$\downarrow$$

$$a_{\pm} = \gamma_{\pm} m_{\pm}$$

As the activities in aqueous electrolyte solutions are defined with respect to the 1 molality standard state (or infinitely dilute solution standard state), the activity of an ionic species becomes equal to its molality as the concentration approaches zero (where the Henry's Law applies).

$$\lim_{m_{A^+} \to 0} \left( \frac{a_{A^+}}{m_{A^+}} \right) = 1$$

$$\lim_{m_{B^-} \to 0} \left( \frac{a_{B^-}}{m_{B^-}} \right) = 1$$

Thus, at infinite dilution, $\gamma_{A^+} = 1$, $\gamma_{B^-} = 1$, and $\gamma_{\pm} = 1$, and hence

$$a_{\pm} = m_{\pm}$$

For a 1-1 electrolyte, $m_{A^+} = m_{B^-} = m^{o}_{AB}$, and hence $a_{\pm} = m^{o}_{AB}$.

We now generalize our discussion with more complex types of electrolyte. Consider the non-symmetrical electrolyte, $A_xB_y$, which dissociates into $A^{z+}$ positive ions and $B^{z-}$ negative ions in an aqueous solution:

$$A_xB_y = xA^{z+} + yB^{z-}$$

For the dissolution of one mole of $A_xB_y$

$$\boldsymbol{G} = x\bar{G}_{A^{z+}} + y\bar{G}_{B^{z-}}$$

$$\downarrow \quad \bar{G}_{A^{z+}} = G^{o}_{A^{z+}} + RT\,ln\,a_{A^{z+}}, \qquad \bar{G}_{B^{z-}} = G^{o}_{B^{z-}} + RT\,ln\,a_{B^{z-}}$$

$$\boldsymbol{G} = xG^{o}_{A^{z+}} + yG^{o}_{B^{z-}} + RT\,ln\,(a^{x}_{A^{z+}} a^{y}_{B^{z-}})$$

$\boldsymbol{G}$ in the above equation is the total free energy of $(x+y)$ moles of ions ($x$ moles of $A^{z+}$ ions and $y$ moles of $B^{z-}$ ions). From the definition of the mean partial molar Gibbs energy of ion ($\overline{G}_{\pm}$),

$$\overline{G}_{\pm} = \frac{\boldsymbol{G}}{(n_+ + n_-)}$$

$$n_+ = x, \quad n_- = y$$

$$\overline{G}_{\pm} = \frac{\boldsymbol{G}}{x+y} = \frac{x\overline{G}_{A^{Z+}} + y\overline{G}_{B^{Z-}}}{x+y} = \frac{xG^{o}_{A^{+}} + yG^{o}_{B^{-}}}{x+y} + RT\,ln(a^{x}_{A^{Z+}}a^{y}_{B^{Z-}})^{\frac{1}{x+y}}$$

Let's define the mean ionic activity as we did for the 1:1 electrolyte:

$$a_{\pm} = (a^{x}_{A^{Z+}}a^{y}_{B^{Z-}})^{\frac{1}{x+y}}$$

$$a_{A^{Z+}} = \gamma_{A^{Z+}}m_{A^{Z+}}$$
$$a_{B^{Z-}} = \gamma_{B^{Z-}}m_{B^{Z-}}$$

$$a_{\pm} = (\gamma^{x}_{A^{Z+}}\gamma^{y}_{B^{Z-}})^{\frac{1}{x+y}}(m^{x}_{A^{Z+}}m^{y}_{B^{Z-}})^{\frac{1}{x+y}}$$

Define $m_{\pm}$ and $\gamma_{\pm}$

$$m_{\pm} = (m^{x}_{A^{Z+}}m^{y}_{B^{Z-}})^{\frac{1}{x+y}}$$

$$\gamma_{\pm} = (\gamma^{x}_{A^{Z+}}\gamma^{y}_{B^{Z-}})^{\frac{1}{x+y}}$$

$$a_{\pm} = \gamma_{\pm}m_{\pm}$$

The total Gibbs energy of the ions in the electrically neutral solution is the sum of the free energies of positive ions and of negative ions:

$$\boldsymbol{G} = xG^{o}_{A^{Z+}} + yG^{o}_{B^{Z-}} + RT\,ln(a^{x}_{A^{Z+}}a^{y}_{B^{Z-}})$$

$$a_{A^{Z+}} = \gamma_{A^{Z+}}m_{A^{Z+}}$$
$$a_{B^{Z-}} = \gamma_{B^{Z-}}m_{B^{Z-}}$$

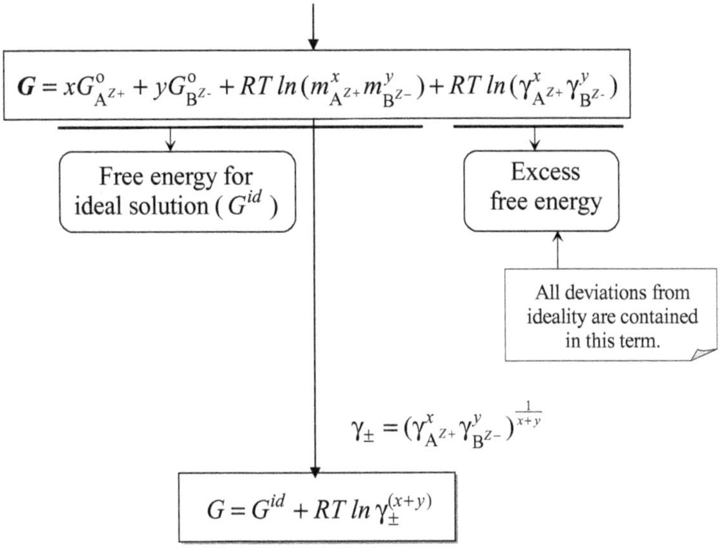

$G = xG^{o}_{A^{Z+}} + yG^{o}_{B^{Z-}} + RT\,ln(m^{x}_{A^{Z+}} m^{y}_{B^{Z-}}) + RT\,ln(\gamma^{x}_{A^{Z+}} \gamma^{y}_{B^{Z-}})$
Free energy for ideal solution ( $G^{id}$ )
Excess free energy
All deviations from ideality are contained in this term.
$\gamma_{\pm} = (\gamma^{x}_{A^{Z+}} \gamma^{y}_{B^{Z-}})^{\frac{1}{x+y}}$
$G = G^{id} + RT\,ln\,\gamma_{\pm}^{(x+y)}$


Observed deviations from ideal behavior are ascribed to electrical interactions between ions. Oppositely charged ions attract each other. As a result, in the immediate neighbourhood of a given ion, an oppositely charged ion is more likely to be found. Overall the solution is electrically neutral, but near any given ion there is an excess of oppositely charged ions. Consequently the chemical potential of an ion is lowered as a result of its electrostatic interaction with its ionic neighbors. This lowering of energy appears as the difference between the Gibbs energy $G$ and the ideal value $G^{id}$ of the solution; *i.e.*, $RT\,ln\,\gamma_{\pm}^{(x+y)}$.

On the assumption that deviations of a dilute solution from ideality are caused entirely by the electrostatic interactions, the activity coefficient can be calculated from the ***Debye-Hückel limiting law***:

$$log\,\gamma_{\pm} = -0.509 \left| z_{+} z_{-} \right| I^{1/2}$$

where

$z_{+}, z_{-}$ : charge numbers of positive and negative ions, respectively

$I$ : ionic strength of the solution which is given by

$$I = \frac{1}{2} \sum_{i} z_i^2 m_i$$

*Example 13.14*

Consider the solution of $La_2(SO_4)_3$ at a molality $m$. Express the mean ionic activity $a_{\pm}$ in terms of $\gamma_{\pm}$ and $m$.

$$La_2(SO_4)_3 = 2La^{3+} + 3SO_4^{2-}$$

$$a_\pm = \gamma_\pm m_\pm$$

$$m_\pm = (m_{La^{3+}}^2 m_{SO_4^{2-}}^3)^{\frac{1}{2+3}}$$

$$m_{La^{3+}} = 2m$$

$$m_{SO_4^{2-}} = 3m$$

$$m_\pm = [(2m)^2(3m)^3]^{1/5} = 108^{1/5} m$$

$$a_\pm = \gamma_\pm (108)^{1/5} m$$

*Example 13.15*

Values of the mean ionic activity coefficients for several electrolytes in water at 25°C are given in the following table:

Mean activity coefficients $\gamma_\pm$ for strong electrolytes at 25°C

| Electro-lytes | Molality($m$) | | | | | | | | | |
|---|---|---|---|---|---|---|---|---|---|---|
| | 0.001 | 0.002 | 0.005 | 0.01 | 0.05 | 0.1 | 0.5 | 1.0 | 2.0 | 4.0 |
| HCl | 0.996 | 0.952 | 0.928 | 0.904 | 0.830 | 0.796 | 0.758 | 0.809 | 1.01 | 1.76 |
| $HNO_3$ | 0.965 | 0.951 | 0.927 | 0.902 | 0.823 | 0.785 | 0.715 | 0.720 | 0.783 | 0.982 |
| $H_2SO_4$ | 0.830 | 0.757 | 0.639 | 0.544 | 0.340 | 0.265 | 0.154 | 0.130 | 0.124 | 0.171 |
| $CaCl_2$ | 0.89 | 0.85 | 0.785 | 0.725 | 0.57 | 0.515 | 0.52 | 0.71 | | |
| $CuCl_2$ | 0.89 | 0.85 | 0.78 | 0.72 | 0.58 | 0.52 | 0.42 | 0.43 | 0.51 | |
| $CuSO_4$ | 0.74 | | 0.53 | 0.41 | 0.21 | 0.16 | 0.068 | 0.047 | | |
| $FeCl_2$ | 0.89 | 0.86 | 0.80 | 0.75 | 0.62 | 0.58 | 0.59 | 0.67 | | |
| KCl | 0.965 | 0.952 | 0.927 | 0.901 | 0.815 | 0.769 | 0.651 | 0.606 | 0.576 | 0.579 |
| $K_2SO_4$ | 0.89 | | 0.78 | 0.71 | 0.52 | 0.43 | | | | |
| $MgCl_2$ | | | | | | 0.56 | 0.52 | 0.62 | 1.05 | |
| NaCl | 0.966 | 0.953 | 0.929 | 0.904 | 0.823 | 0.780 | 0.68 | 0.66 | 0.67 | 0.78 |
| $PbCl_2$ | 0.86 | 0.80 | 0.70 | 0.61 | | | | | | |
| $ZnCl_2$ | 0.88 | 0.84 | 0.77 | 0.71 | 0.56 | 0.50 | 0.38 | 0.33 | | |
| NaOH | | | | | 0.82 | | 0.69 | 0.68 | | |
| KOH | | | 0.92 | 0.90 | 0.82 | 0.8 | 0.73 | 0.76 | | |
| NaBr | 0.966 | | 0.934 | 0.914 | 0.844 | 0.800 | 0.695 | 0.686 | | |
| $MgSO_4$ | | | | 0.40 | 0.22 | 0.18 | 0.088 | 0.064 | | |
| $ZnSO_4$ | 0.70 | 0.61 | 0.48 | 0.39 | | 0.15 | 0.065 | 0.045 | 0.036 | |

Calculate the mean activities of HCl and $CuCl_2$ in the aqueous solution at 25°C. The concentration of both electrolytes is 0.01 molality.

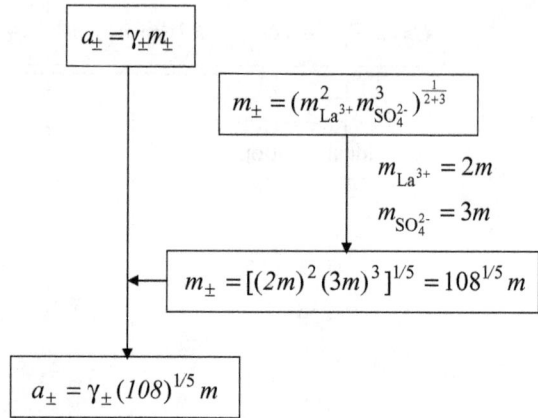

$$HCl \rightarrow H^+ + Cl^-$$

$$a_\pm = \gamma_\pm m_\pm$$

$$m_\pm = (m^x_{A^{z+}} m^y_{B^{z-}})^{\frac{1}{x+y}}$$

$$m_{H^+} = 0.01,\ m_{Cl^-} = 0.01$$

$$x = 1,\ y = 1$$

$$m_\pm = 0.01$$

$\gamma_\pm = 0.904$ from the table

$$a_\pm = 9.04 \times 10^{-3}$$

$$CuCl_2 \rightarrow Cu^{2+} + 2Cl^-$$

$$a_\pm = \gamma_\pm m_\pm$$

$$m_\pm = (m^x_{A^{z+}} m^y_{B^{z-}})^{\frac{1}{x+y}}$$

$$m_{Cu^{2+}} = 0.01,\ m_{Cl^-} = 0.02$$

$$x = 1,\ y = 2$$

$$m_\pm = 0.016$$

$\gamma_\pm = 0.72$ from the table

$$a_\pm = 0.0114$$

*Example 13.16*

Estimate the mean activity coefficient of the aqueous solution of KCl at 0.005 molal at 25°C.

Applying the Debye-Hückel limiting law,

$$\log \gamma_\pm = -0.509 |z_+ z_-| I^{\frac{1}{2}}$$

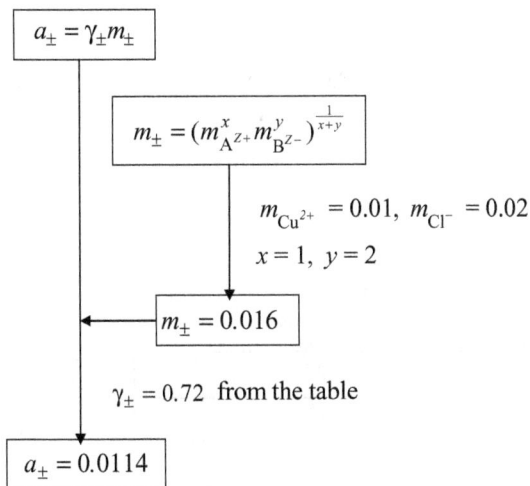

$$I = \frac{1}{2}\sum_i z_i^2 m_i$$

$z_+ = 1, \quad z_- = -1$

$m_{K^+} = 0.005,$

$m_{Cl^-} = 0.005$

$$I = 0.005$$

$$\gamma_\pm = 0.920$$

*Example 13.17*

The solubility of AgCl in water is $1.274\times10^{-5}$ molal at 25°C. Calculate the standard Gibbs energy change of reaction

$$\text{AgCl}(s) \rightarrow \text{Ag}^+(aq) + \text{Cl}^-(aq)$$

$$\Delta G^o = -RT\, ln\left(\frac{a_{\text{Ag}^+} a_{\text{Cl}^-}}{a_{\text{AgCl}}}\right)$$

$$a_\pm = (a_{\text{Ag}^+} a_{\text{Cl}^-})^{1/2}$$

$$\Delta G^o = -RT\, ln\left(\frac{a_\pm^2}{a_{\text{AgCl}}}\right)$$

$a_{\text{Ag}^+} = \gamma_{\text{Ag}^+} m_{\text{Ag}^+}$

$a_{\text{Cl}^-} = \gamma_{\text{Cl}^-} m_{\text{Cl}^-}$

$$a_\pm = (\gamma_{\text{Ag}^+}\gamma_{\text{Cl}^-})^{1/2}(m_{\text{Ag}^+} m_{\text{Cl}^-})^{1/2}$$

$\gamma_\pm = (\gamma_{\text{Ag}^+}\gamma_{\text{Cl}^-})^{1/2}$

$m_\pm = (m_{\text{Ag}^+} m_{\text{Cl}^-})^{1/2}$

$$a_\pm = \gamma_\pm m_\pm$$

$m_{\text{Ag}^+} = 1.274\times10^{-5}$

$m_{\text{Cl}^-} = 1.274\times10^{-5}$

$m_\pm = 1.274\times10^{-5}$

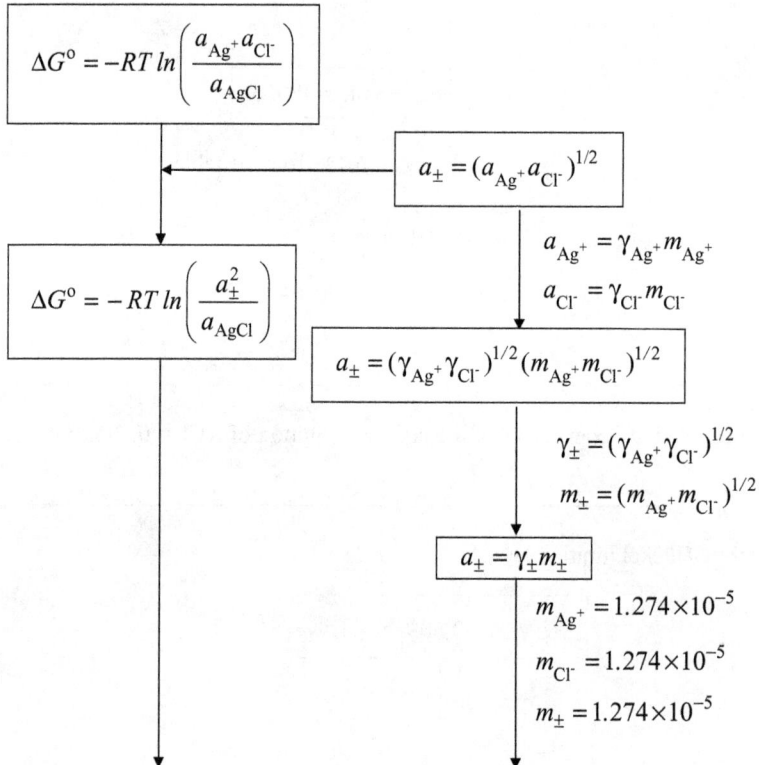

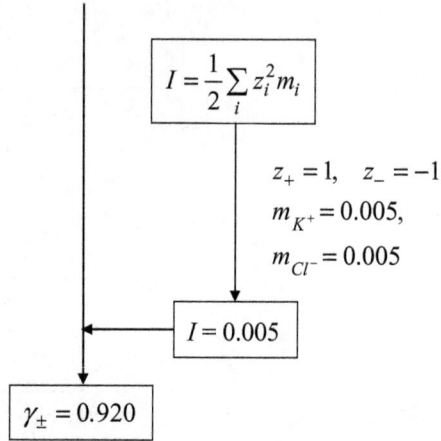

$$a_{\pm} = (1.274 \times 10^{-5})\gamma_{\pm}$$

$$T = 298\text{K}$$
$$R = 8.314 \text{ J mol}^{-1} \text{ K}^{-1}$$
$$a_{\text{AgCl}} = 1 \text{ (saturation)}$$

$$\log \gamma_{\pm} = -0.509 |z_+ z_-| I^{1/2}$$

$$I = \frac{1}{2}\sum_i z_i^2 m_i$$

$$I = 1.274 \times 10^{-5}$$

$$\gamma_{\pm} = 0.996$$

$$a_{\pm} = 1.269 \times 10^{-5}$$

$$\Delta G^{\text{o}} = 55{,}870 \text{ J mol}^{-1}$$

## 13.6 Solubility Products

Consider the dissolution of a weakly soluble salt or electrolyte $A_xB_y$. Saturation of the aqueous solution occurs when $A_xB_y$ has dissolved to the extent that the activity of $A_xB_y$ in the solution, with the respect to the solid $A_xB_y$ as the standard state, is *unity*.

$$\text{A}_x\text{B}_y = x\text{A}^{z+}(aq) + y\text{B}^{z-}(aq)$$

$$K_{sp} = a^x_{\text{A}^{z+}} a^y_{\text{B}^{z-}}$$

The equilibrium constant $K_{sp}$ is called the ***solubility constant*** or ***solubility product*** of the salt.

$$K_{sp} = (\gamma^x_{\text{A}^{z+}} \gamma^y_{\text{B}^{z-}})(m^x_{\text{A}^{z+}} m^y_{\text{B}^{z-}})$$

$$K_{sp} = (\gamma_{\pm} m_{\pm})^{(x+y)}$$

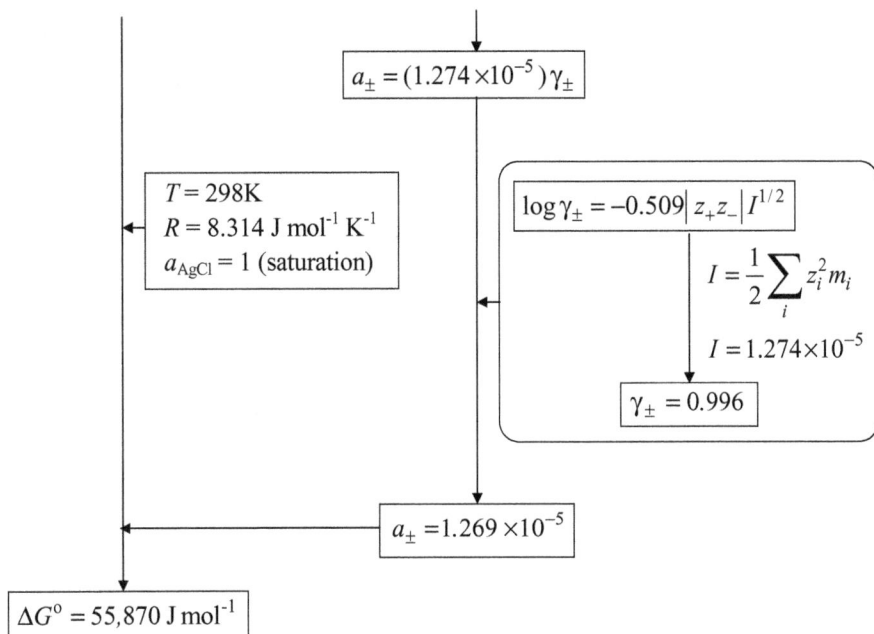

*Example 13.18*

Evaluate the solubility constant $K_{sp}$ of NaCl at 25°C using cell potential data and the free energy of formation of NaCl:

$$2Cl^- = Cl_2 + 2e \qquad \varepsilon^o = -1.360V$$

$$Na = Na^+ + e \qquad \varepsilon^o = 2.714V$$

$$Na(s) + \tfrac{1}{2}Cl_2(g) = NaCl(s) \qquad \Delta G^o = -383{,}880\ J\ mol^{-1} \text{ at } 298K$$

1) $2Cl^- = Cl_2 + 2e \qquad \Delta G^o_{Cl} = -nF\varepsilon^o = -(1)(96487)(-1.360) = 131{,}220$ J

2) $Na = Na^+ + e \qquad \Delta G^o_{Na} = -nF\varepsilon^o = -(1)(96487)(2.714) = -261{,}870$ J

3) $Na(s) + \tfrac{1}{2}Cl_2(g) = NaCl(s) \qquad \Delta G^o_{NaCl} = -383{,}880$ J

Combination of the above reactions yields

$$NaCl(s) = Na^+(aq) + Cl^-(aq) \qquad \Delta G^o = -9{,}210\ J$$

$$K_{sp} = (\gamma_\pm m_\pm)^2 = exp\left(\frac{-\Delta G^o}{RT}\right) = 41.2$$

For NaCl,

$$m_\pm = m_{Na^+} = m_{Cl^-} = m_{NaCl}.$$

$$(\gamma_\pm m_\pm)^2 = (\gamma_\pm m_{NaCl})^2 = 41.2$$

or

$$\gamma_\pm m_{NaCl} = 6.42$$

*Example 13.19*

Calculate the concentrations of $H^+$ and $OH^-$ in water at 25°C. Use cell potential data.

From the table of the standard half-cell potentials,

$$H_2 + 2OH^- = 2H_2O + 2e \qquad \varepsilon^o = 0.828V$$

$$H_2 = 2H^+ + 2e \qquad \varepsilon^o = 0\ V$$

Combination yields

$$H_2O = H^+ + OH^- \qquad \varepsilon^o = -0.828V$$

$$K_{sp} = \frac{a_{H^+}a_{OH^-}}{a_{H_2O}} = exp\left(-\frac{\Delta G^o}{RT}\right) = exp\left(-\frac{-nF\varepsilon^o}{RT}\right) = 9.9\text{x}10^{-15}$$

Since

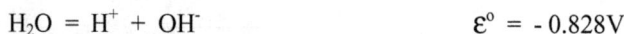

$$a_{H^+} = \gamma_{H^+} m_{H^+}, \quad a_{OH^-} = \gamma_{OH^-} m_{OH^-}, \quad a_{H_2O} = 1, \quad \gamma_{\pm} = (\gamma_{H^+}\gamma_{OH^-})^{\frac{1}{2}}$$

$$\gamma_{\pm}^2 m_{H^+} m_{OH^-} = 9.9\text{x}10^{-15}$$

↓ $\gamma_{\pm} \cong 1$ (very dilute)
$m_{H^+} = m_{OH^-}$ (from the stoichiometry)

$$m_{H^+} = 10^{-7}$$

↓

pH = 7 for water

> *Definition*: **pH = - *log*[H$^+$]**
>
> where [H$^+$] : molarity of H$^+$
>
> ***Molarity***: Number of moles of solute per liter of water.
>
> For dilute solutions, molarity ≅ molality

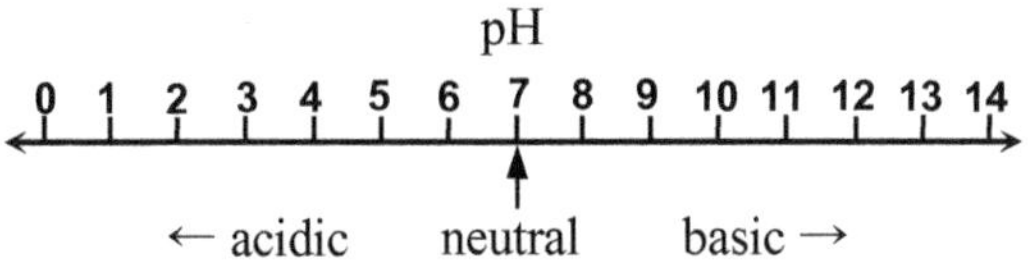


*Example 13.20*

> The thermodynamic behavior of weak electrolytes is based on the conditions for equilibrium between dissociated ions and undissociated portion in the solution. Prove that, for most weak electrolytes, the degree of dissociation increases as the electrolyte concentration decreases.

Consider the weak electrolyte $A_xB_y$

$$A_xB_y(aq) = xA^{z+}(aq) + yB^{z-}(aq)$$

$$K_C = \frac{a_{A^{z+}}^x a_{B^{z-}}^y}{a_{A_xB_y}}$$

↓ $a_i = \gamma_i m_i$

$$K_C = \left(\frac{\gamma_{A^{z+}}^x \gamma_{B^{z-}}^y}{\gamma_{A_xB_y}}\right)\left(\frac{m_{A^{z+}}^x m_{B^{z-}}^y}{m_{A_xB_y}}\right)$$

↓

In dilute solutions, the activity coefficients, $\gamma_i$, approach 1. Thus

$$\left(\frac{\gamma^x_{A^{z+}}\gamma^y_{B^{z-}}}{\gamma_{A_xB_y}}\right) \cong 1$$

↓

$$K_C = \frac{m^x_{A^{z+}} m^y_{B^{z-}}}{m_{A_xB_y}}$$

↓

Let
$m$ : molality of the electrolyte dissolved
$\xi$ : fraction of dissociation of the dissolved electrolyte

$$m_{A^{z+}} = x\xi m,$$
$$m_{B^{z-}} = y\xi m,$$
$$m_{A_xB_y} = (1-\xi)m$$

↓

$$K_C = \frac{(x^x y^y)\xi^{(x+y)} m^{(x+y-1)}}{(1-\xi)}$$

↓ For dilute solutions, $\xi \ll 1$.

$$\xi = \left(\frac{K_C}{x^x y^y}\right)^{\frac{1}{x+y}} m^{(\frac{1}{x+y}-1)}$$

As $m$ increases, $m^{(\frac{1}{x+y}-1)}$ decreases, and hence $\xi$ decreases.

*Example 13.21*

A strong electrolyte is defined as a compound that, when added to a solvent, dissociates completely into its ionic components. If a strong acid such as HCl is dissolved in water, it dissociates completely into $H^+$ ions and $Cl^-$ ions. Calculate the pH of a 0.1 molal solution of HCl in water.

$$HCl \rightarrow H^+ + Cl^-$$

↓ Every HCl molecule dissociates forming one $H^+$ ion.

$$m_{HCl} = m_{H^+} = 0.1 \cong [H^+]$$

↓

$$pH = -\log[H^+] = 1$$

*Example 13.22*

A solution is prepared by mixing two liters of 0.02 molal HCl and one liter of 0.05 molal NaOH. Calculate the pH of the solution. HCl and NaOH are both strong electrolytes.

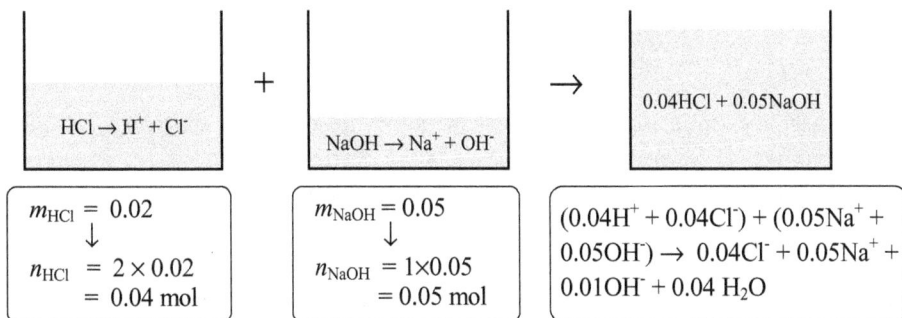

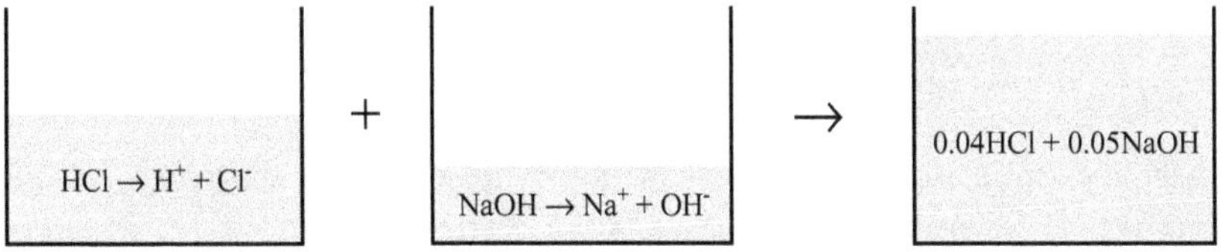


$m_{HCl} = 0.02$ ↓ $n_{HCl} = 2 \times 0.02 = 0.04$ mol

$m_{NaOH} = 0.05$ ↓ $n_{NaOH} = 1\times 0.05 = 0.05$ mol

$(0.04H^+ + 0.04Cl^-) + (0.05Na^+ + 0.05OH^-) \rightarrow 0.04Cl^- + 0.05Na^+ + 0.01OH^- + 0.04\ H_2O$

Upon mixing, 0.04 mole of $H^+$ reacts with 0.04 mole of $OH^-$ to form 0.04 mole of $H_2O$, leaving 0.01 mole of $OH^-$ unreacted. The total volume of the solution becomes 3 liters.

$$m_{OH^-} \cong [OH^-] = \frac{0.01}{3} = 0.0033 \text{ mol l}^{-1}$$

↓ $[H^+][OH^-] = 10^{-14}$

$$[H^+] = 3.03\ x\ 10^{-12} \text{ mol l}^{-1}$$

↓

$$pH = 11.5$$

*Exercises*

13.13 The solubility product of $Cu_2S$ is $3 \times 10^{-48}$. Calculate the solubility of this salt.

13.14 5 grams of $Pb(NO_3)_2$ is added to 1 *liter* of 0.01molal NaCl solution. Would $PbCl_2$ precipitate? The solubility product for $PbCl_2$ is $2 \times 10^{-5}$.

13.15 A solution contains 0.10 molal each of $Ba^{2+}$ and $Sr^{2+}$. The solubility products of $BaCO_3$ and $SrCO_3$ are $2.0\times10^{-9}$ and $5.2\times10^{-10}$, respectively. Describe what happens as each of the solids is added to the solution.

13.16 Magnesium hydroxide is slightly soluble in water. If the pH of a saturated solution of $Mg(OH)_2$ is 10.38, find the solubility product of $Mg(OH)_2$.

13.17 Find the concentration of $Ca^{2+}$ and $CO_3^{2-}$ in air-saturated water (the partial pressure of $CO_2$ is $3\times10^{-4}$ *atm.*) assuming $K_{sp} = 3.84\times10^{-9}$ for $CaCO_3$ and the equilibrium constant $K = 1.4 \times 10^{-6}$ for the reaction

$$CO_2(g) + H_2O(l) + CaCO_3(s) = Ca^{2+} + 2HCO_3^-(aq)$$

## 13.7 Pourbaix Diagrams

Phase diagrams (in particular, phase stability or predominance diagrams) provide a convenient means of finding stable phases of a thermodynamic system at particular compositions, temperature and pressure. It is sometimes useful to look at the system in another way by considering what combinations of electrochemical potential (cell *emf*, $\varepsilon$) and the acidity of the solution (electrolyte pH) allow a particular species to exist in thermodynamic and electrochemical equilibria in aqueous systems. This information is most usefully represented by means of a ***ε-pH diagram***, also known as a ***Pourbaix diagram***. Construction of the diagram is based upon repeated applications of the Nernst equation to all the competing reactions that may occur in a given system (cell).

Let's develop the Pourbaix diagram for the system of water. We analyze the system in the thermodynamic and electrochemical ways:

- Number of components: $H_2O$, $O_2$, $H_2$, $H^+$, $OH^-$
- Other variables: T, P, $\varepsilon$

For a graphical presentation of the system, we need to choose two variables for two axes. The choices are,

- pH: Both $H^+$ and $OH^-$ are related to each other and can be represented by pH.
- $\varepsilon$: The *emf* experienced by the system

In order to relate these variables the Nernst equation is employed to possible electrochemical reactions of the system.

First, consider the hydrogen half-cell reaction:

$$2H^+ + 2e = H_2(g) \qquad \varepsilon^o = 0.0000 \text{ volts}$$

Adding $2OH^-$ to both sides,

$$2H_2O + 2e = H_2(g) + 2OH^-$$

The above two reactions are identical, and can be considered to represent the competition between $H_2O$ and $H_2(g)$ as to which one is stable under given thermodynamic and electrochemical conditions. Applying the Nernst Equation,

$$\boxed{\varepsilon = \varepsilon^o - \frac{RT}{2F} ln\left(\frac{P_{H_2}}{a_{H^+}^2}\right)}$$

↓

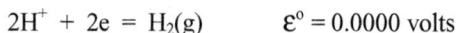

$$\varepsilon = \varepsilon^{o} - \frac{2.303RT}{2F} \log\left(\frac{P_{H_2}}{a_{H^+}^2}\right)$$

$\varepsilon^{o} = 0$ for $H_2/H^+$ half-cell
Let's fix the temperature at
$T = 25°C$ (298 K)

$$\varepsilon = -0.02958 \log\left(\frac{P_{H_2}}{a_{H^+}^2}\right)$$

For a dilute solution of $H^+$,
$a_{H^+} = [H^+]$
and pH = $-log[H^+]$

$$\varepsilon = -0.0591\text{pH} - 0.0296 \log P_{H_2}$$

This equation defines conditions at which both $H_2O$ and $H_2(g)$ coexist as stable phases. Using the above equation, let's examine how $P_{H_2}$ affects the stability of $H_2O$ and $H_2$ for a given pH. For pH = 0 (*i.e.*, $[H^+] = 1$), the above equation reduces to $\varepsilon = -0.0296 \log P_{H_2}$ or $\log P_{H_2} = -33.784\varepsilon$. The graphical representation of this equation will show,

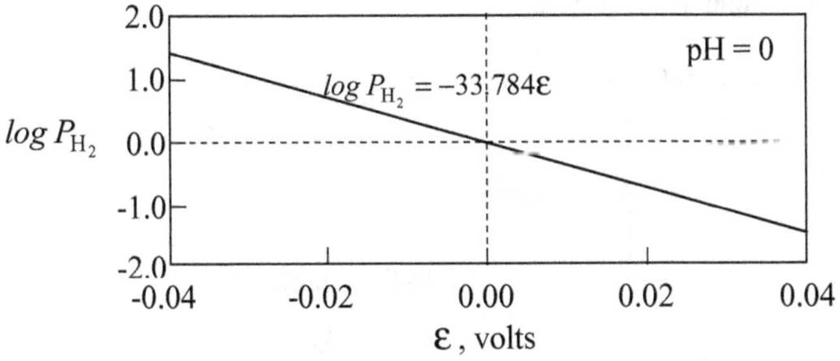


The above diagram may be interpreted as follows:

- The hydrogen pressure ($P_{H_2}$) in equilibrium with $H_2O$ increases with decrease in $\varepsilon$.
- When the $P_{H_2}$ of the system is chosen to be 1 *atm.*, *i.e.*, $\log P_{H_2} = 0$,
  - The half-cell reaction of
    $2H^+ + 2e = H_2(g)$, or $2H_2O + 2e = H_2(g) + 2OH$
    is electrochemically at equilibrium when $\varepsilon = 0.0$ volt.

- However, if $\varepsilon$ is lower than the above, *i.e.*, $\varepsilon < 0$, the equilibrium $P_{H_2}$ is greater than 1 *atm.*, and hence the above cell reaction proceeds in the direction which increases $P_{H_2}$ :

$$2H^+ + 2e \rightarrow H_2(g), \text{ or}$$
$$2H_2O + 2e \rightarrow H_2(g) + 2OH$$

This implies that, when $\varepsilon < 0$, $H_2O$ is unstable and dissociates to form $H_2(g)$.

- If $\varepsilon$ is greater than the above, *i.e.*, $\varepsilon < 0$, the equilibrium $P_{H_2}$ is smaller than 1 atm., and hence the cell reaction proceeds in the opposite direction which $H_2$ dissolves to form $H_2O$.

- In summary, at pH = 0 and $P_{H_2}$ = 1 atm.,

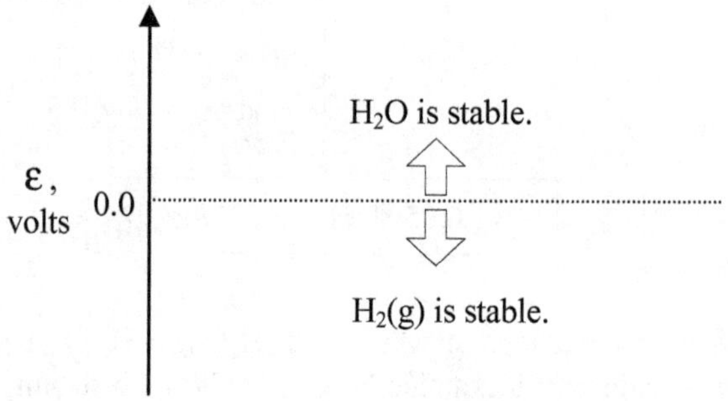


Similar analysis can be made at other conditions.

For instance, at pH = 7 and $P_{H_2}$ = 1 atm.,

$$\varepsilon = -0.0591\text{pH} - 0.0296 \log P_{H_2} = -0.414 \text{ volts}$$

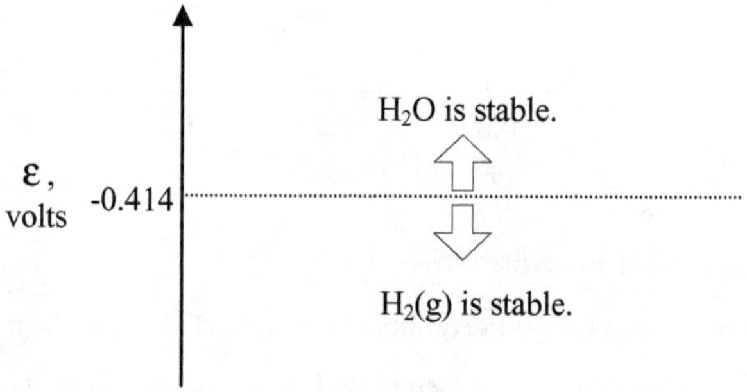


The $\varepsilon$-pH relationship on the stability of $H_2$ and $H_2O$ for different values of pH can be conveniently plotted as follows:

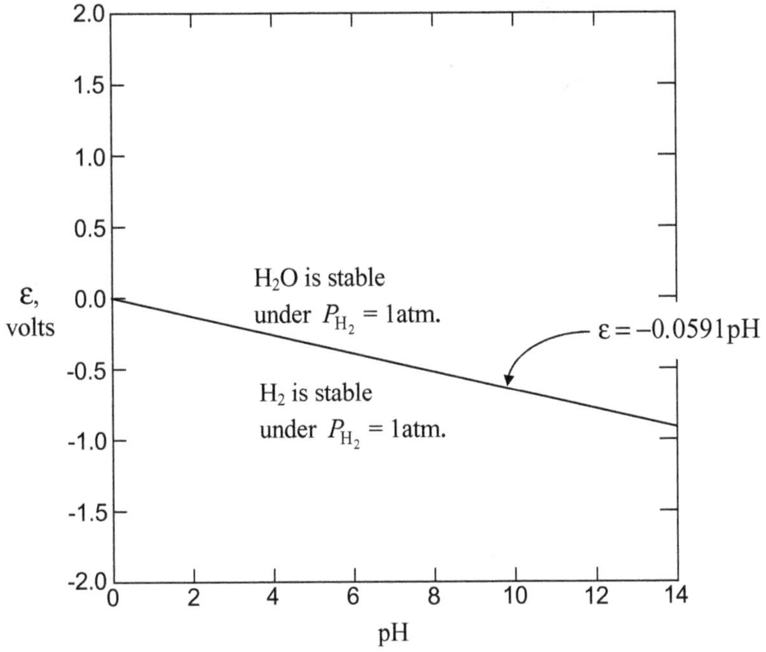


There is another half-cell reaction which represents the competition of $O_2$ and $H_2O$ for stability:

$$O_2 + 4H^+ + 4e = 2H_2O \qquad \varepsilon^o = +1.229 \text{ volts}$$

Application of the Nernst equation yields,

$$\boxed{\varepsilon = 1.229 - 0.0591\text{pH} + 0.0148 \log P_{O_2}}$$

As was done for $H_2/H_2O$, the above equation may be plotted in the ε-pH space under the oxygen pressure at 1 atm., $P_{O_2}$ = 1 atm. The result is given in the diagram given below. This is the Pourbaix diagram for water.

- The Pourbaix diagram divides the ε-pH space into three regions of predominance:
  - At high *emf* values (above the line 1) water decomposes to form oxygen under $P_{O_2}$ = 1 atm.: Oxygen is the stable phase. It is said that $H_2O$ is electrolyzed *anodically* to $O_2$.
  - At low *emf* values (below the line 2), water decomposes to form hydrogen under $P_{H_2}$ = 1 atm.: Hydrogen is the stable phase. It is said that $H_2O$ is electrolyzed *cathodically* to $H_2$.
  - At intermediate *emf* values (between the lines 1 and 2) water is stable over either

oxygen or hydrogen under $P_{H_2}$ = 1 atm. and $P_{O_2}$ = 1 atm. $H_2$ is oxidized to $H_2O$ or $H^+$, and $O_2$ is reduced to $H_2O$ or $OH^-$.

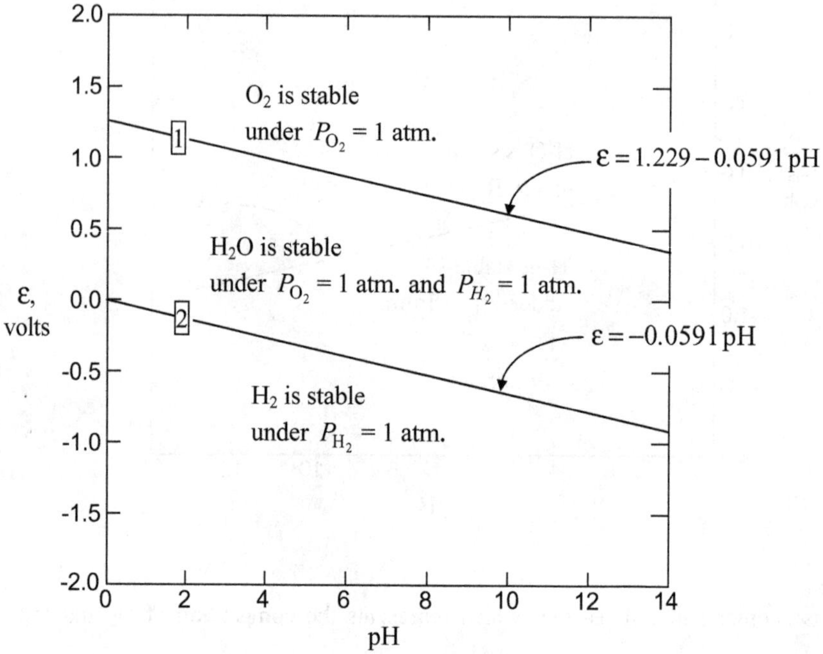


Pourbaix diagrams are convenient tools in determining the corrosion behavior of a metal in water solutions *i.e.* the direction of electrochemical processes and the equilibrium state of the metal at a certain electrode potential in a water solution at a certain value of pH.

Aluminum provides a good example for illustrating the general strategy for construction of a Pourbaix diagram. The Al-containing species in the aluminum-water system at 25°C are two solids (Al and $Al_2O_3$) and two ions ($Al^{+3}$ and $AlO_2^-$). The number of half-cell reactions need to be considered will be (4x3)/2 = 6. These are the combinations of ($Al^{3+}$/Al), ($Al_2O_3$/Al), ( $AlO_2^-$ /Al), ($Al^{3+}$/$Al_2O_3$), ( $AlO_2^-$ /$Al_2O_3$), and ($Al^{3+}$/ $AlO_2^-$ ), all with the aqueous solution.

First consider the half-cell reaction of the (**$Al^{3+}$/Al**) combination:

$$Al^{3+} + 3e = Al$$

↓ Applying the Nernst equation

$$\varepsilon = \varepsilon^{o} - \frac{2.303RT}{3F} log \frac{a_{Al}}{a_{Al^{3+}}}$$

↓

$\varepsilon^o$ = -1.66 volts
$a_{Al} = 1$ (pure Al)
$a_{Al^{3+}} = [Al^{3+}]$ (dilute at the 1 molality standard state)
$T = 298$ K

$$\varepsilon = -1.66 + 0.0197\,log[Al^{3+}]$$

This equation does not include pH and thus the $Al/Al^{3+}$ half-cell electrode potential is independent of pH, because no $H^+$ is involved in the $Al/Al^{+3}$ reaction. The potential depends only on the activity of aluminum in the aqueous solution. If $[Al^{3+}]$ is unity, $\varepsilon = -1.66$ volts for all pH. Normally the Poubaix diagrams are built for the water solutions with the concentrations of metal ions $m = 10^{-6}M$ and at the temperature 25°C. Then the above equation is reduced to,

$$\varepsilon = -1.78 \text{ volts}$$

The half-cell reaction of (**$Al_2O_3/Al$**) is,

$$Al_2O_3 + 6H^+ + 6e = 2Al + 3H_2O \qquad \varepsilon^o$$

If the standard *emf* of this cell ($\varepsilon^o$) is not readily available, it may be found from the Gibbs energy change of the reaction using the relationship of

$$\Delta G^o = -nF\varepsilon^o$$

$$2Al + 3/2\,O_2 = Al_2O_3 \qquad \Delta G^o = -1609000 \text{ J}$$

$$H_2 + 1/2\,O_2 = H_2O \qquad \Delta G^o = -237180 \text{ J}$$

$$2H^+ + 2e = H_2 \qquad \Delta G^o = 0 \text{ J}$$

Combining these three reactions,

$$Al_2O_3 + 6H^+ + 6e = 2Al + 3H_2O \qquad \Delta G^o = 897500 \text{ J}$$

$\Delta G^o = -nF\varepsilon^o$

$$Al_2O_3 + 6H^+ + 6e = 2Al + 3H_2O \qquad \varepsilon^o = -1.55 \text{ volts}$$

Applying the Nernst equation,

$$\varepsilon = \varepsilon^o - \frac{2.303RT}{6F} log \frac{a_{Al}^2 a_{H_2O}^3}{a_{Al_2O_3}[H^+]^6}$$

$a_{Al} = a_{Al_2O_3} = a_{H_2O} = 1$

$pH = -log[H^+]$

$$\varepsilon = -1.55 - 0.0591\,pH\text{, volts}$$

The reaction of ($\mathbf{AlO_2^-/Al}$) is

$$AlO_2^- + 4H^+ + 3e = Al + 2H_2O$$

$Al + O_2 + e = AlO_2^-$:

$\varepsilon^o = 8.7$ volts $\rightarrow \Delta G^o = -839760$ J

$H_2 + 1/2O_2 = H_2O(l)$

$\Delta G^o = -237180$ J

Therefore,

$AlO_2^- + 4H^+ + 3e = Al + 2H_2O$

$$\Delta G^o = 365300\text{ J}$$

$\Delta G^o = -nF\varepsilon^o$

$$\varepsilon^o = -1.262\text{ volts}$$

$$\varepsilon = \varepsilon^o - \frac{2.303RT}{3F} log \frac{a_{Al} a_{H_2O}^2}{[AlO_2^-][H^+]^4}\text{, volts}$$

$$\varepsilon = -1.262 - 0.0788pH + 0.0197\,log[AlO_2^-]\text{, volts}$$

For $[AlO_2^-] = 10^{-6} M$

$$\varepsilon = -1.380 - 0.0788pH\text{ volts}$$

The reaction of ($\mathbf{Al^{3+}/Al_2O_3}$) is,

$$6H^+ + Al_2O_3 = 2Al^{3+} + 3H_2O \qquad \Delta G^o = -87650\text{ J}$$

This reaction does not involve electrons and thus the equilibrium of this reaction is independent of the *emf* of the electrode. From the equilibrium constant,

$$K = exp\left(-\frac{\Delta G^o}{RT}\right) = \frac{[Al^{3+}]^2}{[H^+]^6}$$

↓

$$5.74 = log\,[Al^{+3}] + 3\,pH$$

↓ For $[Al^{3+}] = 10^{-6} M$

$$pH = 3.91$$

The reaction of ($\mathbf{AlO_2^-/Al_2O_3}$) is,

$$Al_2O_3 + H_2O = 2\,AlO_2^- + 2H^+ \quad \Delta G^o = 166700\text{ J}$$

$$K = exp\left(-\frac{\Delta G^o}{RT}\right) = [H^+]^2[AlO_2^-]^2$$

↓

$$14.61 = pH - log[AlO_2^-]$$

↓ For $[AlO_2^-] = 10^{-6} M$

$$pH = 8.61$$

Finally the reaction of ($\mathbf{Al^{3+}/AlO_2^-}$) is,

$$Al^{3+} + 2H_2O = AlO_2^- + 4H^+, \qquad \Delta G^o = 115100\text{ J}$$

$$K = exp\left(-\frac{\Delta G^o}{RT}\right) = \frac{[AlO_2^-][H^+]^4}{[Al^{3+}]}$$

↓

$$20.18 = log[Al^{3+}] - log[AlO_2^-] + 4pH$$

↓ For $[AlO_2^-] = [Al^{3+}] = 10^{-6} M$, or

$[Al^{3+}]/[AlO_2^-] = 1$

$$pH = 5.04$$

We now have 6 equations, each of which determines the stability boundary of two competing species. The following are the equations and the Pourbaix diagram of the aluminium-water system at 25°C with the concentration of all ions in solutions kept at $10^{-6}M$.

| | |
|---|---|
| $Al/Al^{3+}$ | $\varepsilon = -1.78$ |
| $Al_2O_3/Al$ | $\varepsilon = -1.55 - 0.0591pH$ |
| $AlO_2^-/Al$ | $\varepsilon = -1.380 - 0.0788pH$ |
| $Al^{3+}/Al_2O_3$ | $pH = 3.91$ |
| $AlO_2^-/Al_2O_3$ | $pH = 8.61$ |
| $Al^{3+}/AlO_2^-$ | $pH = 5.04$ |

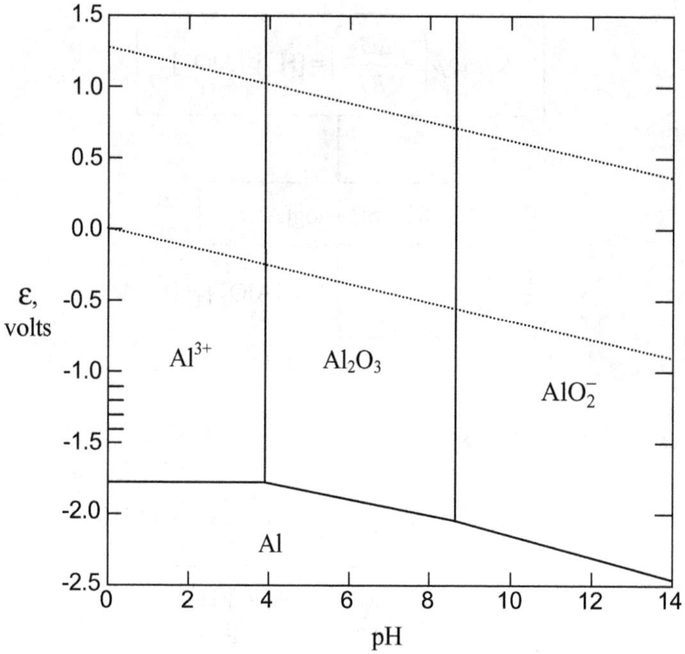


The Dashed lines in the above diagram are for water. They enclose the practical region of stability of the aqueous solvent to oxidation or reduction *i.e.* the region of interest in aqueous systems. Outside this region, it is the water that breaks down, not aluminum.

*Example 13.23*

The Pourbaix diagram is sometimes divided into three zones, *namely*,

*Immunity zone*: The electrochemical reactions in this zone proceed in the direction of reduction of metallic ions. No corrosion occurs in this zone.

*Corrosion zone*: Metallic iron oxidizes in this zone.

*Passivation zone*: Metal oxidizes (corrodes) in this zone however the resulted oxide film depresses the oxidation process causing passivation (corrosion protection of the metal due to formation of a film of a solid product of the oxidation reaction).

Determine the above-mentioned three zones in the Pourbaix diagram of aluminum.

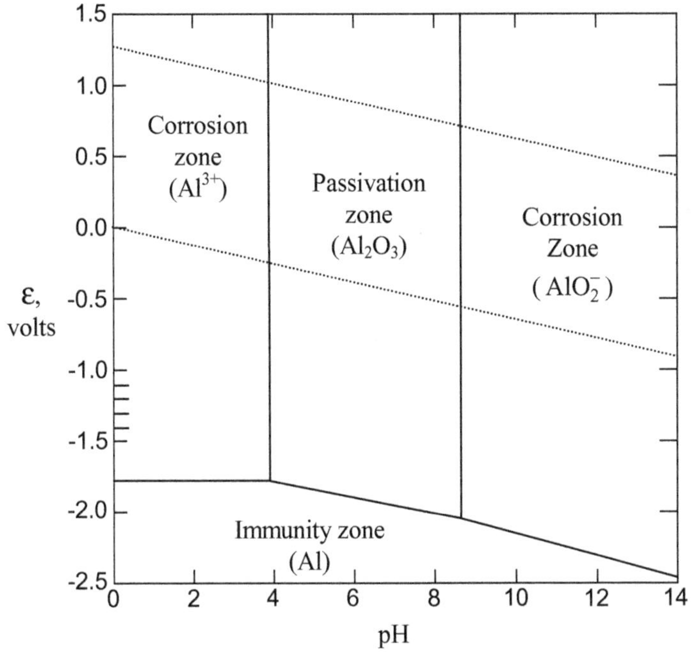


*Example 13.24*

Discuss significances of the vertical lines, horizontal lines and diagonal lines in the Pourbaix diagram.

Horizontal lines correspond to the redox reactions, which are independent of PH.
Vertical lines correspond to the non-redox reactions (electrons are not involved), which are dependent on PH only.
Diagonal lines correspond to the redox reactions, which are dependent on both PH and *emf*.

# Appendices

| | |
|---|---|
| Appendix I | Heats of formation, standard entropies and heat capacities |
| Appendix II | Standard Gibbs energies of formation |
| Appendix III | Properties of selected elements |
| Appendix IV | Standard half-cell potentials in aqueous solution |

# APPENDIX I

## Heats of formation, standard entropies and heat capacities

| Substance | Phase | $\Delta H^{o}_{f,298}$ (kJ mol$^{-1}$) | $S^{o}_{298}$ (J mol$^{-1}$K$^{-1}$) | $C_P = a+bT+cT^{-2}$ (J mol$^{-1}$K$^{-1}$) | | |
|---|---|---|---|---|---|---|
| | | | | a | b × $10^3$ | c × $10^{-5}$ |
| Ag | Solid | 0 | 42.70 | 21.30 | 8.54 | 1.51 |
| Ag | Liquid | 8.94 | 47.19 | 30.54 | | |
| AgCl | Solid | -127.1 | 96.28 | 62.26 | 4.18 | -11.30 |
| AgBr | Solid | -100.7 | 107.2 | 33.19 | 64.46 | |
| AgI | Solid, α | -61.95 | 115.5 | 24.36 | 100.9 | |
| $Ag_2O$ | Solid | -30.56 | 121.8 | 59.36 | 40.81 | -4.19 |
| $Ag_2S$ | Solid, α | -31.31 | 143.6 | 42.40 | 110.5 | |
| Al | Solid | 0 | 28.34 | 20.67 | 12.39 | |
| Al | Liquid | 8.23 | 34.74 | 31.80 | | |
| $AlF_3$ | Solid, α | -1511.1 | 66.52 | 72.29 | 45.88 | -9.63 |
| AlCl | Gas | -51.49 | 227.9 | 37.67 | | -2.85 |
| $AlCl_3$ | Solid | -706.0 | 109.3 | 55.46 | 117.2 | |
| $AlCl_3$ | Gas | -584.8 | 314.5 | 82.88 | | -11.05 |
| $Al_2O_3$ | Solid | -1678.2 | 51.07 | 106.7 | 17.79 | -28.55 |
| $Al_2S_3$ | Solid | -723.8 | 123.5 | 102.2 | 36.08 | |
| AlN | Solid | -318.6 | 20.18 | 34.40 | 16.95 | -8.37 |
| As | Solid | 0 | 35.71 | 23.18 | 5.52 | |
| $As_4$ | Gas | 153.4 | 327.5 | 82.94 | 0.13 | -5.13 |
| $As_2O_3$ | Solid | -653.6 | 122.8 | 59.83 | 175.7 | |
| $As_2S_3$ | Solid | -167.4 | 163.7 | 105.7 | 36.46 | |
| Au | Solid | 0 | 47.39 | 23.69 | 5.19 | |
| B | Solid | 0 | 2.99 | 19.82 | 5.78 | -9.21 |
| $B_2O_3$ | Solid | -1272.5 | 54.0 | 57.06 | 73.05 | -14.06 |
| $B_2O_3$ | Liquid | -1253.7 | 78.49 | 127.7 | | |
| BN | Solid | -252.4 | 14.32 | 33.91 | 14.73 | -23.06 |
| $B_4C$ | Solid | -71.58 | 27.13 | 96.24 | 22.60 | -44.87 |
| Ba | Solid | 0 | 62.46 | -44.43 | 158.4 | 22.49 |
| $BaCl_2$ | Solid, α | -1207.7 | 96.40 | 92.93 | 3.18 | -16.74 |
| BaO | Solid | -553.8 | 70.45 | 53.33 | 4.35 | -8.31 |
| Be | Solid | 0 | 9.50 | 21.22 | 5.69 | -5.88 |
| BeO | Solid | -608.6 | 14.15 | 41.61 | 10.21 | -17.37 |
| Bi | Solid | 0 | 56.72 | 18.80 | 22.60 | |
| $Bi_2O_3$ | Solid, α | -571.0 | 151.5 | 103.6 | 33.49 | |
| $Bi_2S_3$ | Solid | -201.8 | 200.5 | 110.0 | 41.02 | |
| Br | Gas | 111.9 | 174.9 | 19.88 | 1.49 | 0.42 |
| $Br_2$ | Gas | 30.9 | 245.4 | 37.36 | 0.46 | -1.30 |
| C(graphite) | Solid | 0 | 5.74 | 17.15 | 4.27 | -8.79 |

| Substance | Phase | $\Delta H^o_{f,298}$ (kJ mol$^{-1}$) | $S^o_{298}$ (J mol$^{-1}$K$^{-1}$) | $C_P$ = a+b$T$+c$T^{-2}$ (J mol$^{-1}$K$^{-1}$) | | |
|---|---|---|---|---|---|---|
| | | | | a | b × 10$^3$ | c × 10$^{-5}$ |
| $CH_4$ | Gas | -74.85 | 186.2 | 23.64 | 47.87 | -1.93 |
| $C_2H_2$ | Gas | 226.8 | 200.9 | 43.66 | 31.67 | -7.51 |
| CO | Gas | -110.5 | 197.6 | 28.41 | 4.10 | -0.46 |
| $CO_2$ | Gas | -393.5 | 213.7 | 44.14 | 9.04 | -8.54 |
| COS | Gas | -138.4 | 231.5 | 47.41 | 2.61 | -7.66 |
| Ca | Solid, α | 0 | 41.6 | 25.51 | 2.61 | |
| Ca | Liquid | 10.90 | 50.65 | 30.13 | | |
| $CaF_2$ | Solid, α | -1220.2 | 68.86 | 59.86 | 30.47 | 1.97 |
| $CaCl_2$ | Solid | -796.2 | 104.7 | 30.06 | 12.73 | -2.51 |
| CaO | Solid | -635. | 38.10 | 49.62 | 4.52 | -7.00 |
| $Ca(OH)_2$ | Solid | -985.8 | 83.43 | 105.4 | 11.95 | -18.98 |
| $CaSO_4$ | Solid | -1434.8 | 106.7 | 70.24 | 98.79 | |
| $CaC_2$ | Solid, α | -59.02 | 70.32 | 68.65 | 11.89 | -8.67 |
| $CaCO_3$ | Solid | -1207.1 | 88.70 | 104.5 | 21.92 | -25.94 |
| CaS | Solid | -476.1 | 56.48 | 45.19 | 7.74 | |
| $CaSiO_3$ | Solid | -1634.9 | 82.01 | 108.2 | 16.49 | -23.64 |
| $Ca_2SiO_4$ | Solid | -136.9 | 120.6 | 151.7 | 36.96 | -30.31 |
| Cd | Solid | 0 | 51.80 | 22.22 | 12.30 | |
| Cd | Liquid | 5.81 | 61.05 | 29.71 | | |
| Cd | Gas | 111.8 | 167.6 | 20.79 | | |
| $CdCl_2$ | Solid | -391.0 | 115.3 | 47.30 | 91.67 | |
| CdO | Solid | -259.4 | 54.80 | 48.24 | 6.38 | -4.90 |
| CdS | Solid | -141.4 | 69.04 | 44.56 | 13.81 | |
| Ce | Solid, α | 0 | 72.00 | 23.50 | 10.40 | |
| $Ce_2O_3$ | Solid | -1822.6 | 150.7 | 107.9 | 41.44 | -9.21 |
| $Cl_2$ | Gas | 0 | 223.0 | 36.90 | 0.25 | -2.84 |
| Co | Solid, α | 0 | 30.04 | 21.39 | 14.31 | -0.88 |
| Co | Solid, β | 1.29 | 32.66 | 13.81 | 24.52 | |
| CoO | Solid | -238.9 | 52.93 | 48.28 | 8.54 | 1.67 |
| CoS | Solid | -121.7 | 63.95 | 44.37 | 10.51 | |
| Cr | Solid | 0 | 23.64 | 24.44 | 9.87 | -3.68 |
| Cr | Liquid | 26.10 | 36.23 | 39.33 | | |
| $Cr_2O_3$ | Solid | -1130.2 | 81.21 | 119.4 | 9.21 | -15.66 |
| $Cr_3C_2$ | Solid | -109.7 | 85.39 | 125.7 | 23.36 | -31.23 |
| $Cr_7C_3$ | Solid | -228.1 | 200.9 | 238.4 | 60.86 | -42.36 |
| Cs | Liquid | 0 | 85.27 | 31.90 | | |
| Cu | Solid | 0 | 33.15 | 22.64 | 6.28 | |
| Cu | Liquid | 9.31 | 36.25 | 31.38 | | |
| Cu | Gas | 336.8 | 166.3 | 9.93 | 5.07 | |
| $Cu_2O$ | Solid | -167.4 | 93.09 | 62.34 | 23.85 | |
| CuO | Solid | -155.2 | 42.68 | 38.79 | 20.08 | |

| Substance | Phase | $\Delta H^{o}_{f,298}$ (kJ mol$^{-1}$) | $S^{o}_{298}$ (J mol$^{-1}$K$^{-1}$) | $C_P$ = a+b$T$+c$T^{-2}$ (J mol$^{-1}$K$^{-1}$) | | |
|---|---|---|---|---|---|---|
| | | | | a | b × 10$^3$ | c × 10$^{-5}$ |
| $Cu_2S$ | Solid, α | -79.50 | 120.9 | 81.59 | | |
| $Cu_2S$ | Solid, β | -79.53 | 121.0 | 97.28 | | |
| $Cu_2S$ | Solid, γ | -79.53 | 121.0 | 85.02 | | |
| CuS | Solid | -52.30 | 66.53 | 44.35 | 11.05 | |
| $CuSO_4$ | Solid | -771.1 | 109.2 | 78.53 | 71.97 | |
| Fe | Solid, α | 0 | 27.28 | 17.49 | 24.77 | |
| Fe | Solid, γ | 6.78 | 33.66 | 26.61 | 6.28 | |
| Fe | Solid, δ | 3.89 | 29.71 | 28.28 | 7.53 | |
| Fe | Liquid | 13.13 | 34.29 | 35.40 | 3.75 | |
| Fe | Gas | 416.3 | 180.4 | 15.72 | 3.47 | |
| $FeCl_2$ | Solid | -342.4 | 118.1 | 79.28 | 8.71 | -4.90 |
| $Fe_{0.947}O$ | Solid | -264.6 | 58.81 | 48.81 | 8.37 | -2.80 |
| FeO | Solid | -264.4 | 58.79 | 51.80 | 6.78 | -1.59 |
| $Fe_3O_4$ | Solid, α | -1116.7 | 155.5 | 91.55 | 201.7 | |
| $Fe_2O_3$ | Solid, α | -821.3 | 87.45 | 98.28 | 77.82 | -14.85 |
| FeS | Solid, α | -100.4 | 60.29 | 21.72 | 110.5 | |
| $Fe_3C$ | Solid, α | 25.12 | 104.7 | 82.21 | 83.72 | |
| Ga | Solid | 0 | 41.02 | 26.10 | | |
| GaAs | Solid | -81.63 | 64.26 | 45.21 | 6.07 | |
| GaN | Solid | -109.7 | 29.72 | 38.09 | 9.00 | |
| $Ga_2O_3$ | Solid | -1083.3 | 84.68 | 112.9 | 15.45 | -21.01 |
| GaP | Solid | -122.2 | 52.33 | 41.86 | 6.82 | 0 |
| GaSb | Solid | -41.86 | 77.36 | 45.63 | 12.56 | 0 |
| Ge | Solid | 0 | 31.10 | 21.60 | 5.86 | 0 |
| $GeO_2$ | Solid | -580.2 | 39.77 | 66.64 | 11.60 | -17.75 |
| $H_2$ | Gas | 0 | 130.6 | 27.28 | 3.26 | 0.50 |
| HBr | Gas | -36.38 | 198.6 | 26.15 | 5.86 | 1.09 |
| HCl | Gas | -92.31 | 186.8 | 26.53 | 4.60 | 1.09 |
| HI | Gas | 26.36 | 206.5 | 26.32 | 5.94 | 0.92 |
| $H_2O$ | Liquid | -285.8 | 69.95 | 75.44 | | |
| $H_2O$ | Gas | -241.8 | 188.7 | 30.00 | 10.71 | 0.34 |
| $H_2S$ | Gas | -20.50 | 205.6 | 32.68 | 12.39 | -1.93 |
| Hf | Solid | 0 | 43.58 | 23.47 | 7.62 | |
| $HfO_2$ | Solid | -1113.5 | 59.44 | 72.79 | 8.71 | -14.57 |
| Hg | Liquid | 0 | 75.93 | 30.39 | -11.47 | |
| $HgCl_2$ | Solid | -228.4 | 140.1 | 63.96 | 43.53 | |
| HgO | S,red | 90.84 | 70.32 | 37.67 | 25.12 | |
| HgS | S,red | 53.37 | 82.46 | 43.79 | 15.57 | |
| $I_2$ | Gas | 62.43 | 260.6 | 37.40 | 0.57 | -0.63 |
| In | Solid | 0 | 57.85 | 24.32 | 10.47 | |
| $In_2O_3$ | Solid | -926.4 | 108.0 | 123.9 | 7.95 | -23.06 |

| Substance | Phase | $\Delta H^{o}_{f,298}$ (kJ mol$^{-1}$) | $S^{o}_{298}$ (J mol$^{-1}$K$^{-1}$) | $C_P$ = a+b$T$+c$T^{-2}$ (J mol$^{-1}$K$^{-1}$) | | |
|---|---|---|---|---|---|---|
| | | | | a | b × $10^3$ | c × $10^{-5}$ |
| K | Solid | 0 | 64.72 | 25.28 | 13.06 | |
| KCl | Solid | -436.9 | 82.59 | 41.40 | 21.77 | 3.22 |
| $K_2O$ | Solid | -363.4 | 94.19 | 95.69 | -4.94 | 11.05 |
| $K_2CO_3$ | Solid | -1150.2 | 155.5 | 80.29 | 109.0 | |
| La | Solid | 0 | 56.93 | 25.83 | 6.70 | |
| $La_2O_3$ | Solid | -1794.1 | 128.1 | 120.8 | 12.89 | -13.73 |
| Li | Solid | 0 | 29.09 | 13.94 | 34.37 | |
| LiF | Solid | -617.2 | 35.58 | 38.26 | 21.73 | |
| $Li_2O$ | Solid | -596.9 | 37.93 | 62.54 | 25.45 | -14.15 |
| Mg | Solid | 0 | 42.51 | 22.30 | 10.25 | -0.43 |
| Mg | Liquid | 9.29 | 148.6 | 22.05 | 10.90 | |
| Mg | Gas | 147.6 | 26.95 | 20.79 | | |
| $MgF_2$ | Solid | -1123.9 | 57.26 | 70.87 | 10.55 | -9.21 |
| $MgCl_2$ | Solid | -641.7 | 89.66 | 79.12 | 5.94 | -8.62 |
| MgO | Solid | -601.5 | 26.95 | 49.00 | 3.14 | -11.72 |
| MgS | Solid | -351.6 | 50.36 | 43.12 | 8.25 | |
| $MgCO_3$ | Solid | -1111.7 | 65.86 | 77.91 | 57.74 | -17.41 |
| Mn | Solid, α | 0 | 32.02 | 23.86 | 14.15 | -1.57 |
| $MnCl_2$ | Solid | -482.2 | 118.3 | 75.52 | 13.23 | -5.73 |
| MnO | Solid | -385.1 | 59.86 | 46.51 | 8.12 | -3.68 |
| $Mn_3O_4$ | Solid, α | -1387.2 | 154.0 | 145.0 | 45.29 | -9.21 |
| $Mn_2O_3$ | Solid | -957.3 | 110.5 | 103.5 | 35.08 | -13.56 |
| $MnO_2$ | Solid | -520.3 | 53.16 | 69.49 | 10.21 | -16.24 |
| MnS | S, green | -213.5 | 80.37 | 47.72 | 7.53 | |
| $MnSO_4$ | Solid | -1065.8 | 112.2 | 122.5 | 37.34 | -29.47 |
| Mo | Solid | 0 | 28.67 | 25.57 | 2.85 | -2.18 |
| $MoO_3$ | Solid | -1164.1 | 77.82 | 84.01 | 24.70 | -15.40 |
| $Mo_2N$ | Solid | -69.49 | 87.91 | 46.84 | 57.77 | |
| $N_2$ | Gas | 0 | 191.5 | 27.87 | 4.27 | |
| $NH_3$ | Gas | -45.94 | 192.7 | 37.32 | 18.66 | -6.49 |
| $N_2O$ | Gas | 82.09 | 220.0 | 45.71 | 8.62 | -8.54 |
| NO | Gas | 90.33 | 210.8 | 29.43 | 3.85 | -0.59 |
| Na | Solid | 0 | 51.28 | 82.51 | -369.5 | |
| NaCl | Solid | -412.8 | 72.17 | 45.96 | 16.33 | |
| $Na_2O$ | Solid, α | -415.3 | 75.10 | 55.51 | 70.24 | -4.14 |
| NaOH | Solid | -426.1 | 64.46 | 71.79 | -110.9 | |
| $Na_2SO_4$ | Solid, α | 1396.0 | 149.6 | 98.37 | 132.9 | |
| $Na_2CO_3$ | Solid | -1130.9 | 138.8 | 58.49 | 227.6 | -13.10 |
| $Na_3AlF_6$ | Solid, α | -82.88 | 238.6 | 192.4 | 123.3 | -11.64 |
| Nb | Solid | 0 | 36.54 | 23.72 | 4.02 | |
| NbO | Solid | -419.9 | 46.05 | 42.03 | 9.84 | -3.27 |

| Substance | Phase | $\Delta H^o_{f,298}$ (kJ mol$^{-1}$) | $S^o_{298}$ (J mol$^{-1}$K$^{-1}$) | $C_P$ = a+b$T$+c$T^{-2}$ (J mol$^{-1}$K$^{-1}$) | | |
|---|---|---|---|---|---|---|
| | | | | a | b × 10$^3$ | c × 10$^{-5}$ |
| $NbO_2$ | Solid, α | -795.3 | 55.67 | 61.45 | 25.79 | -10.13 |
| Ni | Solid, α | 0 | 29.87 | 12.54 | 35.82 | 2.47 |
| Ni | Solid, β | 0.63 | 30.95 | 25.10 | 7.53 | |
| Ni | Liquid | 8.32 | 27.38 | 38.91 | | |
| $NiCl_2$ | Solid | -305.6 | 97.74 | 73.26 | 13.23 | -4.98 |
| NiO | Solid, α | -240.6 | 38.08 | -20.88 | 157.2 | 16.28 |
| NiO | Solid, β | -240.6 | 38.08 | 58.10 | | |
| NiO | Solid, γ | -240.6 | 38.08 | 46.80 | 8.46 | |
| NiS | Solid, α | -94.14 | 52.93 | 43.76 | 22.20 | -2.90 |
| $NiSO_4$ | Solid | -873.6 | 103.9 | 126.0 | 41.53 | |
| $O_2$ | Gas | 0 | 205.0 | 29.96 | 4.18 | -1.67 |
| P | S, white | 0 | 41.11 | 19.13 | 15.82 | |
| $P_2O_5$ | Solid | -1492.7 | 114.5 | 74.93 | 162.4 | -15.61 |
| Pb | Solid | 0 | 64.80 | 23.56 | 9.75 | |
| Pb | Liquid | 4.29 | 71.72 | 32.43 | -3.10 | |
| Pb | Gas | 195.6 | 175.3 | 6.67 | 8.96 | |
| $PbCl_2$ | Solid | -359.6 | 136.1 | 67.39 | 16.74 | |
| $PbCl_2$ | Liquid | -174.1 | 317.3 | 118.1 | | |
| PbO | S, red | -219.3 | 65.27 | 38.20 | 25.52 | |
| PbO | S, yellow | -217.9 | 67.36 | 45.09 | 12.23 | |
| PbO | Liquid | -195.4 | 85.97 | 65.00 | | |
| PbS | Gas | 131.9 | 251.5 | 44.62 | 16.41 | |
| S | S, ortho | 0 | 32.05 | 14.81 | 24.06 | 0.73 |
| S | S, mono | 0.34 | 32.97 | 68.35 | -118.6 | |
| $S_2$ | Gas | 128.5 | 228.1 | 36.49 | 0.67 | -3.77 |
| $SO_2$ | Gas | -296.8 | 248.1 | 43.43 | 10.63 | -5.94 |
| $SO_3$ | Gas | -395.7 | 256.7 | 57.32 | 26.86 | -13.05 |
| Sb | Solid | 0 | 45.54 | 23.06 | 7.28 | |
| $SbCl_3$ | Gas | -313.7 | 337.4 | 43.12 | 239.0 | |
| $Sb_2O_3$ | Solid | -708.9 | 141.1 | 79.95 | 71.58 | |
| $Sb_2S_3$ | Solid | -205.1 | 182.1 | 101.9 | 60.57 | |
| Se | Solid | 0 | 42.28 | 17.90 | 25.12 | |
| $SeO_2$ | Solid | -225.2 | 66.72 | 69.61 | 3.90 | -11.05 |
| Si | Solid | 0 | 18.84 | 23.94 | 2.47 | -4.14 |
| Si | Liquid | 455.9 | 167.9 | 25.62 | | |
| SiC | Solid | -66.98 | 16.53 | 50.78 | 1.97 | -49.23 |
| $SiCl_2$ | Gas | -167.4 | 282.0 | 57.60 | 0.38 | -5.65 |
| $SiCl_4$ | Gas | -663.1 | 331.0 | 101.5 | 6.87 | -11.51 |
| $SiI_4$ | Solid | -199.3 | 265.6 | 82.00 | 87.49 | |
| $SiO_2$ | Solid | -190.7 | 41.46 | 46.95 | 34.31 | -11.30 |
| $Si_3N_4$ | Solid | -745.1 | 113.0 | 43.91 | 1.00 | -6.03 |

| Substance | Phase | $\Delta H^o_{f,298}$ (kJ mol$^{-1}$) | $S^o_{298}$ (J mol$^{-1}$K$^{-1}$) | $C_P = a+bT+cT^{-2}$ (J mol$^{-1}$K$^{-1}$) | | |
|---|---|---|---|---|---|---|
| | | | | a | b × $10^3$ | c × $10^{-5}$ |
| Sm | Solid | 0 | 69.57 | 52.62 | | |
| $Sm_2O_3$ | S, cubic | | | 128.3 | 21.26 | -16.58 |
| Sn | S, white | 0 | 51.21 | 21.59 | 18.16 | |
| Sn | Liquid | 6.63 | 63.97 | 34.69 | -9.21 | |
| $SnO_2$ | Solid | -580.7 | 52.30 | 73.89 | 10.04 | -21.59 |
| $SnS_2$ | Solid | -153.6 | 87.49 | 64.92 | 17.58 | |
| Ta | Solid | 0 | 41.53 | 27.84 | -2.18 | -1.88 |
| $Ta_2O_5$ | Solid | -2047.0 | 143.2 | 154.9 | 27.46 | -24.78 |
| Te | Solid | 0 | 49.52 | 19.17 | 21.98 | |
| $TeO_2$ | Solid | -323.6 | 74.09 | 65.22 | 14.57 | -5.02 |
| Ti | Solid, α | 0 | 30.63 | 22.09 | 10.04 | |
| Ti | Solid, β | 6,59 | 38.38 | 19.83 | 7.95 | |
| TiO | Solid, α | -542.9 | 34.74 | 44.25 | 15.07 | -7.79 |
| TiO | Solid, β | -542.9 | 34.74 | 49.60 | 12.56 | |
| $Ti_2O_3$ | Solid, α | -1521.6 | 77.27 | 30.60 | 224.0 | |
| $Ti_2O_3$ | Solid, β | -1521.6 | 77.27 | 145.2 | 5.44 | -42.70 |
| $Ti_3O_5$ | Solid, α | -2460.5 | 129.5 | 148.5 | 123.5 | |
| $Ti_3O_5$ | Solid, β | -2460.5 | 129.5 | 174.1 | 33.49 | |
| $TiO_2$ | S, rutile | -944.8 | 50.33 | 75.19 | 1.17 | -18.20 |
| Tl | Solid, α | 0 | 64.21 | 15.66 | 25.28 | 2.80 |
| $Tl_2O_3$ | Solid | -390.6 | 137.3 | 131.9 | 3.56 | -22.27 |
| U | Solid, α | 0 | 50.20 | 10.92 | 37.45 | 4.90 |
| $UO_2$ | Solid | -1085.0 | 77.03 | 80.33 | 6.78 | -16.57 |
| V | Solid | 0 | 28.95 | 20.50 | 10.80 | 0.84 |
| VN | Solid | -217.3 | 37.30 | 45.79 | 8.79 | -9.25 |
| VO | Solid | -432.0 | 39.01 | 47.39 | 13.48 | -5.27 |
| $VO_2$ | Solid, α | -713.7 | 51.78 | 62.62 | | |
| $VO_2$ | Solid, β | -713.7 | 51.78 | 74.72 | 7.12 | -16.53 |
| $V_2O_3$ | Solid | -1219.4 | 98.12 | 122.9 | 19.93 | -22.69 |
| $V_2O_5$ | Solid | -1551.3 | 130.6 | 194.8 | -16.33 | -55.34 |
| W | Solid | 0 | 32.64 | 23.81 | 3.26 | |
| WC | Solid | -38.09 | 41.86 | 43.41 | 8.62 | -9.33 |
| $WO_3$ | Solid | -843.3 | 75.93 | 73.17 | 28.42 | |
| Y | Solid, α | 0 | 44.50 | 23.94 | 7.56 | 0.33 |
| YN | Solid | -299.3 | 37.67 | 45.63 | 6.49 | -7.33 |
| Zn | Solid | 0 | 41.63 | 22.38 | 10.04 | |
| $ZnCl_2$ | Solid, α | -415.3 | 111.5 | 60.70 | 23.02 | |
| ZnO | Solid | -350.6 | 43.64 | 49.00 | 5.11 | -9.12 |
| ZnS | Solid | -205.2 | 57.66 | 50.88 | 5.19 | -5.69 |
| $ZnSO_4$ | Solid | -981.8 | 110.6 | 91.67 | 76.19 | |

| Substance | Phase | $\Delta H^o_{f,298}$ (kJ mol$^{-1}$) | $S^o_{298}$ (J mol$^{-1}$K$^{-1}$) | $C_P = a+bT+cT^{-2}$ (J mol$^{-1}$K$^{-1}$) | | |
|---|---|---|---|---|---|---|
| | | | | a | b × 10$^3$ | c × 10$^{-5}$ |
| Zr | Solid, α | 0 | 39.00 | 21.97 | 11.63 | |
| ZrN | Solid | -368.4 | 38.89 | 46.46 | 7.03 | -7.20 |
| $ZrO_2$ | Solid, α | -1100.8 | 50.71 | 69.62 | 7.53 | -14.06 |

Source: Data are largely from *Metallurgical Thermochemistry*, O. Kubaschewski and C.B. Alcock, 5th ed. Pergamon Press, 1979.

# APPENDIX II

## Standard Gibbs energies of formation

$$\Delta G^o = \Delta H^o - T\Delta S^o$$

| Reactions | $\Delta H^o$ (kJ mol$^{-1}$) | $\Delta S^o$ (J mol$^{-1}$ K$^{-1}$) | Temperature Range (K) | Ref |
|---|---|---|---|---|
| $Ag(s) + \frac{1}{2}Br_2(g) = AgBr(l)$ | -97.3 | -27.7 | 715~838 | 11 |
| $Ag(s) + \frac{1}{2}Cl_2(g) = AgCl(l)$ | -106.0 | -25.4 | 803~1193 | 11 |
| $Ag(s) + \frac{1}{2}F_2(g) = AgF(s)$ | -200.8 | 54.8 | 298~708(M) | 4,7 |
| $Ag(s) + \frac{1}{2}I_2(g) = AgI(l)$ | -74.1 | -24.1 | 873~973 | 11 |
| $2Ag(s) + \frac{1}{2}O_2(g) = Ag_2O(s)$ | -28.1 | -60.6 | 298~1000 | 1 |
| $2Ag(s) + \frac{1}{2}S_2(g) = Ag_2S(s)$ | -161.3 | 168.6 | 298~1103(M) | 4 |
| $Al(l) + \frac{1}{2}Cl_2(g) = AlCl(g)$ | -77.4 | 58.2 | 933(m)~2273 | 4 |
| $Al(l) + 1\frac{1}{2}Cl_2(g) = AlCl_3(g)$ | -602.1 | -67.9 | 933(m)~2273 | 4 |
| $Al(l) + 1\frac{1}{2}F_2(g) = AlF_3(s)$ | -1,507.7 | -257.9 | 933(m)~1549(s) | 4 |
| $Al(l) + 1\frac{1}{2}F_2(g) = AlF_3(g)$ | -1,227.8 | -78.1 | 933(m)~2000 | 2 |
| $Al(l) + \frac{1}{2}N_2(g) = AlN(s)$ | -327.1 | -115.5 | 933(m)~2273 | 4 |
| $Al(l) + 3Na(l) + 3F_2(g) = Na_3AlF_6(l)$ | -3,378.2 | -623.4 | 1285(M)~2273 | 4 |
| $2Al(l) + \frac{1}{2}O_2(g) = Al_2O(g)$ | -170.7 | 49.4 | 933~2273 | 15 |
| $2Al(s) + 1\frac{1}{2}O_2(g) = Al_2O_3(s)$ | -1,675.1 | -313.2 | 298~933(m) | 2 |
| $2Al(l) + 1\frac{1}{2}O_2(g) = Al_2O_3(s)$ | -1,682.9 | -323.2 | 933~2315(M) | 2 |
| $2Al(l) + 1\frac{1}{2}O_2(g) = Al_2O_3(l)$ | -1,574.1 | -275.0 | 2315~2767(b) | 2 |
| $2Al(g) + 1\frac{1}{2}O_2(g) = Al_2O_3(l)$ | -2,106.4 | -468.6 | 2767~3500 | 2 |
| $Al_2O_3(s) + SiO_2(s) = Al_2O_3{\cdot}SiO_2(s)$ | -8.8 | -3.9 | 298~1973 | 4 |
| $3Al_2O_3(s) + 2SiO_2(s) = 3Al_2O_3{\cdot}2SiO_2(s)$ | 8.6 | 17.4 | 293~2023(M) | 15 |
| $Al_2O_3(s) + TiO_2(s) = Al_2O_3{\cdot}TiO_2(s)$ | -25.3 | -3.9 | 298~2133(M) | 8 |
| $Al_2O_3(s) + CaO(s) = CaO{\cdot}Al_2O_3(s)$ | -18.0 | 19.0 | 773~1878(M) | 18 |
| $Al_2O_3(s) + 3CaO(s) = 3CaO{\cdot}Al_2O_3(s)$ | -12.6 | 24.7 | 773~1808(M) | 18 |
| $2Al_2O_3(s) + CaO(s) = CaO{\cdot}2Al_2O_3(s)$ | -17.0 | 25.5 | 773~2023(M) | 18 |

| Reactions | $\Delta H^{o}$ (kJ mol$^{-1}$) | $\Delta S^{o}$ (J mol$^{-1}$ K$^{-1}$) | Temperature Range (K) | Ref |
|---|---|---|---|---|
| $Al(l) + 3Li(l) + 3F_2(g) = Li_3AlF_6(l)$ | -3,240.0 | -400 | 1058(M)~1615(M) | 15 |
| $Al_2O_3(s) + MgO(s) = MgO{\cdot}Al_2O_3(s)$ | -35.6 | 2.1 | 298~1673 | 15, 18 |
| $Al_2O_3(s) + MnO(s) = MnO{\cdot}Al_2O_3(s)$ | -48.1 | -7.3 | 773~1473 | 18 |
| $Al_2O_3(s) + Na_2O(s) = Na_2O{\cdot}Al_2O_3(s)$ | -185.0 | 2.9 | 773~1404 | 18 |
| $Al_2O_3(s) + CaO(s) + SiO_2(s) = CaO{\cdot}Al_2O_3{\cdot}SiO_2(s)$ | -105.9 | -14.2 | 298~1673 | 4 |
| $Al_2O_3(s) + CaO(s) + 2SiO_2(s) = CaO{\cdot}Al_2O_3{\cdot}2SiO_2(s)$ | -139.0 | -17.2 | 298~1826 | 8 |
| $Al_2O_3(s) + 2CaO(s) + SiO_2(s) = 2CaO{\cdot}Al_2O_3{\cdot}SiO_2(s)$ | -170.0 | -8.8 | 298~1773 | 8 |
| $Al_2O_3(s) + 3CaO(s) + 3SiO_2(s) = 3CaO{\cdot}Al_2O_3{\cdot}3SiO_2(s)$ | -389.0 | -100.0 | 298~1673 | 8 |
| $\frac{1}{4}As_4(g) + Ga(l) = GaAs(s)$ | -115.0 | -72.0 | 303~1238 | 12 |
| $\frac{1}{4}As_4(g) + In(l) = InAs(s)$ | -99.3 | -64.8 | 430~1215 | 12 |
| $4B(s) + C(s) = B_4C(s)$ | -41.5 | -5.6 | 298~2303 | 15 |
| $B(s) + \frac{1}{2}N_2(g) = BN(s)$ | -250.6 | -87.6 | 298~2453(m) | 4 |
| $B(s) + \frac{1}{2}O_2(g) = BO(g)$ | -3.8 | 88.8 | 298~2303 | 15 |
| $2B(s) + 1\frac{1}{2}O_2(g) = B_2O_3(g)$ | -1,229.0 | -210 | 723~2316(B) | 15 |
| $Ba(l) + C(s) + 1\frac{1}{2}O_2(g) = BaCO_3(s)$ | -1,203.3 | -249.2 | 1073~1333 | 4 |
| $Ba(l) + Cl_2(g) = BaCl_2(l)$ | -811.3 | -121.5 | 1235(M)~1235(M) | 15 |
| $Ba(l) + F_2(g) = BaF_2(l)$ | -1,154.0 | 129 | 1641~1895 | 15 |
| $Ba(s) + H_2(g) + O_2(g) = Ba(OH)_2(s)$ | -941.3 | -291.1 | 298~681(M) | 7,8 |
| $Ba(l) + H_2(g) + O_2(g) = Ba(OH)_2(l)$ | -918.4 | -248.4 | 1002(m)~1263 | 7,8 |
| $Ba(s) + \frac{1}{2}O_2(g) = BaO(s)$ | -568.2 | -97.0 | 298~1002(m) | 3 |
| $Ba(l) + \frac{1}{2}O_2(g) = BaO(s)$ | -557.2 | -102.7 | 1002~1895(b) | 4 |
| $Ba(l) + \frac{1}{2}S_2(g) = BaS(s)$ | -543.9 | -123.4 | 1002(m)~1895(b) | 4 |
| $3Be(l) + N_2(g) = Be_3N_2(s)$ | -616.0 | -203.0 | 1560~2473(M) | 15 |
| $Be(s) + \frac{1}{2}O_2(g) = BeO(s)$ | -608.0 | -97.7 | 298~1560 | 15 |
| $Be(l) + \frac{1}{2}O_2(g) = BeO(s)$ | -613.6 | -100.9 | 1560~2273 | 15 |
| $Be(s) + \frac{1}{2}S_2(g) = BeS(s)$ | -300.0 | -86.6 | 298~1560 | 8 |

| Reactions | $\Delta H^\circ$ (kJ mol$^{-1}$) | $\Delta S^\circ$ (J mol$^{-1}$ K$^{-1}$) | Temperature Range (K) | Ref |
|---|---|---|---|---|
| $2Bi(l) + 1\frac{1}{2}O_2(g) = Bi_2O_3(s)$ | -590.2 | -292.6 | 545~1097(M) | 1 |
| $2Bi(l) + 1\frac{1}{2}O_2(g) = Bi_2O_3(l)$ | -445.2 | -159.6 | 1097(M)~1773 | 4 |
| $2Bi(l) + 1\frac{1}{2}S_2(g) = Bi_2S_3(s)$ | -360.0 | -274.0 | 545~1050(M) | 8 |
| $C(s) + 2Cl_2(g) = CCl_4(g)$ | -89.1 | -129.2 | 298~2273 | 15 |
| $C(s) + 2F_2(g) = CF_4(g)$ | -933.2 | -151.5 | 298~2273 | 15 |
| $C(s) + 2H_2(g) = CH_4(g)$ | -87.4 | -108.7 | 298~2500 | 2 |
| $C(s) + \frac{1}{2}N_2(g) = CN(g)$ | 433.5 | 99.6 | 298~2273 | 15 |
| $C(s) + \frac{1}{2}O_2(g) = CO(g)$ | -112.9 | 86.5 | 298~2500 | 2 |
| $C(s) + O_2(g) = CO_2(g)$ | -394.8 | -0.836 | 298~2500 | 2 |
| $C(s) + \frac{1}{2}S_2(g) = CS(g)$ | 163.2 | 87.9 | 298~2273 | 4 |
| $C(s) + S_2(g) = CS_2(g)$ | -11.4 | 6.5 | 298~2273 | 4 |
| $Ca(l) + 2C(s) = CaC_2(s)$ | -60.3 | 26.3 | 1115~1755 | 15 |
| $Ca(s) + C(s) + 1\frac{1}{2}O_2(g) = CaCO_3(s)$ | -1,196.3 | -242.1 | 298~1112(m) | 3 |
| $Ca(l) + C(s) + 1\frac{1}{2}O_2(g) = CaCO_3(s)$ | -1,196.2 | -245.0 | 1112~1473 | 4 |
| $Ca(s) + Cl_2(g) = CaCl_2(s)$ | -794.5 | -142.3 | 298~1045(M) | 3 |
| $Ca(l) + Cl_2(g) = CaCl_2(l)$ | -798.6 | -146.0 | 1112(m)~1764(b) | 4 |
| $Ca(s) + F_2(g) = CaF_2(s)$ | -1,221.8 | -164.9 | 298~1112(m) | 2 |
| $Ca(l) + F_2(g) = CaF_2(s)$ | -1,212.6 | -156.7 | 1112(m)~1691(M) | 2 |
| $Ca(s) + H_2(g) + O_2(g) = Ca(OH)_2(s)$ | -983.1 | -285.2 | 298~1000 | 7,8 |
| $Ca(s) + \frac{1}{2}O_2(g) = CaO(s)$ | -633.1 | -99.0 | 298~1112(m) | 3 |
| $Ca(l) + \frac{1}{2}O_2(g) = CaO(s)$ | -640.2 | -108.6 | 1112~1764(b) | 4 |
| $Ca(g) + \frac{1}{2}O_2(g) = CaO(s)$ | -795.4 | -195.1 | 1764~2500 | 3 |
| $3Ca(s) + P_2(g) = Ca_3P_2(s)$ | -650.0 | -216.0 | 298~1115 | 8 |
| $Ca(s,a) + \frac{1}{2}S_2(g) = CaS(s)$ | -541.6 | -95.4 | 298~721 | 3 |
| $Ca(s,\beta) + \frac{1}{2}S_2(g) = CaS(s)$ | -542.6 | -96.1 | 721~1112(m) | 3 |
| $Ca(l) + \frac{1}{2}S_2(g) = CaS(s)$ | -548.1 | -103.8 | 1112~1764(b) | 4 |
| $CaO(s) + SO_2(g) + \frac{1}{2}O_2(g) = CaSO_4(s,\alpha)$ | -454.0 | -232.0 | 1468~1638(M) | 16 |
| $CaO(s) + SO_2(g) + \frac{1}{2}O_2(g) = CaSO_4(s,\beta)$ | -460.0 | -238.0 | 1223~1468 | 16 |
| $CaO(s) + Al_2O_3(s) = CaO{\cdot}Al_2O_3(s)$ | -18.0 | 19.0 | 773~1878(M) | 18 |

| Reactions | $\Delta H^o$ (kJ mol$^{-1}$) | $\Delta S^o$ (J mol$^{-1}$ K$^{-1}$) | Temperature Range (K) | Ref |
|---|---|---|---|---|
| $CaO(s) + 2Al_2O_3(s) = CaO \cdot 2Al_2O_3(s)$ | -17.0 | 25.5 | 773~2023(M) | 18 |
| $3CaO(s) + Al_2O_3(s) = 3CaO \cdot Al_2O_3(s)$ | -12.6 | 24.7 | 773~1808(M) | 18 |
| $CaO(s) + Fe_2O_3(s) = CaO \cdot Fe_2O_3(s)$ | -30.0 | 4.8 | 973~1489(M) | 18 |
| $2CaO(s) + Fe_2O_3(s) = 2CaO \cdot Fe_2O_3(s)$ | -53.1 | 2.5 | 973~1723(M) | 18 |
| $2CaO(s) + P_2(g) + 2\frac{1}{2}O_2(g) = 2CaO \cdot P_2O_5(s)$ | -2,190.0 | -586.0 | 298~1626(M) | 8 |
| $2CaO(s) + SiO_2(s) = 2CaO \cdot SiO_2(s)$ | -120.0 | 11.3 | 298~2403(M) | 8 |
| $3CaO(s) + 2SiO_2(s) = 3CaO \cdot 2SiO_2(s)$ | -237.0 | -9.6 | 298~1773 | 8 |
| $CaO(s) + TiO_2(s) = CaO \cdot TiO_2(s)$ | -80.0 | -3.4 | 298~1673 | 18 |
| $3CaO(s) + 2TiO_2(s) = 3CaO \cdot 2TiO_2(s)$ | -270.0 | 11.5 | 298~1673 | 18 |
| $4CaO(s) + 3TiO_2(s) = 4CaO \cdot 3TiO_2(s)$ | -293.0 | 17.6 | 298~1673 | 18 |
| $3CaO(s) + P_2(g) + 2\frac{1}{2}O_2(g) = 3CaO \cdot P_2O_5(s)$ | -2,314.0 | -556.0 | 298~2003(M) | 8 |
| $3CaO(s) + SiO_2(s) = 3CaO \cdot SiO_2(s)$ | -118.8 | 6.7 | 298~1773 | 4 |
| $2CaO(s) + SiO_2(s) = 2CaO \cdot SiO_2(s)$ | -118.8 | 11.3 | 298~2403(M) | 4 |
| $CaO(s) + SiO_2(s) = CaO \cdot SiO_2(s)$ | -92.8 | -2.5 | 298~1813(M) | 4 |
| $3CaO(s) + 2SiO_2(s) = 3CaO \cdot 2SiO_2(s)$ | -236.8 | -9.6 | 298~1773 | 4 |
| $Cd(l) + Cl_2(g) = CdCl_2(s)$ | -389.6 | -153.0 | 594(m)~8421(M) | 4 |
| $Cd(l) + Cl_2(g) = CdCl_2(l)$ | -352.6 | -110.3 | 841(M)~1040(b) | 14 |
| $Cd(l) + \frac{1}{2}O_2(g) = CdO(s)$ | -263.2 | -104.9 | 594(m)~1040(b) | 4 |
| $Cd(g) + \frac{1}{2}O_2(g) = CdO(s)$ | -356.7 | -198.5 | 1040~1500 | 5 |
| $Ce(s) + O_2(g) = CeO_2(s)$ | -1,084.0 | -212.0 | 298~1071 | 20 |
| $2Ce(s) + 1\frac{1}{2}O_2(g) = Ce_2O_3(s)$ | -1,788.0 | -286.6 | 298~1071(m) | 4 |
| $Ce(l) + \frac{1}{2}S_2(g) = CeS(s)$ | -534.9 | -91.0 | 1071(m)~2723(M) | 4 |
| $Co(s) + \frac{1}{2}O_2(g) = CoO(s)$ | -233.9 | -70.7 | 298~1400 | 3 |
| $Co(s) + S_2(g) = CoS_2(s)$ | -280.3 | 182.4 | 298~872 | 4 |

| Reactions | $\Delta H^o$ (kJ mol$^{-1}$) | $\Delta S^o$ (J mol$^{-1}$ K$^{-1}$) | Temperature Range (K) | Ref |
|---|---|---|---|---|
| $3Cr(s) + 2C(s) = Cr_3C_2(s)$ | -79.1 | 17.7 | 298~2130(m) | 4 |
| $7Cr(s) + 3C(s) = Cr_7C_3(s)$ | -153.6 | 37.2 | 298~2130(m) | 4 |
| $23Cr(s) + 6C(s) = Cr_{23}C_6(s)$ | -309.6 | 77.4 | 298~1773 | 4 |
| $Cr(s) + ½O_2(g) = CrO(l)$ | -334.2 | 63.8 | 1938~2023(M) | 4 |
| $2Cr(s) + 1½O_2(g) = Cr_2O_3(s)$ | -1,110.1 | -249.3 | 1173~1923 | 4 |
| $3Cr(s) + 2O_2(g) = Cr_3O_4(s)$ | -1,355.0 | -265.0 | 1923~1938(M) | 21 |
| $Cu(s) + ½O_2(g) = CuO(s)$ | -152.5 | -85.3 | 298~1356(m) | 2 |
| $2Cu(s) + ½O_2(g) = Cu_2O(s)$ | -168.4 | -71.2 | 298~1356(m) | 4 |
| $2Cu(l) + ½O_2(g) = Cu_2O(s)$ | -181.7 | -80.6 | 1356~1509(M) | 2 |
| $2Cu(l) + ½O_2(g) = Cu_2O(l)$ | -118.7 | -39.5 | 1509~2273 | 2 |
| $Cu(s) + ½S_2(g) = CuS(s)$ | -115.6 | -76.1 | 298~708 | 4 |
| $2Cu(s) + ½S_2(g) = Cu_2S(s,\gamma)$ | -140.7 | -43.3 | 298~708 | 4 |
| $2Cu(s) + ½S_2(g) = Cu_2S(s,\mu)$ | -131.8 | -30.8 | 708~1356(m) | 4 |
| $Cu(s) + Fe(s) + S_2(g) = CuFeS_2(s,a)$ | -278.6 | -115.3 | 830~973 | 4 |
| $3Fe(s,\alpha) + C(s) = Fe_3C(s)$ | 29.0 | 28.0 | 298~1000 | 4 |
| $3Fe(s,\gamma) + C(s) = Fe_3C(s)$ | 11.2 | 11.0 | 1000~1410 | 4 |
| $Fe(s,\alpha) + Cl_2(g) = FeCl_2(s)$ | -339.4 | -119.2 | 298~950(M) | 3 |
| $Fe(s) + Cl_2(g) = FeCl_2(l)$ | -286.4 | -63.7 | 950(M)~1297(B) | 3 |
| $Fe(s) + Cl_2(g) = FeCl_2(g)$ | -169.6 | 26.5 | 1297(B)~1809(m) | 3 |
| $Fe(s) + 1½Cl_2(g) = FeCl_3(s)$ | -396.5 | -210.4 | 298~577(M) | 2 |
| $Fe(s) + 1½Cl_2(g) = FeCl_3(g)$ | -261.3 | -28.0 | 605(B)~1809(m) | 2 |
| $0.947Fe(s) + ½O_2(g) = Fe_{0.947}O(s)$ | -263.7 | -64.4 | 298~1643(M) | 15 |
| $Fe(s) + ½O_2(g) =$ "FeO"$(s)$ | -264.0 | -64.6 | 298~1650 | 2 |
| $Fe(l) + ½O_2(g) = FeO(l)$ | -256.0 | -53.7 | 1644(M)~2273 | 4 |
| $2Fe(s) + 1½O_2(g) = Fe_2O_3(s)$ | -815.0 | -251.1 | 298~1735 | 2 |
| $3Fe(s) + 2O_2(g) = Fe_3O_4(s)$ | -1,103.1 | -307.4 | 298~1870(M) | 2 |
| $Fe(s,\gamma) + ½S_2(g) = FeS(s)$ | -154.9 | -56.9 | 1179~1261 | 4 |
| $Fe(s) + ½S_2(g) = FeS(l)$ | -164.0 | -61.1 | 1261~1468(M) | 4 |
| $Fe(s) + S_2(g) = FeS_2(s)$ | -336.9 | -244.5 | 903~1033 | 4 |
| $Cu(s) + Fe(s) + S_2(g) = CuFeS_2(s,\alpha)$ | -278.6 | -115.3 | 830~973 | 4 |

| Reactions | $\Delta H^o$ (kJ mol$^{-1}$) | $\Delta S^o$ (J mol$^{-1}$ K$^{-1}$) | Temperature Range (K) | Ref |
|---|---|---|---|---|
| $Ga(l) + \frac{1}{4}As_4(g) = GaAs(s)$ | -115.0 | -72.0 | 303~1238 | 12 |
| $Ga(l) + \frac{1}{2}Cl_2(g) = GaCl(g)$ | -79.6 | 52.4 | 303~2000 | 12 |
| $Ga(l) + 1\frac{1}{2}Cl_2(g) = GaCl_3(g)$ | -442.0 | 84.7 | 351(M)~575(B) | 3,8 |
| $2Ga(l) + 1\frac{1}{2}O_2(g) = Ga_2O_3(s)$ | -1,089.9 | -323.6 | 303(m)~2068(M) | 4 |
| $Ga(l) + \frac{1}{4}P_4(g) = GaP(s)$ | -142.3 | -77.0 | 303~1790(M) | 7,8 |
| $Ga(l) + \frac{1}{2}S_2(g) = GaS(s)$ | -276.0 | -111.0 | 303~1233(M) | 8 |
| $2Ga(l) + 1\frac{1}{2}S_2(g) = Ga_2S_3(s)$ | -719.6 | -318.4 | 303(m)~1363(M) | 4 |
| $Ge(s) + O_2(g) = GeO_2(s)$ | -575.0 | -188.0 | 298~1210 | 8 |
| $\frac{1}{2}H_2(g) + \frac{1}{2}Br_2(g) = HBr(g)$ | -53.6 | 6.9 | 298~2273 | 4 |
| $\frac{1}{2}H_2(g) + \frac{1}{2}Cl_2(g) = HCl(g)$ | -94.1 | 6.4 | 298~2273 | 4 |
| $\frac{1}{2}H_2(g) + \frac{1}{2}F_2(g) = HF(g)$ | -274.5 | 3.5 | 298~2273 | 4 |
| $\frac{1}{2}H_2(g) + \frac{1}{2}I_2(g) = HI(g)$ | -4.2 | 8.8 | 298~2273 | 4 |
| $H_2(g) + \frac{1}{2}O_2(g) = H_2O(g)$ | -247.4 | -55.8 | 298~2500 | 2 |
| $\frac{1}{2}H_2(g) + \frac{1}{2}S_2(g) = HS(g)$ | 79.7 | 15.5 | 718~2273 | 2 |
| $H_2(g) + \frac{1}{2}S_2(g) = H_2S(g)$ | -91.6 | -50.6 | 298~2273 | 4 |
| $Hf(s) + O_2(g) = HfO_2(s,a)$ | -1,060.0 | -174.0 | 298~1973 | 7 |
| $Hg(l) + \frac{1}{2}O_2(g) = HgO(s,red)$ | -90.8 | -70.3 | 298~773 | 15 |
| $Hg(l) + \frac{1}{2}S_2(g) = HgS(s,red)$ | -53.0 | -82.0 | 298~618 | 3 |
| $HgS(s,black) = HgS(s,red)$ | -4.0 | -6.3 | 618~618 | 3 |
| $In(l) + \frac{1}{4}As_4(g) = InAs(s)$ | -99.3 | -64.8 | 430~1215 | 12 |
| $In(l) + \frac{1}{2}Cl_2(g) = InCl(l)$ | -169.1 | -37.1 | 498(M)~881(B) | 7,8 |
| $In(l) + \frac{1}{2}Cl_2(g) = InCl(g)$ | -87.2 | 56.3 | 430(m)~2000 | 12 |
| $In(l) + 1\frac{1}{2}Cl_2(g) = InCl_3(s)$ | -533.4 | -242.4 | 430(m)~856(M) | 7,8 |
| $In(l) + 1\frac{1}{2}Cl_2(g) = InCl_3(g)$ | -375.0 | -36.7 | 500~800 | 3,8 |
| $In(l) + \frac{1}{4}P_4(g) = InP(s)$ | -92.4 | -74.1 | 430~1328(M) | 7,8 |
| $Ir(s) + O_2(g) = IrO_2(s)$ | -234.0 | -170.0 | 298~1273 | 7 |
| $Ir(s) + S_2(g) = IrS_2(s)$ | -268.0 | -190.0 | 298~1273 | 8 |
| $2K(l) + C(s) + 1\frac{1}{2}O_2(g) = K_2CO_3(s)$ | -1,149.3 | -288.5 | 336(m)~1037(b) | 2 |
| $2K(g) + C(s) + 1\frac{1}{2}O_2(g) = K_2CO_3(s)$ | -1,277.1 | -410.4 | 1037~1174(M) | 2 |

| Reactions | $\Delta H^o$ (kJ mol$^{-1}$) | $\Delta S^o$ (J mol$^{-1}$ K$^{-1}$) | Temperature Range (K) | Ref |
|---|---|---|---|---|
| $2K(g) + C(s) + 1½O_2(g) = K_2CO_3(l)$ | -1,204.8 | -353.7 | 1174~2500 | 2 |
| $K(l) + ½Cl_2(g) = KCl(s)$ | -438.9 | -100.4 | 336(m)~1037(b) | 3 |
| $K(g) + ½Cl_2(g) = KCl(l)$ | -474.0 | -131.8 | 1044(M)~1710(B) | 4 |
| $K(g) + ½Cl_2(g) = KCl(g)$ | -306.3 | -35.5 | 1710(B)~2273 | 4 |
| $K(l) + ½H_2(g) + ½O_2(g) = KOH(l)$ | -402.3 | -118.0 | 673(M)~1037(b) | 2 |
| $K(g) + ½H_2(g) + ½O_2(g) = KOH(l)$ | -469.6 | -182.9 | 1037~1600(B) | 2 |
| $2K(l) + ½O_2(g) = K_2O(s)$ | -363.2 | -140.4 | 336(m)~1037(b) | 2 |
| $2K(g) + ½O_2(g) = K_2O(s)$ | -478.7 | -253.0 | 1037~2000 | 2 |
| $2K(l) + ½S_2(g) = K_2S(s)$ | -481.2 | -143.5 | 336~1037(b) | 4 |
| $2K(g) + ½S_2(g) = K_2S(s)$ | -633.1 | -289.8 | 1037~1221(M) | 4 |
| $2K(g) + ½S_2(g) = K_2S(l)$ | -616.9 | -276.6 | 1221~2000 | 4,8 |
| $2La(s) + 1½O_2(g) = La_2O_3(s)$ | -1,790.0 | -278.0 | 298~1193 | 8 |
| $La(s) + ½S_2(g) = LaS(s)$ | -527.0 | -104.0 | 1193~1773 | 8 |
| $2La(s) + 1½S_2(g) = La_2S_3(s)$ | -1,420.0 | -286.0 | 1193~1773 | 8 |
| $Li(l) + ½F_2(g) = LiF(l)$ | -583.4 | -66.8 | 1121(M)~1620(b) | 4 |
| $Li(g) + ½F_2(g) = LiF(g)$ | -500.0 | -43.0 | 1990(B)~2273 | 15 |
| $2Li(l) + ½O_2(g) = Li_2O(s)$ | -603.8 | -136.6 | 454(m)~1620(b) | 2 |
| $2Li(g) + ½O_2(g) = Li_2O(s)$ | -854.7 | -290.8 | 1620~1843(M) | 2 |
| $2Li(l) + ½S_2(g) = Li_2S(s)$ | -514.6 | -121.3 | 454(m)~1273 | 4,8 |
| $Mg(l) + Cl_2(g) = MgCl_2(l)$ | -596.8 | -114.2 | 987(M)~1378(b) | 2 |
| $Mg(g) + Cl_2(g) = MgCl_2(l)$ | -649.0 | -157.7 | 987(M)~1710(B) | 15 |
| $Mg(s) + Cl_2(g) = MgCl_2(s)$ | -637.0 | -155.4 | 298~923 | 15 |
| $Mg(l) + F_2(g) = MgF_2(s)$ | -1,126.8 | -177.8 | 922(m)~1276 | 3 |
| $Mg(g) + F_2(g) = MgF_2(l)$ | -1,172.0 | -215.6 | 1536(M)~2536(B) | 15 |
| $Mg(s) + F_2(g) = MgF_2(s)$ | -1,120.0 | -171.2 | 298~922(m) | 15 |
| $Mg(s) + C(s) + 1½O_2(g) = MgCO_3(s)$ | -1,109.5 | -274.4 | 298~922(m) | 2 |
| $Mg(s) + H_2(g) + O_2(g) = Mg(OH)_2(s)$ | -922.9 | -300.8 | 298~922(m) | 2 |
| $Mg(s) + ½O_2(g) = MgO(s)$ | -601.2 | -107.6 | 298~922(m) | 2 |
| $Mg(l) + ½O_2(g) = MgO(s)$ | -609.6 | -116.5 | 922~1378(b) | 2 |

| Reactions | $\Delta H^o$ (kJ mol$^{-1}$) | $\Delta S^o$ (J mol$^{-1}$ K$^{-1}$) | Temperature Range (K) | Ref |
|---|---|---|---|---|
| $Mg(g) + ½O_2(g) = MgO(s)$ | -732.7 | -206.0 | 1378~2000 | 2 |
| $Mg(s) + ½S_2(g) = MgS(s)$ | -409.6 | -94.4 | 298~922(m) | 4,8 |
| $Mg(l) + ½S_2(g) = MgS(s)$ | -408.9 | -98.0 | 922~1378(b) | 2,4 |
| $Mg(g) + ½S_2(g) = MgS(s)$ | -539.7 | -193.0 | 1378~1973 | 4,8 |
| $MgO(s) + Al_2O_3(s) = MgO{\cdot}Al_2O_3(s)$ | -35.6 | 2.1 | 298~1673 | 15 |
| $MgO(s) + SiO_2(s) = MgO{\cdot}SiO_2(s)$ | -41.1 | -6.1 | 298~1850(M) | 4 |
| $2MgO(s) + SiO_2(s) = 2MgO{\cdot}SiO_2(s)$ | -67.2 | -4.3 | 298~2171(M) | 4 |
| $2MgO(s) + TiO_2(s) = 2MgO{\cdot}TiO_2(s)$ | -26.4 | -1.3 | 298~1773 | 15 |
| $3Mn(s) + C(s) = Mn_3C(s)$ | -13.9 | -1.1 | 298~1310 | 4,7 |
| $7Mn(s) + 3C(s) = Mn_7C_3(s)$ | -127.6 | -21.1 | 298~1473 | 4,8 |
| $Mn(s) + Cl_2(g) = MnCl_2(s)$ | -478.2 | -127.7 | 298~923(M) | 4 |
| $Mn(s) + Cl_2(g) = MnCl_2(l)$ | -440.6 | -86.9 | 923(M)~1200 | 4 |
| $Mn(s) + ½O_2(g) = MnO(s)$ | -388.9 | -76.3 | 298~1517(m) | 4 |
| $3Mn(s) + 2O_2(g) = Mn_3O_4(s)$ | -1,384.9 | -344.4 | 298~1517(m) | 4 |
| $2Mn(s) + 1½O_2(g) = Mn_2O_3(s)$ | -953.9 | -255.2 | 298~1517(m) | 4 |
| $Mn(s) + O_2(g) = MnO_2(s)$ | -519.0 | -181.0 | 298~783 | 18 |
| $Mn(s) + ½S_2(g) = MnS(s)$ | -296.0 | -76.7 | 973~1473 | 4 |
| $Mn(s) + Si(s) = MnSi(s)$ | -61.5 | -6.3 | 298~1519 | 8 |
| $MnO(s) + Al_2O_3(s) = MnO{\cdot}Al_2O_3(s)$ | -48.1 | -7.3 | 773~1473 | 18 |
| $MnO(s) + SiO_2(s) = MnO{\cdot}SiO_2(s)$ | -28.0 | -2.8 | 298~1564(M) | 4 |
| $2MnO(s) + SiO_2(s) = 2MnO{\cdot}SiO_2(s)$ | -53.6 | -24.7 | 298~1618(M) | 4 |
| $Mo(s) + C(s) = MoC(s)$ | -7.5 | 5.4 | 298~973 | 4,7 |
| $2Mo(s) + C(s) = Mo_2C(s)$ | -45.6 | 4.2 | 298~1373 | 4,7 |
| $Mo(s) + 2Cl_2(g) = MoCl_4(s)$ | -472.4 | -236.4 | 298~590(M) | 2 |
| $2Mo(s) + ½N_2(g) = Mo_2N(s)$ | -60.7 | -14.6 | 298~773 | 8 |
| $Mo(s) + O_2(g) = MoO_2(s)$ | -578.2 | -166.5 | 298~2273 | 4 |
| $Mo(s) + O_2(g) = MoO_2(g)$ | -18.4 | 33.9 | 298~2273 | 4 |
| $Mo(s) + 1½O_2(g) = MoO_3(s)$ | -740.2 | -246.7 | 298~1074(m) | 4 |

| Reactions | $\Delta H^o$ (kJ mol$^{-1}$) | $\Delta S^o$ (J mol$^{-1}$ K$^{-1}$) | Temperature Range (K) | Ref |
|---|---|---|---|---|
| $Mo(s) + 1\frac{1}{2}O_2(g) = MoO_3(l)$ | -664.5 | -176.6 | 1074~2000 | 2 |
| $Mo(s) + 1\frac{1}{2}O_2(g) = MoO_3(g)$ | -359.8 | -59.4 | 298~2273 | 4 |
| $2Mo(s) + 1\frac{1}{2}S_2(g) = Mo_2S_3(s)$ | -594.1 | -265.3 | 298~1473 | 4,8 |
| $Mo(s) + S_2(g) = MoS_2(s)$ | -397.5 | -182.0 | 298~1458(M) | 4,8 |
| $Na(l) + \frac{1}{2}Cl_2(g) = NaCl(s)$ | -411.6 | -93.1 | 371(m)~1074(M) | 4 |
| $Na(g) + \frac{1}{2}Cl_2(g) = NaCl(l)$ | -464.4 | -133.9 | 1074(M)~1738(B) | 4 |
| $Na(l) + \frac{1}{2}F_2(g) = NaF(s)$ | -576.6 | -105.5 | 371(m)~1269(M) | 4 |
| $Na(g) + \frac{1}{2}F_2(g) = NaF(l)$ | -624.3 | -148.2 | 1269(M)~2060(B) | 4 |
| $3Na(g) + Al(l) + 3F_2(g) = Na_3AlF_6(l)$ | -3,378.2 | -623.4 | 1285(M)~2273 | 4 |
| $2Na(l) + C(s) + 1\frac{1}{2}O_2(g) = Na_2CO_3(s)$ | -1,127.5 | -273.6 | 371(m)~1156(b) | 2 |
| $2Na(g) + C(s) + 1\frac{1}{2}O_2(g) = Na_2CO_3(l)$ | -1,229.6 | -362.5 | 1123(M)~2500 | 2 |
| $Na(l) + \frac{1}{2}H_2(g) + \frac{1}{2}O_2(g) = NaOH(l)$ | -408.1 | -125.7 | 592(M)~1156(b) | 2 |
| $Na(g) + \frac{1}{2}H_2(g) + \frac{1}{2}O_2(g) = NaOH(l)$ | -486.6 | -192.5 | 1156(b)~1663(B) | 2 |
| $2Na(l) + \frac{1}{2}O_2(g) = Na_2O(s)$ | -421.6 | -141.3 | 371(m)~1156(b) | 4 |
| $2Na(g) + \frac{1}{2}O_2(g) = Na_2O(s)$ | -571.7 | -269.8 | 1156~1405(M) | 2 |
| $2Na(g) + \frac{1}{2}O_2(g) = Na_2O(l)$ | -519.8 | -234.3 | 1405~2223 | 2 |
| $2Na(l) + \frac{1}{2}S_2(g) = Na_2S(s)$ | -394.0 | -83.7 | 371(m)~1156(b) | 2 |
| $2Na(g) + \frac{1}{2}S_2(g) = Na_2S(s)$ | -521.2 | -200.1 | 1156~1223(M) | 2 |
| $2Na(g) + \frac{1}{2}S_2(g) = Na_2S(l)$ | -610.9 | -274.7 | 1223~2000 | 2 |
| $\frac{1}{2}N_2(g) + 1\frac{1}{2}H_2(g) = NH_3(g)$ | -53.7 | -116.5 | 298~2273 | 2,4 |
| $\frac{1}{2}N_2(g) + \frac{1}{2}O_2(g) = NO(g)$ | 90.4 | 12.7 | 298~2273 | 4 |
| $\frac{1}{2}N_2(g) + O_2(g) = NO_2(g)$ | 32.3 | -63.3 | 298~2273 | 15 |
| $2Nb(s) + C(s) = Nb_2C(s)$ | -194.0 | -11.7 | 298~1773 | 7 |
| $Nb(s) + 0.98C(s) = NbC_{0.98}(s)$ | -137.0 | -2.4 | 298~1773 | 15 |
| $Nb(s) + O_2(g) = NbO_2(s)$ | -784.0 | -167 | 298~2175(M) | 15 |
| $2Nb(s) + 2\frac{1}{2}O_2(g) = Nb_2O_5(s)$ | -1,888.0 | -420 | 298~1785(M) | 15 |
| $Nb(s) + \frac{1}{2}O_2(g) = NbO(s)$ | -415.0 | -87.0 | 298~2210 | 15 |
| $3Ni(s) + C(s) = Ni_3C(s)$ | 39.7 | 17.1 | 298~773 | 4,8 |
| $Ni(s) + Cl_2(g) = NiCl_2(s)$ | -305.4 | -146.4 | 298~1260 | 4 |

| Reactions | $\Delta H^o$ (kJ mol$^{-1}$) | $\Delta S^o$ (J mol$^{-1}$ K$^{-1}$) | Temperature Range (K) | Ref |
|---|---|---|---|---|
| $Ni(s) + ½S_2(g) = NiS(l)$ | -111.7 | -43.6 | 1067~1728(m) | 9,13 |
| $Ni(s) + ½O_2(g) = NiO(s)$ | -235.8 | -86.2 | 298~1728(m) | 6 |
| $3Ni(s) + S_2(g) = Ni_3S_2(s)$ | -336.7 | -162.9 | 298~1064(M) | 9,13 |
| $3Ni(s) + S_2(g) = Ni_3S_2(l)$ | -237.3 | -62.4 | 1064~1728(m) | 9,13 |
| $Ni(s) + ½S_2(g) = NiS(s)$ | -153.6 | -83.6 | 298~1067(M) | 9,13 |
| $NiO(s) + SO_2(g) + ½O_2(g) = NiSO_4(s)$ | -347.5 | -293.2 | 873~1133 | 17 |
| $¼P_4(g) = P(s, red)$ | -32.1 | -45.6 | 298~704 | 15 |
| $4P_{red}(s) = P_4(g)$ | 128.5 | 182.6 | 298~704 | 4 |
| $½P_4(g) = P_2(g)$ | 108.6 | 69.5 | 298~1973 | 4 |
| $½P_2(g) + 1½H_2(g) = PH_3(g)$ | -71.5 | -108.2 | 298~1973 | 4,8 |
| $2P_2(g) + 5O_2(g) = P_4O_{10}(g)$ | -3,155.0 | -1011 | 631~1973 | 15 |
| $½P_2(g) + O_2(g) = PO_2(g)$ | -386.0 | -60.3 | 298~1973 | 15 |
| $½P_2(g) + ½O_2(g) = PO(g)$ | -77.8 | 11.6 | 298~1973 | 15 |
| $Pb(l) + Cl_2(g) = PbCl_2(l)$ | -324.6 | -103.5 | 774(M)~1226(B) | 4 |
| $Pb(l) + Cl_2(g) = PbCl_2(g)$ | -188.3 | 7.5 | 1226(B)~2023(b) | 4 |
| $Pb(l) + ½O_2(g) = PbO\ (s, red)$ | -221.5 | -104.6 | 600(m)~762(t) | 2 |
| $Pb(l) + ½O_2(g) = PbO\ (s, yellow)$ | -218.1 | -100.2 | 762~1170(M) | 2 |
| $Pb(l) + ½O_2(g) = PbO(s)$ | -219.1 | -101.2 | 600~1170(M) | 4 |
| $Pb(l) + ½O_2(g) = PbO(l)$ | -185.1 | -72.0 | 1170~1789(B) | 15 |
| $3Pb(l) + 2O_2(g) = Pb_3O_4(s)$ | -702.5 | -368.9 | 600~1473 | 4 |
| $Pb(l) + ½S_2(g) = PbS(s)$ | -163.2 | -88.0 | 600(m)~1386(M) | 4 |
| $Pb(l) + ½S_2(g) = PbS(g)$ | 59.0 | 54.0 | 1100~1400 | 5 |
| $PbO(s) + SO_2(g) + ½O_2(g) = PbSO_4(s)$ | -401.0 | -262.0 | 298~1363(M) | 7 |
| $S(l) = S(s)$ | -1.7 | -4.4 | 388(M)~388(M) | 15 |
| $½S_2(g) = S(l)$ | -58.6 | -68.3 | 388(M)~718(B) | 15 |
| $½S_2(g) = S(g)$ | 217.0 | 59.6 | 298~1973 | 2 |
| $2S(l) = S_2(g)$ | 120.0 | 139.6 | 388(m)~718(b) | 10 |
| $2S(g) = S_2(g)$ | -469.3 | -161.3 | 298~1973 | |
| $1½S_2(g) = S_3(g)$ | -56.3 | -80.4 | 298~800 | 10 |
| $2S_2(g) = S_4(g)$ | -117.9 | -154.7 | 298~800 | 10 |

| Reactions | $\Delta H^o$ (kJ mol$^{-1}$) | $\Delta S^o$ (J mol$^{-1}$ K$^{-1}$) | Temperature Range (K) | Ref |
|---|---|---|---|---|
| $2S_2(g) = S_4(g)$ | -62.8 | -115.5 | 298~1973 | 4 |
| $2\frac{1}{2}S_2(g) = S_5(g)$ | -203.0 | -240.0 | 298~800 | 10 |
| $3S_2(g) = S_6(g)$ | -276.1 | -305.0 | 298~1973 | 4 |
| $3\frac{1}{2}S_2(g) = S_7(g)$ | -331.6 | -374.1 | 298~800 | 10 |
| $4S_2(g) = S_8(g)$ | -397.5 | -448.1 | 298~1973 | 4 |
| $\frac{1}{2}S_2(g) + \frac{1}{2}O_2(g) = SO(g)$ | -57.8 | 5.0 | 718(b)~2273 | 4 |
| $\frac{1}{2}S_2(g) + O_2(g) = SO_2(g)$ | -361.7 | -72.7 | 718(b)~2273 | 4 |
| $\frac{1}{2}S_2(g) + 1\frac{1}{2}O_2(g) = SO_3(g)$ | -457.9 | -163.3 | 718(b)~2273 | 4 |
| $2Sb(s) + 1\frac{1}{2}O_2(g) = Sb_2O_3(s)$ | -687.6 | -241.1 | 298~904(m) | 3 |
| $2Sb(l) + 1\frac{1}{2}O_2(g) = Sb_2O_3(l)$ | -660.7 | -198.1 | 929(M)~1860(b) | 4 |
| $\frac{1}{2}Se_2(g) + O_2(g) = SeO_2(g)$ | -178.0 | -66.1 | 958~1973 | 8 |
| $\frac{1}{2}Se_2(g) + \frac{1}{2}O_2(g) = SeO(g)$ | -9.2 | -4.2 | 958~1973 | 8 |
| $Si(s) + C(s) = SiC(s_{,\beta})$ | -73.1 | -7.7 | 29 8~1685(m) | 4 |
| $Si(l) + C(s) = SiC(s_{,\beta})$ | -122,6 | -37.0 | 1685~2273 | 4 |
| $Si(s) + 2Cl_2(g) = SiCl_4(g)$ | -660.2 | -128.8 | 334(B)~1685(m) | 4 |
| $3Si(s) + 2N_2(g) = Si_3N_4(s_{,\alpha})$ | -723.8 | -315.1 | 298~1685(m) | 4 |
| $3Si(l) + 2N_2(g) = Si_3N_4(s_{,\alpha})$ | -874.5 | -405.0 | 1685(m)~1973 | 4 |
| $Si(s) + \frac{1}{2}O_2(g) = SiO(g)$ | -104.2 | 82.5 | 298~1685(m) | 4 |
| $Si(s) + O_2(g) = SiO_2(s_{,quartz})$ | -907.1 | -175.7 | 298~1685(m) | 4 |
| $Si(s) + O_2(g) = SiO_2(s_{,\beta\text{-}cristobalite})$ | -904.8 | -173.8 | 298~1685(m) | 2 |
| $Si(l) + O_2(g) = SiO_2(s_{,\beta\text{-}cristobalite})$ | -946.3 | -197.6 | 1685~1996(M) | 2 |
| $Si(l) + O_2(g) = SiO_2(l)$ | -921.7 | -185.9 | 1996~3514(b) | 2 |
| $Si(s) + \frac{1}{2}S_2(g) = SiS(g)$ | 51.8 | 81.6 | 973~1685(m) | 2 |
| $Si(s) + S_2(g) = SiS_2(s)$ | -326.4 | -139.0 | 298~1363(M) | 4 |
| $Sn(l) + Cl_2(g) = SnCl_2(l)$ | -333.0 | -118.4 | 520~925(B) | 3 |
| $Sn(l) + Cl_2(g) = SnCl_2(g)$ | -225.9 | -13.0 | 925(B)~1473 | 4 |
| $Sn(l) + 2Cl_2(g) = SnCl_4(g)$ | -512.5 | -150.6 | 500~1200 | 3 |
| $Sn(l) + \frac{1}{2}O_2(g) = SnO(g)$ | 6.3 | 50.9 | 505(m)~1973 | |
| $Sn(l) + O_2(g) = SnO_2(s)$ | -574.9 | -198.4 | 505~1903(M) | |
| $Sn(l) + \frac{1}{2}S_2(g) = SnS_2(g)$ | 26.0 | 49.4 | 505~1973 | 8 |

| Reactions | $\Delta H^o$ (kJ mol$^{-1}$) | $\Delta S^o$ (J mol$^{-1}$ K$^{-1}$) | Temperature Range (K) | Ref |
|---|---|---|---|---|
| $Ta(s) + \frac{1}{2}O_2(g) = TaO(g)$ | 188.0 | 87.0 | 298~2273 | 15 |
| $Ta(s) + O_2(g) = TaO_2(g)$ | -209.0 | 20.5 | 298~2273 | 15 |
| $\frac{1}{2}Te_2(g) + \frac{1}{2}O_2(g) = TeO(l)$ | -7.1 | 6.0 | 1282~1973 | 8 |
| $Ti(s) + B(s) = TiB(s)$ | -163.0 | -5.9 | 298~1939 | 15 |
| $Ti(s) + 2B(s) = TiB_2(s)$ | -285.0 | -20.5 | 298~1939 | 15 |
| $Ti(s) + O_2(g) = TiO_2\ (s_{,rutile})$ | -941.0 | -177.6 | 298~1943(m) | 4 |
| $Ti(s) + 2Cl_2(g) = TiCl_4(g)$ | -764.0 | -121.5 | 298~1943(m) | 4 |
| $3Ti(s) + 2\frac{1}{2}O_2(g) = Ti_3O_5(s)$ | -2,435.1 | -420.5 | 298~1943(m) | 4 |
| $2Ti(s) + 1\frac{1}{2}O_2(g) = Ti_2O_3(s)$ | -1,502.1 | -258.1 | 298~1943(m) | 4 |
| $Ti(s) + \frac{1}{2}O_2(g) = TiO\ (s_{,\beta})$ | -514.6 | -74.1 | 298~1943(m) | 4 |
| $U(l) + C(s) = UC(s)$ | -109.6 | -1.8 | 1405(m)~1800 | 4,7 |
| $U(s) + \frac{1}{2}N_2(s) = UN(s)$ | -292.9 | -80.8 | 298~1405(m) | 4,7 |
| $U(s) + O_2(g) = UO_2(s)$ | -1,079.5 | -167.4 | 298~1405(m) | 3 |
| $U(l) + O_2(g) = UO_2(s)$ | -1,086.6 | -172.3 | 1405~2273 | 4 |
| $V(s) + B(s) = VB(s)$ | -138.0 | -5.9 | 298~2273 | 8 |
| $V(s) + 0.23N_2(g) = VN_{0.23}(s)$ | -130.0 | -44.4 | 298~1973 | 15 |
| $V(s) + \frac{1}{2}N_2(g) = VN(s)$ | -214.6 | -82.4 | 298~2619 | 4 |
| $V(s) + O_2(g) = VO_2(s)$ | -706.0 | -155.0 | 298~1633(M) | 15 |
| $2V(s) + 1\frac{1}{2}O_2(g) = V_2O_3(s)$ | -1,203.0 | -238.0 | 298~2343 | 15 |
| $V(s) + \frac{1}{2}O_2(g) = VO(s)$ | -425.0 | -80.0 | 298~2073 | 15 |
| $W(s) + C(s) = WC(s)$ | -42.3 | -5.0 | 1173~1575 | 4 |
| $2W(s) + C(s) = W_2C(s)$ | -30.5 | 2.3 | 1575~1673 | 19 |
| $Zn(l) + \frac{1}{2}S_2(g) = ZnS(s)$ | -277.8 | -107.9 | 693(m)~1180(b) | 4,8 |
| $Zn(g) + \frac{1}{2}S_2(g) = ZnS(s)$ | -375.4 | -191.6 | 1120~2000 | 3 |
| $Zn(g) + \frac{1}{2}S_2(g) = ZnS(g)$ | 5.0 | -30.5 | 1180~1973 | 4,8 |
| $Zr(s) + 2B(s) = ZrB_2(s)$ | -328.0 | -23.4 | 298~2125 | 15 |
| $Zr(s) + C(s) = ZrC(s)$ | -196.6 | -9.2 | 298~2125(m) | 4,7 |
| $Zr(s) + \frac{1}{2}N_2(g) = ZrN(s)$ | -363.0 | -92.0 | 298~2125(m) | 4,7 |
| $Zr(s) + \frac{1}{2}S_2(g) = ZrS(g)$ | 237.2 | 78.2 | 298~2125(m) | 4,8 |
| $Zr(s) + S_2(g) = ZrS_2(s)$ | -698.7 | -178.2 | 298~1823(M) | 4,8 |

Note : (m) = melting point of metal, (M) = melting point of compound
(b) = boiling point of metal, (B) = boiling point of compound

References :


1. J.P. Coughlin, *Contributions to the Data on Theoretical Metallurgy. XII. Heats and Free Energies of Formation of Inorganic Oxides*, Bulletin 542, Bureau of Mines, U.S. Department of the Interior, Washington, D.C., 1954.
2. D.R. stull and H. prophet, *JANAF Thermochemical Tables*, 2nd ed., NSRDS-NBS 37, U.S. Department of Commerce, Washington, D.C., 1971.
3. O. Kubaschewski and C. B. Alcock, *Metallurgical Thermochemistry*, 5th ed., Pergamon Press, New York, 1979.
4. E. T. Turkdogan, *Physical Chemistry of High Temperature Technology*, Academic Press, New York, 1980.
5. H. H. Kellogg, *Trans. Met. Soc. AIME*, 236:602, 1966.
6. H. H. Kellogg, *J. Chem. Eng. Data*, 14:41, 1969.
7. I. Barin and O. Knacke, *Thermodynamical Properties of Inorganic Substances*, Springer-Verlag, New York, 1973.
8. I. Barin, O. Knacke, and O. Kubaschewski, *Thermodynamical Properties of Inorganic Substances*, Supplement, Springer-Verlag, New York, 1977.
9. M. Nagamori and T. R. Ingraham, *Metall. Trans*., 1:1821, 1970.
10. H. H. Kellogg, *Metall. Trans*., 2:2161, 1971.
11. G. J. Janz and G.M. Dijkhuis, in *Molten Salts*, vol.2, NBS, U.S. Department of Commerce, Washington, D.C., 1969.
12. H. Nagai, *J. Electrochem. Soc*., 126:1400, 1979.
13. A. D. Mah and L.B. Pankratz, *Contributions to the Data on Theoretical Metallurgy, XVI. Thermodynamic properties of Nickel and Its Inorganic Compounds*, Bulletin 668, Bureau of Mines, U.S. Department of the Interior, Washington, D.C. 1976.
14. L. B. Pankratz, *Thermodynamic Properties of Halides*, Bulletin 674, Bureau of Mines, U.S. Department of the interior, Washington, D.C. 1984.
15. M.W. chase, *JANAF Thermochemical Tables*, 3rd ed., 1985, *American chem. Soc. And the American institute of physics for the national Bureau of standard. J. phys. Chem. Ref. Data Vol 14*, Supplement NO.1, Michigan 48674 USA, 1985.
16. Turkdogan, E. T. Rice, B. B., and Vinters, J. V. *Sulfide and Sulfate Solid Solubility in Lime, Metall. Trans*., 5, 1527~35., 1974
17. Skeaff, J. M. and Espelund, A. W., *An E.M.F. Method for sulfate-oxide equilibria results for the Mg, Mn, Fe, Ni, Ca and Zn system*., Can. Metall., Q., 12, 445~58., 1973
18. Kubaschewski, O., *The thermodynamic properties of double oxides High temperature – High pressure*, 4, 1~12., 1972.
19. Gupta, D. K. and seigle. L. L., *Free energies of formation of WC and $W_2C$, and the thermodynamic properties of carbon in solid tungsten,* Metall. Trans. A6, 1939~44, 1975.
20. Baker, F. B. and Holley, *C. E., Enthalpy of formation of cerium sesquioxide*. J. chem. Eng. Data, 13, 405~8, 1968.
21. Toker, N., ph. D., *Thesis*., Pennsylvania state university, 1977.

# APPENDIX III

## Properties of Selected Elements

| Atomic number | Element Symbol | Element Name | Atomic weight | Density* ($kg\ m^{-3}$) | Melting point, K | Boiling point, K |
|---|---|---|---|---|---|---|
| 1 | H | Hydrogen | 1.0079 | (0.090) | 14.025 | 20.268 |
| 2 | He | Helium | 4.00260 | (0.179) | 0.95 | 4.215 |
| 3 | Li | Lithium | 6.941 | 530 | 453.7 | 1615 |
| 4 | Be | Beryllium | 9.01218 | 1850 | 1560 | 2745 |
| 5 | B | Boron | 10.81 | 2340 | 2300 | 4275 |
| 6 | C | Carbon | 12.011 | 2620 | 4100 | 4470 |
| 7 | N | Nitrogen | 14.0067 | (1.250) | 63.14 | 77.35 |
| 8 | O | Oxygen | 15.9994 | (1.429) | 50.35 | 90.18 |
| 9 | F | Fluorine | 18.9984 | (1.696) | 53.48 | 84.95 |
| 10 | Ne | Neon | 20.179 | (0.901) | 24.553 | 27.096 |
| 11 | Na | Sodium | 22.9898 | 970 | 371.0 | 1156 |
| 12 | Mg | Magnesium | 24.305 | 1740 | 922 | 1363 |
| 13 | Al | Aluminium | 26.9815 | 2700 | 933.25 | 2793 |
| 14 | Si | Silicon | 28.0855 | 2330 | 1685 | 3540 |
| 15 | P | Phosphorus | 30.9738 | 1820 | 317.3 | 550 |
| 16 | S | Sulphur | 32.06 | 2070 | 388.36 | 717.75 |
| 17 | Cl | Chlorine | 35.453 | (3.17) | 172.16 | 239.1 |
| 18 | Ar | Argon | 39.948 | (1.784) | 83.81 | 87.30 |
| 19 | K | Potassium | 39.0983 | 860 | 336.35 | 1032 |
| 20 | Ca | Calcium | 40.08 | 1550 | 1112 | 1757 |
| 21 | Sc | Scandium | 44.9559 | 3000 | 1812 | 3104 |
| 22 | Ti | Titanium | 47.90 | 4500 | 1943 | 3562 |
| 23 | V | Vanadium | 50.9415 | 5800 | 2175 | 3682 |
| 24 | Cr | Chromium | 51.996 | 7190 | 2130 | 2945 |
| 25 | Mn | Manganese | 54.9380 | 7430 | 1517 | 2335 |
| 26 | Fe | Iron | 55.847 | 7860 | 1809 | 3135 |
| 27 | Co | Cobalt | 58.9332 | 8900 | 1768 | 3201 |
| 28 | Ni | Nickel | 58.70 | 8900 | 1726 | 3187 |

| Atomic number | Element Symbol | Element Name | Atomic weight | Density* (kg m$^{-3}$) | Melting point, K | Boiling point, K |
|---|---|---|---|---|---|---|
| 29 | Cu | Copper | 63.546 | 8960 | 1357.6 | 2836 |
| 30 | Zn | Zinc | 65.38 | 7140 | 692.73 | 1180 |
| 31 | Ga | Gallium | 69.72 | 5910 | 302.90 | 2478 |
| 32 | Ge | Germanium | 72.59 | 5320 | 1210.4 | 3107 |
| 33 | As | Arsenic | 74.9216 | 5720 | | 876 |
| 34 | Se | Selenium | 78.96 | 4800 | 494 | 958 |
| 35 | Br | Bromine | 79.904 | 3120 | 265.9 | 332.25 |
| 36 | Kr | Krypton | 83.80 | (3.74) | 115.78 | 119.80 |
| 37 | Rb | Rubidium | 85.4678 | 1530 | 312.64 | 961 |
| 38 | Sr | Strontium | 87.62 | 2600 | 1041 | 1650 |
| 39 | Y | Yttrium | 88.9059 | 4500 | 1799 | 3611 |
| 40 | Zr | Zirconium | 91.22 | 6490 | 2125 | 4682 |
| 41 | Nb | Niobium | 92.9064 | 8550 | 2740 | 5017 |
| 42 | Mo | Molybdenum | 95.94 | 10200 | 2890 | 4912 |
| 43 | Tc | Technetium | 98 | 11500 | 2473 | 4538 |
| 44 | Ru | Ruthenium | 101.07 | 12200 | 2523 | 4423 |
| 45 | Rh | Rhodium | 102.9055 | 12400 | 2236 | 3970 |
| 46 | Pd | Palladium | 106.4 | 12000 | 1825 | 3237 |
| 47 | Ag | Silver | 107.868 | 10500 | 1234 | 2436 |
| 48 | Cd | Cadmium | 112.41 | 8650 | 594.18 | 1040 |
| 49 | In | Indium | 114.82 | 7310 | 429.76 | 2346 |
| 50 | Sn | Tin | 118.69 | 7300 | 505.06 | 2876 |
| 51 | Sb | Antimony | 121.75 | 6680 | 904 | 1860 |
| 52 | Te | Tellurium | 127.60 | 6240 | 722.65 | 1261 |
| 53 | I | Iodine | 126.9045 | 4920 | 386.7 | 458.4 |
| 54 | Xe | Xenon | 131.30 | (5.89) | 161.36 | 165.03 |
| 55 | Cs | Cesium | 132.9054 | 1870 | 301.55 | 944 |
| 56 | Ba | Barium | 137.33 | 3500 | 1002 | 2171 |
| 57 | La | Lanthanum | 138.9055 | 6700 | 1193 | 3730 |
| 58 | Ce | Cerium | 140.12 | 6780 | 1071 | 3699 |

| Atomic number | Element Symbol | Element Name | Atomic weight | Density* (kg $m^{-3}$) | Melting point, K | Boiling point, K |
|---|---|---|---|---|---|---|
| 60 | Nd | Neodymium | 144.24 | 7000 | 1289 | 3341 |
| 62 | Sm | Samarium | 150.4 | 7540 | 1345 | 2064 |
| 72 | Hf | Hafnium | 178.49 | 13100 | 2500 | 4876 |
| 73 | Ta | Tantalum | 180.9479 | 16600 | 3287 | 5731 |
| 74 | W | Tungsten | 183.85 | 19300 | 3680 | 5828 |
| 75 | Re | Rhenium | 186.207 | 21000 | 3453 | 5869 |
| 76 | Os | Osmium | 190.2 | 22400 | 3300 | 5285 |
| 77 | Ir | Iridium | 192.22 | 22500 | 2716 | 4701 |
| 78 | Pt | Platinum | 195.09 | 21400 | 2045 | 4100 |
| 79 | Au | Gold | 196.9665 | 19300 | 1337.58 | 3130 |
| 80 | Hg | Mercury | 200.59 | 13530 | 234.28 | 630 |
| 81 | Tl | Thallium | 204.37 | 11850 | 577 | 1746 |
| 82 | Pb | Lead | 207.2 | 11400 | 600.6 | 2023 |
| 83 | Bi | Bismuth | 208.9804 | 9800 | 544.52 | 1837 |
| 84 | Po | Polonium | 209 | 9400 | 527 | 1235 |
| 86 | Rn | Radon | 222 | (9.91) | 202 | 610 |
| 88 | Ra | Radium | 226.0254 | 5000 | 973 | 1809 |
| 89 | Ac | Actium | 227.0278 | 10070 | 1323 | 3473 |
| 90 | Th | Thorium | 232.0381 | 11700 | 2028 | 5061 |
| 92 | U | Uranium | 238.029 | 18900 | 1405 | 4407 |
| 94 | Pu | Plutonium | 244 | 19800 | 913 | 3503 |

* Density at 300*K* for solids and liquids, and at 273*K* for gases indicated by ( ).

# APPENDIX IV

## Standard half-cell potentials in aqueous solutions

(T = 298 *K*, Standard state = 1 molal)

| Electrode reaction | Potentials (V) | |
|---|---|---|
| | Oxidation | Reduction |
| Acid solutions | | |
| $Li = Li^+ + e$ | 3.045 | -3.045 |
| $K = K^+ + e$ | 2.925 | -2.925 |
| $Cs = Cs^+ + e$ | 2.923 | -2.923 |
| $Ba = Ba^{2+} + 2e$ | 2.90 | -2.90 |
| $Ca = Ca^{2+} + 2e$ | 2.87 | -2.87 |
| $Na = Na^+ + e$ | 2.714 | -2.714 |
| $La = La^{3+} + 3e$ | 2.52 | -2.52 |
| $Mg = Mg^{2+} + 2e$ | 2.37 | -2.37 |
| $2H^- = H_2 + 2e$ | 2.25 | -2.25 |
| $Th = Th^{4+} + 4e$ | 1.90 | -1.90 |
| $U = U^{3+} + 3e$ | 1.80 | -1.80 |
| $Al = Al^{3+} + 3e$ | 1.66 | -1.66 |
| $Mn = Mn^{2+} + 2e$ | 1.18 | -1.18 |
| $Zn = Zn^{2+} + 2e$ | 0.763 | -0.763 |
| $Cr = Cr^{3+} + 3e$ | 0.74 | -0.74 |
| $U^{3+} = U^{4+} + e$ | 0.61 | -0.61 |
| $O_2^- = O_2 + e$ | 0.56 | -0.56 |
| $S^{2-} = S + 2e$ | 0.48 | -0.48 |
| $Ni + 6NH_3(aq) = Ni(NH_3)_6^{2+} + 2e$ | 0.47 | -0.47 |
| $Fe = Fe^{2+} + 2e$ | 0.44 | -0.44 |
| $Cu + CN^- = CuCN_2^- + e$ | 0.43 | -0.43 |
| $Cr^{2+} = Cr^{3+} + e$ | 0.41 | -0.41 |
| $Cd = Cd^{2+} + 2e$ | 0.403 | -0.403 |
| $Pb + SO_4^{2-} = PbSO_4 + 2e$ | 0.356 | -0.356 |

| Electrode reaction | Potentials (V) | |
|---|---|---|
| | Oxidation | Reduction |
| $Tl = Tl^{+} + e$ | 0.336 | -0.336 |
| $Co = Co^{2+} + 2e$ | 0.277 | -0.277 |
| $Pb + 2Cl^{-} = PbCl_2 + 2e$ | 0.268 | -0.268 |
| $Ni = Ni^{2+} + 2e$ | 0.250 | -0.250 |
| $Ag + I^{-} = AgI + e$ | 0.151 | -0.151 |
| $Sn = Sn^{2+} + 2e$ | 0.136 | -0.136 |
| $Pb = Pb^{2+} + 2e$ | 0.126 | -0.126 |
| $Cu + 2NH_3(aq) = Cu(NH_3)_2^{+} + e$ | 0.12 | -0.12 |
| $H_2 = 2H^{+} + 2e$ | 0.000 | 0.000 |
| $2S_2O_3^{2-} = S_4O_6^{2-} + 2e$ | -0.08 | 0.08 |
| $Ag + Br^{-} = AgBr + e$ | -0.095 | 0.095 |
| $H_2S = S + 2H^{+} + e$ | -0.141 | 0.141 |
| $Sn^{2+} = Sn^{4+} + 2e$ | -0.15 | 0.15 |
| $Ag + Cl^{-} = AgCl + e$ | -0.222 | 0.222 |
| $2Hg + 2Cl^{-} = Hg_2Cl_2 + 2e$ | -0.2677 | 0.2677 |
| $Cu = Cu^{2+} + 2e$ | -0.337 | 0.337 |
| $Cu = Cu^{+} + e$ | -0.521 | 0.521 |
| $2I^{-} = I_2 + 2e$ | -0.5355 | 0.5355 |
| $Fe^{2+} = Fe^{3+} + e$ | -0.771 | 0.771 |
| $2Hg = Hg_2^{2+} + 2e$ | -0.789 | 0.789 |
| $Ag = Ag^{+} + e$ | -0.7991 | 0.7991 |
| $Hg_2^{2+} = 2Hg^{2+} + 2e$ | -0.920 | 0.920 |
| $Au + 4Cl^{-} = AuCl_4^{-} + 3e$ | -1.0 | 1.0 |
| $2Br^{-} = Br_2(l) + 2e$ | -1.065 | 1.065 |
| $H_2O = \frac{1}{2}O_2 + 2H^{+} + 2e$ | -1.229 | 1.229 |
| $Mn^{2+} + 2H_2O = MnO_2 + 4H^{+} + 2e$ | -1.23 | 1.23 |
| $2Cr^{3+} + 7H_2O = Cr_2O_7^{2-} + 14H^{+} + 6e$ | -1.33 | 1.33 |
| $2Cl^{-} = Cl_2 + 2e$ | -1.3595 | 1.3595 |
| $Pb^{2+} + 2H_2O = PbO_2 + 4H^{+} + 2e$ | -1.455 | 1.455 |

| Electrode reaction | Potentials (V) | |
|---|---|---|
| | Oxidation | Reduction |
| $Ce^{3+} = Ce^{4+} + e$ | -1.61 | 1.61 |
| $PbSO_4 + 2H_2O = PbO_2 + SO_4^{2-} + 4H^+ + 2e$ | -1.685 | 1.685 |
| $Co^{2+} = Co^{3+} + e$ | -1.82 | 1.82 |
| $Ag^+ = Ag^{2+} + e$ | -1.98 | 1.98 |
| $2SO_4^{2-} = S_2O_8^{2-} + 2e$ | -2.01 | 2.01 |
| $O_2 + H_2O = O_3 + 2H^+ + 2e$ | -2.07 | 2.07 |
| $2F^- = F_2 + 2e$ | -2.89 | 2.89 |
| $SO_4^{2-} = S(s) + 2O_2 + 2e$ | -3.8587 | 3.8587 |
| Basic solutions | | |
| $Ca + 2OH^- = Ca(OH)_2 + 2e$ | 3.03 | -3.03 |
| $Cr + 3OH^- = Cr(OH)_3 + 3e$ | 1.3 | -1.3 |
| $Zn + 4OH^- = ZnO_2^{2-} + 2H_2O + 2e$ | 1.126 | -1.126 |
| $CN^- + 2OH^- = CNO^- + H_2O + 2e$ | 0.97 | -0.97 |
| $SO_3^{2-} + 2OH^- = SO_4^{2-} + H_2O + 2e$ | 0.93 | -0.93 |
| $Fe + 2OH^- = Fe(OH)_2 + 2e$ | 0.877 | -0.877 |
| $H_2 + 2OH^- = 2H_2O + 2e$ | 0.828 | -0.828 |
| $Ni + 2OH^- = Ni(OH)_2 + 2e$ | 0.72 | -0.72 |
| $Fe(OH)_2 + OH^- = Fe(OH)_3 + e$ | 0.56 | -.56 |
| $2Cu + 2OH^- = Cu_2O + H_2O + 2e$ | 0.358 | -0.358 |
| $2Ag + 2OH_- = Ag_2O + H_2O + 2e$ | -0.344 | 0.344 |
| $4OH^- = O_2 + 2H_2O + 4e$ | -0.401 | 0.401 |
| $Hg_2^{2+} = 2Hg^{2+} + 2e$ | -0.920 | 0.920 |
| $O_2 + 2OH^- = O_3 + H_2O + 2e$ | -1.24 | 1.24 |

Sources: W.M. Latimer, The oxidation states of the elements and their potentials in aqueous solutions, 2nd ed., Prentice-Hall, Englewood Cliffs, N.J., 1952

# Index


Activity
Activity 141, 143
aqueous solution 403
coefficient 145, 216
coefficient, mean ionic 404
Henrian..coefficient 167
Mean..of ion 404
negative deviation 162
positive deviation 162
quotient 185
Adiabatic flame temperature 127, 128
Adsorption
adsorption 227
coefficient 236
equilibria 227
Gibbs..equation 233
Gibbs..isotherm 234
Langmuir...isotherm 236
relative 229, 230
Alkemade line 344
Alkemade theorem 344
Alkemade triangle 346
Allotrope 294,
Allotropy 294
Alpha function 174
Anode 369
Aqueous solution 403
Azeotrope 303,

BET isotherm 238
Binary
eutectic 308
liquid system 297
solution 317
system 297
system, with solid solution 317
total solubility 317
partial solubility 318
system, without solid solution 308
Bivariant 341
Boiling
boiling 248
boiling point 169, 170
temperature 291
Boltzmann's constant 55
Boltzmann distribution 67
Boltzmann equation 55
Boundary curve 343

Capillary effect 273
on melting point 276
on solubility 278
on vapor pressure 274
Carnot cycle 33, 41
Cathode 369
Cell
chemical 390
concentration 390
Daniell 380
electrochemical 370, 378

electrode concentration 390
electrolyte concentration 390, 394
with transference 390
without transference 390, 395
galvanic 369
Chemical cell 390
Chemical potential 135, 143, 146
Clapeyron equation 251
Clausius-Clapeyron equation 252
Clustering 166
Coarsening 280
Colligative property 168
Common tangent 268, 279,
Common tangent plane 359
Component 240
Composition triangle 335
Compound
formation 166
intermetallic 269
stoichiometric 269
Compressibility
compressibility 109
factor 138
Compression
adiabatic 33, 35, 37, 44, 45, 47, 52
isothermal 33, 35, 37,44, 45, 46, 52
Concentration cell 390
Configuration 54, 64, 31
Congruent melting 311, 322
Conjugate pair 360
Conjugate phase 359
Conjugation line 346
Contact angle 284
Cooling curve 292
Coordination number 178
Coring 324
Corrosion zone 425
Criterion of equilibrium 62, 100, 189
Critical pressure 290
Critical point 289
Critical temperature 290, 304
Crystallization path 343, 346, 347, 348

Daniell cell 380
Darken's quadratic formalism 182
Debye-Huckel limiting law 407
Degradation 29, 31, 53, 55, 58
Degree of freedom 239, 242
Degree of irreversibility 62, 87, 95,
Dendrite growth 278
Dilute solution 167
Distillation 303, 306

Efficiency of engine 32, 39, 48
Electrochemical cell 370, 378
Electrochemical equilibrium 369
Electrochemical reaction 368
Electrochemistry 367
Electrode
electrode 369, 378
hydrogen 378
negative 379
positive 379
reference 382
standard hydrogen 382, 383
standard..potential 383
Electrolysis 376
Electrolyte 379
Electromotive force 372
Electronegativity 264
Electroneutrality 368
Ellingham diagram 218, 219
Emf 372
Endothermic 15, 124, 166
Energy
ability 28
bond 9
energy 2
chemical 2, 9
degradation 29, 31, 58
dispersion 53, 54
grade 27, 28
internal 3, 70
kinetic 2, 3, 17
macroscopic..transfer 1
motional 88
potential 2, 3, 17
quality 27, 28, 31,

surface 271
thermal 3, 17, 27, 62
translational 61,
units 5
useful 28
Enthalpy
enthalpy 15, 16
excess partial 163
change 23, 43, 119, 121
excess 163
of formation 121
of mixing 135
standard.. of formation 121
of reaction 122
Entropy
entropy 32, 39, 40, 42, 53, 54, 62, 71, 80, 112
absolute 108, 115
change 61, 72, 73, 74, 78, 90, 91
excess 163
generation 49, 89
maximum 59, 67
mixing 135
partial 154
production 49, 73, 86, 89, 90
relative 154
transfer 44, 49, 90
ε-pH diagram 416
Equilateral composition triangle 336
Equilibrium
equilibrium 62, 67, 98, 101, 103, 106
adsorption 227
constant 185, 186
criteria 62, 100, 189
diagram 309
effect of pressure 203
effect of temperature 201
electrochemical 369
local 278
phase 239, 255
reaction 185
state 97, 98
Eutectic
binary 308
hyper- 319
hypo- 319
point 309
reaction 310
structure 311
ternary 345
Eutectoid
point 319
reaction 319
Excess integral molar Gibbs energy 164
Excess Gibbs energy of mixing 178
Excess partial molar enthalpy 163
Excess partial molar entropy 163
Excess partial molar Gibbs energy 163
Excess property 161, 163, 227
Exothermic 2, 15, 73, 124, 166
Expansion
adiabatic 21, 33, 34, 37, 44, 49, 50
coefficient 112
isothermal 33, 37, 44, 45, 46, 49, 50
Extensive property 13
Extremum principle 190

Faraday 368
Faraday's law 367
First law of thermodynamics 7
Fractional distillation 306
Free energy
free energy 97
minimization method 197
Gibbs 98, 101,
Helmholtz 98, 99
Freezing point 169, 171
Freundich isotherm 238
Fugacity 136, 142
Fusion curve 289

Galvanic cell 369
Gibbs adsorption equation 233
Gibbs adsorption isotherm 234

Gibbs dividing surface 227, 228
Gibbs-Duhem equation 151, 160, 171
Gibbs energy
  Gibbs energy 98, 101, 255
  change 103, 119, 129
  effect of pressure 104
  effect of temperature 106
  excess 163
  excess integral 164
  excess..of solution 164
  excess partial 163
  integral 151
  mean partial ..of ion 404
  Mean standard..of ion 404
  of mixing 135, 257
  of mixture 134
  partial 135, 143, 146, 153
  relative 135, 153
  standard ..of formation 129
  standard..of reaction 130
Gibbs-Helmholtz equations 106
Gibbs phase rule 242
Gibbs Thomson effect 273

Half-cell 369
  Standard..potential 383
Heat
  heat 4, 5
  of formation 121,
  of fusion 22, 249
  of reaction 122
  of transformation 120
  of vaporization 22, 23
  sensible 127
  sign convention 5
Heat capacity
  heat capacity 16, 17, 113
  constant pressure 16
  constant volume 16
Heat engine 32
Heat pump
  heat pump 51
  efficiency 51
Heat reservoir 33,

Henry's law 167, 168
Henrian activity coefficient 167
Hess's law 26, 123
Hydrogen electrode 378
Hypereutectic 319
Hypoeutectic 319

Ideal gas 84, 133
Ideal gas equation 133
Ideal solution 157, 158
Immiscibility 297, 303
Immunity zone 424
Incongruent melting 312, 322
Inflection point 267
Integral molar volume 150
Integral property 151
Intensive property 13
Interaction coefficient 215, 216
Interaction parameter 177
Interface 271
Intermediate phase 269,
Internal energy 3, 70
Invariant 315, 341
Irreversibility 31, 62, 78, 87, 89, 90, 95
Irreversible process 11, 42, 49, 78, 87, 90
Isobar 287
Isotherm 287, 343
Isothermal plane 339, 342
Isothermal section 353

Join 346
Junction potential 396

Kirchhoff's law 125

Langmuir adsorption isotherm 236
Latent heat 249
Le Chatelier's principle 205
Lever rule 259
Liquidus 255, 302, 309
Liquidus isotherm 343
Lost work 86, 87, 88

Macrostate 53

Margules formalism
    Margules formalism 181
    three suffix 182
Maxwell's equations 108
Mean ionic activity coefficient 404
Mean activity of ion 404
Mean ionic molality 404
Mean molar Gibbs energy of ion 403,
Mean thermodynamic property 403
Melting
    congruent 311, 322
    incongruent 312, 322
    point 248
Meta-stable phase 269
Method of intercept 152
Microstate 53, 54, 63, 64, 67
Miscibility gap 259
Miscibility
    miscibility 259, 260,
    partial 297, 303
    total 297
Molality 403
Molar Gibbs energy of mixing 135, 257
Molarity 413
Monotectic
    point 314
    reaction 314
    system 314

Natural process 27
Negative deviation 161, 162
Nernst equation 373, 374
Non-ideal solution 161
Non-wetting 284

One component system 287
Order-disorder transition 324
Ordered structure 324
Ostwald ripening 274, 280
Oxygen sensor 393

Partial miscibility 297
Partial property 130, 148, 149
Partition function 67
Passivation zone 425
Perfect gas 133
Perfect gas equation 133
Performance coefficient 51
Peritectic
    point 313, 315
    reaction 313
    system 312
    ternary 345
pH 413
Phase
    phase 240
    boundary 250
    conjugate 359
    diagram 250, 287, 309
    equilibrium 239, 255
    Gibbs..rule 239, 242
    intermediate 268, 269
    meta-stable 269
    primary 343
    reaction 310
    rule 242
    separation 166
    terminal 268
    transformation 83, 248
    transition 22, 249
Phase rule 242
Polymorphic transformation 294
Polymorph 294
Polythermal projection 342
Positive deviation 161, 162
Pourbaix diagram 416
Predominance diagram 363, 364
Primary field 309, 343
Primary phase 343
Probability 67

Raoult's law 158
Real gas 136
Redox reaction
Reference electrode 382
Reference state 119
Refrigerator 51
Regular solution model 176, 266

Relative integral Gibbs energy 153
Relative partial molar Gibbs energy 135
Reversibility 12, 62, 75, 76, 77
Reversible process 11, 76, 77, 78, 80, 98
Reversible work 8, 12
Rotation 2

Sessile drop technique 283
Solid electrolyte 392
Solid solubility
  solid solubility 317
  partial 318
  total 317
Solid solution
  intermediate 322
  interstitial 317
  substitutional 317
  terminal 322
Solidus 302, 309
Solubility constant 411
Solubility product 411
Solution model 175
Solution
  aqueous 403
  dilute 167
  ideal 157, 158
  interstitial 317
  model 175
  regular 176
  solid 264, 317
    terminal 322
    intermediate 322
  substitutional 317
Spontaneous process 27, 55
Stability diagram 363, 364
Standard electrode potential 383
Standard enthalpy of formation 121
Standard Gibbs energy of formation 129
Standard half-cell potential 383
Standard hydrogen electrode 383
Standard state
  standard state 119
  alternative 207
  1 molality 404
  1wt% 210
  Henrian 208
  infinitely dilute solution 210
  Raoultian 207
State 53
State function 4, 13
Steam distillation 303
Stirling's approximation 64
Sublimation curve 289
Surface active 234, 235
Surface energy 271
Surface excess concentration 229, 234
Surface tension 231
Surroundings 5

Taylor series expansion 217
Temperature 3
Temperature coefficient 375
Temkin isotherm 238
Terminal phase 268
Terminal solid solution 322
Ternary eutectic 345
Ternary peritectic 345
Ternary system 335
Theoretical plate 306
Thermal expansion coefficient 112
Thermodynamic model
  Darken's quadratic formalism 182
  ideal solution 326
  non-ideal solution 329
  regular solution 331
  Margules formalism 181
Thermodynamics
  chemical 2
  classical 1
  electrochemical 370
  first law of 7
  second law of 27, 94
  statistical 1
  third law of 111, 112
Tie line 309, 360
Tie triangle 360
Transition temperature 249
Translation 2

Triple point 250, 289
Trivariant 341

Unary system 287
Undetermined multiplier 66
Univariant 316, 341

van der Waals equation 139
van't Hoff equation 201
Vapor pressure 162
Vaporisation curve 289
Vaporus 302
Vibration 2
Virial coefficient 138
Voltage 372

Work
  work 4
  additional 99, 102
  capacity 31
  electric 27
  irreversible 12
  lost 86,88
  lost capacity 31
  maximum 32, 87
  mechanical 27
  reversible 8, 12

www.ingramcontent.com/pod-product-compliance
Lightning Source LLC
Chambersburg PA
CBHW061927190326
41458CB00009B/2672
* 9 7 8 9 8 1 4 3 6 8 0 5 6 *